Transition Metal-Catalyzed Carbene Transformations

Transition Metal-Catalyzed Carbene Transformations

Edited by Jianbo Wang, Chi-Ming Che and Michael P. Doyle

WILEY-VCH

Editors

Prof. Jianbo Wang
Peking University
College of Chemistry
100871 Beijing
P. R. China

Prof. Chi-Ming Che
The University of Hong Kong
Department of Chemistry
Pokfulam Road
Hong Kong SAR
P. R. China

Prof. Michael P. Doyle
The University of Texas at San Antonio
Department of Chemistry
San Antonio, Texas 78249-0698
United States

Cover Image: Kai Chen

Library of Congress Card No.: applied for

British Library Cataloguing-in-Publication Data
A catalogue record for this book is available from the British Library.

Bibliographic information published by The Deutsche Nationalbibliothek The Deutsche Nationalbibliothek lists this publication in the Deutsche Nationalbibliografie; detailed bibliographic data are available on the Internet at <http://dnb.d-nb.de>.

Print ISBN: 978-3-527-34799-5
ePDF ISBN: 978-3-527-82916-3
ePub ISBN: 978-3-527-82915-6
oBook ISBN: 978-3-527-82917-0

Cover Design: ADAM DESIGN, Weinheim, Germany
Typesetting Straive, Chennai, India
Printing and Binding CPI Group (UK) Ltd, Croydon, CR0 4YY

Printed on acid-free paper

C107542_091221

Contents

Preface

Transition-metal-catalyzed carbene transfer has been one of the major topics in the arena of organic synthesis and organometallics for many decades. It remains as a dynamic research area. In the past decade, remarkable progress has been made in transition-metal-catalyzed carbene transformations, particularly in asymmetric catalysis, new carbene-based reactions, new catalytic systems with engineered enzymes, and so on. While numerous reviews are available on these developments, the purpose of this book, written by the leading experts, is to provide a collective summary of the most recent advances in this ever-growing field.

The book has 12 chapters covering the major advances in transition-metal-catalyzed carbene transformations. As a typical carbene reaction, C—H bond insertion has been applied in the challenging topic of inert C—H bond functionalization, and Chapter 1 covers the applications of carbene C—H bond insertion in alkane functionalization, including methane C—H bond functionalization. Another typical carbene reaction is cyclopropanation. The catalytic radical approach for selective carbene cyclopropanation is summarized in Chapter 2. In carbene transfer reactions, a long-standing challenge is to achieve high enantioselectivity in X–H insertions (X = heteroatom). With newly developed chiral ligands, significant advances have been made in this arena, which are summarized in Chapter 3. A remarkable advance in asymmetric carbene transfer is the application of engineered enzymes, which are generated by directed evolution. The concepts and practices of those bio-inspired carbene transformations are presented in Chapter 4. In addition, Chapter 10 introduces chiral *N,N′*-dioxide and chiral guanidine-based catalysts, which have proven to be highly effective for some asymmetric carbene transfer reactions.

Cycloaddition reactions with metal carbenes is a highly efficient approach for rapidly generating molecular complexity. The recent progress in this domain is summarized in Chapter 5. Though diazo compounds are the most commonly used carbene precursors, efforts have been devoted to the search for other safer carbene precursors. In Chapter 6, a metal-catalyzed decarbenation by retro-cyclopropanation, which generates metal carbene species, is presented. In Chapter 7, metal carbenes are generated by oxidation of alkynes. Both chapters focus on transformations with gold catalysis.

Polymerization with carbene transformations is a unique approach toward polymer synthesis. The C1 polymerization with carbenes, as well as polymerization with classic carbene reactions, is summed up in Chapter 8. Metal-quinoid carbene complexes represent a unique metal carbene species, which are introduced along with their transformations in Chapter 9. Finally, the exploration of new transformations based on metal carbenes has been flourishing in the past decades. Chapter 11 outlines the multi-component reactions through *gem*-difunctionalization. The enormous progress in transition-metal-catalyzed cross-coupling using carbene precursors is summarized in Chapter 12.

Having read all the chapters, we hope that readers are left with the notion that transition-metal-catalyzed carbene transfer is still a rapidly growing area with constant emergence of novel catalytic systems, transformations, and applications. Thus, further advances can be expected in the coming years.

We would like to appreciate Dr. Elke Maase and Ms. Katherine Wong for their assistance in the whole process. We would also like to thank the referees for their comments and suggestions. Finally, this book would not have been possible without the participation of the authors, to whom we are deeply indebted for their hard work and enduring patience throughout this project.

Beijing
Hong Kong
San Antonio
01 June 2021

Jianbo Wang
Chi-Ming Che
Michael P. Doyle

1

Alkane Functionalization by Metal-Catalyzed Carbene Insertion from Diazo Reagents

María Álvarez, Ana Caballero, and Pedro J. Pérez

Universidad de Huelva, Laboratorio de Catálisis Homogénea, Unidad Asociada al CSIC – CIQSO-Centro de Investigación en Química Sostenible and Departamento de Química, 21007, Huelva, Spain

1.1 Introduction

The metal-catalyzed decomposition of diazo compounds, known from the beginning of the twentieth century [1], occurs through the formation of metal-carbene intermediates which display electrophilic nature, mainly located at the carbenic carbon atom (**MC**, Scheme 1.1) [2]. Such feature is the origin of the reactivity toward nucleophiles triggering the transfer of the carbene group and subsequent functionalization of the latter. A variety of powerful nucleophiles have been employed along the years, such as unsaturated molecules (olefins, alkynes, or arenes) or amines and alcohols, for which carbene incorporation, either added to an unsaturated bond or inserted into a saturated one, was favored. At variance with that, the functionalization of the less nucleophilic carbon–hydrogen bonds by this methodology has developed at a lower rate [3]. Intramolecular transformations were described with a substantial degree of success, albeit many examples took advantage of the existence of carbon–hydrogen bonds vicinal to heteroatoms (N, O), of being located at benzylic positions, and/or of the formation of very stable five- or six-member rings. But when facing the modification of non-activated carbon–hydrogen bonds, and under intermolecular conditions, this reaction becomes a challenge that only in the last decade has been significantly achieved and with some goals yet to be reached.

In this chapter, an account of the state of the art on the functionalization of the carbon–hydrogen bonds of alkanes by carbene incorporation from diazo compounds is presented. In this sense, only saturated hydrocarbons C_nH_{2n+2} as well as their cyclic partners C_nH_{2n} are considered. Other C—H bonds in substrates containing activating groups (heteroatoms, aryl, or olefin) are not considered herein, unless a substantial degree of novelty is implicit. As shown in Figure 1.1, the strengths of the C—H bonds of the alkanes are considerably higher than those of substrates containing activating groups [4]. As a representative comparison, the C—H bond of cyclohexane displays a bond dissociation energy (BDE) of 99.5 kcal mol^{-1}, whereas the

Transition Metal-Catalyzed Carbene Transformations, First Edition.
Edited by Jianbo Wang, Chi-Ming Che, and Michael P. Doyle.

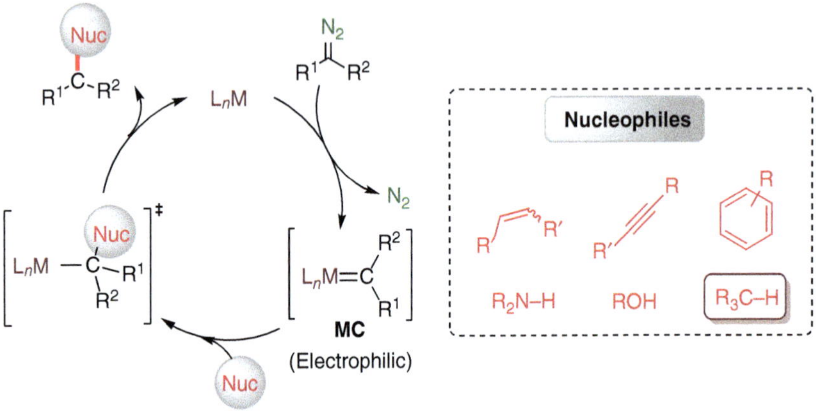

Scheme 1.1 General catalytic cycle for several nucleophiles functionalization by carbene transfer from diazo compounds.

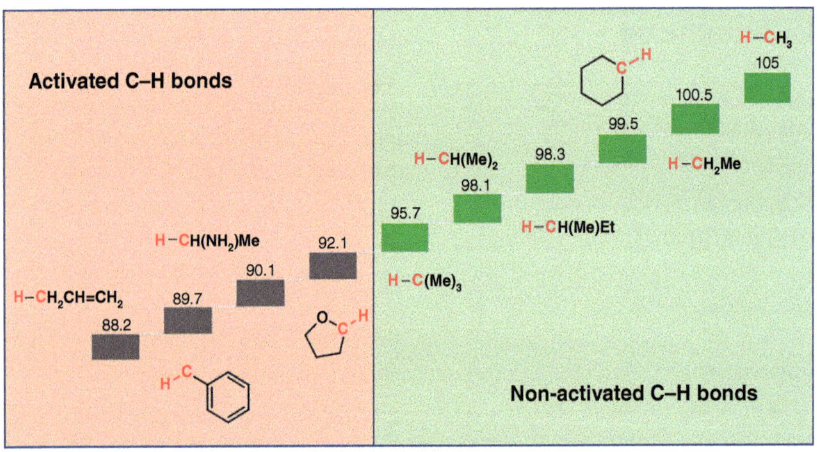

Figure 1.1 Bond dissociation energies of representative carbon–hydrogen bonds. Values in kcal mol^{-1}.

methyl C—H bonds of toluene are nearly 10 kcal mol^{-1} below. Interestingly, some literature reports refer to the latter as "alkane functionalization," although it is clearly far of being and behaving as an alkane.

Based on the low nucleophilicity of the targeted C—H bonds, the success of the carbene transfer depends on exerting a high electrophilicity in the metallocarbene intermediate (**MC**, Scheme 1.1) generated from the catalyst precursor and the diazo reagent. The carbenic carbon needs to be drained of electron density; therefore, reactivity toward the weak electrophile is increased. Several tactics can be employed to achieve such effect (Scheme 1.2). On one hand, electron donation from the ancillary ligand(s) bonded to the metal ion should be as low as possible while always ensuring that coordination is maintained. The use of a very poor coordinating ligand could be a problem if it is de-coordinated during catalysis. Additionally, the nature of the

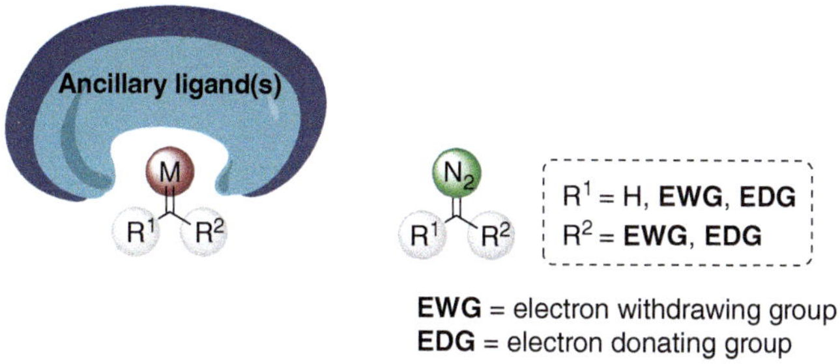

Scheme 1.2 Factors affecting the electrophilicity of the carbenic carbon atom.

substituents at the diazo reagent may also help since electron withdrawing, electron donating, or the neutral H can be employed at those positions. It could be thought that using two electron-withdrawing groups could be the best option to enhance the reactivity. However, this is not the case since other side reactions can also occur (see Section 1.2), and very reactive metallocarbenes could enhance the formation of undesired products. A balance between the ancillary ligand, the metal center and the diazo substituent must be found to optimize the reaction outcome.

1.2 Chemo- and Regioselectivity

1.2.1 Definitions

The metallocarbene intermediates are highly reactive and interact with available nucleophiles in the reaction mixture. In addition to the substrate employed, other side, non-desired reactions might occur, decreasing the selectivity of the process. Scheme 1.3 shows the most common byproducts derived from such behavior. Olefins derived from the catalytic coupling of two diazo molecules are the most frequent byproducts, in a process which usually is highly favored [5]. The use of slow addition techniques to maintain a low diazo concentration is a tactic to decrease their formation. Such olefins can also be transformed into cyclopropanes with a third diazo molecule, a reaction which may occur with highly active catalysts and high concentration of the olefins. When adventitious water is present, the O—H bond can also be modified upon incorporation of a carbene moiety. Because all these possible processes decrease the yields into the targeted alkane C—H bond functionalization product, the chemoselectivity is usually defined as the amount of the latter referred to the initial amount of the diazo reagent, which can also be transformed into the other byproducts. The formation of the olefins and cyclopropanes involve two and three, respectively, molecules of diazo reagent, a fact that needs to be accounted for the chemoselectivity value.

Regioselectivity in the context of alkane functionalization by this methodology refers to the distinct reactivity shown by the different C—H bonds available in the alkane molecule employed as substrate. Considering *n*-hexane as an example, three distinct potential sites are present, i.e. the primary C—H bonds at C1 and the secondary sites at C2 and C2′ (Scheme 1.4). Once the reaction outcome is quantified, a

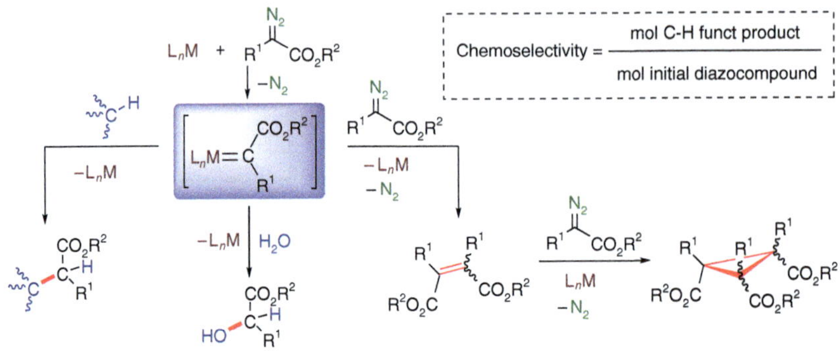

Scheme 1.3 Chemoselectivity in C—H bond functionalization by carbene insertion, with potential undesired side reactions.

distribution of the three products derived from the carbene insertion in such those three sites is obtained. Albeit those numbers can be employed to define the selectivity of the catalyst, it is more convenient to correct them employing the number of C—H bonds existing at each site, therefore eliminating statistic effects. In this manner, six C—H bonds are available at C1 sites, whereas C2 and C2′ contain four C—H bonds each. In the case of 2,3-dimethylbutane, there are 12 primary sites at C1 and 2 tertiary ones at C3. Scheme 1.4 shows a comparison with hypothetical values for the distributions of products and the regioselectivity defined as the latter corrected by the number of available C—H bonds. The selectivity toward C1 and C2 in hexane corresponds to a 1 : 9 ratio, which may be misinterpreted focusing only on the distribution of products (10 : 60). The case of 2,3-dimethylbutane is more pronounced: a 60 : 40 ratio for the distribution of products, which provides more of the primary site functionalized product, corresponds to a 1 : 4 regioselectivity favoring the tertiary site.

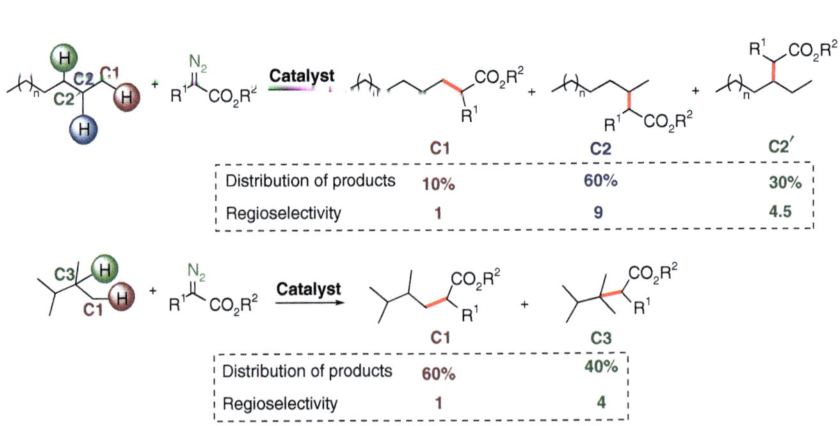

Scheme 1.4 Examples for the comparison of the distribution of products and the regioselectivity calculated employing the number of C—H bonds of each class in the substrate.

1.2.2 Catalysts

A number of catalysts have been described for the functionalization of the carbon–hydrogen bonds of alkanes and cycloalkanes through this methodology, many of them consisting of transition metal complexes bearing somewhat elaborated ligands. Scheme 1.5 displays all those commented along this chapter, either showing the structure of the catalyst or, in some cases, the ligand used as an additive

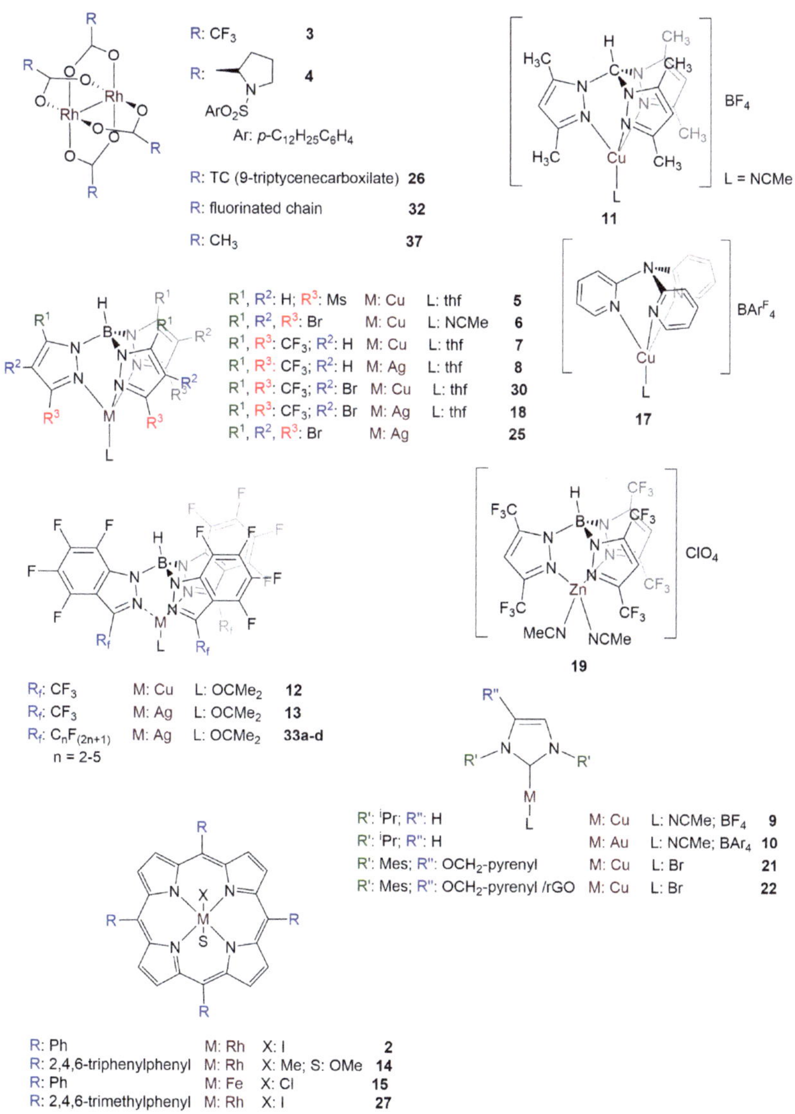

Scheme 1.5 Catalysts employed for the functionalization of alkanes or cycloalkanes by carbene insertion from diazo compounds.

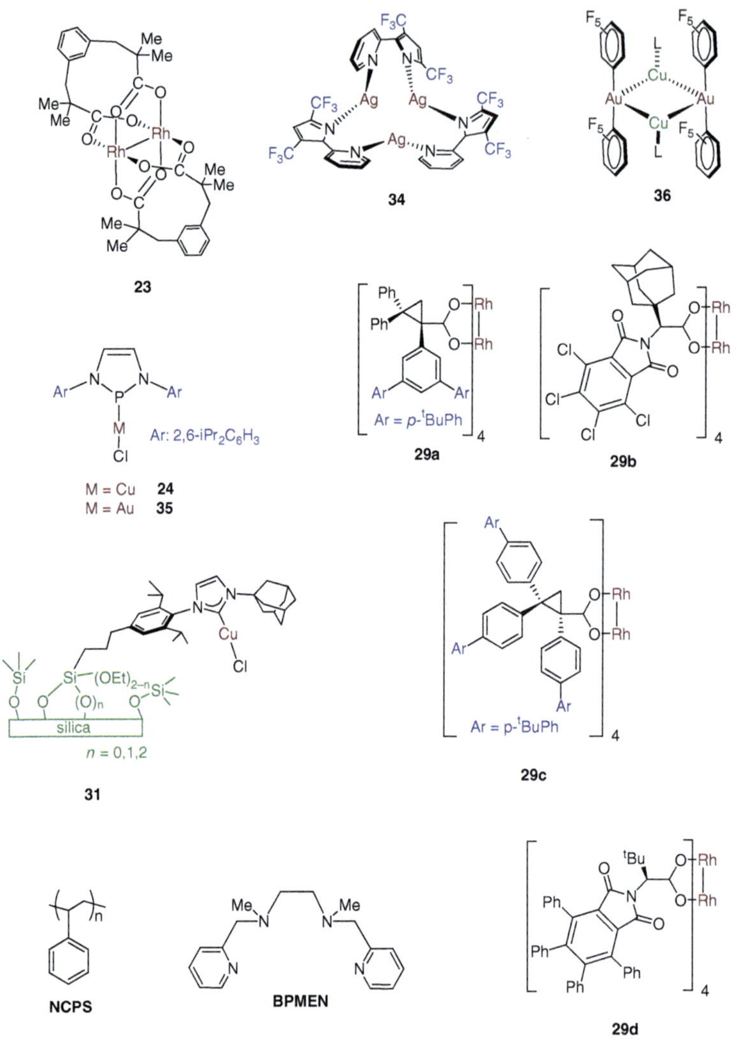

Scheme 1.5 (*Continued*)

along with a simple metal salt. Each catalyst is given a number which is later employed in the discussion of results.

1.2.3 Chemoselectivity

Cyclohexane is very often employed as the probe substrate to evaluate the potential of a catalyst toward the carbene transfer from a diazo compound and subsequent alkane C—H bond functionalization. With a substantial BDE value (99 kcal mol^{-1}), it is an appropriate reactant since only one product is formed, at variance with linear alkanes, and thus provides a measure of the chemoselectivity induced by

Table 1.1 Cyclohexane functionalization by metal-catalyzed carbene insertion.

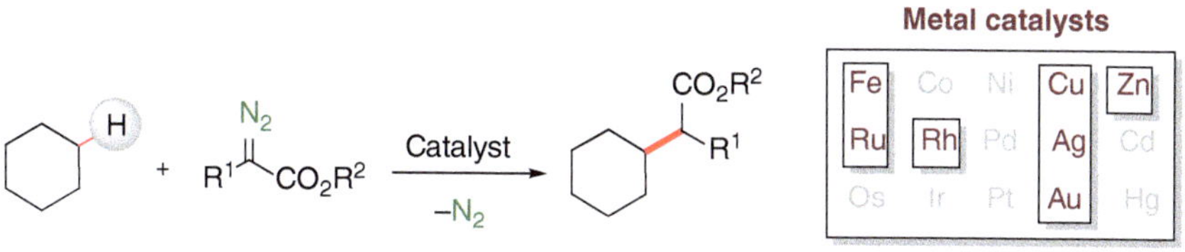

Catalyst		R^1	R^2	Yield (%)	Diazo addition[a]	Reference
$CuSO_4$	1	H	CO_2Et	24	sa	[6]
$CuSO_4$	1	H	CO_2Ph	17	sa	[6]
RhTPPI	2	H	CO_2Et	71	sa	[7]
$Rh_2(OOCCF_3)_4$	3	H	CO_2Et	75	sa	[8]
Rh_2(S-DOSP)$_4$	4	Ar	CO_2Et	80	sa	[9]
Tp^{Ms}Cu(thf)	5	H	CO_2Et	54	sa	[10]
Tp^{Br3}Cu(NCMe)	6	H	CO_2Et	90	sa	[11]
$Tp^{(CF3)2}$Cu(thf)	7	H	CO_2Et	58	sa	[12]
$Tp^{(CF3)2}$Ag(thf)	8	H	CO_2Et	99	sa	[12]
[IPrCu(NCMe)]BF_4	9	H	CO_2Et	80	sa	[13]
[IPrAu(NCMe)]BAr^F_4	10	H	CO_2Et	84	sa	[13]
*[Tpm*Cu(NCMe)]BF_4*	11	H	CO_2Et	80	sa	[14]
Tp^{F21}Cu($OCMe_2$)	12	H	CO_2Et	99	sa	[15]
Tp^{F21}Ag($OCMe_2$)	13	H	CO_2Et	97	op	[15]
Rh(ttppp)(Me)(MeOH)	14	Ar	CO_2Et	80	op	[16]
Fe(TPP)Cl	15	Ar	CO_2Me	78	op	[17]
NCPS-Ru	16	Ph	CO_2Me	60	op	[18]
[(TPN)Cu(thf)]+BAr^F_4	17	H	CO_2Et	85	sa	[19]
$Tp^{(CF3)2Br}$Ag(thf)	18	H	CO_2Et	99	op	[20]
{LZn$(NCMe)_2$}ClO_4	19	H	CO_2Et	91	sa	[21]
Fe$(ClO_4)_2$/BPMEN	20	Ph	CO_2Me	84	op	[22]
L^{Mes}CuBr	21	H	CO_2Et	15	sa	[23]
L^{Mes}CuBr-rGO	22	H	CO_2Et	35	sa	[23]
[$Rh_2(esp)_2$]	23	CONHPh	CO_2Me	89	sa	[24]
(ADAP)CuCl	24	H	CO_2Et	98	sa	[25]
[Tp^{Br3}Ag]$_2$	25	N-triftosylhydrazones		99	op	[26]

[a] sa = slow addition; op = one portion.

the catalyst chosen. In this section, an overview of the catalytic systems described for the conversion of cyclohexane into the corresponding derivatives upon carbene insertion is given. Table 1.1 contains a list of such catalysts (see their structures in Scheme 1.5). The substituents at the diazo group are also detailed, as well as the temperature and the mode employed for the incorporation of the diazo reagent, either in one portion addition (op) or employing slow addition (sa) devices. Diazo-based yields are given as chemoselectivity values.

In spite of the interest of this transformation, only seven metals have been found as productive in terms of catalytic activity for this reaction: Fe, Ru, Rh, Cu, Ag, Au, and Zn. The first report from 1974 [6] employed simple copper(II) sulfate to generate ethyl 2-cyclohexylacetate in 24% yield. Since then, several metal complexes have been described toward that end, most of them providing high yields into the functionalized product. Such degree of chemoselectivity has been frequently reached employing the slow addition technique which maintains the diazo concentration low enough to disfavor the formation of carbene-coupling products. Catalysts **6**, **8**, **12**, **19**, and **24** induce >90% yields into the functionalized product using such technique. Interestingly, some of the catalysts (**13**, **25**) induced very high yields adding the diazo reactant in one unique portion at the beginning of the experiment, demonstrating a very high chemoselectivity toward the C—H bond functionalization.

From the array of catalysts shown in Table 1.1, several trends can be extracted. First, the nature of the diazo reagent seems to be coupled with the metal of choice. Thus, the monosubstituted ethyl diazoacetate (EDA), with one acceptor substituent (CO_2Et), is largely employed with group 11 metal-based catalysts. EDA usually generates a highly reactive metallocarbene intermediate which therefore is disposed to verifying some of the non-desired reactions already commented in Scheme 1.3. In this sense, the observation of very high selectivities with some catalysts is of note, most of them with tridentate ligands surrounding the metal center (**6–8**, **11–13**, **17**, **18**, **25**). In other examples, monodentate ligands of type N-heterocyclic carbene (**9**, **10**, **21**, **22**) or alkoxydiaminophosphines (**24**) also give high yields into the functionalized product.

The other group of catalysts are rhodium-based, either mononuclear with porphyrin ligands or dinuclear derived from the parent dirhodium tetraacetate. These $Rh_2(L\text{–}L)_4$ catalysts are the most relevant in the chemistry of carbene transfer reactions in view of their extensive use with outstanding performance in many reactions. Regarding alkanes as substrates, donor–acceptor diazo reagents are preferred, mainly with aryl (donor) and carboxylate (acceptor) groups. Their use will be developed in Sections 1.2.4 and 1.3.

Other metals such as iron, ruthenium, and zinc have been described for this transformation, with good to high yields. The iron catalyst **15** contains a porphyrin ligand, resembling the similar rhodium catalyst **2**. At variance with that, the ruthenium catalyst **16** is formed by nanoparticles of this metal supported onto a polystyrene material. In the case of zinc (**19**), albeit the complex is required for diazo decomposition, the authors could not rule out that carbene dissociation from the metal center could exist prior to attack to the C—H bond.

1.2.4 Regioselectivity

Except for methane and ethane, which will be discussed in a separate section, all acyclic alkanes display a number of distinct C—H bonds, providing several reaction sites. Because of this, catalysts exerting a certain control in the regioselectivity (see Scheme 1.4 for definition) are desired. In this sense, a tremendous advance in this area has taken place in the last decade. To discuss such development, three representative alkanes have been chosen: *n*-pentane, *n*-hexane, and 2,3-dimethyl butane, exemplifying the potential of a given catalyst to direct the carbene insertion to the different primary, secondary, or tertiary C—H bonds of these non-activated sites.

Table 1.2 contains the catalytic systems reported to date for effective *n*-pentane or *n*-hexane functionalization by carbene insertion from a diazo compound. A few are far of being considered regioselective, despite their good catalytic activity in terms of yields. Most are based on the coinage metals, which also employ EDA as the carbene source. Copper catalysts display very low or null capabilities for the functionalization of primary sites, whereas silver and gold provide mixtures of the three possible products at variable extent, with noticeable amounts of that derived from primary sites functionalization.

At variance with group 11 metals, rhodium-based catalysts have provided the best results to date for the selective functionalization to C1 or C2 positions in *n*-pentane or *n*-hexane. Catalyst Rh(ttppp)(Me)(MeOH) (**14**) induced the highly selective incorporation of the C(Ph)CO_2Me carbene group into the primary C—H bonds of hexane, with a 10.5 : 1 regioselectivity for C1:C2 sites, respectively, and no incorporation into the C2′ site. The dirhodium complex **4** promoted the preferential functionalization of the secondary C—H bond at C2 of pentane, with no activation of that at C1 and very minor at C2′, the C2:C2′ regioselectivity being 29 : 1.

The regioselectivity of this transformation seems to be governed by the steric hindrance of the catalytic pocket, defined by the geometry of the ancillary ligand(s). This is clearly observed when a series of $Tp^FAg(L)$ catalyst with fluorinated ponytails variable in length, but similar electronic properties at the silver center were employed with several alkanes. Additionally, electronic effects are of note, particularly if the highly reactive EDA is employed, the control of the regioselectivity becoming more difficult. With less reactive, donor–acceptor diazo reagents, the regioselectivity is enhanced toward a site which is mainly defined by the steric pressure around the metallocarbene unit.

Carboxylate $CO_2CH_2CCl_3$ groups are the most effective substituents of the diazo group in this case. Extension of the latter system has led to more regioselective catalysts of composition $Rh_2(L\text{–}L)_4$. Due to their chiral nature and the enantioselectivity induced, they are discussed in Section 1.3.

2,3-Dimethylbutane constitutes the prototypical example to discern between tertiary and primary C—H bonds in terms of regioselectivity. Table 1.3 shows the most relevant catalysts employed with this substrate (see Scheme 1.5 for catalyst structures). Very high selectivity toward the tertiary C3 site can be achieved with $Rh_2(OOCCH_3)_4$ (**37**), $Rh_2(OOCCF_3)_4$ (**3**), or $Tp^{Br3}Cu(NCMe)$ (**6**). On the other hand, such degree of regioselectivity has not yet been induced onto the primary sites of this substrate. Best results to date correspond to Rh(ttppp)(Me)(MeOH) (**14**, 1 : 1.3 for C1:C3 regioselectivity), several silver catalysts of formulation Tp^xAgL (**8**, **13**, **28**; C1:C3 within the range 1 : 1.5–2), the gold catalyst $[IPrAu(NCMe)]BAr^F{}_4$ (**10**, with C1:C3 as 1 : 1.2), and the zinc catalyst (**19**, with C1:C3 as 1 : 1.9).

1.3 Enantioselectivity

In line with the continuously pursued great challenge of enantioinduction in catalytic reactions, the development of chiral complexes as catalysts for the

Table 1.2 Functionalization of *n*-pentane/*n*-hexane.

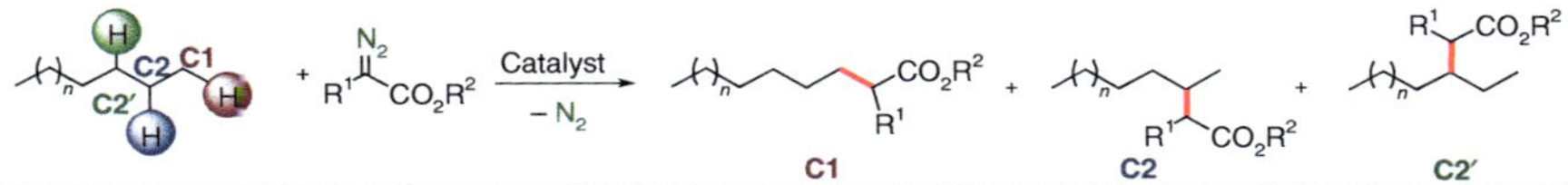

Catalyst		R^1	R^2	*n*	Yield (%)	Selectivity			References
						C1	C2	C2'	
$Rh_2(OOCCF_3)_4$	**3**	H	CO_2Et	1	n.r.	1	14.1	11.6	[8, 27]
$Rh_2(OOCTC)_4$	**26**	H	CO_2Et	1	n.r.	1	3.1	0.9	[27]
RhTPPI	**2**	H	CO_2Et	2	46.	1	13.4	4	[7]
RhTMPI	**27**	H	CO_2Et	2	36	1	3.7	0.8	[7]
$Rh_2(OOCR_F)_4$	**32**	H	CO_2Me	2	92	1	14.4	6.9	[28]
Rh(ttppp)(Me)(MeOH)	**14**	Ph	CO_2Me	2	66	10.5	1		[16]
$Rh_2(S\text{-}DOSP)_4$	**4**	Ar	$CO_2(CH_2CCl_3)$	1	98	n.d.	29	1	[9]
Dirhodium TPCP catalysts	**29**	Ar	$CO_2(CH_2CCl_3)$	1	99	1	30	n.d.	[29]
$Tp^{Br3}Cu(NCMe)$	**6**	H	CO_2Et	1	60	n.d.	3	1	[11]

Tp^{F21}Cu(OCMe$_2$)	**12**	H	CO_2Et	1	47	1	18.5	7	[15]
Tp$^{(CF3)2Br}$Cu(THF)	**30**	H	CO_2Et	2	>99	1	15.8	4.1	[30]
(TPN)Cu(THF)BAr$^F_4$	**17**	H	CO_2Et	2	62	n.d.	1	1	[19]
(ADAP)CuCl	**24**	H	CO_2Et	1	34	1	35	27	[25]
(NHC)Cu-SiO$_2$	**31**	H	CO_2Et	2	45	1	31	17.5	[31]
Tp^{Br3}Ag(thf)	**25**	H	CO_2Et	2	98	1	3.7	1.4	[32]
Tp$^{(CF3)2}$Ag(thf)	**8**	H	CO_2Et	1	81	1	1.7	0.9	[12]
Tp^{Fn}Ag(THF)	**33**	H	CO_2Et	2	>99	2.5–3	3.6–4.5	1	[33]
Tp^{F21}Ag(OCMe$_2$)	**13**	H	CO_2Et	2	>99	1.6	3.3	1	[15]
Ag$_3$(3,5-(CF$_3$)$_2$PyrPy)$_3$	**34**	H	CO_2Et	2	46	1	5	2	[34]
[IPrAu(NCMe)]BAr$^F_4$	**10**	H	CO_2Et	2	78	1	2.9	1.8	[13]
(ADAP)AuCl	**35**	H	CO_2Et	2	29	1	2	1	[35]
[Au$_2$Cu$_2$(C$_6$F$_5$)$_4$(NCMe)$_2$]$_n$	**36**	H	CO_2Et	1	90	n.d.	1.5	1	[36]
Fe(ClO$_4$)$_2$/BPMEN	**20**	Ph	CO_2Et	2	66	1.4	1	1.05	[22]
NCPS-Ru	**16**	Ph	CO_2Et	2	50	1	1[a]		[18]

n.r., no reaction observed; n.d., not detected.

Table 1.3 Functionalization of 2,3-dimethylbutane.

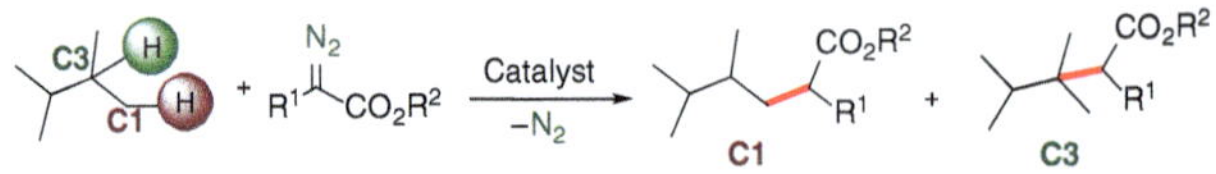

Catalyst		R^1	R^2	Yield (%)	Selectivity C1	Selectivity C3	References
$Rh_2(OOCCH_3)_4$	**37**	H	CO_2Et	n.r.	1	114	[8]
$Rh_2(OOCCF_3)_4$	**3**	H	CO_2Et	n.r.	1	44	[8]
$Rh_2(OOCTC)_4$	**26**	H	CO_2Et	n.r.	1	12	[8]
Rh_2(S-DOSP)$_4$	**4**	Ar	CO_2Et	27	n.d.	1	[9]
Rh(ttppp)(Me)(MeOH)	**14**	Ph	CO_2Me	48	1	1.3	[16]
$Tp^{Br3}Cu(NCMe)$	**6**	H	CO_2Et	56	n.d.	1	[11]
$Tp^{F21}Cu(OCMe_2)$	**12**	H	CO_2Et	16	1	15	[15]
$[IPrCu(NCMe)]BF_4$	**9**	H	CO_2Et	48	1	40	[13]
$(TPN)Cu(THF)BAr'_4$	**17**	H	CO_2Et	69	n.d.	1	[19]
$Tp^{(CF3)2Br}Ag(thf)$	**18**	H	CO_2Et	66	1	1.5	[20]
$Tp^{F21}Ag(OCMe_2)$	**13**	H	CO_2Et	96	1	2	[15]
$Tp^{Br3}Ag(thf)$	**25**	H	CO_2Et	98	1	2	[32]
$Tp^{(CF3)2}Ag(thf)$	**8**	H	CO_2Et	85	1	1.5	[12]
$Ag(3,5\text{-}(CF_3)_2PyrPy)$	**34**	H	CO_2Et	30	1	29	[34]
$[IPrAu(NCMe)]BAr'_4$	**10**	H	CO_2Et	90	1	1.2	[13]
$[Au_2Cu_2(C_6F_5)_4(NCMe)_2]_n$	**36**	H	CO_2Et	88	1	94	[36]

functionalization of alkanes by carbene insertion has undergone an outstanding development in the last two decades. Seminal work [9] using Rh_2(S-DOSP)$_4$ (**4**) with cycloalkanes and a series of donor–acceptor aryldiazoacetates provided good yields and high levels of asymmetric induction (60–93% ee, Scheme 1.6a), which varied when the reaction conditions were modified using degassed cycloalkanes as solvents at lower temperatures (88–96% ee, Scheme 1.6a). Also, acyclic alkanes were used, such as isobutane, 2,3-dimethylbutane, and 2-methylbutane. In this case, insertion in tertiary sites was favored, forming the corresponding products with variable yields and good to high enantioselectivities (Scheme 1.6b).

The rhodium–porphyrin complex **14** was later reported for the same transformation employing $N_2C(Ph)CO_2Me$ as the carbene source [8]. The use of cyclopentane, cyclohexane, and adamantane provided the corresponding products in good yields (64–80%) and with high enantioselectivities (88–93% ee) (Scheme 1.6a). This catalyst promoted a preferential selectivity toward the primary C—H bonds using *n*-hexane

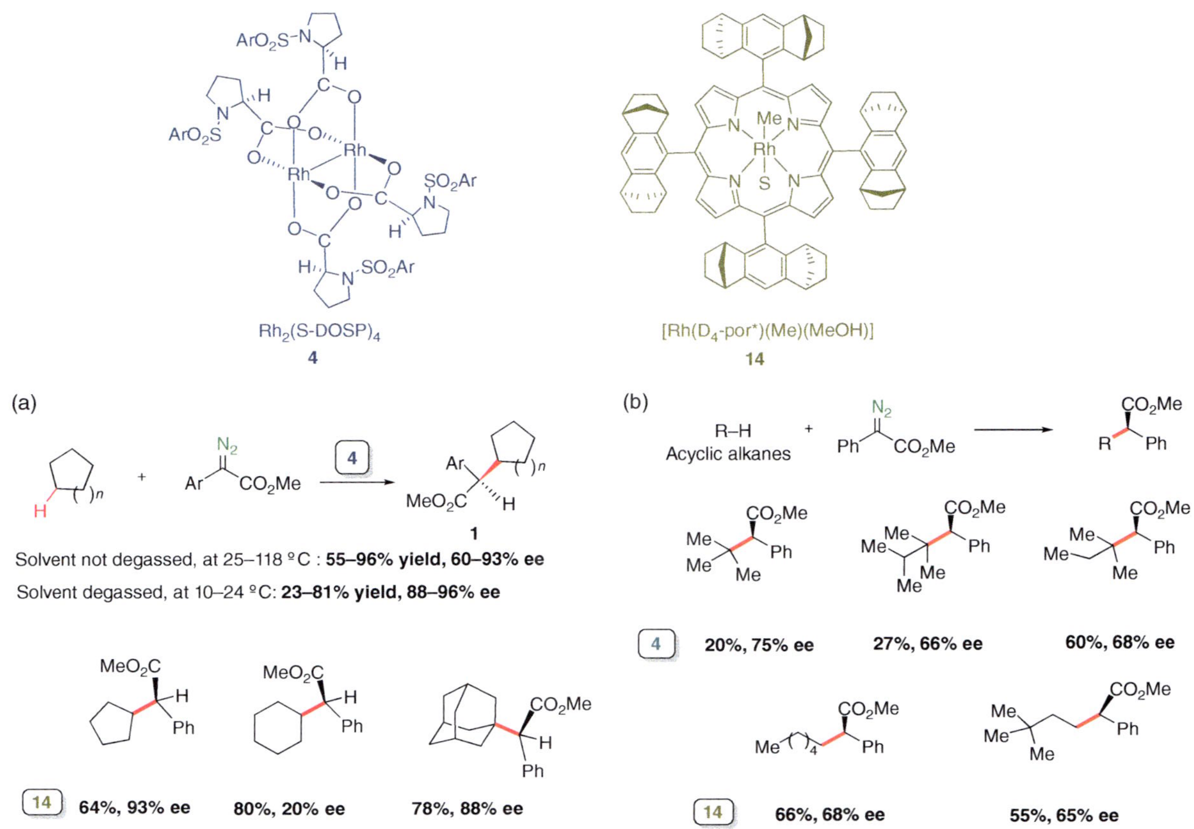

Scheme 1.6 First examples of asymmetric functionalization of alkanes by carbene insertion.

and 2,2-dimethylbutane, with moderate enantioselection (65–68% ee, Scheme 1.6c). A significant feature of **14** is its ability to be recycled up to five times, without significant change in selectivity and enantioselectivity values.

Novel chiral dirhodium catalysts **29** with excellent regio-, diastereo-, and enantioselectivities have been disclosed in the last decade. Their capabilities are maximized with the use of trihaloethyl aryldiazoacetates as the carbene precursor, giving rise to much cleaner outcomes, due to the suppression of adverse reactions. It has been also proposed that this trihaloethyl group provides a slight increase in the electrophilicity of metallocarbene intermediate. In chronological order, the first member of the series was Rh_2[R-3,5-di(p-^{t}BuC_6H_4)TPCP]$_4$ (**29a**, Scheme 1.7) [37], which promoted the preferential functionalization of secondary C—H bonds, with high values of diastereo and enantiomeric ratios.

Turning this selectivity into tertiary C—H bonds of alkanes was achieved with the catalyst Rh_2(TCPTAD)$_4$ (**29b**, Scheme 1.7) [38], with lower steric demand. The corresponding products were obtained with good yields and high stereoselectivity, in both the regio- and the enantio- cases. And the remaining and more challenging primary sites were reached using the complex Rh_2[*R*-tris(p-^{t}BuC_6H_4)TPCP]$_4$ (**29c**, Scheme 1.7) [39], which contains largely bulkier substituents to reduce the catalytic pocket and thus preventing the functionalization of both tertiary and secondary sites. The preferential functionalization of primary C—H bonds was thus reached with an excellent degree of asymmetric induction (Scheme 1.7).

Further advancement for the controlled C—H bond functionalization was accomplished in the case of a variety of mono- and disubstituted cyclohexanes [40], using catalyst Rh_2(S-TPPTTL)$_4$ (**29d**, Scheme 1.8). A collection of mono-substituted cyclohexanes were synthesized, predominantly functionalized in the equatorial position C3, with good yields (41–89%), as well as regio- (9.9 : 1–50 : 1 rr), diastereo- (3.3 : 1–26 : 1 dr), and enantioselectivities (47–98% ee) (Scheme 1.8).

1.4 Methane and Gaseous Alkanes as Substrates

The previous sections 1.2 and 1.3 enumerate catalytic systems for the chemo-, regio, diastereo-, and enantioselective functionalization of *liquid* alkanes C_nH_{2n+2} ($n \geq 5$) upon metal-mediated carbene transfer from a diazo compound onto a carbon–hydrogen bond. The lighter members of the alkane series which are gaseous at room temperature, i.e. those with n = 1–4, were not satisfactorily modified with this methodology until a novel strategy was developed [41]. Such a tactic consists in the use of supercritical carbon dioxide as the reaction medium [42], where those gaseous hydrocarbons are soluble. In this manner, the only C—H bonds available in the reaction mixture are those of the hydrocarbons, which display the highest values of BDE of the series [4]. Because of this, any other solvent containing C—H bonds would be activated before the desired gaseous alkanes.

The mixtures scCO_2-alkane form a fluid at 40 °C, using at least 80 bar of carbon dioxide, and provide a homogeneous medium where the catalyst and the diazo compound must dissolve. While EDA is completely soluble, only

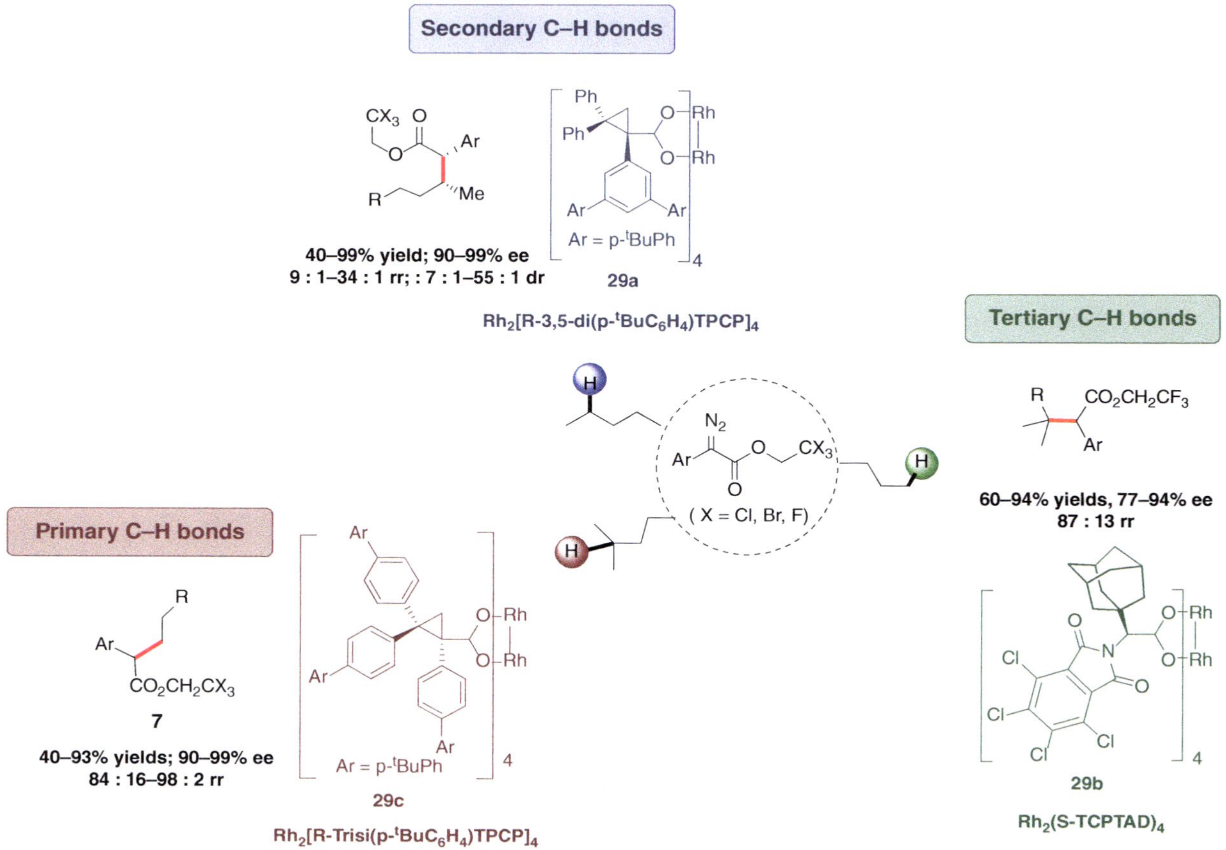

Scheme 1.7 Dirhodium-based catalysts for the control of the reaction site: primary, secondary, and tertiary C—H bond preferential functionalization (regioisomeric ratio, rr; diastereoisomeric ratio, dr).

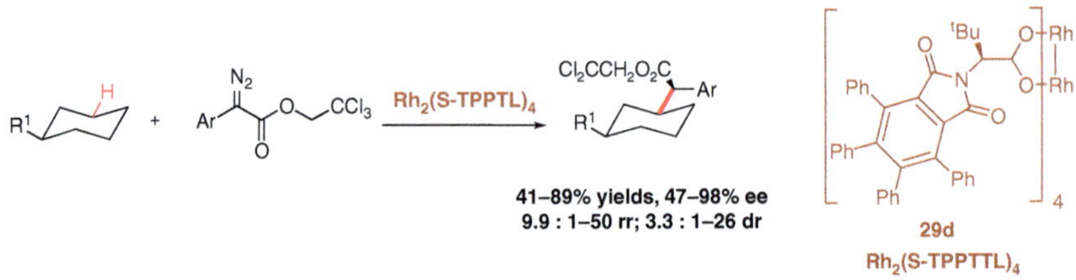

Scheme 1.8 Functionalization of mono- and disubstituted C—H bonds in cyclohexanes.

fluorine-containing catalysts are soluble in such mixture. Thus, a family of perfluorinated hydrotris(indazolyl)borate ligands (Scheme 1.9) was developed and the corresponding silver complexes synthetized [33], affording the first example of the functionalization of methane by incorporation of the $CHCO_2Et$ unit from EDA: ethyl propionate was obtained in 29% yield, based on initial EDA, using catalyst **13** [41, 43]. Further development led to the synthesis of $Tp^{(CF3)2,Br}ML$ (M = Cu, **30**; Ag, **18**) [44], which gave an improved yield of 33% with **18**. Additionally, catalyst **30** gave 4% of ethyl propionate, in the first example of copper for methane functionalization by carbene transfer.

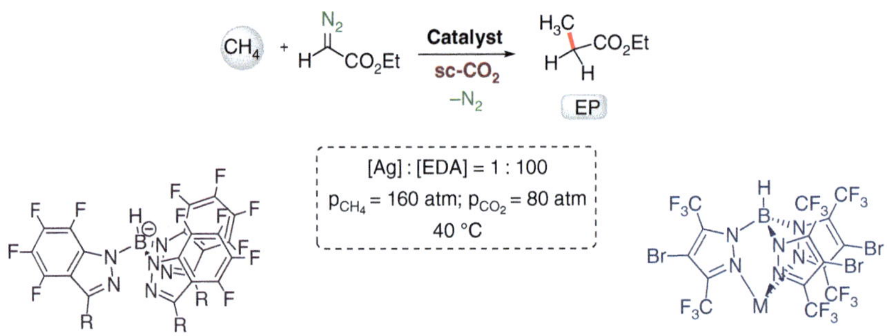

R	Tp^F	Tp^FAgL	Yield EP
CF_3	Tp^{F21}	13	29
CF_2CF_3	Tp^{F27}	33a	22
$CF_2CF_2CF_3$	Tp^{F33}	33b	19
$CF_2CF_2CF_2CF_3$	Tp^{F39}	33c	18
$CF_2CF_2CF_2CF_2CF_2CF_3$	Tp^{F51}	33d	18

M	Catalyst	Yield EP
Cu	30	4
Ag	18	33

Scheme 1.9 Methane functionalization by carbene insertion using silver-based catalysts.

Ethane, propane, and butane have also been employed as the substrates, with the above silver complexes bearing indazolyl- or pyrazolylborate ligands as well as the trinuclear silver complex **34** [34], containing pyrazolyl bridges between the metal ions. Scheme 1.10 displays the comparison between three catalysts, observing moderate to high yields (EDA-based) in most cases. The presence of two different C—H bonds in propane and butane leads to the formation of mixtures of two products, the values of regioselectivities being similar for **18** and **33a**, and significantly different for

Alkane	Products		Yield (%) 33a	18	34
Ethane	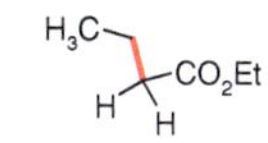		84%	99%	17%
Propane	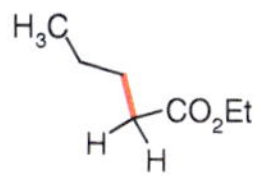	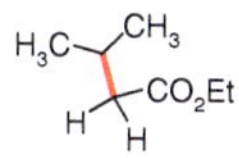	95% (1 : 3.6)	75% (1 : 3.8)	37.5% (1 : 14.5)
Butane	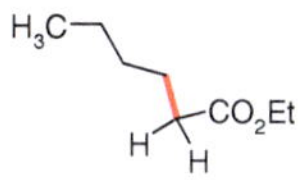	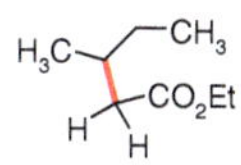	95% (1 : 2.1)	99% (1 : 1.9)	66% (1 : 10.1)

Regioselectivity (1^{ry} : 2^{ry})

Scheme 1.10 Alkane C2–C4 functionalization by carbene insertion using silver-based catalysts.

the trinuclear complex **34**. It is worth mentioning that the latter catalyst is employed under homogeneous conditions, using an organic solvent along with a high pressure of the alkane, whereas the former **18** and **33a** were tested under the abovementioned conditions employing scCO_2 as the reaction medium.

To complete this section, a unique example of the catalytic functionalization of methane and gaseous alkanes at room temperature in water deserves some comment. Catalyst **18** was employed in the reaction of EDA and methane using water as the solvent and in the presence of surfactants capable of generating micelles (Scheme 1.11) [45]. In the absence of such additives, no reaction took place, whereas when added above their critical micellar concentration, significant amounts of ethyl propionate were formed, reaching 14% yield when PFOS (perfluorooctane sulfonate) was employed. The formation of micelles constitutes the critical point in this tactic since methane accumulates in the inner volume of such aggregates, reaching 2 M values, nearly 4 orders of magnitude higher than the solubility of methane in water (c. 10^{-2} M). The other gaseous alkanes were also functionalized under the same conditions. Interestingly, the regioselectivity found in this micellar system was identical to that in the scCO_2 reaction medium. Such similarity is interpreted as due to a null effect of the micelle in the catalytic reaction, serving exclusively as a reaction vessel concentrating all the reactants.

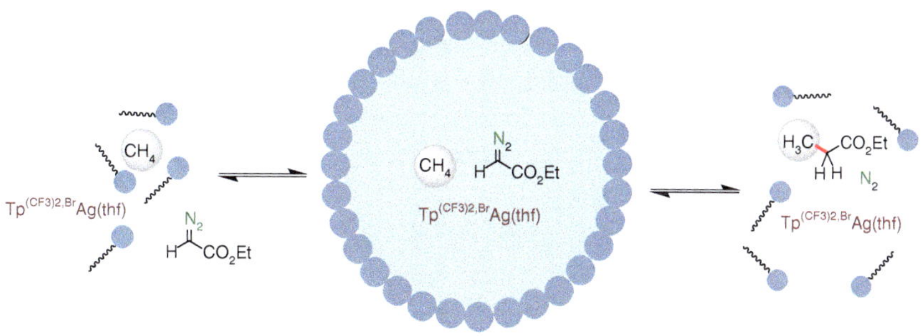

Scheme 1.11 The room temperature functionalization of methane in water surfactant as reaction medium. Source: Based on Gava et al. [45].

1.5 Alkane Nucleophilicity Scale

The development of catalysts capable of transferring a carbene onto the C—H bond of methane provided an opportunity to carry out competition experiments between different alkanes and evaluate the relative reactivity of the distinct reaction sites. Thus, a complete study with 14 alkanes and cycloalkanes and 29 carbon–hydrogen bonds was performed [20], providing the first large set of intermolecular comparative reactivity using methane as the reference. The procedure consisted of experiments using exactly measured amounts of two alkanes and in a careful analysis of the final reaction mixture (Scheme 1.12).

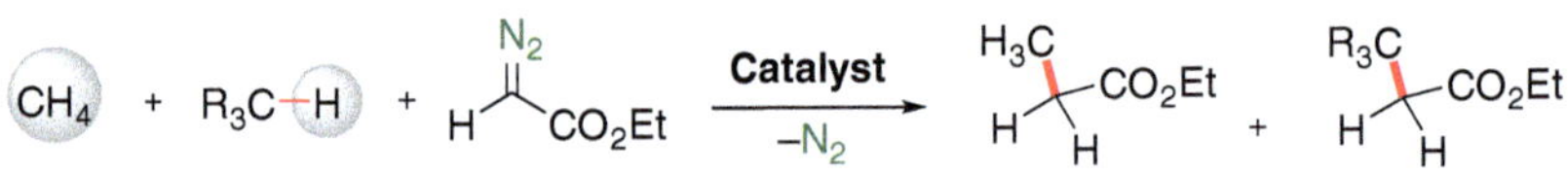

Scheme 1.12 Alkane competition experiments.

Scheme 1.13 displays the reaction profile. The rate determining step in this transformation is the formation of the metallocarbene, the barrier for the interaction of the latter with the C—H bond being lower in energy. Therefore, the rate determining step is not that where selectivity is decided. In addition, the product is formed in one final step which is irreversible. Because of this, the ratio of products derived from the functionalization of each bond can be connected to the difference in activation energies of such bonds.

The array of alkanes employed are displayed in Scheme 1.14, along with a plot of the values of the relative reactivity referred to methane, in log scale. It is noteworthy the observance that the usually claimed order of reactivity 1ry < 2ry < 3ry C—H bonds is not followed in a general manner. This is a consequence of the origin of such trend, developed by Olah using small electrophiles (H^+, NO_2^+, etc.) [46], at variance with the bulkier organometallic metallocarbene complexes acting as electrophiles in this case.

Further work with this experimental data allowed proposing a model for the quantitative prediction of such values employing a set of topological descriptors of the carbon–hydrogen bond being functionalized [47]. After a statistical treatment of experimental data, the new method, named as Q-DEAN (Quantitative DEscriptor-based Alkane Nucleophilicity), contains an equation with six descriptors, from which the observed experimental reactivity can be obtained. When the model is applied, a very good fitting between predicted and experimental data is found. Scheme 1.15 shows the general equation, the definition of the six descriptors, and a radial plot where the quality of the fitting is clearly visible.

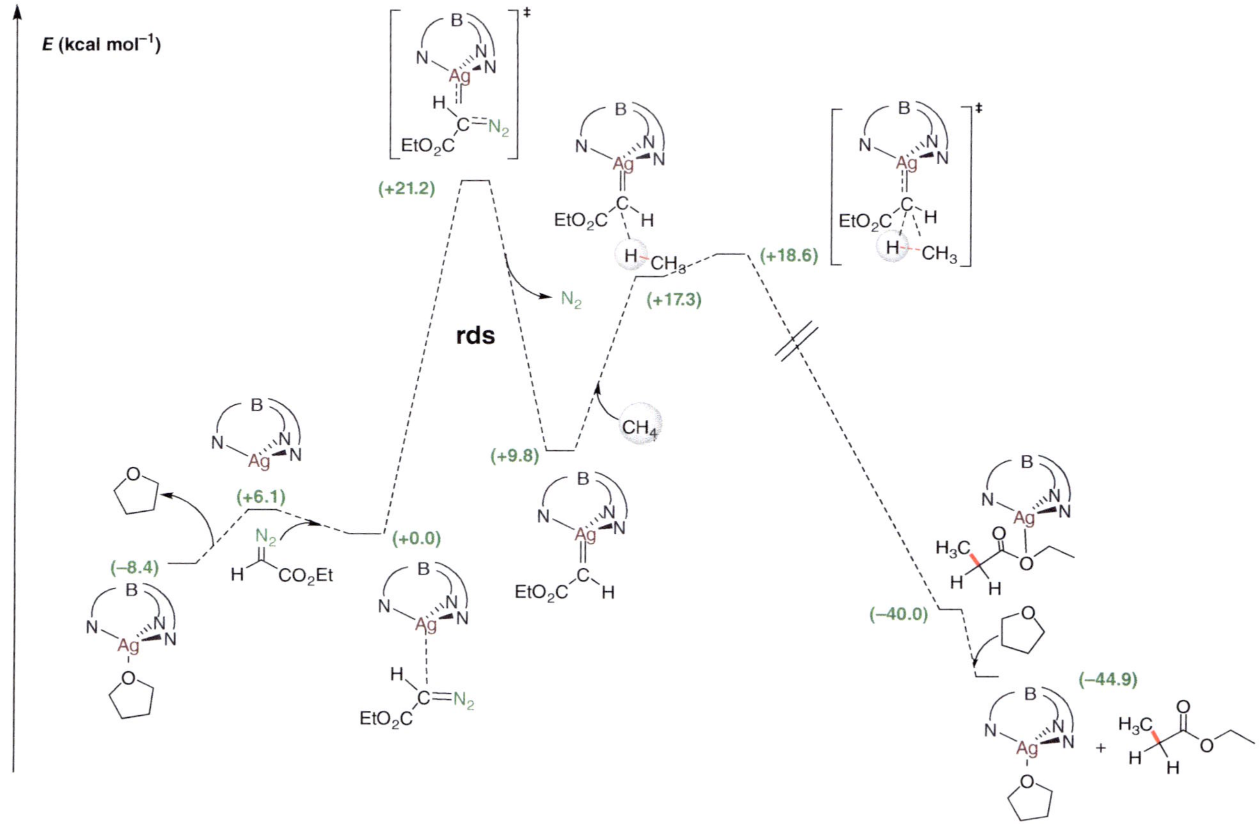

Scheme 1.13 DFT calculations for the functionalization of methane with EDA using catalyst **33a**. Source: Based on Olmos et al. [20].

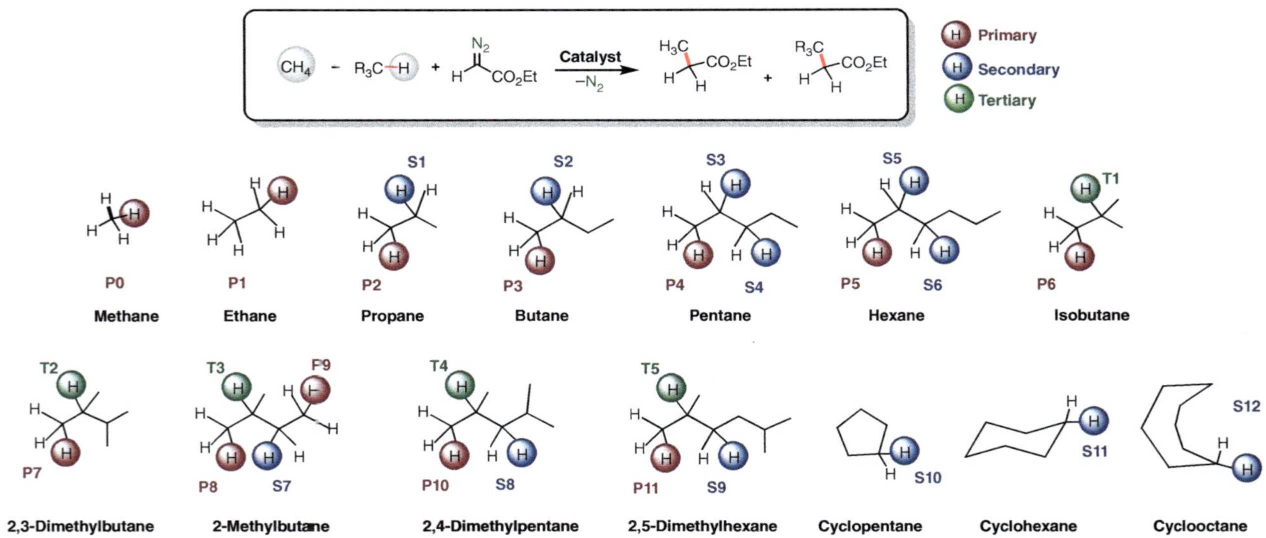

Scheme 1.14 Competition experiments carried out with the alkanes shown. The plot at the bottom shows the variation of $\log(R_r)$.

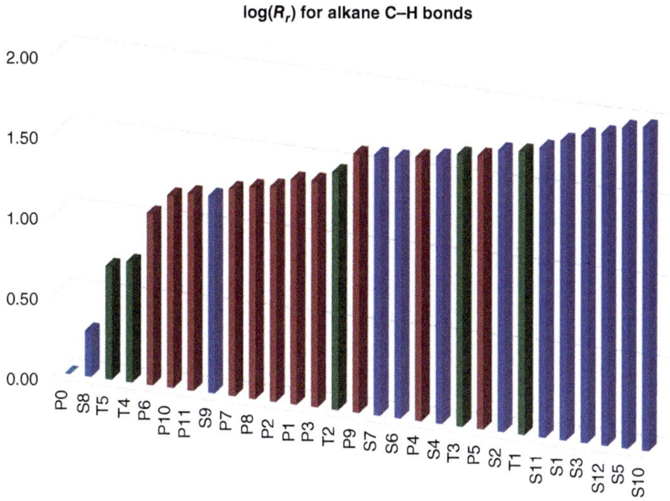

Scheme 1.14 (*Continued*)

$$\log(R_r) = A + c_{type}D_{Ctype} + \Sigma(c_{Ri} \times D_{Ri}) + c_{\beta}D_{\beta} + c_{st}D_{st}$$

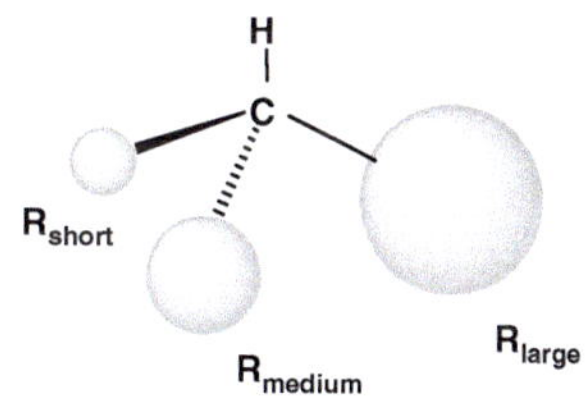

Descriptors

D_{Ctype}	Degree of substitution (1, 2 or 3)	$D_{Rmedium}$	Number of carbons in R_{medium}
D_{Rshort}	Number of carbons in R_{short}	D_{Rlarge}	Number of carbons in R_{large}
D_{st}	The steric factor (1 or 0)	D_{β}	Number of Me groups in C_{β}

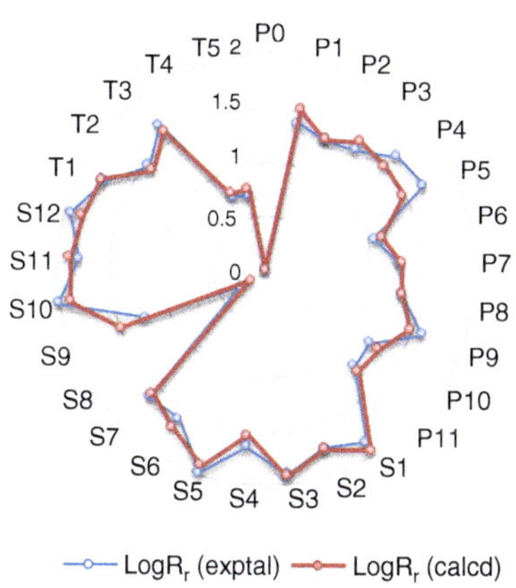

Scheme 1.15 The Q-DEAN model, showing the good approach of the calculated data to the experimental values of relative reactivities (methane as reference; log values).

1.6 Conclusions and Outlook

In nearly two decades, the modification of unreactive alkanes upon carbene insertion into their carbon–hydrogen bonds has moved from nearly non-existent, far from utility examples to an array of catalytic systems exerting a high control in the selectivity. Very active catalysts induce very high yields, based on the diazo reagent, whereas a set of carefully designed metal complexes allows the selective functionalization of the distinct reaction sites, particularly in linear or branched alkanes. However, there are yet several drawbacks to surpass. Among them, the use of base metals, non-expensive and widely available, is not frequent, with the exception of copper. Expansion to other first row elements as well as improvement with copper-based catalysts should be developed. A second target would consist in the use of diazo reagents with a low degree of elaboration, therefore making the transformation of interest for a potential use. Finally, the development of strategies in which the catalyst is readily removed and, preferentially, reused, is nearly unknown in this particular area and should be also targeted in the next future.

Acknowledgments

We thank the support of *MINECO (CTQ2017-82893-C2-1-R), Junta de Andalucía (P18-RT-1536) and PO FEDER 2014-2020 (UHU-1260216).*

References

1 Silberrad, O. and Roy, C.S. (1906). *J. Chem. Soc.* 89: 179.

2 (a) Doyle, M.P., McKervey, M.A., and Ye, T. (1998). *Modern Catalytics Methods for Organic Synthesis with Diazo Compounds*. John Wiley & Sons. (b) Doyle, M.P., Duffy, R., Ratnikov, M., and Zhou, L. (2010). *Chem. Rev.* 110: 704. (c) Doyle, M.P. (1986). *Chem. Rev.* 86: 919.

3 (a) Caballero, A., Díaz-Requejo, M.M., Fructos, M.R. et al. (2015) *Dalton Trans.* 44: 20295. (b) Díaz-Requejo, M.M., Belderrain, T.R., Nicasio, M.C., and Pérez, P.J. (2006). *Dalton Trans.* 47: 5559. (c) Díaz-Requejo, M.M. and Pérez, P.J. (2008). *Chem. Rev.* 108: 3379. (d) Davies, H.M.L. and Dick, A.R. (2010). *Top. Curr. Chem.* 292: 303. (e) Davies, H.M.L. and Manning, J.R. (2008). *Nature* 451: 417. (f) Davies, H.M.L. and Beckwith, R.E.J. (2003). *Chem. Rev.* 103: 2861. (g) Davies, H.M.L. and Morton, D. (2011). *Chem. Soc. Rev.* 40: 1857. (h) Cheng, Q.Q. and Doyle, M.P. (2016). *Adv. Organomet. Chem.* 66: 1. (i) Chen, Z., Rong, M.Y., Nie, J. et al. (2019). *Chem. Soc. Rev.* 48: 4921.

4 Luo, Y.R. (2007). *Comprehensive Handbook of Chemical Bond Energies*. Boca Raton: CRC.

5 Rivilla, I., Sameera, W.M.C., Álvarez, E. et al. (2013). *Dalton Trans.* 42: 4132.

6 Scott, L.T. and DeCicco, G.J. (1974). *J. Am. Chem. Soc.* 96: 322.

7 (a) Callot, H.J. and Metz, F. (1982). *Tetrahedron Lett.* 23: 4321. (b) Callot, H.J. and Metz, F. (1985). *Nouv. J. Chim.* 9: 167.

8 (a) Demonceau, A., Noels, A.F., and Hubert, A. (1989). *J. Mol. Catal.* 57: 149. (b) Demonceau, A., Noels, A.F., Costa, J.L., and Hubert, A. (1990). *J. Mol. Catal.* 58: 21.

9 (a) Davies, H.M.L., Hansen, T., and Churchill, M.R.J. (2000). *J. Am. Chem. Soc.* 122: 3063. (b) Davies, H.M.L. and Hansen, T. (1997). *J. Am. Chem. Soc.* 119: 9075. (c) Davies, H.M.L. and Antoulinakis, E.G. (2001). *J. Organomet. Chem.* 617–618: 47.

10 Díaz-Requejo, M.M., Belderrain, T.R., Nicasio, M.C. et al. (2002). *J. Am. Chem. Soc.* 124: 896.

11 (a) Caballero, A., Díaz-Requejo, M.M., Belderrain, T.R. et al. (2003). *J. Am. Chem. Soc.* 125: 1446. (b) Caballero, A., Díaz-Requejo, M.M., Belderrain, T.R. et al. (2003). *Organometallics* 22: 4145.

12 (a) Dias, H.V.R., Browning, R.G., Richey, S.A., and Lovely, C.J. (2004). *Organometallics* 23: 1200. (b) Lovely, C.J., Flores, J.A., Meng, X., and Dias, H.V.R. (2009). *Synlett*: 129.

13 Fructos, M.R., de Frémont, P., Nolan, S.P. et al. (2006). *Organometallics* 25: 2237.

14 Rodríguez, P., Álvarez, E., Nicasio, M.C., and Pérez, P.J. (2007). *Organometallics* 26: 6661.

15 Despagnet-Ayoub, E., Jacob, K., Vendier, L. et al. (2008). *Organometallics* 27: 4779.

16 Thu, H.Y., Tong, G.S., Huang, J.S. et al. (2008). *Angew. Chem. Int. Ed.* 47: 9747.

17 Mbuvi, H.M. and Woo, L.K. (2008). *Organometallics* 27: 637.

18 Choi, M.K.W., Yu, W.Y., So, M.H. et al. (2008). *Chem. Asian J.* 3: 1256.

19 Pérez, J., Morales, D., García Escudero, L.A. et al. (2007). *Dalton Trans.*: 379.

20 Olmos, A., Gava, R., Noverges, B. et al. (2018). *Angew. Chem. Int. Ed.* 57: 1.

21 Kulkarni, N.V., Dash, C., Jayaratna, N.B. et al. (2015). *Inorg. Chem.* 54: 11043.

22 Cheng, Q.Q., Yang, J.M., Xu, H., and Zhu, S.F. (2017). *Synlett* 20: 1327.

23 Ballestin, P., Ventura Espinosa, D., Martín, S. et al. (2019). *Chem. Eur. J.* 25: 9534.

24 McLarney, B.D., Hanna, S., Musaev, D.G., and France, S. (2019). *ACS Catal.* 9: 4526.

25 Pizarro, J.D., Molina, F., Fructos, M.R., and Pérez, P.J. (2020). *Chem. Eur. J.* 26: 10330.

26 Liu, Z., Cao, S., Yu, W. et al. (2020). *Chem* 6: 2110.

27 Demonceau, A., Noels, A.F., Hubert, A.J., and Teyssie, P. (1984). *Bull. Soc. Chim. Belg.* 93: 945.

28 Endres, A. and Maas, G. (2002). *J. Organomet. Chem.* 643: 174.

29 Liao, K., Negretti, S., Musaev, D.G. et al. (2016). *Nature* 533: 230.

30 Gava, R., Olmos, A., Noverges, B. et al. (2015). *ChemCatChem* 7: 3254.

31 Sarmiento, J.T., Olmos, A., Belderrain, T.R. et al. (2019). *ACS Sustainable Chem. Eng.* 7: 7346.

32 Urbano, J., Belderraín, T.R., Nicasio, M.C. et al. (2005). *Organometallics* 24: 1528.

33 Fuentes, M.A., Muñoz, B.K., Jacob, K. et al. (2013). *Chem. Eur. J.* 19: 1327.

34 Flores, J.A., Komine, N., Pal, K. et al. (2012). *ACS Catal.* 2: 2066.
35 Pizarro, J.D., Molina, F., Fructos, M.R., and Pérez, P.J. (2020). *Organometallics* 39: 2553.
36 Fuentes, M.A., Rodríguez-Castillo, M., Monge, M. et al. (2011). *Inorg. Chim. Acta* 369: 146.
37 (a) Rackl, D., Yoo, C.J., Jones, C.W., and Davies, H.M.L. (2017). *Org. Lett.* 19: 3055. (b) Liao, K., Liu, W., Niemeyer, Z.L. et al. (2018). *ACS Catal.* 8: 678. (c) Liu, W., Ren, Z., Bosse, A.T. et al. (2018). *J. Am. Chem. Soc.* 140: 12247.
38 Liao, K., Pickel, T.C., Boyarskikh, V. et al. (2017). *Nature* 551: 609.
39 Liao, K., Yang, Y.F., Li, Y. et al. (2018). *Nat. Chem.* 10: 1048.
40 Fu, J., Ren, Z., Bacsa, J. et al. (2018). *Nature* 564: 395.
41 Caballero, A., Despagnet-Ayoub, E., Díaz-Requejo, M.M. et al. (2011). *Science* 332: 835.
42 Olmos, A., Asensio, G., and Pérez, P.J. (2016). *ACS Catal.* 6: 4265.
43 Fuentes, M.A., Olmos, A., Muñoz, B.K. et al. (2014). *Chem. Eur. J.* 20: 11013.
44 Gava, R., Olmos, A., Noverges, B. et al. (2015). *ACS Catal.* 5: 3726.
45 Gava, R., Ballestin, P., Prieto, A. et al. (2019). *Chem. Commun.* 55: 11243.
46 Olah, G.A. (1973). *Angew. Chem. Int. Ed. Engl.* 12: 173.
47 Besora, M., Olmos, A., Gava, R. et al. (2020). *Angew. Chem. Int. Ed.* 59: 3112.

2

Catalytic Radical Approach for Selective Carbene Transfers via Cobalt(II)-Based Metalloradical Catalysis

Xiaoxu Wang and X. Peter Zhang

Merkert Chemistry Center, Boston College, Department of Chemistry, Chestnut Hill, MA 02467, USA

2.1 Introduction

Radicals are highly reactive intermediates that have been found to engage in many types of reactions, including radical addition, radical substitution, atom abstraction, and radical scission [1]. Owing to their rich reactivities and unique attributes, there has been an unprecedented growth in the development of radical transformations as powerful tools for modern organic synthesis over the past decades [2]. In comparison with the ionic counterparts, radical reactions have a number of inherent synthetic advantages [3]. For example, radical reactions typically proceed at fast reaction rates under mild and neutral conditions and feature high degree of functional group tolerance. Moreover, radical processes have the capability to undergo multiple reactions in a cascade fashion, allowing for rapid construction of complex molecules from simple and readily available building blocks [4]. Despite the vast synthetic potential of radical chemistry, the control of reactivity and stereoselectivity of the highly reactive radical species has been largely unaddressed and poses a formidable challenge in the field [5].

Among recent advancements to address this enduring issue in radical chemistry [6], metalloradical catalysis (MRC) has emerged as a conceptually new approach in which metal-centered radicals are explored as open-shell catalysts for catalytic initiation and regulation of homolytic radical reactions [7, 8]. To this end, Co(II) complexes of porphyrins [Co(Por)], as stable 15e-metalloradicals, have been demonstrated as effective catalysts for homolytically activating diazo compounds to generate the fundamentally new α-Co(III)-alkyl radicals (also known as Co(III)-carbene radicals) [9]. While these carbon-centered organic radicals are stabilized by the cobalt complex, they retain their radical nature to undergo common radical reactions, such as radical addition and H-atom abstraction as well as radical substitution. More importantly, through the support of a family of chiral amidoporphyrins with diverse electronic, steric, and chiral environments, the reactivity and stereoselectivity of these radical processes can be effectively controlled, leading

Transition Metal-Catalyzed Carbene Transformations, First Edition.
Edited by Jianbo Wang, Chi-Ming Che, and Michael P. Doyle.

to the development of a variety of stereoselective radical transformations. This review aims to provide an account of the recent development of Co(II)-based MRC (Co(II)-MRC) as a one-electron approach for selective carbene transfer processes with diazo compounds. Herein, we will specifically focus on selective carbene transfer reactions with diazo compounds catalyzed by [Co(Por)] that are known to proceed with radical pathways. Representative examples that are covered in this overview include radical cyclopropanation of alkenes, radical cyclopropenation of alkynes, and radical alkylation of C–H bonds. For carbene transfer reactions catalyzed by Co complexes of other types of ligands such as [Co(salen)], which will not be covered in this chapter, the reader is directed to some excellent reviews on this subject [7b, 10].

2.2 Intermolecular Radical Cyclopropanation of Alkenes

Cyclopropane derivatives are ubiquitous structural motifs found in many biologically important molecules and can serve as versatile synthons in organic synthesis [11]. Among different methods, transition-metal-catalyzed olefin cyclopropanation with diazo compounds as carbene sources represents the most general approach for stereoselective construction of the three-membered carbocycles [12]. To date, a number of catalytic systems based on Cu [13], Rh_2 [14], and Ru [15] complexes have been successfully developed for asymmetric olefin cyclopropanation with diazo compounds involving electrophilic metallocarbene intermediates. While the existing catalytic systems have been shown to work well with styrene derivatives and some electron-rich olefins, cyclopropanation of electron-deficient olefins has proven to be difficult presumably due to the electrophilic nature of the metallocarbene intermediates [16]. In addition to the limited scope of olefin substrates, several important types of diazo compounds, such as acceptor/acceptor- and donor-substituted diazo compounds, remain challenging for asymmetric cyclopropanation [16, 17]. Moreover, excess olefins and slow addition of diazo compounds are usually required to minimize unwanted carbene dimerization, which is a common side reaction in electrophilic metallocarbene-mediated cyclopropanation systems. In view of the aforementioned limitations, it is therefore highly desirable to develop a more general and efficient catalytic system that could incorporate a wide array of both olefins and diazo compounds for stereoselective synthesis of various cyclopropane derivatives.

Radical olefin cyclopropanation with diazo compounds via Co(II)-MRC has recently emerged as an alternative strategy for asymmetric synthesis of cyclopropanes. Distinct from the electrophilic metallocarbene-mediated cyclopropanation, the Co(II)-based metalloradical system has been demonstrated to possess a unique profile of reactivity and selectivity proceeding through a stepwise radical mechanism. As illustrated in Scheme 2.1, the general mechanism involves the metalloradical activation of diazo compound by [Co(Por)] to generate α-Co(III)-alkyl radical intermediate **I**. Subsequent radical addition to olefin can break the π bond and translocate the radical character to the olefin carbon, forming the typical

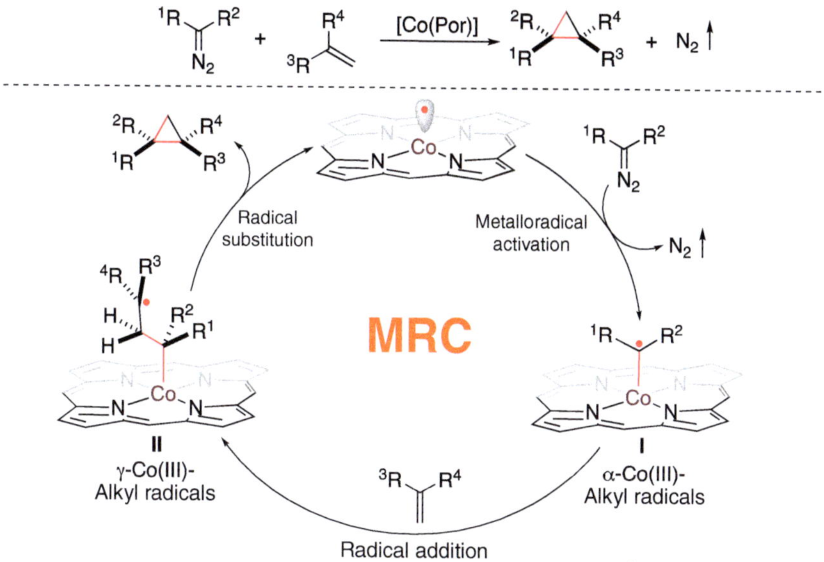

Scheme 2.1 General catalytic cycle of intermolecular olefin cyclopropanation via Co(II)-MRC.

carbon radical intermediate, namely γ-Co(III)-alkyl radical **II**. The highly reactive carbon-centered radical can then undergo radical substitution (3-*exo-tet* radical cyclization) with homolysis of the weak Co–C bond and concomitantly regenerate the catalyst to turn over the catalytic cycle. The following session will describe examples of [Co(Por)]-catalyzed intermolecular olefin cyclopropanation on the basis of different classes of diazo compounds as carbene sources (Scheme 2.2), with emphasis placed upon stereoselective olefin cyclopropanation processes.

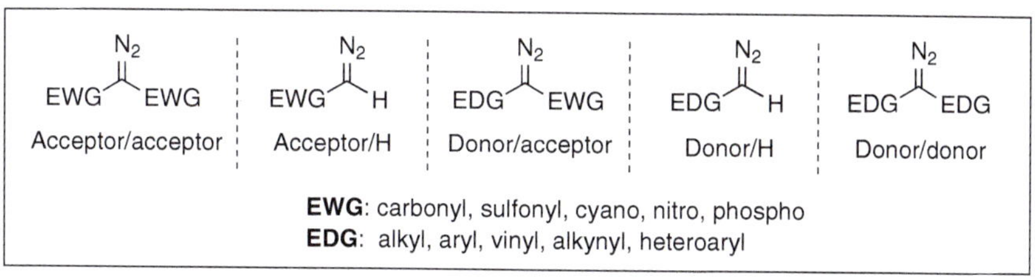

Scheme 2.2 Common classes of diazo compounds.

2.2.1 Cyclopropanation with Acceptor-Substituted Diazo Compounds

The first examples of [Co(Por)]-catalyzed intermolecular olefin cyclopropanation were independently reported by the groups of Cenini and Zhang in 2003 [18, 19]. Cenini and coworkers revealed that [Co(Por)] complexes are efficient catalysts for selective cyclopropanation of styrene and its derivatives with ethyl diazoacetate (EDA), affording the corresponding cyclopropanes in moderate-to-excellent yields with *trans*-diastereoselectivity (Scheme 2.3) [18].

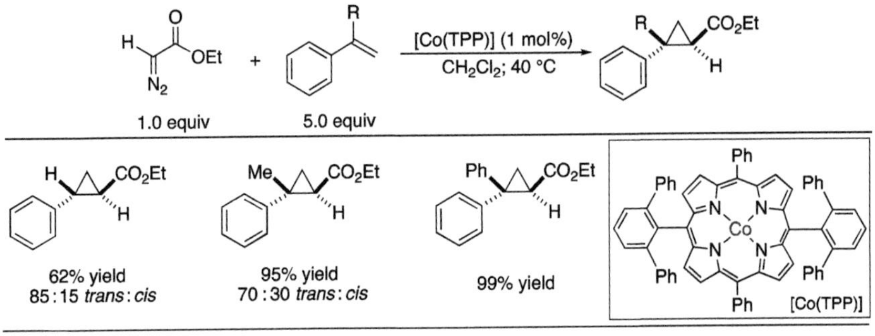

Scheme 2.3 [Co(TPP)]-catalyzed *trans*-selective cyclopropanation with diazoacetates. Source: Based on Penoni et al. [18].

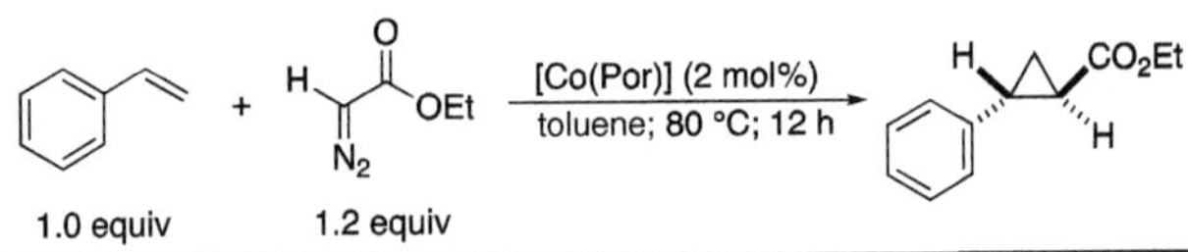

Entry	[Co(Por)]	Yield (%)	*trans* : *cis*	*trans* (% ee)	*cis* (% ee)
1	[Co(**P1**)]	73	36 : 64	62	77
2	[Co(**P2**)]	84	52 : 48	10	31

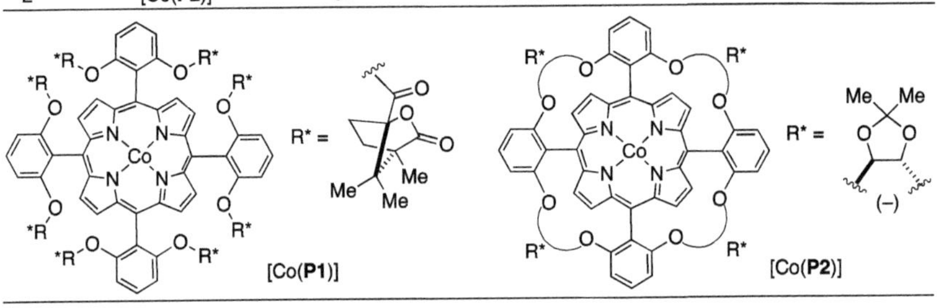

Scheme 2.4 [Co(Por)]-catalyzed asymmetric cyclopropanation of styrene with EDA. Source: Based on Huang et al. [19].

In the same year, Zhang and coworkers demonstrated that [Co(Por)] complexes exhibit superior reactivity and selectivity for stereoselective cyclopropanation in comparison with other common metalloporphyrin catalysts [19]. In addition, the [Co(Por)]-based catalytic system can operate with alkenes as the limiting reagents without the need for slow addition of the diazo compounds. It was further shown that both the diastereoselectivities and enantioselectivities of [Co(Por)]-catalyzed cyclopropanation reactions could be tuned by using different chiral porphyrin ligands (Scheme 2.4). With chiral camphanic acid-based porphyrin **P1** as the supporting ligand, asymmetric cyclopropanation of styrene with EDA resulted in the formation of the desired cyclopropane as *cis*-dominant product in good yields with enantioselectivities of 77% ee and 62% ee for *cis*- and *trans*-cyclopropane, respectively. Although stereoselectivity is limited in this initial catalytic system, the results provide the first indication that asymmetric induction in catalytic cyclopropanation by Co(II) complexes of porphyrins is possible.

In 2004, inspired by the extraordinary potential of Vitamin B_{12} derivatives as natural catalysts for asymmetric olefin cyclopropanation [20], Zhang and coworkers pioneered the development of structurally well-defined Co(II) complexes of D_2-symmetric chiral amidoporphyrins [Co(D_2-Por*)] as stable metalloradical catalysts (Scheme 2.5) [21]. The modular construction of chiral amidoporphyrins is realized by robust palladium-catalyzed amidations using 5,10-bis(2′,6′-dibromophenyl)porphyrins as versatile synthons. Through different combinations of both chiral R* and *meso*-R groups of the porphyrin ligands, a family of D_2-symmetric chiral amidoporphyrins with diverse electronic, steric, and chiral properties can be synthesized in high yields.

The first application of [Co(D_2-Por*)] was showcased in the asymmetric cyclopropanation of styrene with diazoacetates using 1 mol% of [Co(**P3**)] (**P3** = 3,5-Di^{t}Bu-ChenPhyrin) (Scheme 2.6) [21]. It was found that the addition of a substoichiometric amount of 4-(dimethylamino)pyridine (DMAP) as additives is crucial to achieve high level of reactivity and stereoselectivity as a consequence of positive *trans*-effect on the metal center. It is worth noting that the reaction can be performed by using styrene as the limiting reagent under a one-pot protocol rather than the common practice of using excess olefins with slow addition of diazo compounds.

In addition to styrene, its derivatives with varied steric and electronic properties were also shown to undergo a highly stereoselective cyclopropanation with diazoacetates (Scheme 2.7) [22]. Remarkably, the exceptional reactivity with electron-deficient olefins distinguishes the Co(II)-based metalloradical system from the typical electrophilic metallocarbene-mediated cyclopropanation (Scheme 2.7) [23].

In 2008, Gallo, Rose, and coworkers reported the synthesis of Co(II) complexes of C_2-symmetric chiral bis-binaphthyl porphyrins and carried out systematic investigations on the efficiency of the catalysts for asymmetric olefin cyclopropanation with EDA (Scheme 2.8) [24]. It was shown that [Co(**C1**)] and [Co(**C2**)] could catalyze the cyclopropanation of various aromatic and nonaromatic olefins in high yields with *trans*-diastereoselectivities and good enantioselectivities. Intriguingly, by tuning the bridge length of the porphyrin ligands, [Co(**C1**)] and [Co(**C2**)] exhibited distinct reactivity and stereoselectivity even under the same reaction conditions. Furthermore, the presence of axial nitrogen ligand *N*-methylimidazole (NMI) can enhance the enantioselectivity of the cyclopropanation process.

In 2019, a new generation of Co(II) complexes of D_2-symmetric chiral porphyrins [Co(HuPhyrin)] were synthesized by Zhang and coworkers, which feature alkyl bridges across two chiral amide units on both sides of the porphyrin plane [25]. The bridging effect on the catalytic performance was exemplified in the asymmetric cyclopropanation of styrene with EDA. Both the bridged catalysts [Co(**P4**)] (**P4** = 3,5-Di^{t}Bu-Hu(C_4)Phyrin) and [Co(**P5**)] (**P5** = 3,5-Di^{t}Bu-Hu(C_6)Phyrin) exhibited superior catalytic reactivity and stereoselectivity than the open analogue [Co(**P3**)] (**P3** = 3,5-Di^{t}Bu-ChenPhyrin), furnishing the corresponding cyclopropane product in 93% yield with 95 : 5 dr and 96% ee (Scheme 2.9). Notably, the catalytic performance can be influenced by simply changing the length of the alkyl bridges

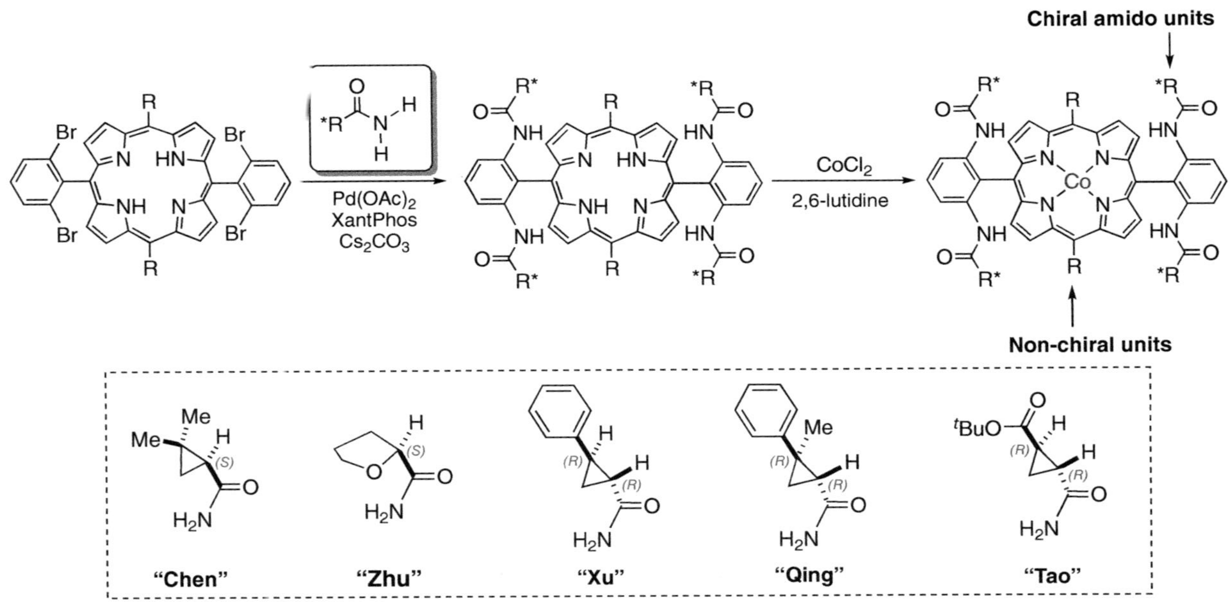

Scheme 2.5 Modular construction of Co(II) complexes of D_2-symmetric chiral amidoporphyrins [Co(D_2-Por*)]. Source: Based on Chen et al. [21].

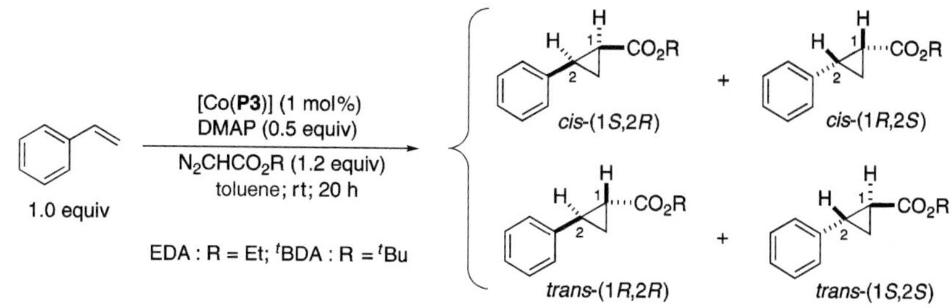

Entry	Diazo	Yield (%)	*trans* : *cis*	ee (%)	Config
1	EDA	82	97 : 3	78	1*R*,2*R*
2	^{t}BDA	84	> 99 : 1	95	1*R*,2*R*

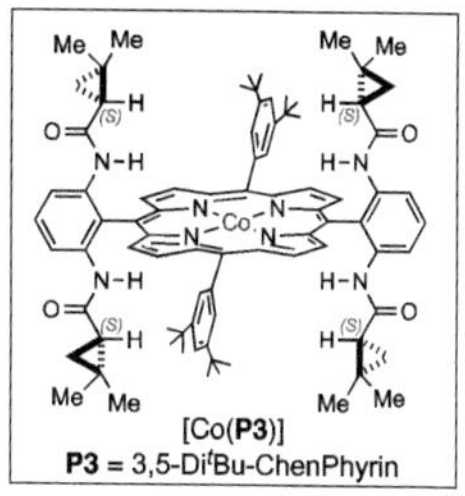

Scheme 2.6 [Co(D_2-Por*)]-catalyzed asymmetric cyclopropanation of styrene with diazoacetates. Source: Based on Chen et al. [21].

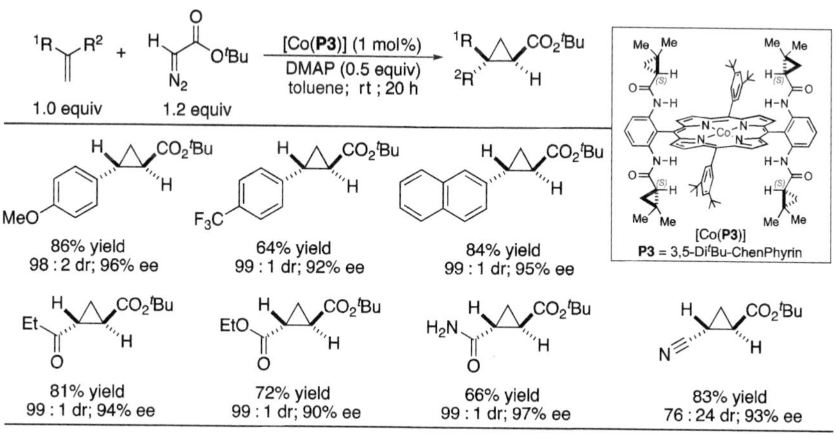

Scheme 2.7 Asymmetric cyclopropanation of different olefins with diazoacetates. Source: Based on Chen et al. [22, 23].

(C4 versus C6-alkyl chain), which may have an important implication in future catalyst design and development.

In addition to alkyl diazoacetates, succinimidyl diazoacetates were also found to be viable carbene sources for Co(II)-catalyzed asymmetric radical cyclopropanation of alkenes (Scheme 2.10) [26]. The resulting enantioenriched succinimidyl cyclopropyl carboxylates could serve as valuable synthons for reactions with different amines, affording a wide range of chiral cyclopropyl carboxamides with retention of the original enantiomeric purity. One useful transformation was demonstrated by the introduction of *D*-(+)-glucosamine moiety to the cyclopropyl core in a two-step approach under mild conditions (Scheme 2.10).

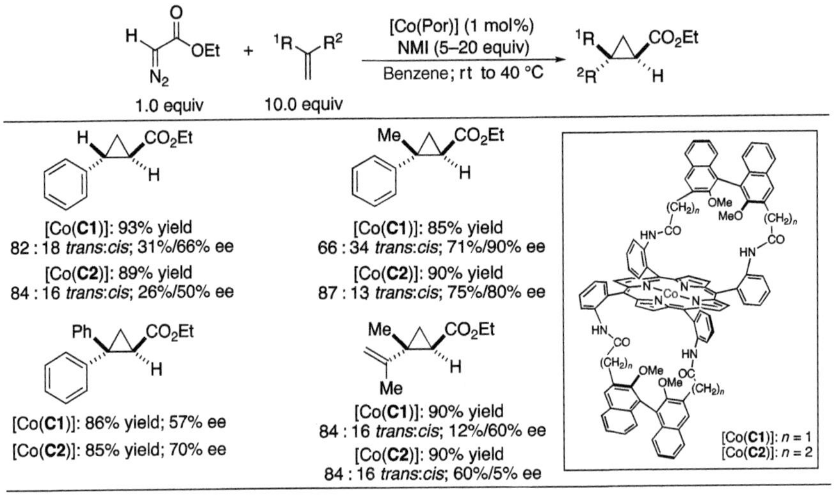

Scheme 2.8 Chiral cobalt(II)-binaphthyl porphyrins-catalyzed asymmetric cyclopropanation with EDA. Source: Based on Fantauzzi et al. [24].

To further extend the generality of Co(II)-based metalloradical cyclopropanation, diazosulfones were successfully employed for asymmetric olefin cyclopropanation (Scheme 2.11) [27]. A new Co(II) complex of D_2-symmetric chiral amidoporphyrins [Co(**P6**)] (**P6** = 2,6-DiMeO-ZhuPhyrin) was synthesized by incorporating both intramolecular O···H–N H-bonding interactions and cyclic tetrahydrofuran structures. As a result of the enhanced rigidity and polarity of the chiral environment for asymmetric induction, the metalloradical system was shown to be generally applicable for a broad range of styrene derivatives as well as electron-deficient olefins, leading to high-yielding formations of cyclopropyl sulfones with excellent control of diastereoselectivities and enantioselectivities.

2.2.2 Cyclopropanation with Acceptor/Acceptor-Substituted Diazo Compounds

While acceptor-substituted diazo compounds have been shown to be effective carbene sources, asymmetric olefin cyclopropanation with acceptor/acceptor substituted diazo compounds bearing two electron-withdrawing groups has been largely underexplored. This underdevelopment could presumably be due to the added stabilization by the second electron-withdrawing group that renders this class of diazo compounds less reactive toward formation of the metallocarbene intermediates. Consequently, the development of more reactive catalysts with high degree of stereocontrol is anticipated to address this challenge.

In 2008, Zhang and coworkers developed the first highly selective catalytic system for asymmetric olefin cyclopropanation with acceptor/acceptor-substituted diazo compounds (Scheme 2.12) [28]. With the employment of D_2-symmetric chiral amidoporphyrin (**P3** = 3,5-Di^{t}Bu-ChenPhyrin) as the optimal supporting ligand, the Co(II)-based metalloradical system enabled the stereoselective synthesis of

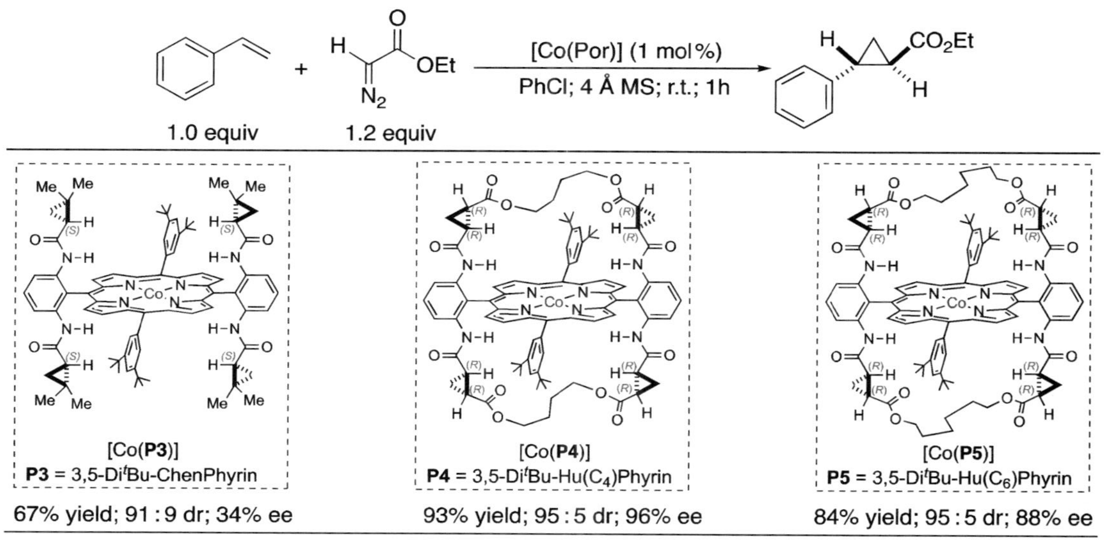

Scheme 2.9 Bridging effect on asymmetric cyclopropanation with diazoacetates. Source: Based on Hu el al. [25].

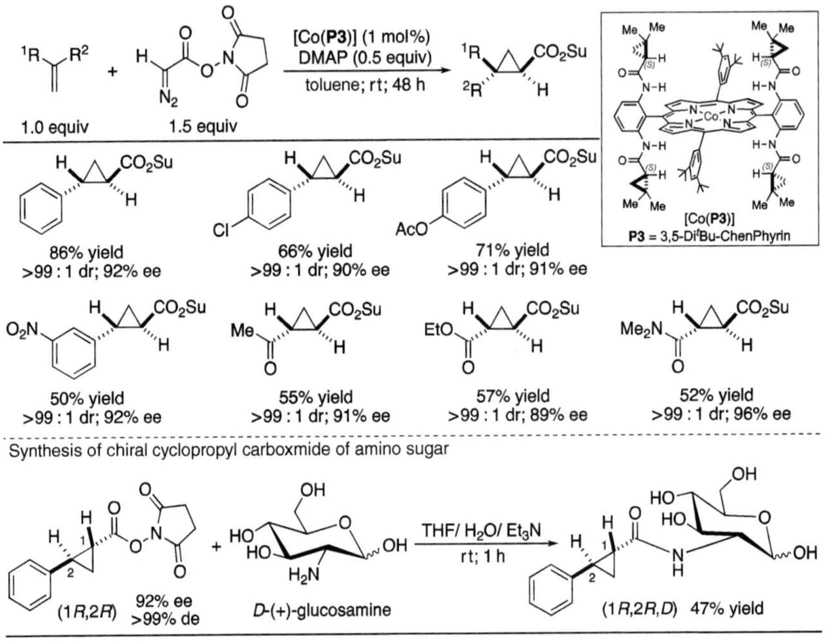

Scheme 2.10 Asymmetric cyclopropanation with succinimidyl diazoacetates. Source: Based on Ruppel et al. [26].

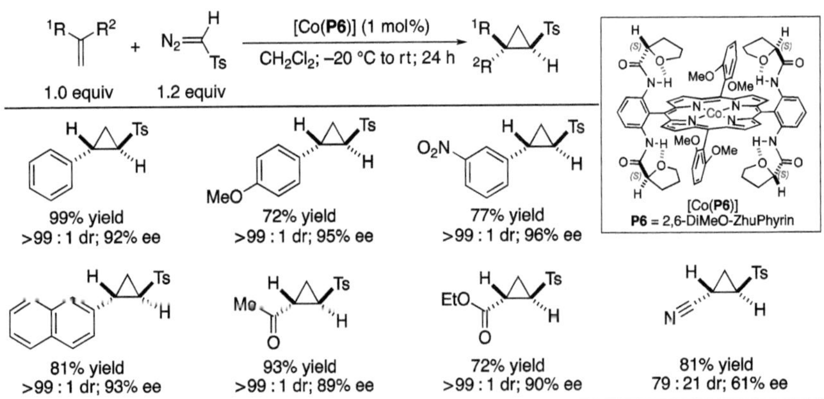

Scheme 2.11 Asymmetric cyclopropanation with diazosulfones. Source: Based on Zhu et al. [27].

(*Z*)-cyclopropane α-nitroesters from α-nitrodiazoacetates, which should allow ready access to valuable cyclopropane α-amino acids and amino cyclopropanes (Scheme 2.12). The effective stereocontrol is rationalized by the potential double H-bonding interactions between the chiral amide N–H units on the porphyrin ligand and the respective N=O and C=O units of the diazo compounds. These non-covalent interactions are postulated to promote the formation of α-Co(III)-alkyl radical intermediates as well as to reinforce their rigidity toward subsequent olefin addition in a more selective fashion. As a unique feature of the Co(II)-based metalloradical

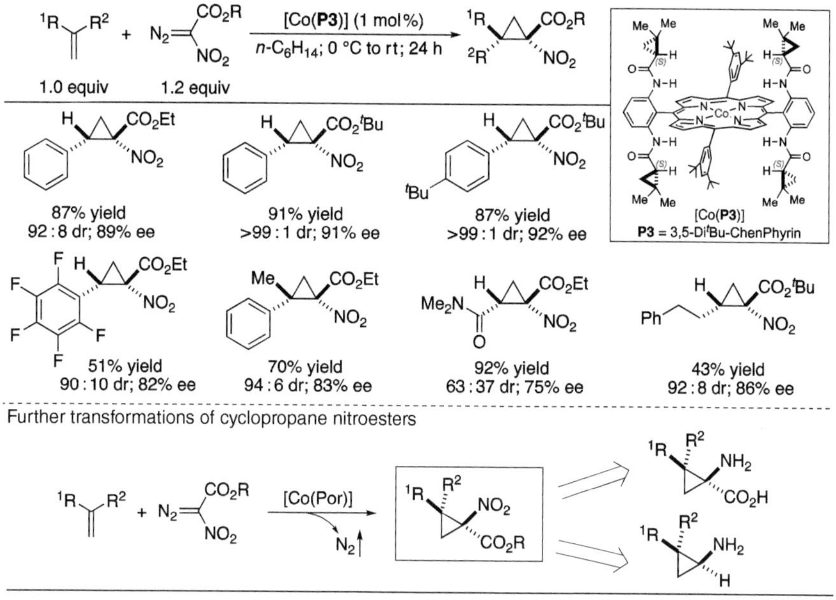

Scheme 2.12 Asymmetric cyclopropanation with nitrodiazoacetates. Source: Based on Zhu et al. [28].

system, electron-deficient pentafluorostyrene, acrylates, and acrylamides were well suited for the catalytic process. Moreover, aliphatic olefin could be successfully cyclopropanated with good diastereoselectivity and enantioselectivity albeit in a relatively lower yield.

Later on, the exceptional potential of [Co(D_2-Por*)] for activating acceptor/acceptor-substituted diazo compounds stimulates further development of asymmetric cyclopropanation with other types of acceptor/acceptor-substituted diazo compounds. In 2010, Zhang and coworkers achieved an unprecedented asymmetric olefin cyclopropanation with *tert*-butyl α-cyanodiazoacetates using [Co(**P3**)] (**P3** = 3,5-Di^{t}Bu-ChenPhyrin) as the optimal catalyst (Scheme 2.13) [29]. It is hypothesized that the existence of similar double H-bonding interactions between the cyano/ester groups and the amide units on the amidoporphyrin ligand can result in the stabilization of α-Co(III)-alkyl radicals, which can presumably enhance the catalytic reactivity and improve the product stereoselectivity. Importantly, this methodology offers a practical approach for preparation of enantioenriched 1,1-cyclopropanenitrile esters, which can serve as versatile synthons for potential synthetic and biological applications.

To explore the scope of acceptor/acceptor-substituted diazo compounds beyond α-nitrodiazoacetates and α-cyanodiazoacetates, Zhang and coworkers showcased another example of asymmetric olefin cyclopropanation by exploiting α-ketodiazoacetates as the carbene source (Scheme 2.14) [30]. Notably, the radical cyclopropanation by [Co(**P3**)] (**P3** = 3,5-Di^{t}Bu-ChenPhyrin) exhibits a distinct sense of diastereoselectivity from the previous Rh_2-catalyzed system [31], enabling a direct synthesis of (*E*)-1,1-cyclopropaneketoesters in high enantiopurity. The scope

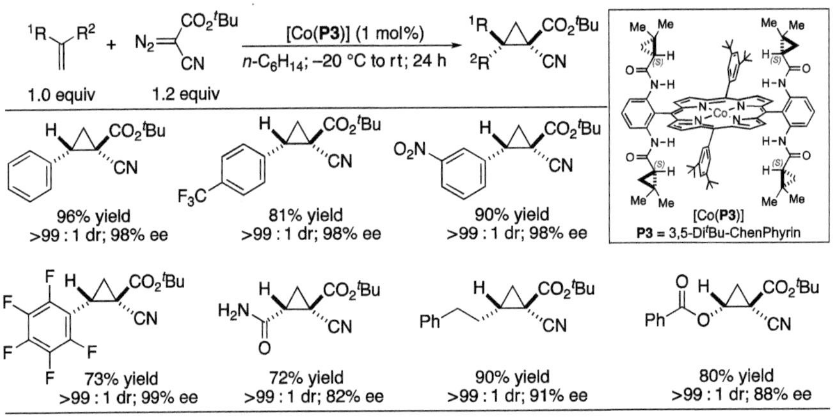

Scheme 2.13 Asymmetric cyclopropanation with cyanodiazoacetates. Source: Based on Zhu et al. [29].

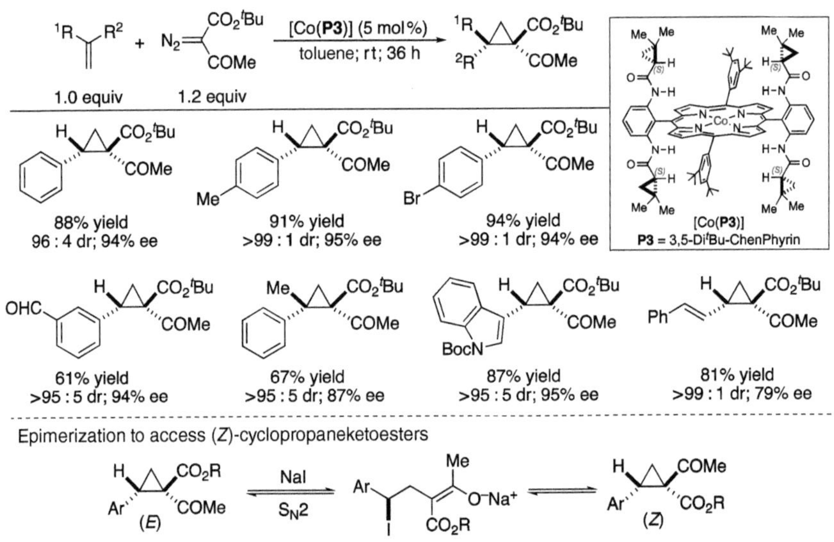

Scheme 2.14 Asymmetric cyclopropanation with ketodiazoacetates. Source: Based on Xu et al. [30].

of the olefin substrates could be extended beyond styrene derivatives, furnishing cyclopropaneketoesters with diversified electronics and functionalities. Furthermore, the opposite (*Z*)-diastereomers can also be accessed through stereospecific epimerization reactions with retention of the original optical purity (Scheme 2.14).

In 2017, Zhang and coworkers reported a highly stereoselective cyclopropanation of alkenes with α-formyldiazoacetates, which represent another challenging class of acceptor/acceptor-substituted diazo compounds (Scheme 2.15) [32]. The under-development is largely attributed to the poor compatibility of aldehyde moieties in the diazo compounds with the existing catalytic systems in addition to their inherent low reactivities. By employing [Co(**P3**)] (**P3** = 3,5-Di^{t}Bu-ChenPhyrin) as the effective metalloradical catalyst, the system enabled the direct synthesis of 1,1-cyclopropaneformylesters from α-formyldiazoacetates. Moreover, the significantly enhanced reactivity with [Co(**P3**)] as compared with [Co(TPP)] (TPP = 5,10,15,20-tetraphenylporphyrin) (<10% product yield) reinforces the hypothesis that the non-covalent H-bonding interactions play an important role in asymmetric inductions.

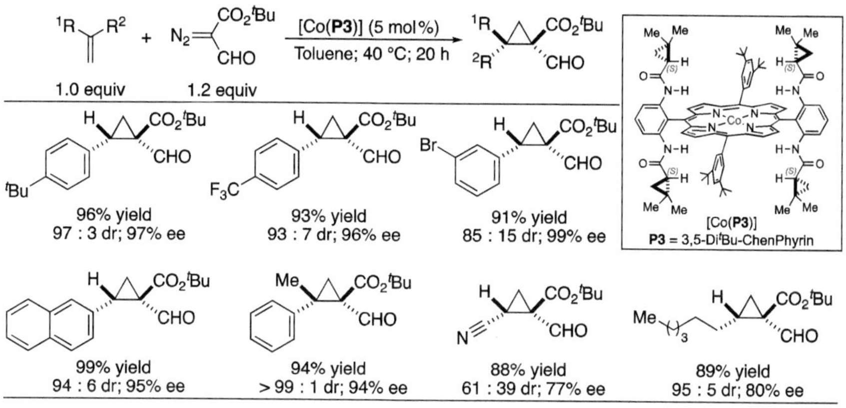

Scheme 2.15 Asymmetric cyclopropanation with formyldiazoacetates. Source: Based on Xu et al. [32].

2.2.3 Cyclopropanation with Donor-Substituted Diazo Compounds

Despite tremendous efforts being devoted to the development of asymmetric olefin cyclopropanation with both acceptor-substituted and acceptor/acceptor-substituted diazo compounds, catalytic systems with donor-substituted diazo compounds have been much less explored due to their intrinsic low stability. To circumvent the issues related with isolation of these less-stabilized diazo compounds, *N*-sulfonylhydrazones have been utilized as stable surrogates for *in situ* generation of the diazo compounds, enabling a broader scope of substrates for carbene transfer reactions [33].

In 2017, Zhang and coworkers developed a new Co(II)-based metalloradical system that is capable of catalyzing asymmetric olefin cyclopropanation with donor-substituted diazo compounds, generated in situ from *N*-sulfonylhydrazones (Scheme 2.16) [34]. The [Co(D_2-Por*)] containing two contiguous stereogenic centers [Co(**P7**)] (**P7** = 3,5-Di^{t}Bu-Xu(2′-Naph)Phyrin) was found to be highly effective for asymmetric radical cyclopropanation with α-aryldiazomethanes. In the previous studies, H-bonding interactions between the carbonyl groups of the diazo compounds and the chiral amide units on the porphyrin ligand have been found to be crucial in enhancing reactivity and controlling stereoselectivity of the cyclopropanation process. To achieve high level of asymmetric induction with donor-substituted diazo compounds lacking of carbonyl functionalities, Zhang and coworkers

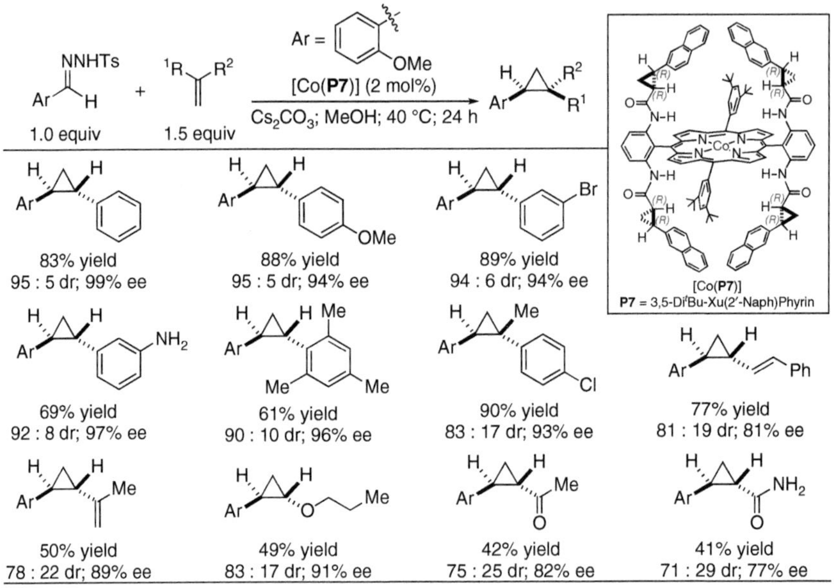

Scheme 2.16 Asymmetric cyclopropanation with *in situ* generated aryldiazomethanes. Source: Based on Wang et al. [34].

considered installing an *ortho*-methoxy group in the α-phenyldiazomethanes, which could engage in potential H-bonding interactions with the amido group of the chiral ligand. It is noted that the unique radical mechanism could enable the utilization of a broad range of olefin substrates, including both electron-rich and electron-deficient nonaromatic olefins, affording the corresponding cyclopropane derivatives in high yields with excellent control of stereoselectivities.

To probe the underlying stepwise radical mechanism, both (*E*)- and (*Z*)-β-deuterostyrenes were employed as substrates for [Co(**P7**)]-catalyzed cyclopropanation with 2,6-difluoroaryltosylhydrazones (Scheme 2.17). Given that a concerted

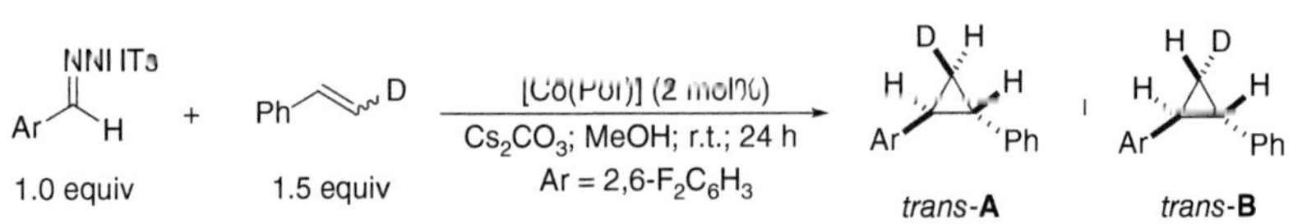

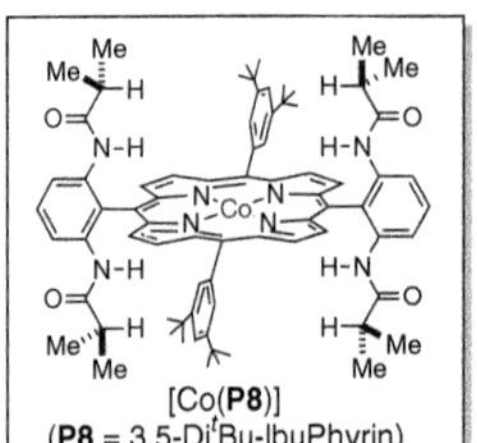

[Co(Por)]	β-D-styrene	*trans* **A** / **B**
[Co(**P7**)]	*E*	93/7
[Co(**P7**)]	*Z*	7/93
[Co(**P8**)]	*E*	80/20
[Co(**P8**)]	*Z*	20/80

Scheme 2.17 Mechanistic study on the cyclopropanation of (*E*)- and (*Z*)-β-deuterostyrene. Source: Based on Wang et al. [34].

mechanism is usually stereospecific, the observation of *trans*-isotopomer **B** and **A** from (*E*)- and (*Z*)-β-deuterostyrene, respectively, indicates a stepwise radical mechanism in which β-C–C bond rotation can occur in γ-Co(III)-alkyl radical intermediate **II** before ring closure. Moreover, the stepwise radical mechanism is further evidenced by the direct detection of α-Co(III)-benzyl radical species **I** with [Co(**P7**)-(2,6-$F_2C_6H_3CH$)]$^+$ (m/z =1361.6515) by high-resolution mass spectrometry (HRMS). Collectively, the authors proposed the following catalytic cycle as shown in Scheme 2.18.

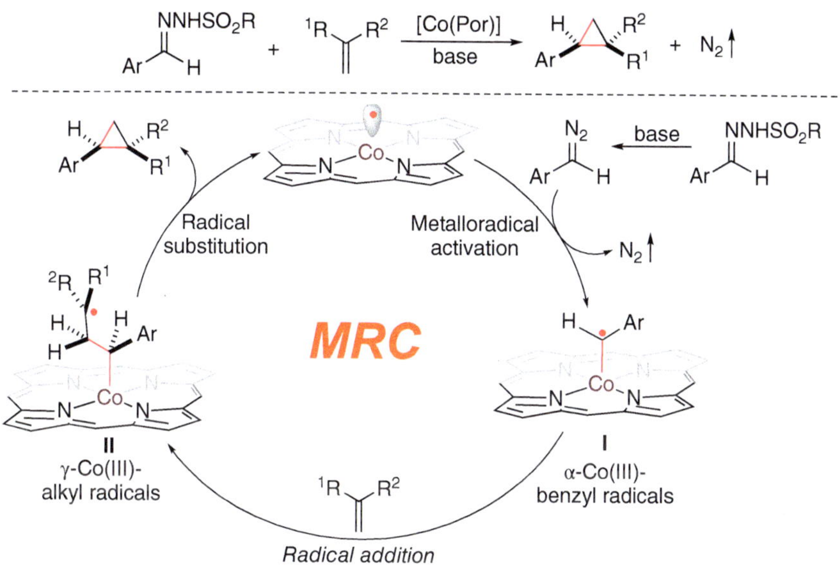

Scheme 2.18 Proposed mechanism of the asymmetric cyclopropanation with *in situ* generated aryldiazomethanes. Source: Based on Wang et al. [34].

Besides aryldiazomethanes as the effective carbene precursors, heteroaryl diazomethanes could also be utilized for the synthesis of biologically important heteroaryl cyclopropanes. In 2018, Chattopadhyay and coworkers described a protocol in which *in situ* generated pyridyl diazomethanes from *N*-tosylhydrazones could be activated by [Co(TPP)] to generate α-Co(III)-pyridyl radical intermediates (Scheme 2.19) [35]. Upon addition to olefins followed by radical substitution could deliver the cyclopropane product with regeneration of the catalyst. This Co(II)-based metalloradical system was shown to be compatible with a range of styrene derivatives with different electronic and steric properties. Moreover, electron-deficient vinyl ketones, acrylates, and vinyl sulfones were well tolerated under the reaction condition.

2.3 Intramolecular Radical Cyclopropanation of Alkenes

In parallel with the development of intermolecular olefin cyclopropanation, considerable efforts have also been devoted to its intramolecular variant of

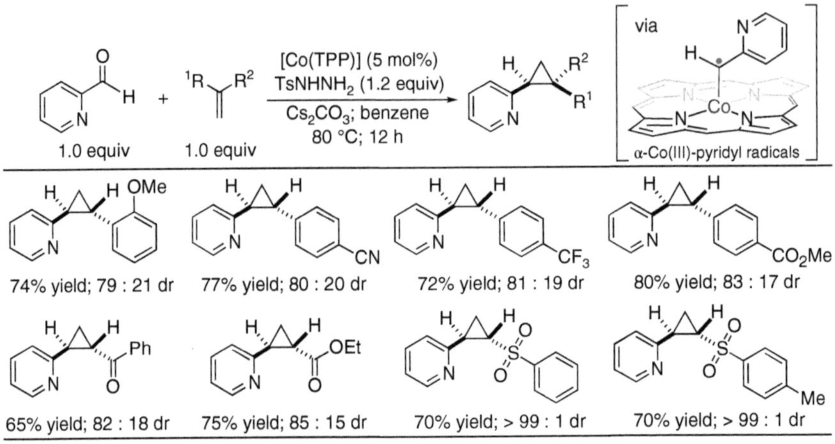

Scheme 2.19 [Co(TPP)]-catalyzed cyclopropanation with *in situ* generated pyridyldiazomethanes. Source: Based on Roy et al. [35].

cyclopropanation, which offers a powerful strategy for the stereoselective construction of [*n*.1.0]bicyclic ring systems [36]. The resulting bicyclic molecules can serve as versatile intermediates in the synthesis of natural products and biologically active compounds. To date, tremendous advancements have been achieved for asymmetric intramolecular cyclopropanation of acceptor-substituted diazo compounds, such as diazoacetates and diazoacetamides [16]. On the other hand, intramolecular cyclopropanation of acceptor/acceptor-substituted diazo compounds faces formidable challenges in the field. In particular, α-acceptor-substituted diazoacetates represent an important class of acceptor/acceptor-substituted diazo compounds that have not been widely explored.

As a fundamentally new approach, the Co(II)-based MRC (Co(II)-MRC) has been demonstrated to be highly effective for asymmetric cyclopropanation of a broad combination of olefin substrates and diazo compounds with excellent control of stereoselectivities. To further extend the application of Co(II)-MRC, Zhang and coworkers developed a novel metalloradical catalyst [Co(**P9**)] (**P9** = 3,5-Di^{t}BuQingPhyrin) through an iterative approach (Scheme 2.20) [37]. This newly synthesized catalyst is capable of catalyzing the asymmetric intramolecular cyclopropanation of various α-acceptor-substituted allyl diazoacetates, affording the 3-oxabicyclo [3.1.0]hexan-2-one derivatives with three contiguous stereogenic centers in high yields with excellent stereoselectivities (Scheme 2.21). The [Co(**P9**)]-catalyzed intramolecular cyclopropanation was shown to be generally applicable for a variety of cinnamyl diazoacetates with different α-substituents, including cyano, nitro, ketone, and ester groups. In addition, the metalloradical system could be extended to acceptor-substituted and donor/acceptor-substituted diazo compounds as exemplified by the cyclopropanation of cinnamyl diazoacetate and cinnamyl α-methyl diazoacetate, respectively. Furthermore, various allyl α-cyanodiazoacetates bearing

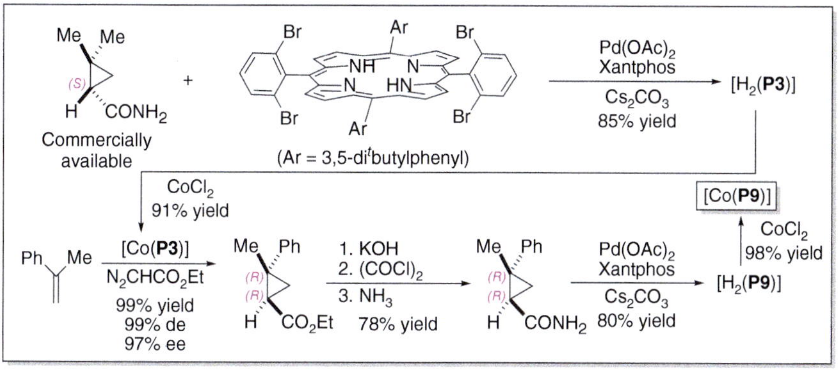

Scheme 2.20 Iterative approach for the synthesis of [Co(P9)]. Source: Based on Xu et al. [37].

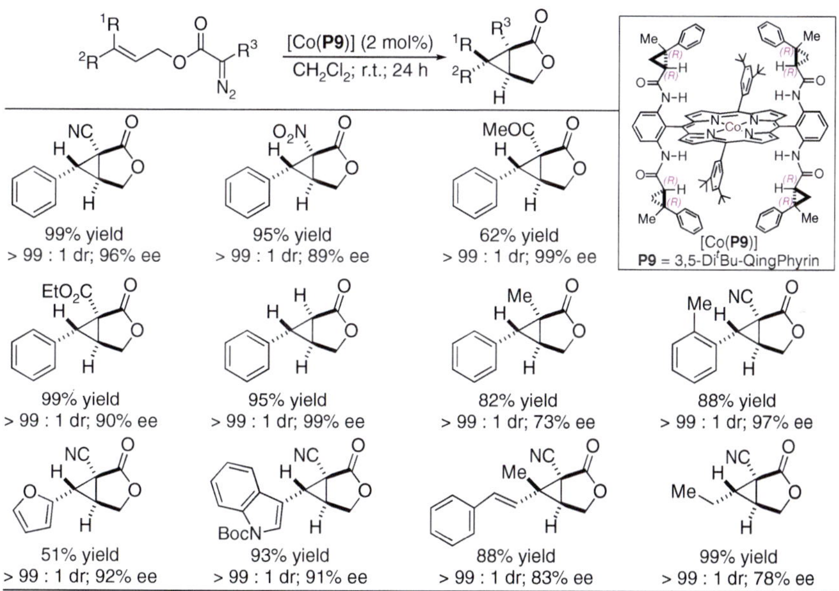

Scheme 2.21 Asymmetric intramolecular cyclopropanation with acceptor-substituted diazoacetates. Source: Based on Xu et al. [37].

different functionalities could be productively cyclopropanated. Remarkably, the catalytic system was also found to be compatible with alkyl-substituted allyl α-cyanodiazoacetate, which formed the corresponding bicyclic products in excellent yields despite the diminished enantioselectivities (Scheme 2.21).

In 2014, Zhang and coworkers described another example of Co(II)-catalyzed asymmetric intramolecular cyclopropanation by utilizing acceptor-substituted diazo compounds (Scheme 2.22) [38]. With DMAP as additives, the readily

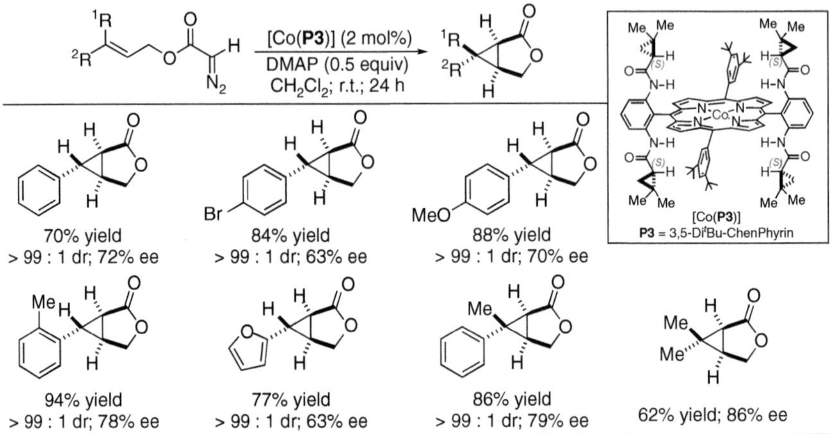

Scheme 2.22 Asymmetric intramolecular cyclopropanation with α-diazoacetates. Source: Based on Ruppel et al. [38].

accessible metalloradical catalyst [Co(**P3**)] (**P3** = 3,5-Di^{t}Bu-ChenPhyrin) could effectively catalyze intramolecular cyclopropanation of allyl α-diazoacetates with varied functional groups and substitution patterns, forming the corresponding [3.1.0]bicyclic products in good-to-high yields with complete diastereocontrol and good enantiocontrol.

In an effort to expand the scope of intramolecular cyclopropanation to less-stabilized donor-substituted diazo compounds, Zhou, Che, and coworkers applied the *in situ* generation protocol using *N*-sulfonylhydrazones as alkyldiazomethane precursors (Scheme 2.23) [39]. This [Co(Por)]-catalyzed protocol allows a chemoselective intramolecular cyclopropanation of *N*-alkyl indoles and pyrroles with alkyldiazomethanes, providing convenient access to a variety of nitrogen-containing polycyclic compounds. The [Co(F_{20}-TPP)] (F_{20}-TPP = *meso*-tetrakis(pentafluorophenyl)porphyrin) can serve as an effective catalyst for intramolecular cyclopropanation featuring a good compatibility with different functionalities, such as halogen,

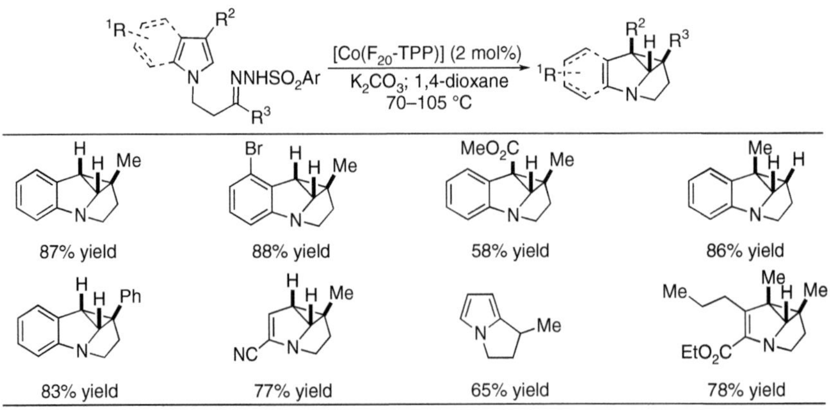

Scheme 2.23 Intramolecular cyclopropanation of *N*-alkyl indoles/pyrroles with alkyldiazomethanes. Source: Based on Reddy et al. [39].

ester, and cyanide groups. Interestingly, the reaction worked equally well with disubstituted and trisubstituted pyrroles, allowing for the rapid construction of multisubstituted polycycles in moderate-to-good yields from readily available starting materials. It is worth noting that pyrrole substrates containing no electron-withdrawing substituent will preferentially undergo facile C–H insertion rather than intramolecular cyclopropanation.

2.4 Intermolecular Radical Cyclopropenation of Alkynes

Optically pure cyclopropenes are important carbocycles with unsaturated, highly strained three-membered ring structures [40]. Among different methods [41], transition-metal-catalyzed cyclopropenation of alkynes with diazo compounds constitutes one of the most effective methods for stereoselective construction of these valuable ring structures [42]. Although a number of Rh_2-catalytic systems have been well established for enantioselective cyclopropenation with both acceptor- and donor/acceptor-substituted diazo compounds [43], asymmetric cyclopropenation with acceptor/acceptor-substituted diazo compounds has been a largely unsolved problem [44]. Given the unique reactivity profile of Co(II)-based metalloradical cyclopropanation, Zhang and coworkers reported a highly enantioselective cyclopropenation system based on a new chiral Co(II) metalloradical catalyst [Co(**P10**)] (**P10** =3,5-DiMes-ChenPhyrin) (Scheme 2.24) [45]. The [Co(**P10**)]-catalyzed cyclopropenation can be successfully applied for a wide

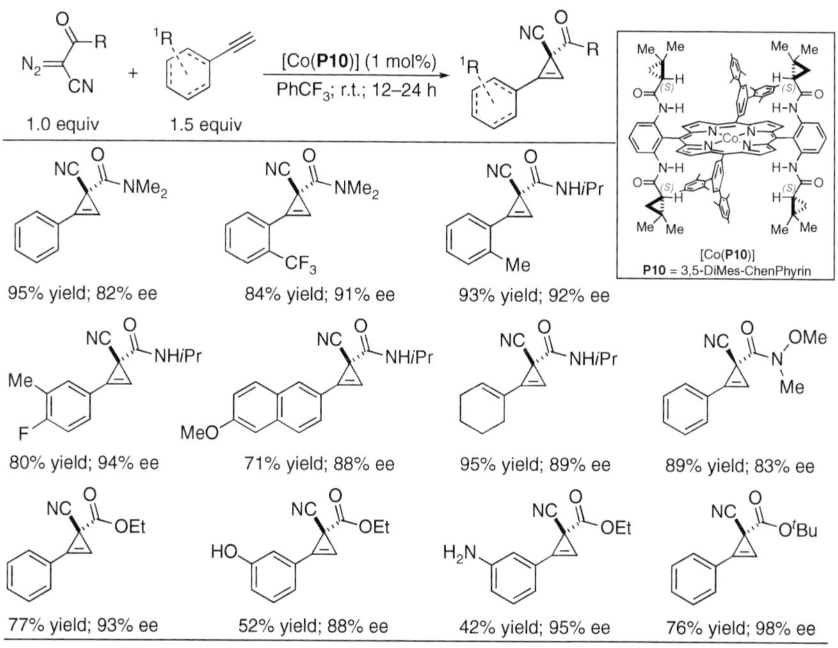

Scheme 2.24 Asymmetric cyclopropenation of aryl/vinyl alkynes with A/A-substituted diazo compounds. Source: Based on Cui et al. [45].

range of aryl- and vinyl-substituted terminal alkynes in combination with different acceptor/acceptor-substituted diazo compounds, and this methodology represents the first examples of utilizing α-cyanodiazoacetates and α-cyanodiazoacetamides for asymmetric cyclopropenation. Furthermore, the metalloradical process is highlighted by an unusual degree of tolerance toward various functional groups as demonstrated by the reactions of phenylacetylene derivatives containing aldehyde, hydroxyl, and amino substituents. The resulting enantioenriched cyclopropenes bearing all-carbon quaternary stereogenic centers can serve as useful chiral intermediates for potential synthetic applications.

2.5 Intramolecular Radical Alkylation of C(sp^3)–H Bonds

C–H alkylation represents a powerful transformation for the formation of various C–C bonds in organic chemistry [46]. Nevertheless, the activation of normally inert C–H bonds while controlling chemo-, regio-, and stereoselectivity is inherently challenging [47]. Significant efforts have been devoted to utilize the C–H alkylation strategy for synthesis of important organic molecules from ubiquitous C–H bonds [48]. Among recent developments, transition-metal-catalyzed carbene insertion is one of the most effective methods for C–C bond formation and has been successfully employed for the synthesis of natural products and pharmaceutically important molecules (Scheme 2.25a) [49]. In particular, catalytic systems based on Rh_2 [50], Cu [51], Ru [52], and Fe [53] complexes have been shown to be highly effective for C–H alkylation with acceptor- and donor/acceptor-substituted diazo compounds as the carbene sources. On the other hand, C–H alkylation with acceptor/acceptor-substituted diazo compounds remains underdeveloped. One of the major challenges lies in the intrinsic low reactivity of acceptor/acceptor-substituted diazo compounds toward formation of the metallocarbene intermediates. Moreover, the subsequent C–H insertion steps are often difficult in terms of controlling regio- and enantioselectivities due to

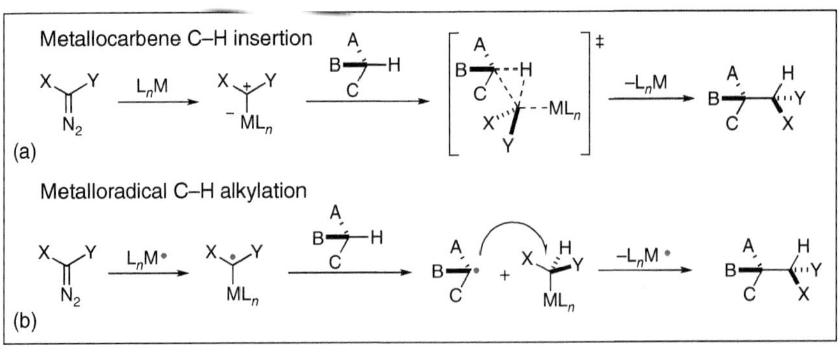

Scheme 2.25 C–H alkylation by (a) electrophilic metallocarbene insertion. Source: Based on Doyle et al. [49] and Xia et al. [49]. (b) Radical C–H alkylation via MRC.

the high electrophilicity of the metallocarbenes. So far, the scope of the existing catalytic systems with acceptor/acceptor-substituted diazo compounds has been limited to mainly electron-rich C–H substrates [54].

2.5.1 Intramolecular C–H Alkylation with Acceptor/Acceptor-Substituted Diazo Compounds

To address these and related challenges, Zhang and coworkers presented a fundamentally different approach based on the concept of MRC for asymmetric C–H alkylation with acceptor/acceptor-substituted diazo compounds (Scheme 2.26) [55]. Unlike the electrophilic Fischer-type carbene systems with concerted insertion pathway (Scheme 2.25a), the Co(II)-based metalloradical system features a stepwise radical pathway that involves the sequential H-atom abstraction and radical substitution (HAA-RS), offering an alternative approach for C–C bond formation via radical C–H alkylation (Scheme 2.25b). Supported by a new D_2-symmetric chiral amidoporphyrin (**P11** = 3,5-DiiPr-(4-^{t}Bu)XuPhyrin), the Co(II) complex [Co(**P11**)] is capable of catalyzing asymmetric intramolecular 1,5-C–H alkylation of α-methoxycarbonyl-α-diazosulfones, leading to high-yielding formations of five-membered sulfolanes with excellent stereoselectivities. Remarkably, this new protocol features low sensitivity to the electronic properties of C–H bonds, allowing for the carbene insertion to electron-deficient benzylic C–H bonds. Moreover, the catalytic system demonstrates other features that are unique to the radical pathway, including high degree of functional group tolerance as well as excellent chemoselectivity toward allylic and allenic C–H bonds.

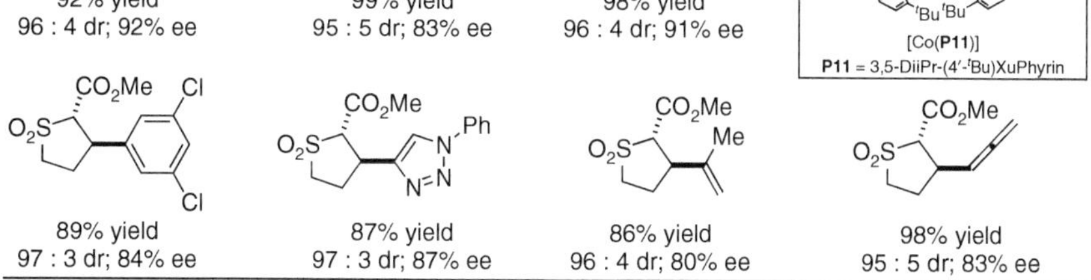

Scheme 2.26 Asymmetric radical C–H alkylation with A/A-substituted diazo compounds. Source: Based on Cui et al. [55].

To probe the underlying radical pathway, the authors also investigated the allylic C–H alkylation of isomerically pure (*E*) and (*Z*)-α-methoxycarbonyl-α-diazosulfones, respectively (Scheme 2.27). The observation of a mixture of (*E*)- and (*Z*)-alkylation products for both substrates supports the existence of allylic radical intermediates, which are presumably formed through H-atom abstraction of allylic C–H bonds by the initial α-Co(III)-alkyl radicals. Moreover, the degree of *E*/*Z* isomerization of the allylic radicals can be influenced by varying the porphyrin ligand environments of the Co(II) complexes. When sterically encumbered [Co(**P11**)] (**P11** = 3,5-DiiPr-(4-^{t}Bu)XuPhyrin) was used the catalyst, only slight isomerization was observed for the alkylation reactions of both (*E*)- and (*Z*)-α-methoxycarbonyl-α-diazosulfones (Scheme 2.27). On the other hand, the use of less sterically hindered [Co(**P8**)] (**P8** = 3,5-Di^{t}Bu-IbuPhyrin) led to the formation of alkylation products with a higher degree of *E*/*Z* isomerization. These experimental observations suggest that the isomerization is kinetically controlled by the ligand environment of the catalyst.

Diazo	[Co(Por)]	Yield (%)	*E* : *Z*	Product
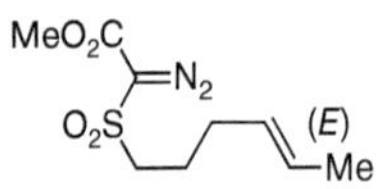	[Co(**P8**)]	96	89 : 11	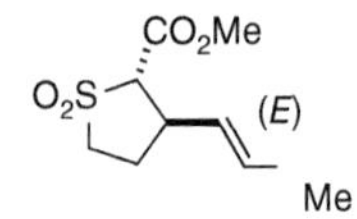
	[Co(**P11**)]	94	95 : 5	
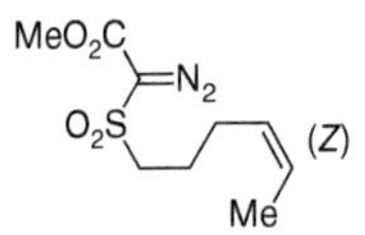	[Co(**P8**)]	94	49 : 51	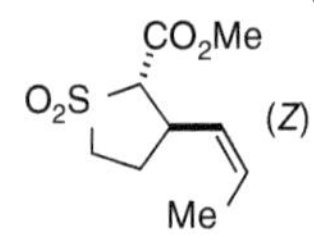
	[Co(**P11**)]	92	18 : 82	

Scheme 2.27 Catalyst-controlled olefin isomerization. Source: Based on Cui et al. [55].

2.5.2 Intramolecular C—H Alkylation with Donor-Substituted Diazo Compounds

Despite intense research efforts in exploring C–H alkylation, the utilization of donor-substituted diazo compounds is still largely underexplored. The intrinsic low stability of this class of diazo compounds essentially limits their application in C–H alkylation. To address this practical challenge, *N*-sulfonyl hydrazones have been employed as user-friendly surrogates for *in situ* generation of donor-substituted diazomethanes [33]. This strategy stimulates significant progresses in the area of C–H alkylation involving less stabilized diazo compounds. As a successful demonstration, Zhang and coworkers expanded the utility of Co(II)-based MRC to C–H alkylation with donor-substituted diazo compounds (Scheme 2.28) [56]. With the design of a new D_2-symmetric chiral amidoporphyrin (**P12** = 2,6-DiPhO-QingPhyrin), the [Co(**P12**)]-based metalloradical system enabled efficient activation of *in situ* generated *ortho*-aminoaryldiazomethanes for enantioselective intramolecular alkylation of different types of C–H bonds, including benzylic, propargylic, and nonactivated aliphatic substrates. A wide array of chiral 2-substituted indolines were obtained in high yields with excellent enantioselectivities at room temperature. In addition to good chemoselectivity and regioselectivity, this catalytic system features tolerance to a wide range of functional groups and exhibits compatibility with heteroaromatic units.

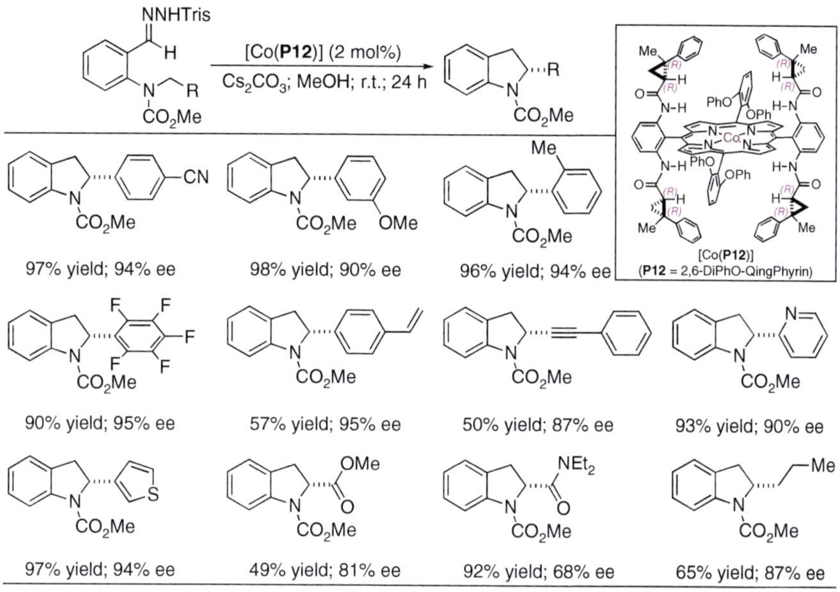

Scheme 2.28 Enantioselective radical C–H alkylation for the synthesis of chiral indolines. Source: Based on Wen et al. [56].

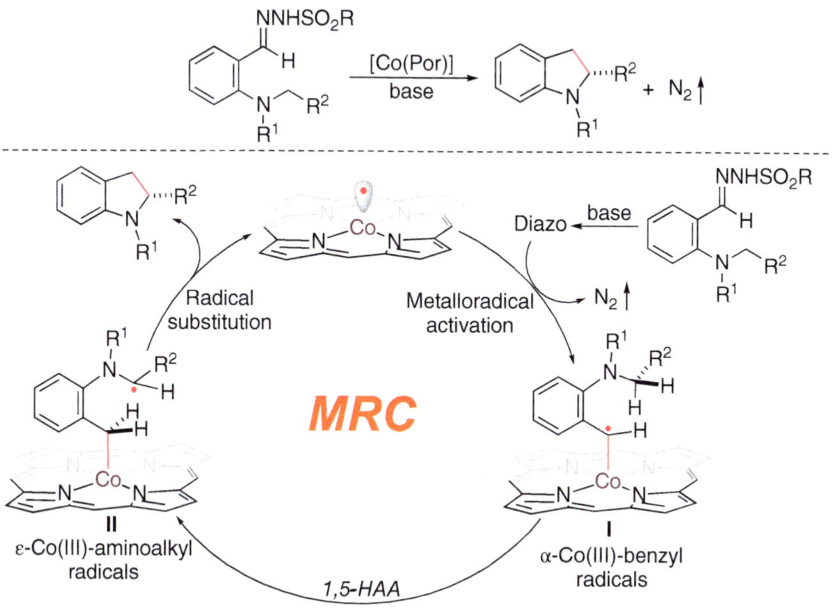

Scheme 2.29 Proposed mechanism for the construction of 2-substituted chiral indolines via Co(II)-MRC. Source: Based on Wen et al. [56].

As shown in Scheme 2.29, upon metalloradical activation, the resulting α-Co(III)-benzyl radical intermediate **I** can undergo intramolecular HAA from the C–H bond at the distal 5-position to form ε-Co(III)-aminoalkyl radical **II**, where the carbon-centered radical is considerably stabilized by the adjacent nitrogen atom. Subsequently, the pendant α-aminoalkyl radical is likely to proceed through 5-*exo-tet* radical cyclization at the α-carbon center to form C–C bond in an asymmetric fashion.

To further elucidate the details of the stepwise radical pathway, a series of mechanistic studies were conducted, including TEMPO-trapping and radical clock experiments (Scheme 2.30). Addition of TEMPO to the reaction of benzylic and allylic C–H substrates resulted in complete inhibition of the C–H alkylation process with the bis-TEMPO-trapped products isolated instead. In the radical clock experiment, the bis-TEMPO-trapped ring-opening products were also generated. Overall, these experimental observations support the proposed stepwise radical mechanism of the Co(II)-based C–H alkylation reaction.

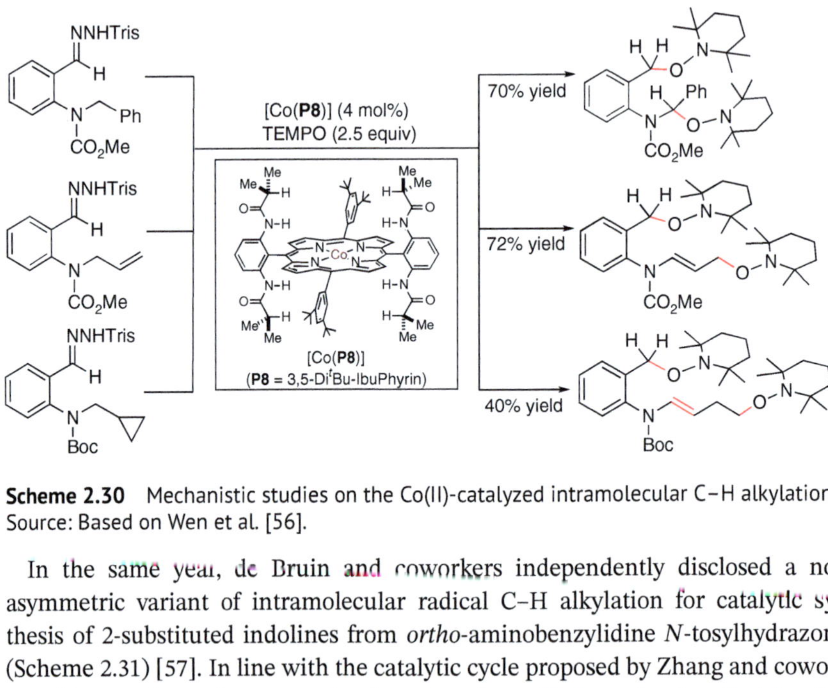

Scheme 2.30 Mechanistic studies on the Co(II)-catalyzed intramolecular C–H alkylation. Source: Based on Wen et al. [56].

In the same year, de Bruin and coworkers independently disclosed a non-asymmetric variant of intramolecular radical C–H alkylation for catalytic synthesis of 2-substituted indolines from *ortho*-aminobenzylidine *N*-tosylhydrazones (Scheme 2.31) [57]. In line with the catalytic cycle proposed by Zhang and coworkers, this transformation proceeds via a stepwise radical mechanism that involves intramolecular 1,5-H-atom abstraction and subsequent radical substitution, affording the desired indoline derivatives with different substituents in 80–98% yields. Moreover, density functional theory (DFT) calculations provide detailed insights into each step of the catalytic cycle, which suggests that the decomposition of tosylhydrazones to the corresponding aryldiazomethanes is more likely the rate-determining step.

To harness the potential of Co(II)-MRC for C–H alkylation, Zhang and coworkers demonstrated that linear aliphatic diazomethanes, generated *in situ* from

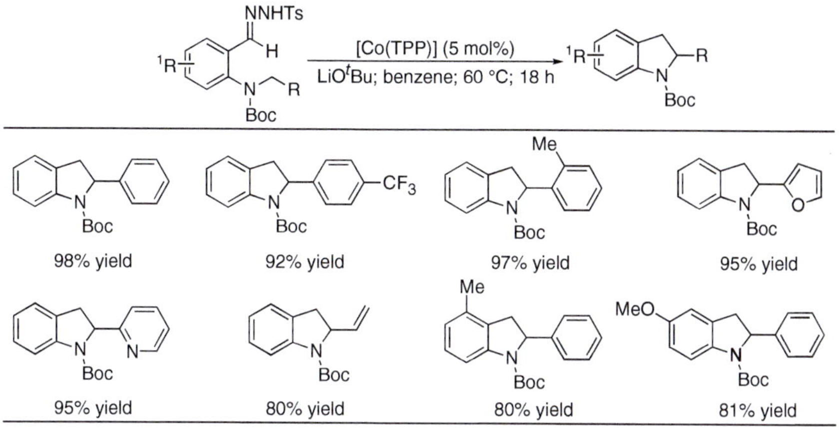

Scheme 2.31 [Co(TPP)]-catalyzed C–H alkylation of *ortho*-aminobenzylidine *N*-tosylhydrazones. Source: Based on Karns et al. [57].

aldehyde-derived *N*-sulfonylhydrazones, could also be utilized for enantioselective C–H alkylation [58]. Different from the traditional radical cyclization system that relies on sequential radical addition to unsaturated bonds and H-atom abstraction (RA-HAA), this metalloradical cyclization approach provides an alternative C–C bond formation strategy by employing H-atom abstraction of saturated C–H bonds followed by radical substitution (HAA-RS) (Scheme 2.32).

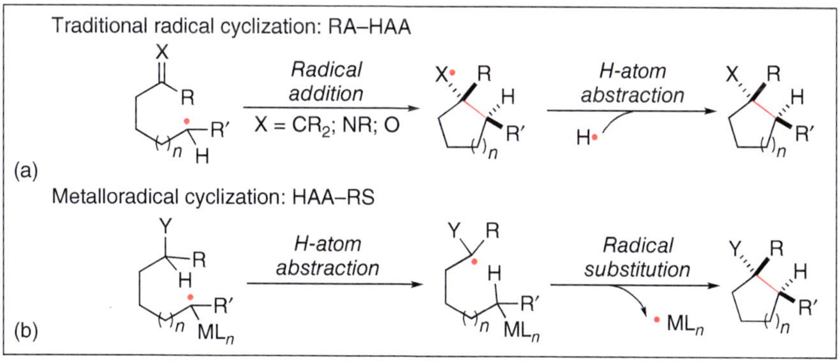

Scheme 2.32 Radical cyclization pathways. Source: Based on Wang et al. [58].

Supported by a new D_2-symmetric chiral amidoporphyrin (**P13** = 2,4,6-TriMe-ZhuPhyrin), [Co(**P13**)] was shown to be capable of activating different aliphatic diazo compounds to generate the corresponding α-Co(III)-alkyl radicals for enantioselective alkylation of both activated and nonactivated C–H bonds (Scheme 2.33). This protocol allows for efficient transformation of readily available aliphatic aldehydes to chiral α-substituted pyrrolidines and other important five-membered ring structures in high yields with excellent enantioselectivities. Interestingly, [Co(**P13**)] could regioselectively alkylate the stronger homobenzylic over the weaker benzylic C–H bond, affording the five-membered pyrrolidine in excellent yield. This regioselectivity suggests that α-Co(III)-alkyl radical intermediates can preferentially undergo 1,5- than 1,6-H-atom abstraction. In addition to chemoselectivity and regioselectivity, the metalloradical C–H alkylation system is highlighted by its compatibility with heteroaryl substrates, as exemplified in the enantioselective synthesis of naturally occurring nicotine and *L*-proline derivatives.

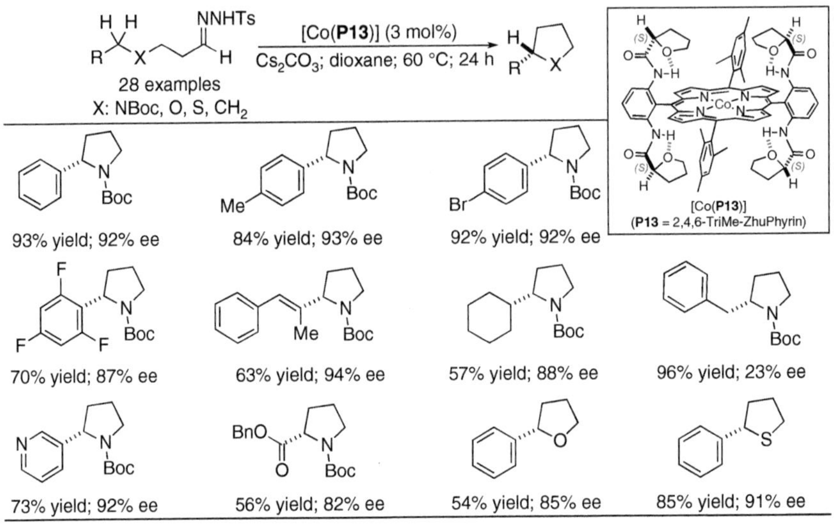

Scheme 2.33 Enantioselective radical cyclization for the synthesis of α-substituted pyrrolidines. Source: Based on Wang et al. [58].

Building on the reported synthesis of five-membered indolines and pyrrolidines via Co(II)-MRC, de Bruin and coworkers in 2019 developed a [Co(TPP)]-catalyzed C–H alkylation system for efficient construction of six-membered piperidine rings directly from aliphatic aldehydes (Scheme 2.34) [59]. The reaction is proposed to

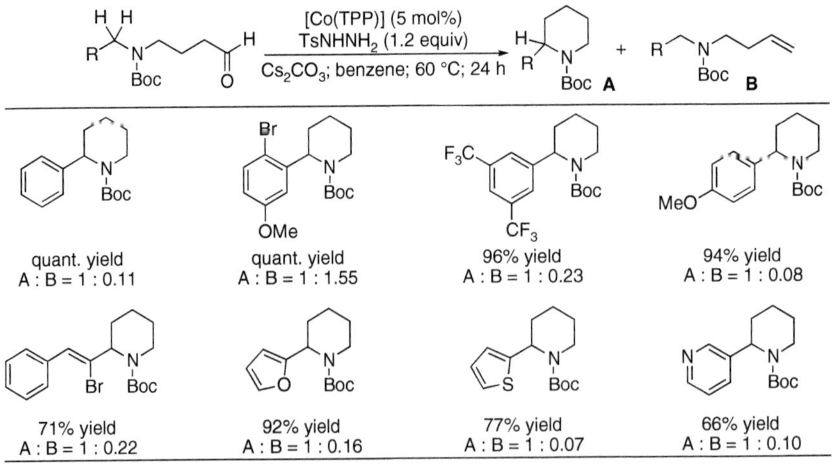

Scheme 2.34 [Co(TPP)]-catalyzed formation of substituted piperidines. Source: Based on Lankelma et al. [59].

proceed through a metalloradical pathway involving formation of α-Co(III)-alkyl radical, followed by 1,6-H-atom abstraction to afford benzylic radical intermediate. The latter can then undergo 6-*exo-tet* radical cyclization to yield the desired six-membered *N*-heterocycles, which are the most prevalent *N*-heterocycles in medicinal chemistry (Scheme 2.35) [60]. In this work, the authors also observed the formation of linear alkenes as side products. Through DFT calculations, this side product formation can be explained by a competitive 1,5-H-atom abstraction followed by β-scission (Scheme 2.35).

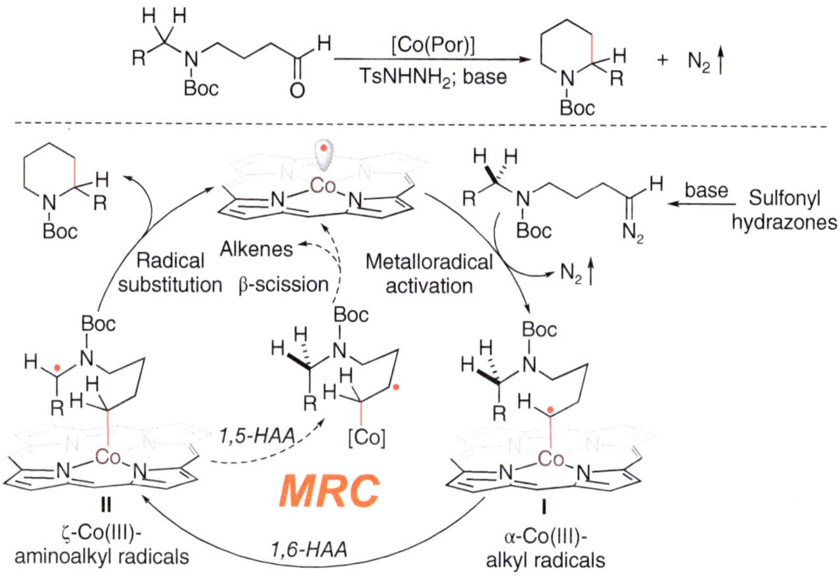

Scheme 2.35 [Co(TPP)]-catalyzed competitive formation of piperidines and alkenes. Source: Based on Lankelma et al. [59].

Besides the synthesis of nitrogen-containing heterocycles, de Bruin and coworkers also demonstrated the utility of Co(II)-MRC for the construction of all carbon-based ring skeletons (Scheme 2.36) [61]. An efficient protocol for the synthesis of substituted 1,2-dihydronaphthalenes was developed by employing *ortho*-styryl *N*-tosylhydrazones as carbene precursors. This [Co(TPP)]-based metalloradical system has been successfully applied to wide-ranging substrates with various R^1 substituents, affording the fused six-membered rings in moderate-to-high yields. Moreover, this radical cyclization protocol is shown to be rather sensitive to the nature of R^2 substituents as changing from the ester group to other substituents can adversely influence the reactivity of the reaction.

Surprisingly, catalytic reactions involving substrates with an alkyl substituent on the allylic position can lead to the selective formation of *E*-aryl-dienes (Scheme 2.37), instead of the expected 1,2-dihydronapthalenes (Scheme 2.36). It is worth mentioning that the presence of the ester substituent at the vinylic position is essential for achieving high yields. Based on these experimental results as well

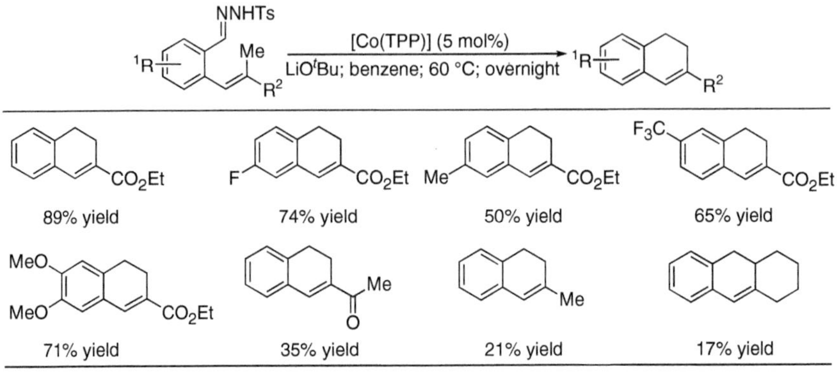

Scheme 2.36 [Co(TPP)]-catalyzed formation of 1,2-dihydronaphthalenes. Source: Based on te Grotenhuis et al. [61].

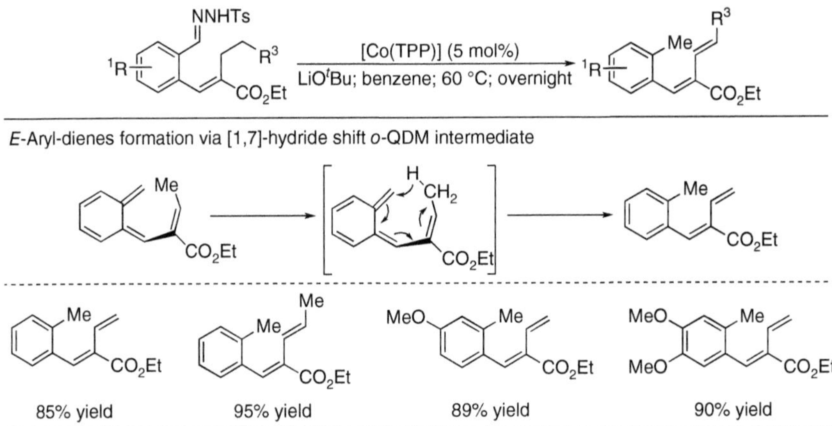

Scheme 2.37 Formation of *E*-aryl-dienes via *o*-quinodimethane. Source: Based on te Grotenhuis et al. [61].

as DFT calculations, the authors proposed a radical pathway involving 1,6-H-atom abstraction of allylic C–H bond to generate benzallylic radical, which is more likely to undergo homolysis of Co C bond to release *ortho*-quinodimethane (*o*-QDM) intermediate (Scheme 2.38). The latter can undergo either 6π-cyclization to produce 1,2-dihydronapthalenes (Scheme 2.36) or [1,7]-hydride shift to afford *E*-aryl-dienes (Scheme 2.37).

Although many methods exist to synthesize five- and six-membered ring structures, cyclization reactions to access medium-sized rings, especially eight-membered cyclic compounds, remain very challenging [62]. Due to the entropic factors and transannular interactions, cyclization of flexible linear molecules can be complicated by unwanted side reactions, such as oligomerization and polymerization. In 2018, de Bruin and coworkers reported an interesting approach to prepare eight-membered cyclic compounds by [Co(TPP)]-catalyzed radical cyclization of *ortho*-benzallylaryl *N*-tosylhydrazones (Scheme 2.39) [63]. A variety of unique dibenzocyclooctenes were obtained in good-to-excellent yields with tolerance of

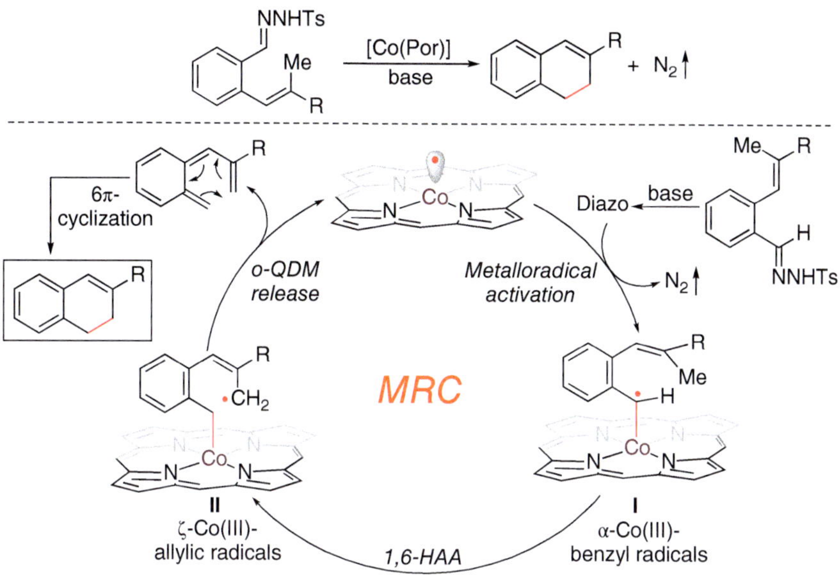

Scheme 2.38 Proposed mechanism for the [Co(TPP)]-catalyzed formation of 1,2-dihydronaphthalenes. Source: Based on et Grotenhuis et al. [61].

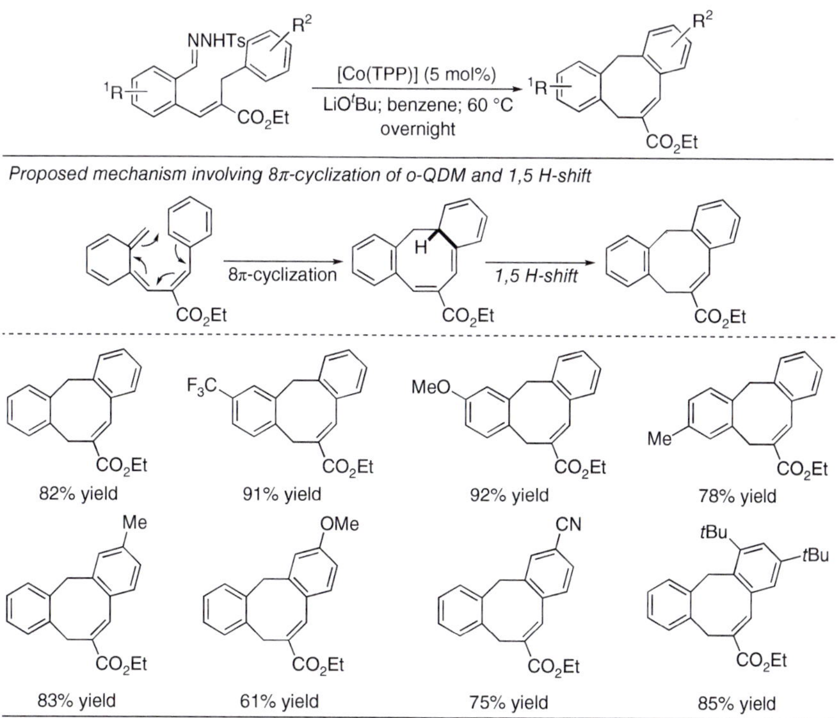

Scheme 2.39 [Co(TPP)]-catalyzed synthesis of dibenzocyclooctenes. Source: Based on te Grotenhuis et al. [63].

diverse substituents, including electron-donating and electron-withdrawing substituents on both aryl rings. The mechanism is proposed to involve intramolecular 1,6-H-atom abstraction followed by dissociation of *o*-QDM intermediate from the catalyst, which is similar to the mechanism reported for dihydronapthalenes and butadienes [61]. Subsequently, the *o*-QDM intermediate can readily undergo 8π-cyclization followed by [1,5]-hydride shift to regain the aromaticity, affording the eight-membered ring product.

In 2020, de Bruin and coworkers attempted to extend the scope of this protocol to allow for the synthesis of other eight-membered ring compounds via Co(II)-MRC [64]. In this work, the authors employed a one-pot approach for the *in situ* generation of sulfonyl hydrazones from the corresponding aldehyde and tosylhydrazide. Under the standard conditions, the reaction was shown to be compatible with substrates containing naphthyl- and pyridyl-substituents at the benzylic position, enabling the synthesis of several new eight-membered dibenzocyclooctenes in high yields (Scheme 2.40).

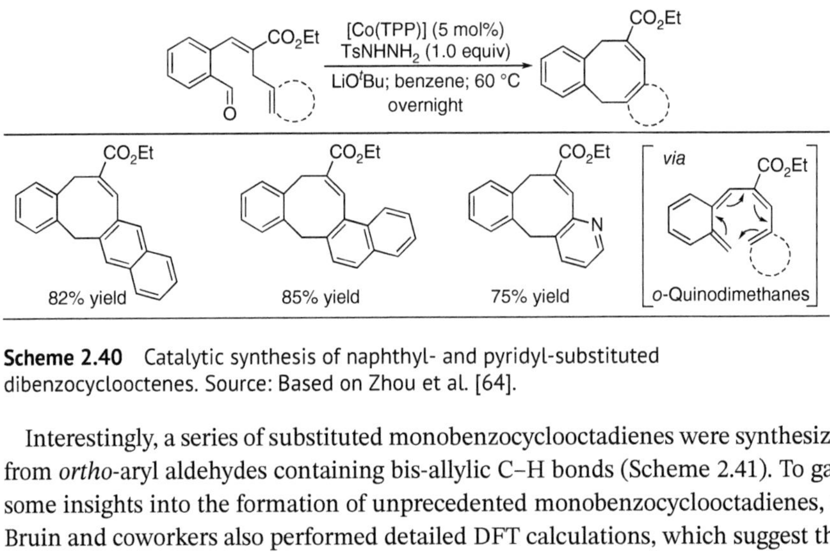

Scheme 2.40 Catalytic synthesis of naphthyl- and pyridyl-substituted dibenzocyclooctenes. Source: Based on Zhou et al. [64].

Interestingly, a series of substituted monobenzocyclooctadienes were synthesized from *ortho*-aryl aldehydes containing bis-allylic C–H bonds (Scheme 2.41). To gain some insights into the formation of unprecedented monobenzocyclooctadienes, de Bruin and coworkers also performed detailed DFT calculations, which suggest that the monobenzocyclooctadienes are likely to be generated from radical substitution rather than proceeding through an *o*-QDM intermediate.

2.6 Other Catalytic Radical Processes for Carbene Transfers

Carbonylation of metal-carbene species represents an attractive method to produce highly reactive ketenes, which are synthetically useful intermediates in various organic transformations [65]. In 2013, Zhang, de Bruin, and coworkers developed a new carbene carbonylation process catalyzed by [Co(**P8**)] (**P8** = 3,5-Di^{t}Bu-IbuPhyrin) to afford ketenes from different carbene precursors and CO under mild conditions (Scheme 2.42) [66]. This Co(II)-based metalloradical carbonylation

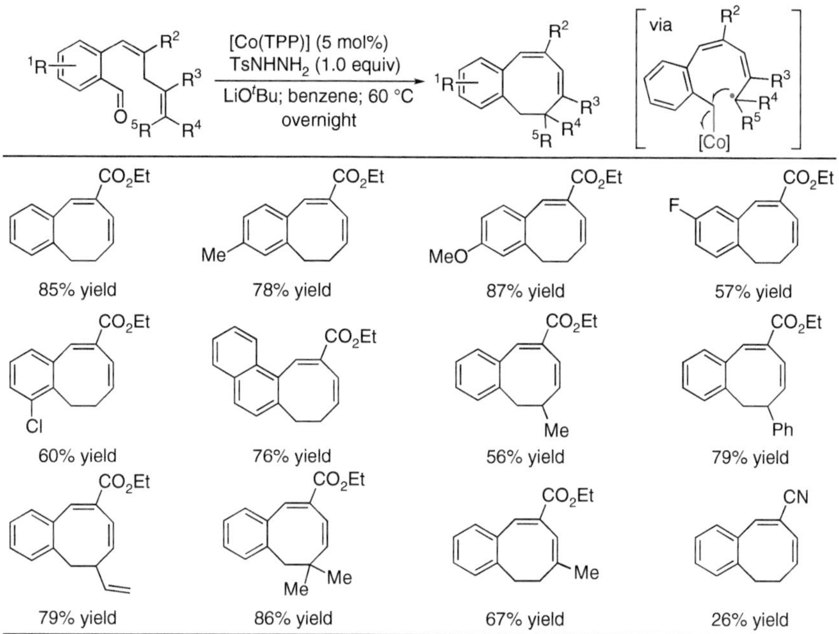

Scheme 2.41 Catalytic synthesis of monobenzocyclooctadienes via Co(II)-MRC. Source: Based on Zhou et al. [64].

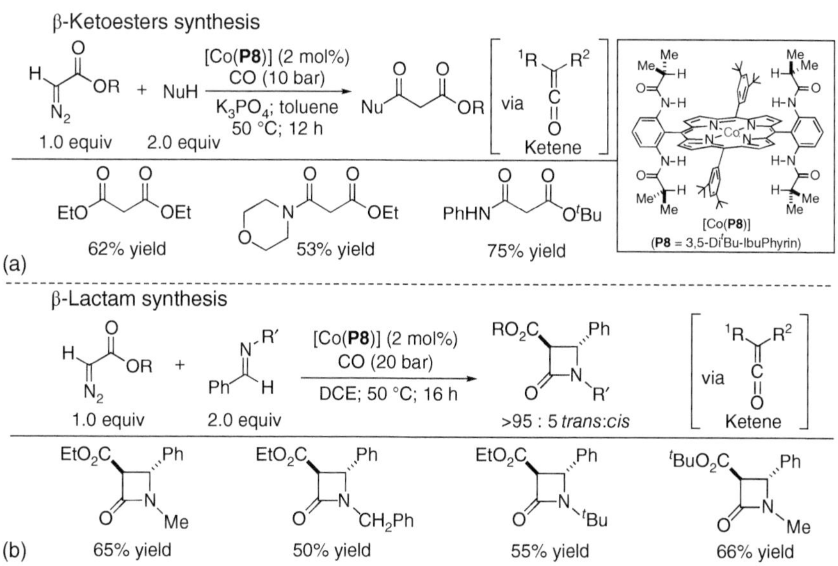

Scheme 2.42 Co(II)-catalyzed carbene carbonylation. Source: Based on Paul et al. [66].

process provides an alternative approach for the formation of ketenes, which can subsequently react with different amines, alcohols, or imines to form esters, amides, or β-lactams in one pot. In addition to α-diazocarbonyls, *N*-tosylhydrazone sodium salts derived from aldehyde and ketones can also be employed as effective carbene sources.

In 2017, de Bruin, Dzik, and coworker disclosed a [Co(Por)]-catalyzed *gem*-difluorocyclopropanation of electron-deficient olefins with Me_3SiCF_3 as the difluorocarbene source (Scheme 2.43) [67]. The metalloradical catalyst [Co(F_{20}-TPP)] proved to be the most effective catalyst for difluorocarbene transfer from Me_3SiCF_3 to *n*-butyl acrylate, delivering the desired *gem*-difluorocyclopropane in up to 40% yield (TON = 8). Based on DFT calculations, the authors proposed a stepwise radical mechanism in which α-Co(III)-difluoroalkyl radical could be generated upon metalloradical activation with release of a fluoride anion. Subsequent radical addition to the acrylate double bond followed by radical substitution would furnish the difluorocyclopropane product to turn over the catalytic cycle. Overall, this methodology illustrates the feasibility of catalytic difluorocarbene transfer to electron-deficient olefins by [Co(Por)]. Future investigations to improve the efficiency of the catalytic system will be highly appealing.

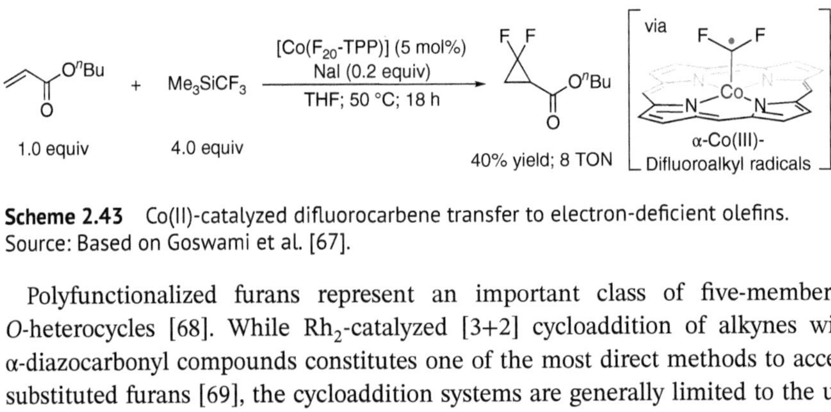

Scheme 2.43 Co(II)-catalyzed difluorocarbene transfer to electron-deficient olefins. Source: Based on Goswami et al. [67].

Polyfunctionalized furans represent an important class of five-membered *O*-heterocycles [68]. While Rh_2-catalyzed [3+2] cycloaddition of alkynes with α-diazocarbonyl compounds constitutes one of the most direct methods to access substituted furans [69], the cycloaddition systems are generally limited to the use of electron-rich arylacetylenes and incompatible with sensitive functionalities. In 2012, Zhang and coworkers developed a new Co(II)-based metalloradical system for direct and regioselective synthesis of multisubstituted furans by cyclization of alkynes with α-diazocarbonyls [70]. As shown in Scheme 2.44 for the catalytic cycle, γ-vinyl radical intermediate **II** generated from radical addition of α-Co(III)-alkyl radical **I** to alkyne can undergo two divergent processes: (i) radical substitution to afford cyclopropene; (ii) a consecutive radical addition to the carbonyl group to form β-Co(III)-alkyl radical **III**, which can then form the furan product via radical β-scission.

Through judicious selection of diazo compounds and careful evaluation of reaction temperatures, [Co(**P8**)] (**P8** = 3,5-Di^{t}Bu-IbuPhyrin) was shown to effectively catalyze the cyclization of terminal alkynes with α-diazocarbonyls, producing polyfunctionalized furans with excellent control of regioselectivities (Scheme 2.45). In addition to its high degree of functional group tolerance, this [Co(**P8**)]-catalyzed

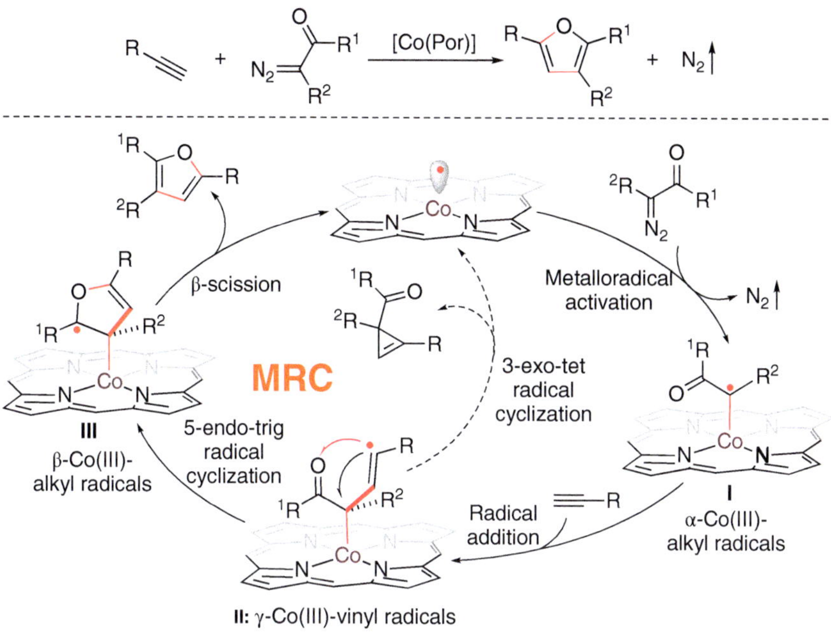

Scheme 2.44 Competitive pathway: cyclopropenation versus furanylation. Source: Based on Cui et al. [70].

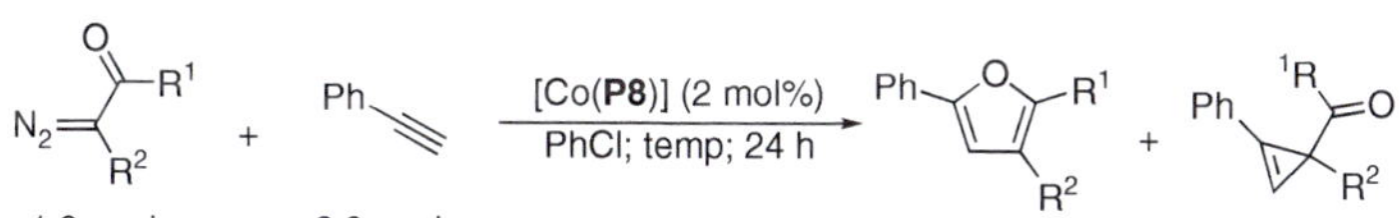

Entry	Diazo	Temp (°C)	**A** yield (%)	**B** yield (%)
1	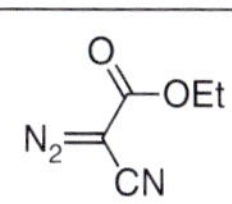	25	31	69
2		80	68	30
3	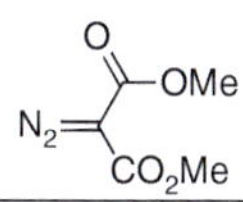	25	13	Trace
4		80	91	Trace

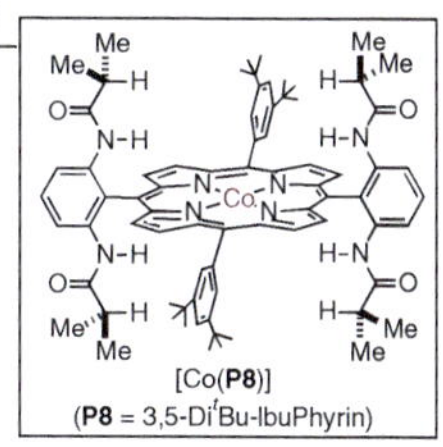

Scheme 2.45 Influence of diazo compounds and temperature on furanylation versus cyclopropenation. Source: Based on Cui et al. [70].

system also enables for the first time the cyclization of electron-deficient alkynes to form furan derivatives in high yields (Scheme 2.46).

In 2013, Zhang, de Bruin, and coworkers extended this tandem radical addition-cyclization protocol for direct synthesis of another important class of *O*-heterocycles, 2*H*-chromenes (Scheme 2.47) [71, 72]. In this process, α-Co(III)-benzyl radical generated from metalloradical activation of salicyl *N*-tosylhydrazones by [Co(**P8**)] (**P8** = 3,5-Di^{t}Bu-IbuPhyrin) can undergo radical addition to terminal alkynes to form salicyl-vinyl radical intermediates. Subsequent H-atom abstraction followed by ring cyclization can lead to the formation of six-membered heterocyclic

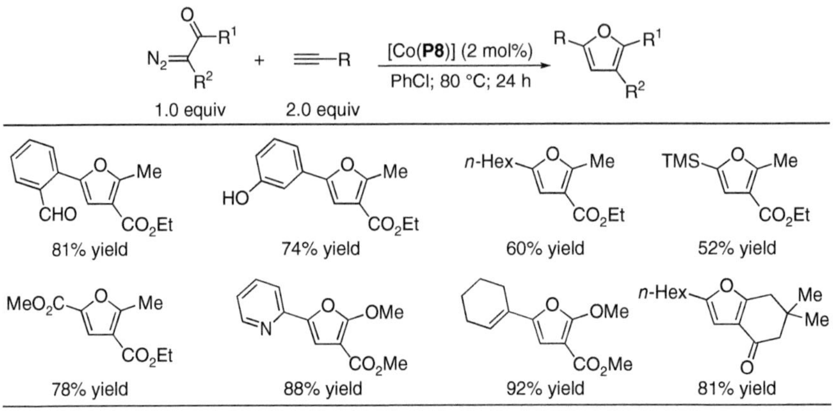

Scheme 2.46 Regioselective synthesis of multisubstituted furans. Source: Based on Cui et al. [70].

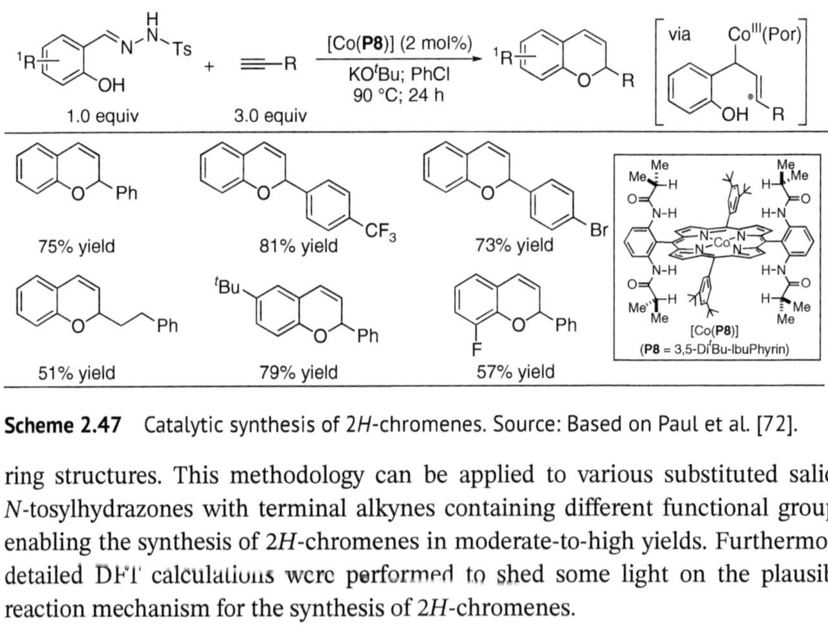

Scheme 2.47 Catalytic synthesis of 2*H*-chromenes. Source: Based on Paul et al. [72].

ring structures. This methodology can be applied to various substituted salicyl *N*-tosylhydrazones with terminal alkynes containing different functional groups, enabling the synthesis of 2*H*-chromenes in moderate-to-high yields. Furthermore, detailed DFT calculations were performed to shed some light on the plausible reaction mechanism for the synthesis of 2*H*-chromenes.

In 2018, Chattopadhyay and coworkers also elaborated the concept of Co(II)-MRC for denitrogenative transannulation of *in situ* generated 2-pyridyldiazomethanes, allowing for the efficient construction of indolizines (Scheme 2.48) [35]. In the same paper, the authors demonstrated that α-Co(III)-pyridyl radical intermediates generated from metalloradical activation could undergo radical addition to olefins to form pyridyl-substituted cyclopropanes in good yields and diastereoselectivities. Interestingly, the resulting α-Co(III)-pyridyl radical intermediates were also capable of undergoing radical addition to terminal alkynes to form γ-Co(III)-vinyl radical intermediates, which subsequently afforded the indolizine derivatives via radical substitution. The [Co(TPP)]-catalyzed denitrogenative transannulation is applicable to various alkynes, including aliphatic- and naphthyl-containing acetylenes.

This method represents the first example of denitrogenative transannulation via a metalloradical approach, which may offer opportunities to address the limitations in the existing ionic systems.

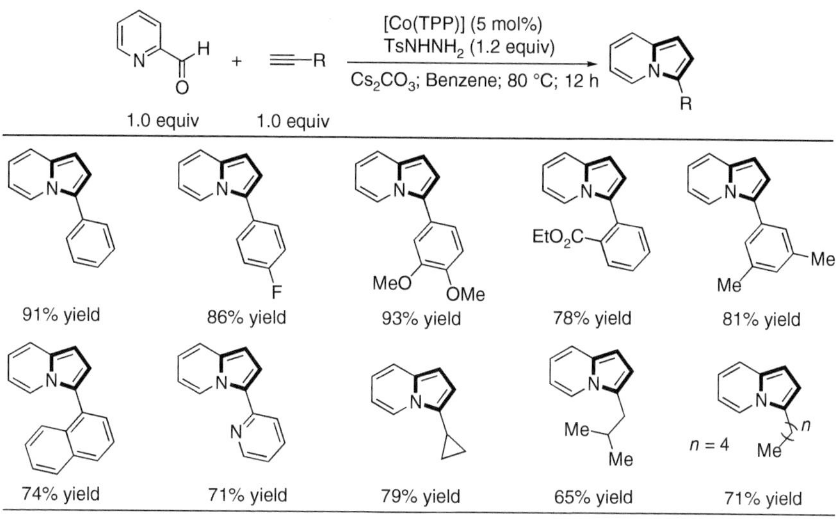

Scheme 2.48 [Co(TPP)]-catalyzed metalloradical transannulation. Source: Based on Roy et al. [35].

2.7 Summary and Outlook

This review has summarized the development of strategies that utilize Co(II) complexes of porphyrins [Co(Por)] as effective metalloradical catalysts to control both reactivity and stereoselectivity in selective carbene transfer reactions. The power of Co(II)-based MRC has been shown in cyclopropanation, C–H alkylation, and other useful organic transformations. Distinct from the traditional two-electron carbene transfer processes, this one-electron catalytic approach has proven to be effective for activating a diversified array of diazo compounds as carbene precursors, particularly acceptor/acceptor-substituted diazo compounds. Moreover, the unique nature of neutral radical intermediates endows the catalytic transformations with an exceptional level of functional group tolerance as exemplified in the cyclopropanation of both electron-rich and electron-deficient olefins. Another feature of Co(II)-MRC is manifested in the regioselective C–H alkylation through pinpointing abstraction of saturated C–H bonds, affording important structural motifs with various ring sizes. In addition, the successful utilization of *in situ* generation protocol in the Co(II)-based metalloradical system further expands the substrate scope to donor-substituted diazo compounds. To address the long-standing challenges of controlling stereoselectivity in radical transformations, a family of chiral amidoporphyrin ligands have been developed and proven to be highly effective for asymmetric induction. The attractive non-covalent interactions, such as H-bonding interactions, are postulated to be responsible for the high level of stereochemical control in these radical processes. Moreover, the combination of experimental results with DFT calculations provides constructive insight into the reaction profile involving a stepwise radical mechanism. Witnessing the surge of development in carbene transfer reactions via Co(II)-MRC, we hope that this tutorial review will further stimulate new discoveries to harness the vast potential of radical chemistry.

Acknowledgment

We are grateful for financial support by National Institutes of Health (R01-GM102554 and R01-GM132471) and National Science Foundation (CHE-1900375).

References

1 For selected books, see: (a) Zard, S.Z. (2003). *Radical Reactions in Organic Synthesis*. Oxford: Oxford University Press. (b) Curran, D.P., Porter, N.A., and Giese, B. (2008). *Stereochemistry of Radical Reactions: Concepts, Guidelines, and Synthetic Applications*. New York: Wiley. (c) Chatgilialoglu, C. and Studer, A. (2012). *Encyclopedia of Radicals in Chemistry, Biology, and Materials*. New York: Wiley.

2 For selected reviews, see: (a) Prier, C.K., Rankic, D.A., and MacMillan, D.W. (2013). Visible light photoredox catalysis with transition metal complexes: applications in organic synthesis. *Chem. Rev.* 113: 5322–5363. (b) Studer, A. and Curran, D.P. (2016). Catalysis of radical reactions: a radical chemistry perspective. *Angew. Chem. Int. Ed.* 55: 58–102.

3 Jasperse, C.P., Curran, D.P., and Fevig, T.L. (1991). Radical reactions in natural product synthesis. *Chem. Rev.* 91: 1237–1286.

4 For selected examples on radical approaches for synthesis of complex molecules, see: (a) Hung, K., Hu, X., and Maimone, T.J. (2018). Total synthesis of complex terpenoids employing radical cascade processes. *Nat. Prod. Rep.* 35: 174–202. (b) Romero, K.J., Galliher, M.S., Pratt, D.A., and Stephenson, C.R.J. (2018). Radicals in natural product synthesis. *Chem. Soc. Rev.* 47: 7851–7866.

5 For selected examples on approaches to controlling radical reactivity and stereoselectivity, see: (a) Sibi, M.P., Manyem, S., and Zimmerman, J. (2003). Enantioselective radical processes. *Chem. Rev.* 103: 3263–3296. (b) Brimioulle, R., Lenhart, D., Maturi, M.M., and Bach, T. (2015). Enantioselective catalysis of photochemical reactions. *Angew. Chem. Int. Ed.* 54: 3872–3890.

6 For selected examples, see: (a) Huo, H., Shen, X., Wang, C. et al. (2014). Asymmetric photoredox transition-metal catalysis activated by visible light. *Nature* 515: 100–103. (b) Kainz, Q.M., Matier, C.D., Bartoszewicz, A. et al. (2016). Asymmetric copper-catalyzed C—N cross-couplings induced by visible light. *Science* 351: 681–684. (c) Zhang, W., Wang, F., McCann, S.D. et al. (2016). Enantioselective cyanation of benzylic C—H bonds via copper-catalyzed radical relay. *Science* 353:

1014–1018. (d) Nakafuku, K.M., Zhang, Z., Wappes, E.A. et al. (2020). Enantioselective radical C–H amination for the synthesis of β-amino alcohols. *Nat. Chem.* 12: 697–704.

7 For selected reviews and highlights on Co(II)-based MRC, see: (a) Lu, H.J. and Zhang, X.P. (2011). Catalytic C–H functionalization by metalloporphyrins: recent developments and future directions. *Chem. Soc. Rev.* 40: 1899–1909. (b) Demarteau, J., Debuigne, A., and Detrembleur, C. (2019). Organocobalt complexes as sources of carbon-centered radicals for organic and polymer chemistries. *Chem. Rev.* 119: 6906–6955. (c) Huang, H.-M., Garduño-Castro, M.H., Morrill, C., and Procter, D.J. (2019). Catalytic cascade reactions by radical relay. *Chem. Soc. Rev.* 48: 4626–4638. (d) Singh, R. and Mukherjee, A. (2019). Metalloporphyrin catalyzed C–H amination. *ACS Catal.* 9: 3604–3617.

8 For selected reviews on Ti(III)-based and other metalloradical-mediated radical processes, see: (a) Zhang, X.-X. and Wayland, B.B. (1994). Rhodium(II) porphyrin bimetalloradical complexes: preparation and enhanced reactivity with CH_4 and H_2. *J. Am. Chem. Soc.* 116: 7897–7898. (b) Gansäuer, A., Hildebrandt, S., Michelmann, A. et al. (2015). Cationic titanocene(III) complexes for catalysis in single-electron steps. *Angew. Chem. Int. Ed.* 54: 7003–7006. (c) Kuo, J.L., Hartung, J., Han, A., and Norton, J.R. (2015). Direct generation of oxygen-stabilized radicals by H• transfer from transition metal hydrides. *J. Am. Chem. Soc.* 137: 1036–1039. (d) Roy, S., Khatua, H., Das, S.K., and Chattopadhyay, B. (2019). Iron(II)-based metalloradical activation: switch from traditional click chemistry to denitrogenative annulation. *Angew. Chem. Int. Ed.* 58: 11439–11443.

9 For experimental evidence of α-Co(III)-alkyl radicals, see: (a) Belof, J.L., Cioce, C.R., Xu, X. et al. (2011). Characterization of tunable radical metal-carbenes: key intermediates in catalytic cyclopropanation. *Organometallics* 30: 2739–2746. (b) Lu, H., Dzik, W.I., Xu, X. et al. (2011). Experimental evidence for cobalt(III)-carbene radicals: key intermediates in cobalt(II)-based metalloradical cyclopropanation. *J. Am. Chem. Soc.* 133: 8518–8521. (c) Intrieri, D., Carminati, D.M., and Gallo, E. (2016). The ligand influence in stereoselective carbene transfer reactions promoted by chiral metal porphyrin catalysts. *Dalton Trans.* 45: 15746–15761.

10 For selected reviews on Co(II)-catalyzed carbene transfer reactions based on other types of ligands, see: (a) Cui, X. and Zhang, X.P. (2013). Cobalt-mediated carbene transfer reactions. In: *Contemporary Carbene Chemistry* (eds. R.A. Moss and M.P. Doyle), 491–525. Hoboken, NJ: Wiley. (b) Pellissier, H. and Clavier, H. (2014). Enantioselective cobalt-catalyzed transformations. *Chem. Rev.* 114: 2775–2823.

11 (a) Talele, T.T. (2016). The "cyclopropyl fragment" is a versatile player that frequently appears in preclinical/clinical drug molecules. *J. Med. Chem.* 59: 8712–8756. (b) Ebner, C. and Carreira, E.M. (2017). Cyclopropanation strategies in recent total syntheses. *Chem. Rev.* 117: 11651–11679.

12 For selected examples on transition metal-catalyzed olefin cyclopropanation with diazo compounds, see: (a) Doyle, M.P. (1986). Electrophilic metal carbenes as reaction intermediates in catalytic reactions. *Acc. Chem. Res.* 19: 348–356. (b) Doyle, M.P. and Forbes, D.C. (1998). Recent advances in asymmetric catalytic metal carbene transformations. *Chem. Rev.* 98: 911–936.

13 Lo, M.M.C. and Fu, G.C. (1998). A new class of planar–chiral ligands: synthesis of a C_2-symmetric bisazaferrocene and its application in the enantioselective Cu(I)-catalyzed cyclopropanation of olefins. *J. Am. Chem. Soc.* 120: 10270–10271.

14 Doyle, M.P., Winchester, W.R., Hoorn, J.A.A. et al. (1993). Dirhodium(II) tetrakis(carboxamidates) with chiral ligands. Structure and selectivity in catalytic metal-carbene transformations. *J. Am. Chem. Soc.* 115: 9968–9978.

15 Che, C.-M., Huang, J.-S., Lee, F.-W. et al. (2001). Asymmetric inter- and intramolecular cyclopropanation of alkenes catalyzed by chiral ruthenium porphyrins. Synthesis and crystal structure of a chiral metalloporphyrin carbene complex. *J. Am. Chem. Soc.* 123: 4119–4129.

16 Lebel, H., Marcoux, J.F., Molinaro, C., and Charette, A.B. (2003). Stereoselective cyclopropanation reactions. *Chem. Rev.* 103: 977–1050.

17 Aggarwal, V.K., de Vicente, J., and Bonnert, R.V. (2001). Catalytic cyclopropanation of alkenes using diazo compounds generated in situ. A novel route to 2-arylcyclopropylamines. *Org. Lett.* 3: 2785–2788.

18 Penoni, A., Wanke, R., Tollari, S. et al. (2003). Cyclopropanation of olefins with diazoalkanes, catalyzed by Co^{II}(porphyrin) complexes – a synthetic and mechanistic investigation and the molecular structure of $Co^{III}(TPP)(CH_2CO_2Et)$ (TPP = dianion of *meso*-tetraphenylporphyrin). *Eur. J. Org. Chem.* 2003: 1452–1460.

19 Huang, L., Chen, Y., Gao, G.-Y., and Zhang, X.P. (2003). Diastereoselective and enantioselective cyclopropanation of alkenes catalyzed by cobalt porphyrins. *J. Org. Chem.* 68: 8179–8184.

20 Chen, Y. and Zhang, X.P. (2004). Vitamin B_{12} derivatives as natural asymmetric catalysts: enantioselective cyclopropanation of alkenes. *J. Org. Chem.* 69: 2431–2435.

21 Chen, Y., Fields, K.B., and Zhang, X.P. (2004). Bromoporphyrins as versatile synthons for modular construction of chiral porphyrins: cobalt-catalyzed highly enantioselective and diastereoselective cyclopropanation. *J. Am. Chem. Soc.* 126: 14718–14719.

22 Chen, Y. and Zhang, X.P. (2007). Asymmetric cyclopropanation of styrenes catalyzed by metal complexes of D_2-symmetrical chiral porphyrin: superiority of cobalt over iron. *J. Org. Chem.* 72: 5931–5934.

23 Chen, Y., Ruppel, J.V., and Zhang, X.P. (2007). Cobalt-catalyzed asymmetric cyclopropanation of electron-deficient olefins. *J. Am. Chem. Soc.* 129: 12074–12075.

24 Fantauzzi, S., Gallo, E., Rose, E. et al. (2008). Asymmetric cyclopropanation of olefins catalyzed by chiral cobalt(II)-binaphthyl porphyrins. *Organometallics* 27: 6143–6151.

25 Hu, Y., Lang, K., Tao, J. et al. (2019). Next-generation D_2-symmetric chiral porphyrins for cobalt(II)-based metalloradical catalysis: catalyst engineering by distal bridging. *Angew. Chem. Int. Ed.* 58: 2670–2674.

26 Ruppel, J.V., Gauthier, T.J., Snyder, N.L. et al. (2009). Asymmetric Co(II)-catalyzed cyclopropanation with succinimidyl diazoacetate: general synthesis of chiral cyclopropyl carboxamides. *Org. Lett.* 11: 2273–2276.

27 Zhu, S., Ruppel, J.V., Lu, H. et al. (2008). Cobalt-catalyzed asymmetric cyclopropanation with diazosulfones: rigidification and polarization of ligand chiral environment via hydrogen bonding and cyclization. *J. Am. Chem. Soc.* 130: 5042–5043.

28 Zhu, S., Perman, J.A., and Zhang, X.P. (2008). Acceptor/acceptor-substituted diazo reagents for carbene transfers: cobalt-catalyzed asymmetric *Z*-cyclopropanation of alkenes with α-nitrodiazoacetates. *Angew. Chem. Int. Ed.* 47: 8460–8463.

29 Zhu, S., Xu, X., Perman, J.A., and Zhang, X.P. (2010). A general and efficient cobalt(II)-based catalytic system for highly stereoselective cyclopropanation of alkenes with alpha-cyanodiazoacetates. *J. Am. Chem. Soc.* 132: 12796–12799.

30 Xu, X., Zhu, S., Cui, X. et al. (2013). Cobalt(II)-catalyzed asymmetric olefin cyclopropanation with α-ketodiazoacetates. *Angew. Chem. Int. Ed.* 52: 11857–11861.

31 Lindsay, V.N.G., Nicolas, C., and Charette, A.B. (2011). Asymmetric Rh(II)-catalyzed cyclopropanation of alkenes with diacceptor diazo compounds: *p*-methoxyphenyl ketone as a general stereoselectivity controlling group. *J. Am. Chem. Soc.* 133: 8972–8981.

32 Xu, X., Wang, Y., Cui, X. et al. (2017). Metalloradical activation of α-formyldiazoacetates for the catalytic asymmetric radical cyclopropanation of alkenes. *Chem. Sci.* 8: 4347–4351.

33 Xia, Y. and Wang, J. (2017). *N*-tosylhydrazones: versatile synthons in the construction of cyclic compounds. *Chem. Soc. Rev.* 46: 2306–2362.

34 Wang, Y., Wen, X., Cui, X. et al. (2017). Asymmetric radical cyclopropanation of alkenes with in situ-generated donor-substituted diazo reagents via Co(II)-based metalloradical catalysis. *J. Am. Chem. Soc.* 139: 1049–1052.

35 Roy, S., Das, S.K., and Chattopadhyay, B. (2018). Cobalt(II)-based metalloradical activation of 2-(diazomethyl)-pyridines for radical transannulation and cyclopropanation. *Angew. Chem. Int. Ed.* 57: 2238–2243.

36 (a) Zhao, W. (2010). Novel syntheses of bridge-containing organic compounds. *Chem. Rev.* 110: 1706–1745. (b) Singh, R. and Vince, R. (2012). 2-azabicyclo[2.2.1]hept-5-en-3-one: chemical profile of a versatile synthetic building block and its impact on the development of therapeutics. *Chem. Rev.* 112: 4642–4686.

37 Xu, X., Lu, H.J., Ruppel, J.V. et al. (2011). Highly asymmetric intramolecular cyclopropanation of acceptor-substituted diazoacetates by Co(II)-based metalloradical catalysis: iterative approach for development of new-generation catalysts. *J. Am. Chem. Soc.* 133: 15292–15295.

38 Ruppel, J.V., Cui, X., Xu, X., and Zhang, X.P. (2014). Stereoselective intramolecular cyclopropanation of α-diazoacetates via Co(II)-based metalloradical catalysis. *Org. Chem. Front.* 1: 515–520.

39 Reddy, A.R., Hao, F., Wu, K. et al. (2016). Cobalt(II) porphyrin-catalyzed intramolecular cyclopropanation of *N*-alkyl indoles/pyrroles with alkylcarbene: efficient synthesis of polycyclic *N*-heterocycles. *Angew. Chem. Int. Ed.* 55: 1810–1815.

40 Fumagalli, G., Stanton, S., and Bower, J.F. (2017). Recent methodologies that exploit C–C single-bond cleavage of strained ring systems by transition metal complexes. *Chem. Rev.* 117: 9404–9432.

41 Marek, I., Simaan, S., and Masarwa, A. (2007). Enantiomerically enriched cyclopropene derivatives: versatile building blocks in asymmetric synthesis. *Angew. Chem. Int. Ed.* 46: 7364–7376.

42 Rubin, M., Rubina, M., and Gevorgyan, V. (2007). Transition metal chemistry of cyclopropenes and cyclopropanes. *Chem. Rev.* 107: 3117–3179.

43 Lou, Y., Remarchuk, T.P., and Corey, E.J. (2005). Catalysis of enantioselective [2+1]-cycloaddition reactions of ethyl diazoacetate and terminal acetylenes using mixed-ligand complexes of the series $Rh_2(RCO_2)_n$ (L^*_{4-n}). Stereochemical heuristics for ligand exchange and catalyst synthesis. *J. Am. Chem. Soc.* 127: 14223–14230.

44 Uehara, M., Suematsu, H., Yasutomi, Y., and Katsuki, T. (2011). Enantioenriched synthesis of cyclopropenes with a quaternary stereocenter, versatile building blocks. *J. Am. Chem. Soc.* 133: 170–171.

45 Cui, X., Xu, X., Lu, H. et al. (2011). Enantioselective cyclopropenation of alkynes with acceptor/acceptor-substituted diazo reagents via Co(II)-based metalloradical catalysis. *J. Am. Chem. Soc.* 133: 3304–3307.

46 For selected reviews, see: (a) Wencel-Delord, J., Dröge, T., Liu, F., and Glorius, F. (2011). Towards mild metal-catalyzed C–H bond activation. *Chem. Soc. Rev.* 40: 4740–4761. (b) Newton, C.G., Wang, S.-G., Oliveira, C.C., and Cramer, N. (2017). Catalytic enantioselective transformations involving C–H bond cleavage by transition-metal complexes. *Chem. Rev.* 117: 8908–8976. (c) Wang, C.-S., Dixneuf, P.H., and Soulé, J.-F. (2018). Photoredox catalysis for building C–C bonds from C(sp^2)–H bonds. *Chem. Rev.* 118: 7532–7585.

47 (a) Giri, R., Shi, B.-F., Engle, K.M. et al. (2009). Transition metal-catalyzed C–H activation reactions: diastereoselectivity and enantioselectivity. *Chem. Soc. Rev.* 38: 3242–3272. (b) Xue, X.-S., Ji, P., Zhou, B., and Cheng, J.-P. (2017). The essential role of bond energetics in C–H activation/functionalization. *Chem. Rev.* 117: 8622–8648.

48 For selected reviews on applications of C–H activation strategy to natural product synthesis, see: (a) McMurray, L., O'Hara, F., and Gaunt, M.J. (2011). Recent developments in natural product synthesis using metal-catalysed C–H bond functionalisation. *Chem. Soc. Rev.* 40: 1885–1898. (b) Abrams, D.J., Provencher, P.A., and Sorensen, E.J. (2018). Recent applications of C–H functionalization in complex natural product synthesis. *Chem. Soc. Rev.* 47: 8925–8967.

49 For selected reviews on carbene insertion into C–H bonds, see: (a) Doyle, M.P., Duffy, R., Ratnikov, M., and Zhou, L. (2010). Catalytic carbene insertion into C–H bonds. *Chem. Rev.* 110: 704–724. (b) Xia, Y., Qiu, D., and Wang, J. (2017).

Transition-metal-catalyzed cross-couplings through carbene migratory insertion. *Chem. Rev.* 117: 13810–13889.

50 Chan, W.-W., Lo, S.-F., Zhou, Z., and Yu, W.-Y. (2012). Rh-catalyzed intermolecular carbenoid functionalization of aromatic C–H bonds by α-diazomalonates. *J. Am. Chem. Soc.* 134: 13565–13568.

51 Díaz-Requejo, M.M., Belderraín, T.R., Nicasio, M.C. et al. (2002). Intermolecular copper-catalyzed carbon–hydrogen bond activation via carbene insertion. *J. Am. Chem. Soc.* 124: 896–897.

52 Choi, M.K.-W., Yu, W.-Y., and Che, C.-M. (2005). Ruthenium-catalyzed stereoselective intramolecular carbenoid C–H insertion for β and γ-lactam formations by decomposition of α-diazoacetamides. *Org. Lett.* 7: 1081–1084.

53 Li, Y., Huang, J.-S., Zhou, Z.-Y. et al. (2002). Remarkably stable iron porphyrins bearing nonheteroatom-stabilized carbene or (alkoxycarbonyl)carbenes: isolation, X-ray crystal structures, and carbon atom transfer reactions with hydrocarbons. *J. Am. Chem. Soc.* 124: 13185–13193.

54 Griffin, J.R., Wendell, C.I., Garwin, J.A., and White, M.C. (2017). Catalytic $C(sp^3)$–H alkylation via an iron carbene intermediate. *J. Am. Chem. Soc.* 139: 13624–13627.

55 Cui, X., Xu, X., Jin, L.-M. et al. (2015). Stereoselective radical C–H alkylation with acceptor/acceptor-substituted diazo reagents via Co(II)-based metalloradical catalysis. *Chem. Sci.* 6: 1219–1224.

56 Wen, X., Wang, Y., and Zhang, X.P. (2018). Enantioselective radical process for synthesis of chiral indolines by metalloradical alkylation of diverse $C(sp^3)$–H bonds. *Chem. Sci.* 9: 5082–5086.

57 Karns, A.S., Goswami, M., and de Bruin, B. (2018). Catalytic synthesis of indolines by hydrogen atom transfer to cobalt(III)-carbene radicals. *Chem. Eur. J.* 24: 5253–5258.

58 Wang, Y., Wen, X., Cui, X., and Zhang, X.P. (2018). Enantioselective radical cyclization for construction of 5-membered ring structures by metalloradical C–H alkylation. *J. Am. Chem. Soc.* 140: 4792–4796.

59 Lankelma, M., Olivares, A.M., and de Bruin, B. (2019). [Co(TPP)]-catalyzed formation of substituted piperidines. *Chem. Eur. J.* 25: 5658–5663.

60 (a) Taylor, R.D., MacCoss, M., and Lawson, A.D.G. (2014). Rings in drugs. *J. Med. Chem.* 57: 5845–5859. (b) Vitaku, E., Smith, D.T., and Njardarson, J.T. (2014). Analysis of the structural diversity, substitution patterns, and frequency of nitrogen heterocycles among U.S. FDA approved pharmaceuticals. *J. Med. Chem.* 57: 10257–10274.

61 te Grotenhuis, C., Das, B.G., Kuijpers, P.F. et al. (2017). Catalytic 1,2-dihydronaphthalene and *E*-aryl-diene synthesis via Co^{III}-carbene radical and *o*-quinodimethane intermediates. *Chem. Sci.* 8: 8221–8230.

62 For selected reviews, see: (a) Yet, L. (2000). Metal-mediated synthesis of medium-
sized rings. *Chem. Rev.* 100: 2963–3008. (b) Hu, Y.-J., Li, L.-X., Han, J.-C. et al. (2020). Recent advances in the total synthesis of natural products containing eight-membered carbocycles (2009–2019). *Chem. Rev.* 120: 5910–5953.

63 te Grotenhuis, C., van den Heuvel, N., van der Vlugt, J.I., and de Bruin, B. (2018). Catalytic dibenzocyclooctene synthesis via cobalt(III)-carbene radical and *ortho*-quinodimethane intermediates. *Angew. Chem. Int. Ed.* 57: 140–145.

64 Zhou, M., Lankelma, M., van der Vlugt, J.I., and de Bruin, B. (2020). Catalytic synthesis of 8-membered ring compounds via cobalt(III)-carbene radicals. *Angew. Chem. Int. Ed.* 59: 11073–11079.

65 For selected highlights on ketene chemistry, see: (a) Tidwell, T.T. (2005). The first century of ketenes (1905–2005): the birth of a versatile family of reactive intermediates. *Angew. Chem. Int. Ed.* 44: 5778–5785. (b) Tidwell, T.T. (2006). Ketene chemistry after 100 years: ready for a new century. *Eur. J. Org. Chem.* 2006: 563–576.

66 Paul, N.D., Chirila, A., Lu, H. et al. (2013). Carbene radicals in cobalt(II)-porphyrin-catalysed carbene carbonylation reactions; a catalytic approach to ketenes. *Chem. Eur. J.* 19: 12953–12958.

67 Goswami, M., de Bruin, B., and Dzik, W.I. (2017). Difluorocarbene transfer from a cobalt complex to an electron-deficient alkene. *Chem. Commun.* 53: 4382–4385.

68 (a) Brown, R.C.D. (2005). Developments in furan syntheses. *Angew. Chem. Int. Ed.* 44: 850–852. (b) Singh, P.K. and Silakari, O. (2018). The current status of *O*-heterocycles: a synthetic and medicinal overview. *ChemMedChem* 13: 1071–1087.

69 Godoi, B., Schumacher, R.F., and Zeni, G. (2011). Synthesis of heterocycles via electrophilic cyclization of alkynes containing heteroatom. *Chem. Rev.* 111: 2937–2980.

70 Cui, X., Xu, X., Wojtas, L. et al. (2012). Regioselective synthesis of multisubstituted furans via metalloradical cyclization of alkynes with α-diazocarbonyls: construction of functionalized α-oligofurans. *J. Am. Chem. Soc.* 134: 19981–19984.

71 Pratap, R. and Ram, V.J. (2014). Natural and synthetic chromenes, fused chromenes, and versatility of dihydrobenzo[*h*]chromenes in organic synthesis. *Chem. Rev.* 114: 10476–10526.

72 Paul, N.D., Mandal, S., Otte, M. et al. (2014). Metalloradical approach to 2*H*-chromenes. *J. Am. Chem. Soc.* 136: 1090–1096.

3

Catalytic Enantioselective Carbene Insertions into Heteroatom–Hydrogen Bonds

Ming-Yao Huang, Shou-Fei Zhu, and Qi-Lin Zhou

Nankai University, State Key Laboratory and Institute of Elemento-Organic Chemistry, College of Chemistry, 94th Weijin Road, Tianjin 300071, China

3.1 Introduction

The development of C—X bond forming methods has long been a vibrant topic in organic chemistry since C—X bonds widely exist in organic compounds [1]. Carbene insertion into heteroatom–hydrogen (X—H) bonds is a straightforward approach to construct C—X bonds, featuring with mild reaction conditions and high activity [2]. The history of transition-metal-catalyzed X—H bond insertions can be traced back to 1950s with the Cu salts as catalysts [3], while Rh(II) acetate and its derivatives become predominant catalysts after they were introduced in 1970s [4]. Generally speaking, X—H (X = N, O, S) bond insertion reactions of metal carbenes go through a stepwise process, including the formation of ylide and subsequent proton transfer (Scheme 3.1A); The Si—H and B—H bond insertion reactions of metal carbenes are a concerted process similar to the C—H bond insertion reactions (Scheme 3.1B).

Although X—H bond insertion reactions have been widely used in the synthesis of amines, alcohols, organosulfur compounds, and organosilanes, their enantioselective versions have long been considered as challenging tasks because of lacking efficient chiral catalysts and poor understanding of the mechanisms. This situation has changed dramatically in the past 15 years, numbers of new chiral catalysts and catalytic strategies were developed, and many highly enantioselective X—H bond insertion reactions and detailed mechanisms have been reported [5]. In this chapter, the research progress of enantioselective X—H bond insertion reactions is reviewed mainly from the perspective of chiral catalysts.

3.2 N—H Bond Insertion Reactions

The N—H bond insertion reaction is an efficient method for preparing α-amino esters, α-amino ketones, and nitrogen-containing heterocycles. The two early works on catalytic asymmetric N—H bond insertion gave only modest enantioselectivity

Transition Metal-Catalyzed Carbene Transformations, First Edition.
Edited by Jianbo Wang, Chi-Ming Che, and Michael P. Doyle.

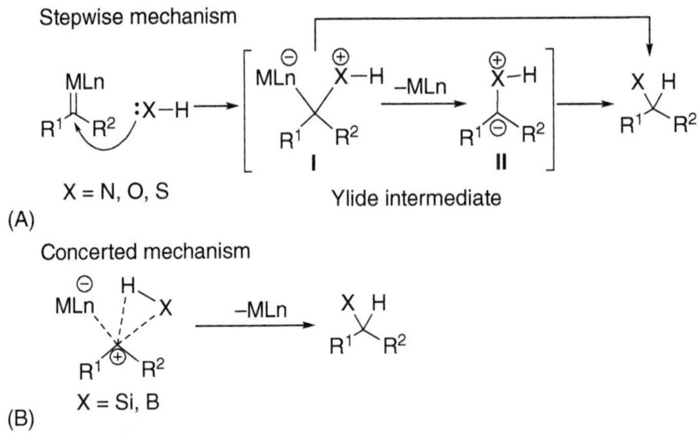

Scheme 3.1 Two distinct mechanisms of X–H bond insertion reactions.

(<50% ee) [6]. Since Zhou's group reported the first highly enantioselective N—H bond insertion reaction using chiral spiro bisoxazoline copper catalysts in 2007 [7], many chiral metal catalysts have been developed, significantly expanding the range of carbene precursors and N—H bond donors of the reaction.

3.2.1 Chiral Metal Catalysts

3.2.1.1 Chiral Cu Catalysts

In 2007, Zhou and coworkers [7] reported the first highly enantioselective N—H bond insertion by using a Cu catalyst *in situ* prepared from CuCl, chiral spiro bisoxazoline ligand (S_a,S,S)-**1**, and $NaBAr_F$ {sodium tetrakis[3,5-bis(trifluoromethyl) phenyl]borate} (Scheme 3.2A). Under optimal reaction conditions, the N—H bond insertion of α-diazopropionates with various primary aniline derivatives afforded α-amino acid derivatives in high yields and high enantioselectivities. With slight variation of the reaction conditions, the insertion reaction of other α-diazocarboxylates with anilines also underwent smoothly, affording the desired N-H insertion products in good yields with high enantioselectivities (Scheme 3.2B) [7]. The potential utility of this reaction was demonstrated by the concise preparation of the chiral arylalanine herbicide (*R*)-flamprop-*M*-isopropyl. The highly enantioselective intramolecular N—H bond insertion was also realized by using chiral spiro bisoxazoline-copper catalyst (Scheme 3.2C) [7]. This reaction provides an efficient access to optically active 2-carboxy tetrahydroquinolines.

The mechanism studies revealed that highly rigid chiral spiro bisoxazoline ligand and bulky and non-coordinating anion BAr_F^- ensure the success of the asymmetric N—H bond insertion reactions. Crystals of Cu(I)-(S_a,S,S)-**1** with different anions, such as PF_6^-, ClO_4^-, or BAr_F^-, have similar unexpected binuclear structure (Scheme 3.3A). In the structure of [Cu(I)-(S_a,S,S)-**1**]$^+$, each of the two Cu(I) atoms is *trans*-coordinated by two nitrogen atoms from the two spiro bisoxazoline ligands and has only 14e. The high resolution mass spectrum (HRMS) analysis and nonlinear effect experiments indicated that the binuclear Cu complex is most likely the

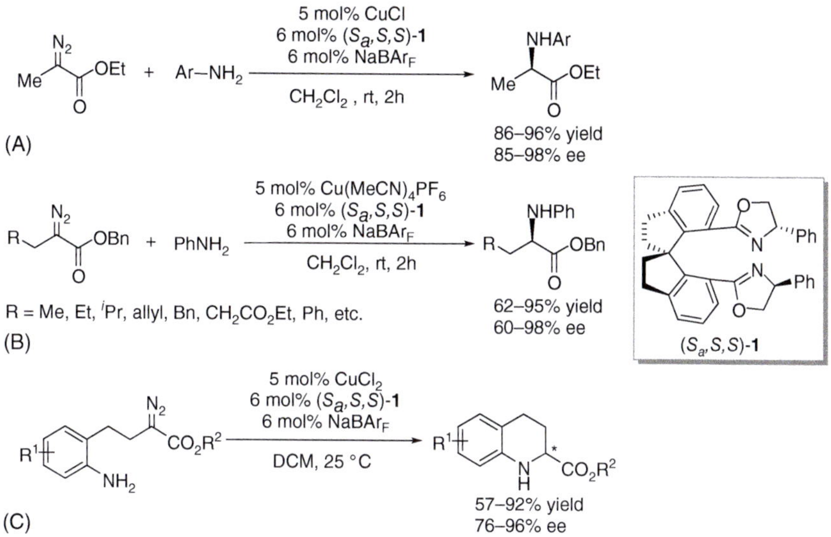

Scheme 3.2 Cu/chiral spiro bisoxazoline catalyzed N–H bond insertion reactions. Source: (B) Based on Zhu et al. [7b]. (C) Based on Song et al. [7c].

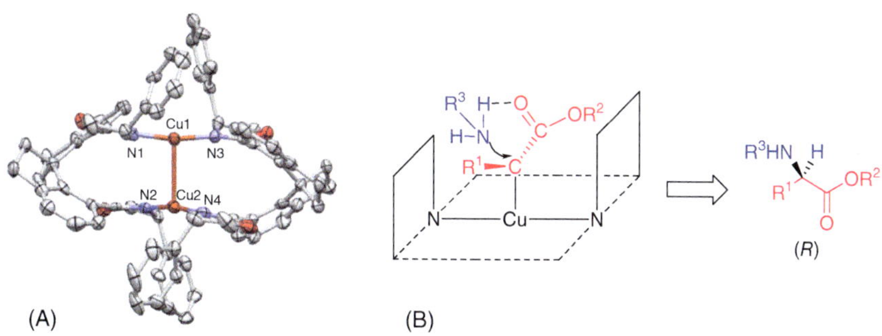

Scheme 3.3 Chiral induction model of Cu/chiral spiro bisoxazoline catalyzed N–H bond insertion reactions.

real catalyst for the N—H bond insertion reaction. A chiral induction model was proposed based on the catalyst structure and absolute configuration of the N—H bond insertion products (Scheme 3.3B): the nitrogen of aniline attacks the electron deficient carbene from less hindered *re* face, affording the insertion product with an *R* configuration.

Soon after, Fu and Lee [8] reported a highly enantioselective N—H bond insertion reaction of α-aryl-α-diazoacetates with carbamates (Scheme 3.4A). Cu complex of chiral bipyridine ligand (+)-**2** having a ferrocene skeleton was proven to be the suitable catalyst for this transformation and afforded up to 95% ee under optimal reaction conditions. Feng and coworkers [9] used a Cu catalyst modified with chiral imidazolidinone ligand having a binaphthyl backbone to catalyze the N—H bond insertion reaction of aniline derivatives and obtained good-to-high

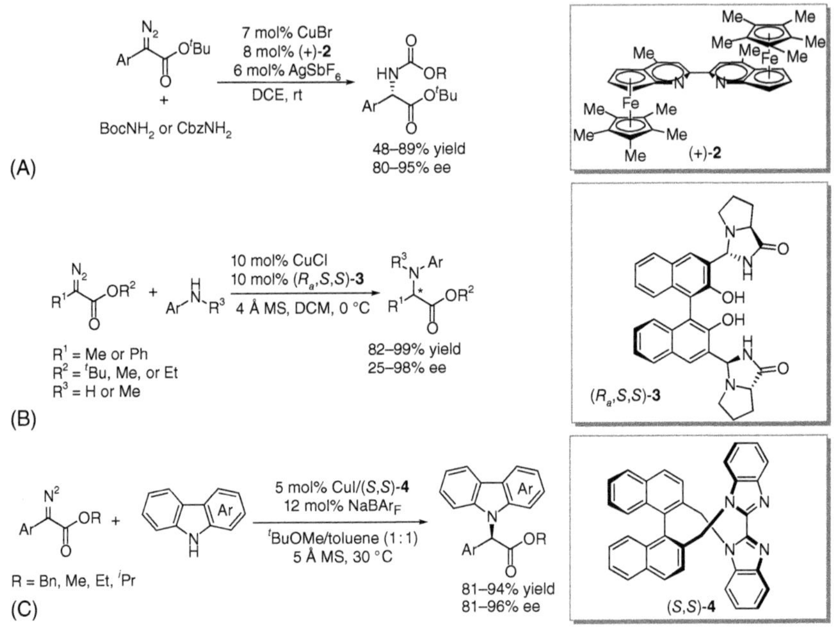

Scheme 3.4 Cu-catalyzed asymmetric N—H bond insertion reactions.

enantioselectivity (Scheme 3.4B). Zhou and coworkers [10] introduced axially chiral ligands with 2,2′-biimidazole linked to binaphthyl skeleton in the Cu(I)-catalyzed N—H bond insertion of carbazoles. Various optically active N-substituted carbazoles were prepared with satisfactory yields and enantioselectivities (Scheme 3.4C).

3.2.1.2 Chiral Pd Catalysts

Liu, Feng and coworkers [11] used Pd complexes of chiral guanidines as catalysts to promote the asymmetric insertion of N—H bond insertions of anilines with either primary or secondary amino moiety (Scheme 3.5A).

van Vranken and coworkers [12] found that the Pd catalysts modified with pyridine-bisoxazoline ligands can catalyze the N—H insertion of aromatic N-heterocycles including carbazoles and C3-substituted indoles. The reaction with α-aryldiazoesters afforded α-(*N*-carbazolyl)-α-arylacetates and α-(*N*-indolyl)-α-arylesters in moderate-to-good yields and high enantioselectivities (Scheme 3.5B). It is noteworthy that the C3 position of indoles must be masked to avoid competing C—H bond insertion reaction.

3.2.1.3 Other Chiral Metal Catalysts

Doyle and coworkers [13] reported an N—H bond insertion of α-vinyldiazoesters with hydrazones under the catalysis of chiral dirhodium carboxylates. In contrast to conventional process, a vinylogous addition took place by nucleophilic attack of hydrazone to the vinyl dirhodium carbene followed by 1,4-proton

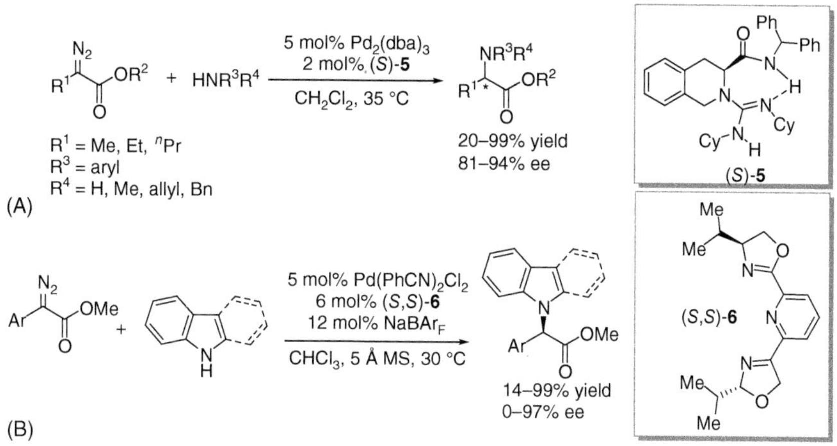

Scheme 3.5 Pd-catalyzed asymmetric N–H bond insertion reactions.

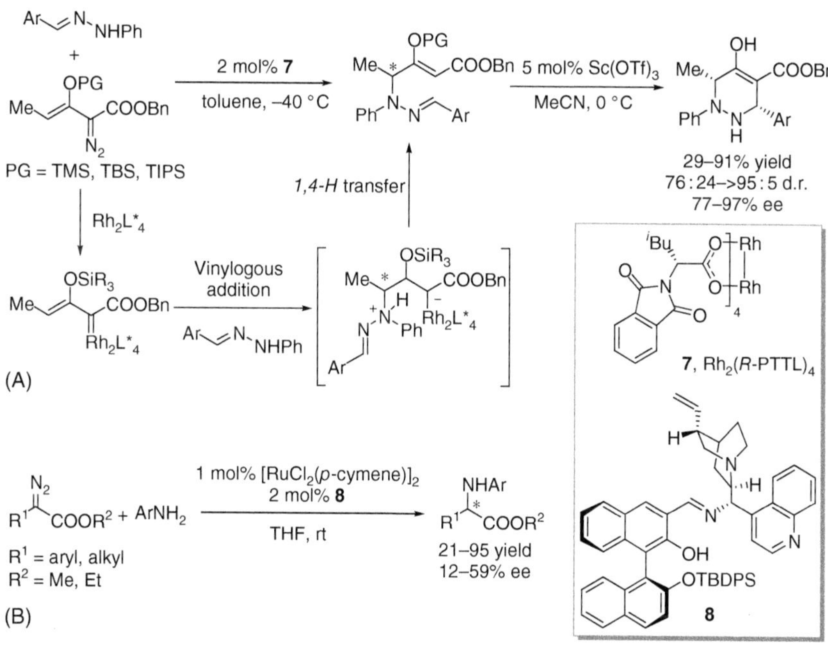

Scheme 3.6 Rh and Ru-catalyzed asymmetric N–H bond insertion reactions.

transfer to form the insertion products. Subsequent Lewis acid–promoted Mannich addition of the vinylogous N-H insertion products smoothly produces 1,2,3,6-tetrahydropyridazines with high diastereoselectivity (Scheme 3.6A).

Xu and coworkers [14] reported an enantioselective Ru-catalyzed N—H bond insertion reaction of α-diazoesters with aromatic amines in the presence of a multifunctional *N,N,O*-ligand, but the enantioselectivities were only moderate (Scheme 3.6B).

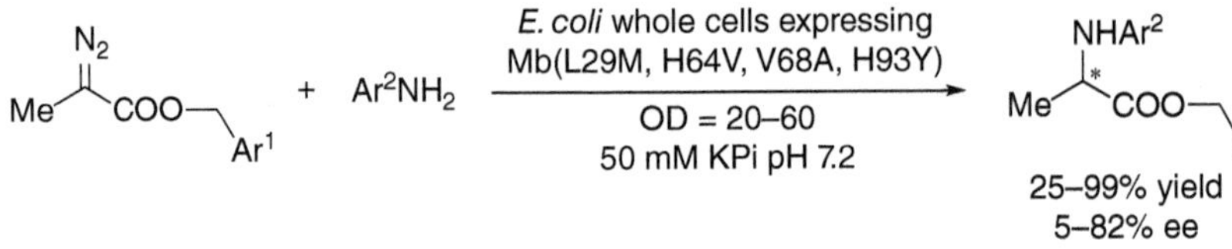

Scheme 3.7 Enzyme-catalyzed N–H bond insertion reactions.

3.2.1.4 Enzymes

Fasan and coworkers [15] reported a biocatalytic strategy for the asymmetric N—H bond insertion of α-diazopropinoates with aromatic amines by using engineered variants of myoglobin (Scheme 3.7). Although only moderate-to-good enantioselectivities were obtained, this reaction can be carried out in whole cell system, which simplifies their applications in organic synthesis.

3.2.1.5 Chiral Proton-Transfer Shuttle Catalysts

Although abovementioned chiral metal catalysts have been applied in various N—H bond insertion reactions, the enantioselectivities of these reactions highly depended on the substrates. For instance, the N—H bond insertion of α-diazoesters gave high enantioselectivity, but the similar insertion of α-diazoketones gave racemic product. This phenomenon mainly attributes to the stepwise mechanism of N—H bond insertion: when the stability of free ylide or enol increased, the metal catalyst is easier to dissociate from the intermediate, and the reaction undergoes through a non-enantioselective proton transfer process, providing a racemic product (Scheme 3.8A).

Chiral phosphoric acid, which has both Brønsted acid and Brønsted base functions, can provide a proton and accept a proton synchronously from ylide or enol through a cyclic transition state (Scheme 3.8B). This chiral phosphoric acid acts as a proton-shuttle to promote the proton-transfer of active intermediates and control the enantioselectivity. Zhu, Zhou and coworkers [16] coined a term of "chiral proton-transfer shuttle (CPTS)" for this type of catalysts. The CPTS catalyst directly controls the enantioselectivity of proton-transfer process without need of the coordination of metal catalyst with the active intermediate and thus provides a new strategy for the enantiocontrol of proton-transfer reactions.

3.2.1.6 Chiral Phosphoric Acids as CPTS Catalysts

The feasibility of the CPTS catalysis was first tested by Zhu, Zhou and coworkers [17] in the N—H bond insertion reaction of α-aryl-diazoesters with *tert*-butyl carbamate. By using chiral spiro phosphoric acid (SPA) (*R*)-**9** and achiral $Rh_2(TPA)_4$ as catalysts, a series of α-aryl-α-amino acid derivatives were prepared in high yields with high enantioselectivities (Scheme 3.9A). It is worth to mention that this combined catalytic system is highly active. Even when the catalyst loading reduced to 0.1 mol%, the reaction still accomplished within 1 minute (turnover frequency is 1000 min^{-1} in this case). α-Alkenyl-α-diazoacetates were also suitable carbene precursors for asymmetric N—H bond insertion of carbamate and afforded α-alkenyl α-amino acid derivatives with high enantioselectivity (Scheme 3.9B) [18].

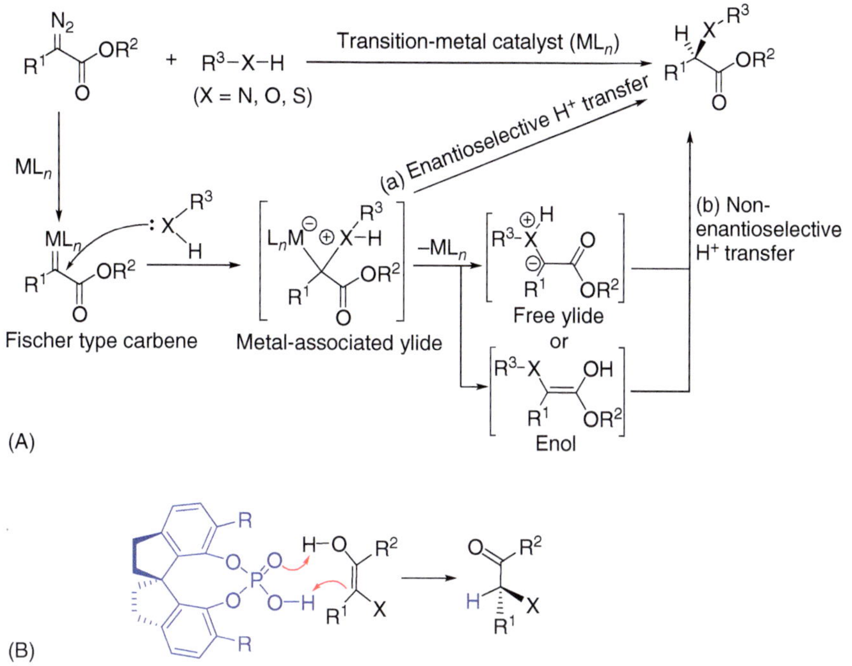

Scheme 3.8 (A) Mechanisms of transition-metal-catalyzed N–H bond insertion reactions; (B) Working model of CPTS catalysis.

In contrast to α-diazoesters, α-diazoketones tend to form free ylide or enol, and the chiral metal catalyst is not involved in the enantio-determining proton transfer step. Thus, the chiral metal catalysts could not control the enantioselectivity of N—H bond insertion reaction of α-diazoketones. However, this problem can be solved by the CPTS catalysts. Zhu, Zhou and coworkers [19] found that the CPTS catalyst (*R*)-**11**, combined with achiral catalyst $Rh_2(TFA)_4$ can promote a highly enantioselective N—H bond insertion reaction of α-diazoketones with carbamate (Scheme 3.9C).

3.2.1.7 Chiral Amino Thioureas as CPTS Catalysts

For a long period of time, the N-H donors of N—H bond insertion reaction have been restricted to aromatic amines or amides. Aliphatic amines are failed to be used in this reaction for the following reasons: (i) strong Lewis base poisons the metal catalysts by coordination, interfering with generation of the metal carbene; (ii) strong basicity neutralizes the chiral phosphoric acids in the CPTS catalysis; (iii) excess aliphatic amines can displace the ylide from metal–ylide intermediates, forming free ylide.

To overcome these barriers, Zhu, Zhou and coworkers [20] developed chiral ATU (amino-thiourea) catalysts for the highly enantioselective N—H bond insertion of aliphatic amines with cocatalysts Tp*Cu [Tp* is hydrotris(3,5-dimethylpyrazolyl) borate] (Scheme 3.10A). Density functional theory (DFT) calculations along with mechanistic experiments revealed that the Tp*Cu-catalyzed decomposition of diazo substrates is the rate-determining step. [Tp*Cu]•**12** complex not only serves as

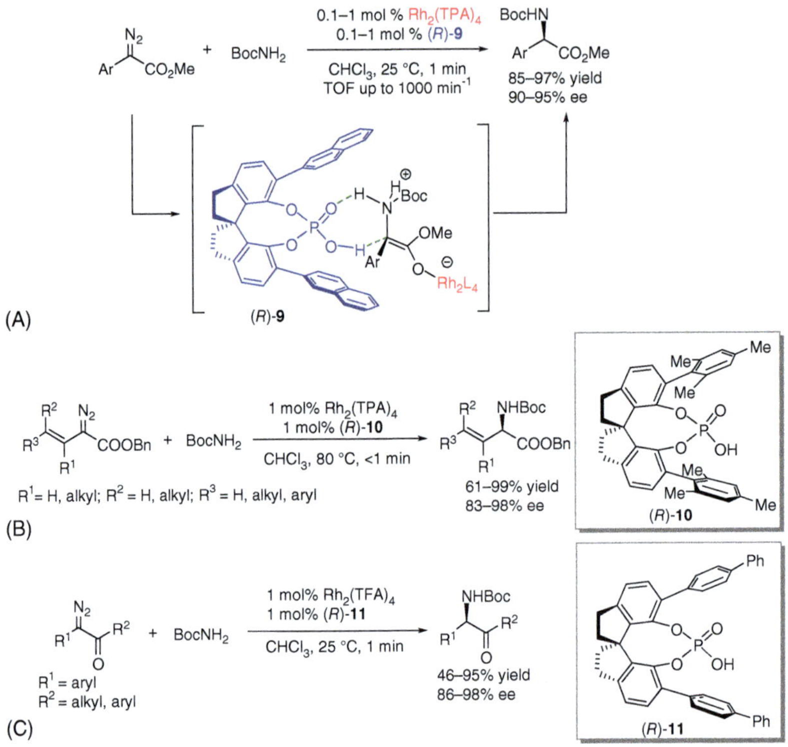

Scheme 3.9 N–H bond insertion reactions catalyzed by CPTS catalysts. Source: (B) Based on Guo et al. [18].

a resting state of the catalyst to release the active Tp*Cu, but also the real CPTS catalyst for the proton transfer, in which the amino moiety accepts a proton from the hydroxy group of the enol while the thiourea moiety donates a proton to the β-carbon of the enol (Scheme 3.10B). It is noteworthy that CPTS Tp*Cu•**12** significantly lowered the transition-state energy of the transformation from enol to ketone, which exits insurmountable barriers in the catalyst-free or H_2O-assisted pathways.

3.3 O–H Bond Insertion Reactions

The O—H bond insertion is an efficient method for preparing α-alkoxy, α-aryloxy, or α-hydroxyesters and oxygen-containing heterocyclic compounds. However, the earlier attempts on catalytic asymmetric O—H bond insertion reactions have poor enantioselectivity [21].

3.3.1 Chiral Metal Catalysts

3.3.1.1 Chiral Cu Catalysts

The first highly enantioselective Cu-catalyzed insertion of α-aryl-α-diazoacetates into O—H bonds of alcohols was accomplished by Fu and Maier [22] by using a

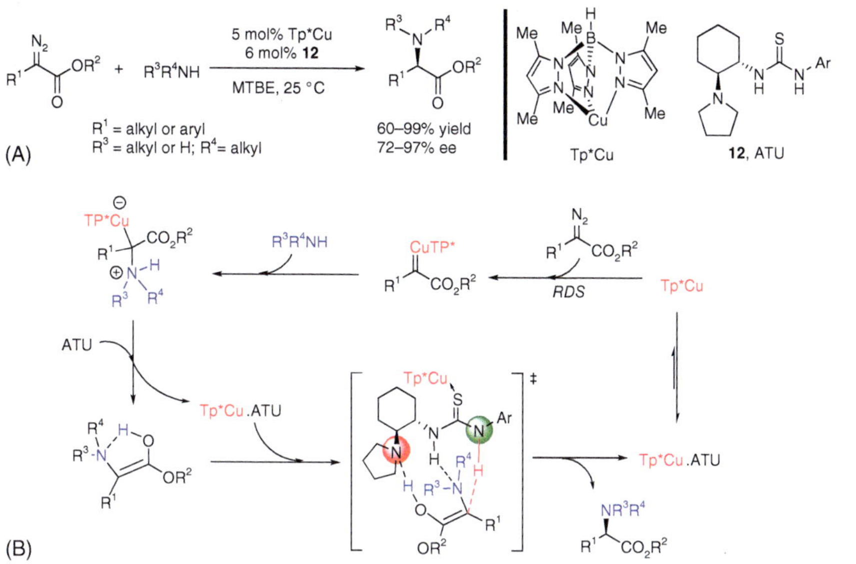

Scheme 3.10 Tp*Cu/chiral ATU-catalyzed N—H bond insertion of aliphatic amines.

chiral bisazaferrocene ligand (Scheme 3.11A). Zhou and coworkers [23] developed the asymmetric O—H bond insertion of α-diazoesters with phenols. By using the catalysts generated *in situ* from CuCl/chiral spiro bisoxazoline/$NaBAr_F$, a wide range of phenol derivatives underwent O—H bond insertion reaction with α-diazopropionates, providing α-aryloxypropionates with high enantioselectivity (Scheme 3.11B). Zhou and coworkers [24] also realized the highly enantioselective O—H bond insertion reaction of α-diazo phosphonates with alcohols by using chiral spiro Cu catalysts (Scheme 3.11C).

The catalytic intramolecular O—H bond insertion is a very useful reaction for the construction of cyclic ethers and esters. The Cu-catalyzed highly enantioselective intramolecular O—H bond insertions of alcohols [25] (Scheme 3.12A) and phenols [26] (Scheme 3.12B) were achieved by using ligand (S_a,S,S)-**14** and (S_a)-**15**, respectively. The substrates of intramolecular O—H bond insertions can be easily modified at side chain and thus provide a useful method for preparing chiral 2-carboxy cyclic ethers with different ring sizes and substitution patterns.

Other Cu complexes with different chiral ligands, such as chiral imidazoindolephosphine [27], planar chiral [2.2]paracyclophane-based bisoxazoline [28, 29], and chiral bicyclo bisoxazoline [30], can also catalyze the O—H bond insertion reactions, but generally have moderate enantioselectivity.

Water can also serve as an O—H bond donor for carbene insertion reactions. Zhou and coworkers [31] realized the Cu-catalyzed highly enantioselective O—H bond insertion reaction of water (Scheme 3.13). Using the chiral spiro ligand (S_a,S,S)-**1**, a series of α-hydroxyesters could be prepared through this protocol with high yields and enantioselectivities.

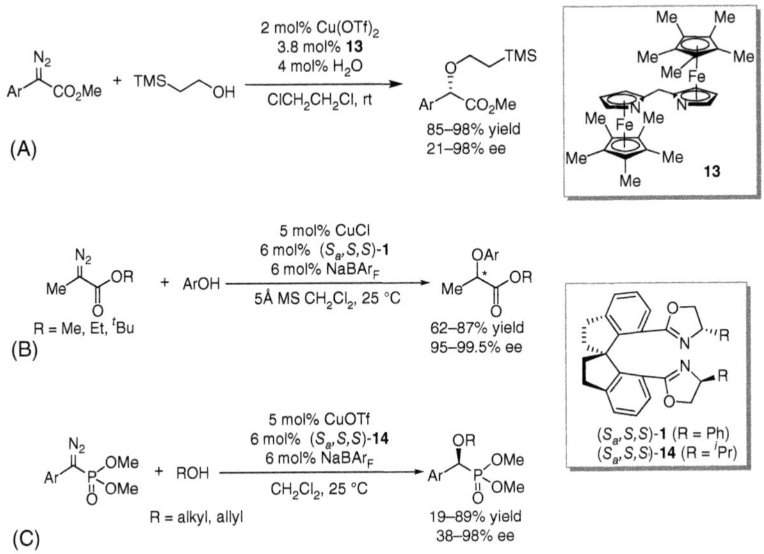

Scheme 3.11 Cu-catalyzed O—H bond insertion of alcohols.

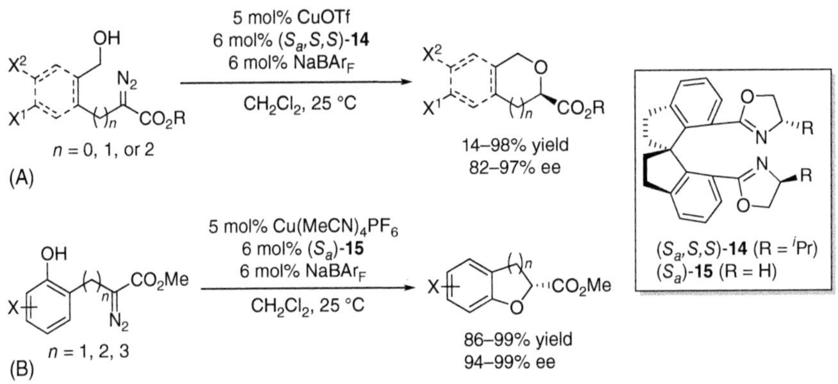

Scheme 3.12 Cu-catalyzed asymmetric intramolecular O—H bond insertions.

3.3.1.2 Chiral Fe Catalysts

The Fe catalysts prepared *in situ* from $FeCl_2{\cdot}4H_2O$ and chiral spiro bisoxazoline ligands (S_a,*S*,*S*)-**1** and (S_a,*S*,*S*)-**14** exhibited high enantioselectivity as well as reactivity for the insertion of O—H bonds of various saturated alcohols, allylic alcohols, and even water (Scheme 3.14A) [32]. The Fe-catalyzed asymmetric O—H bond insertions of alcohols and water provided a highly efficient method for the preparations of chiral α-alkyloxy, α-allyloxy, and α-hydroxyl arylacetates, which are useful building blocks in organic synthesis. In the O—H bond insertions of alcohols and water, the Fe catalyst generally showed higher yields and higher enantioselectivities than Cu and other metal catalysts, for instance, in the O–H insertion of α-aryl-α-diazoacetates with water (Scheme 3.14B). The excellent performance of Fe catalysts bearing chiral

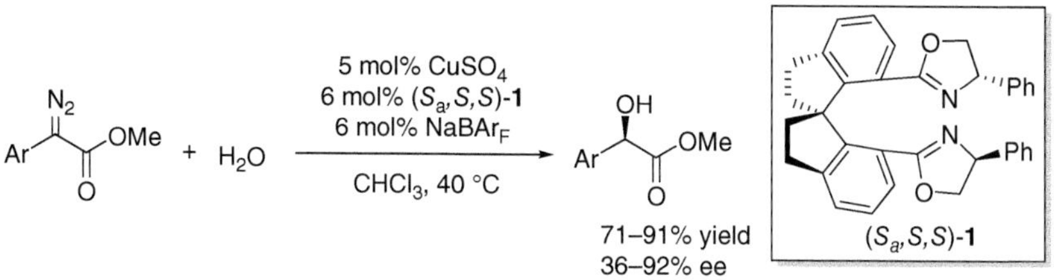

Scheme 3.13 Cu-catalyzed asymmetric O–H bond insertion of water.

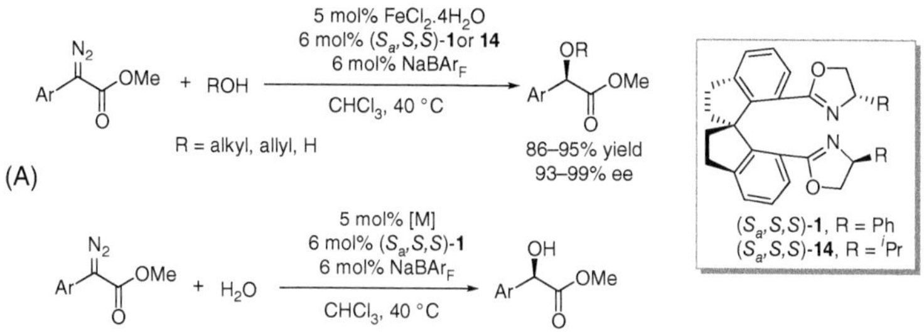

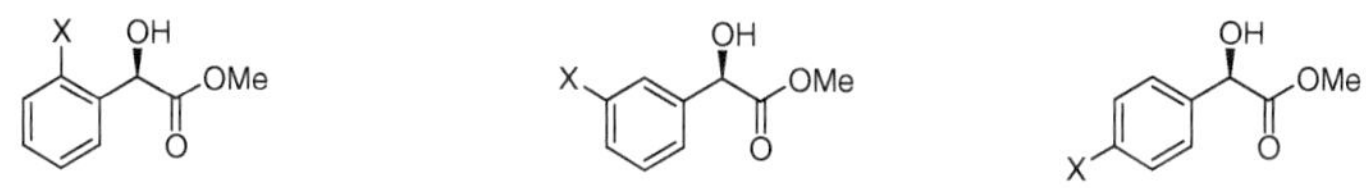

X	$FeCl_2.4H_2O$	$CuSO_4$
Me	87%, 94% ee	81%, 89% ee
MeO	72%, 92% ee	71%, 50% ee
Cl	92%, 95% ee	90%, 36% ee
Br	94%, 94% ee	

X	$FeCl_2.4H_2O$	$CuSO_4$
Me	93%, 92% ee	87%, 92% ee
MeO	92%, 94% ee	89%, 91% ee
Cl	92%, 91% ee	88%, 88% ee
Br	93%, 90% ee	92%, 88% ee

X	$FeCl_2.4H_2O$	$CuSO_4$
H	90%, 95% ee	91%, 90% ee
Me	89%, 94% ee	83%, 92% ee
Cl	90%, 94% ee	83%, 92% ee
Br	93%, 92% ee	86%, 91% ee

(B)

Scheme 3.14 Fe-catalyzed asymmetric O–H bond insertion reactions. Source: (A) Based on Zhu et al. [32].

spiro bisoxazoline ligands encourages the use of environmentally benign Fe catalysts in the catalytic asymmetric synthesis.

3.3.1.3 Chiral Pd Catalysts

The Pd/(S_a,S,S)-**1**-catalyzed asymmetric O—H bond insertion reaction between α-aryl-α-diazoacetates and phenols enabled the first enantioselective method for the preparation of chiral α-aryloxy-α-arylacetates, which are ubiquitous in biologically active molecules [33] (Scheme 3.15A). The reaction has broad substrate scope and thus fulfills the diverse requirements of the optically active α-aryloxy-α-arylacetates.

Zhou and coworkers [35] used Pd complexes of axially chiral 2,2′-bipyridyl ligands to catalyze the O–H insertion of α-aryl-α-diazoesters with phenols, affording various α-aryl-α-aryloxyacetates in high yield and high enantioselectivity. Chiral 2,2′-biimidazole ligands were also efficient for this reaction, giving satisfactory results (Scheme 3.15B) [34].

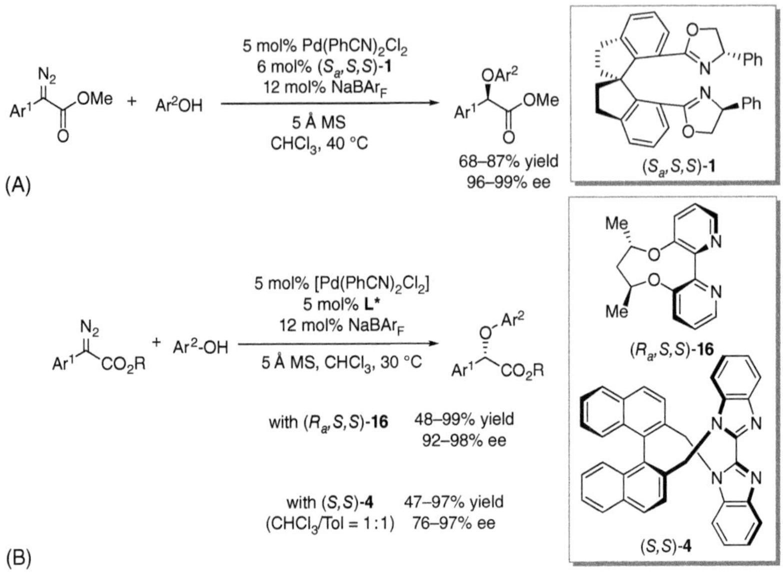

Scheme 3.15 Pd-catalyzed asymmetric O—H bond insertion reactions. Source: (A) Based on Xie et al. [33]. (B) Based on Shen et al. [34].

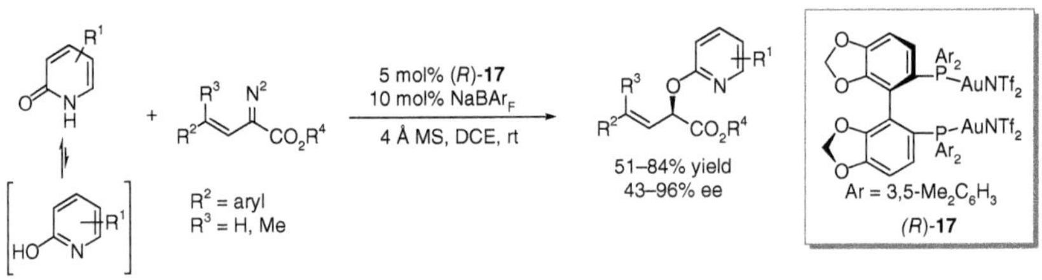

Scheme 3.16 Au-catalyzed asymmetric O—H bond insertion reactions.

3.3.1.4 Chiral Au Catalysts

Sun and coworkers [36] reported an Au-catalyzed regio- and enantioselective carbene insertion into O—H bond of 2-pyridones with α-styryldiazoesters by using chiral Segphos biphosphine ligand and obtained α-styryl-α-(2-pyridyl)oxyacetates with good yields and moderate-to-high enantioselectivities (Scheme 3.16). The reaction likely proceeds via an insertion of gold carbene into the O—H bond of 2-hydroxylpridine (tautomer of 2-pyridones).

3.3.1.5 Chiral Bases as CPTS Catalysts

Saito et al. [37] used a kind of alkaloid cinchona derivative **18** as the additive to realize the asymmetric insertion of phenyldiazomethyl ester into O—H bond of water under the catalysis of achiral dirhodium complex. However, the enantioselectivity of the reaction was only moderate (Scheme 3.17A). Liu and coworkers [38] found that the chiral guanidine **19** was an efficient catalyst for the asymmetric O—H bond insertion reaction of carboxylic acids with α-diazoketones or α-diazoesters, affording optically active α-acyloxy carbonyl compounds with up to

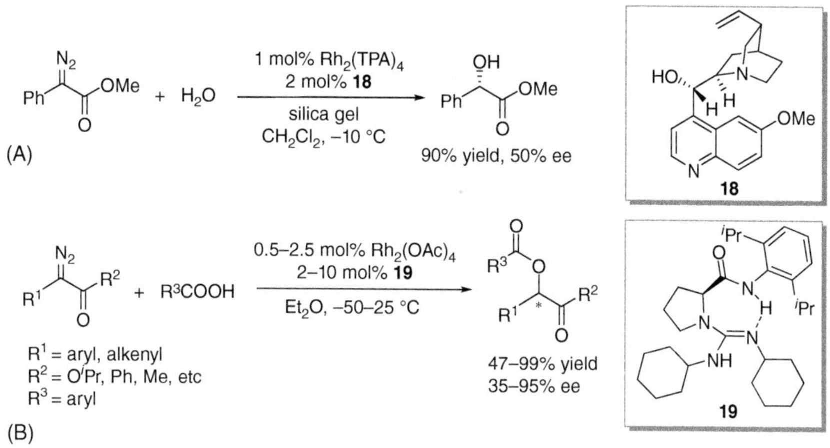

Scheme 3.17 Rh/chiral organo base-catalyzed asymmetric O–H bond insertions.

95% ee (Scheme 3.17B). Although the role of cinchona derivative and guanidine remains unclear, they likely play as CPTS catalyst to promote the protonation of enolate intermediates.

3.3.1.6 Chiral Phosphoric Acids as CPTS Catalysts

By using chiral binaphthyl phosphoric acid/Rh(II) catalysts, Shi and coworkers [39] realized a highly enantioselective O—H bond insertion of α-aryl-diazoesters with carboxylic acids (Scheme 3.18A). Zhu, Zhou and coworkers [40] introduced SPA to the Rh(II)-catalyzed O—H bond insertion reaction of α-diazoketones with alcohols and achieved good yields and high enantioselectivities (Scheme 3.18B).

Zhu and coworkers [41] further accomplished asymmetric O—H bond insertion reaction of water with α-alkyl-diazoesters and α-alkenyl-diazoesters using the SPA/Rh(II) catalysts and prepared various chiral α-hydroxyl esters (Scheme 3.19A).

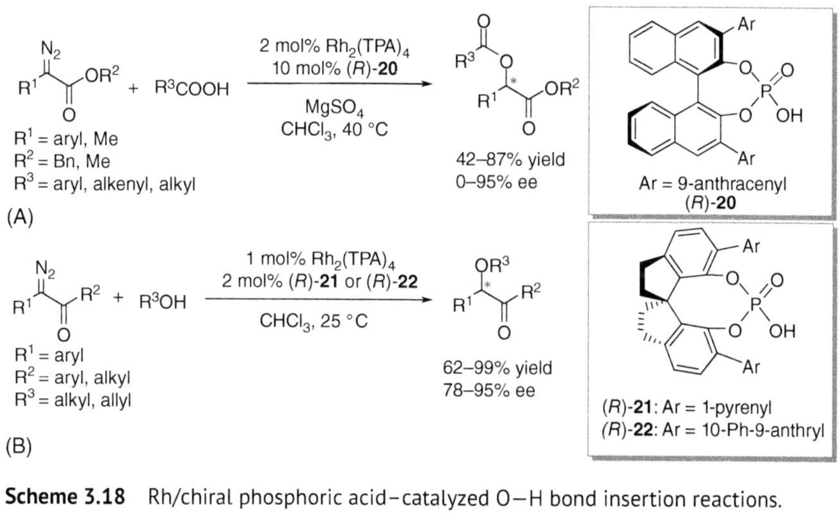

Scheme 3.18 Rh/chiral phosphoric acid-catalyzed O–H bond insertion reactions.

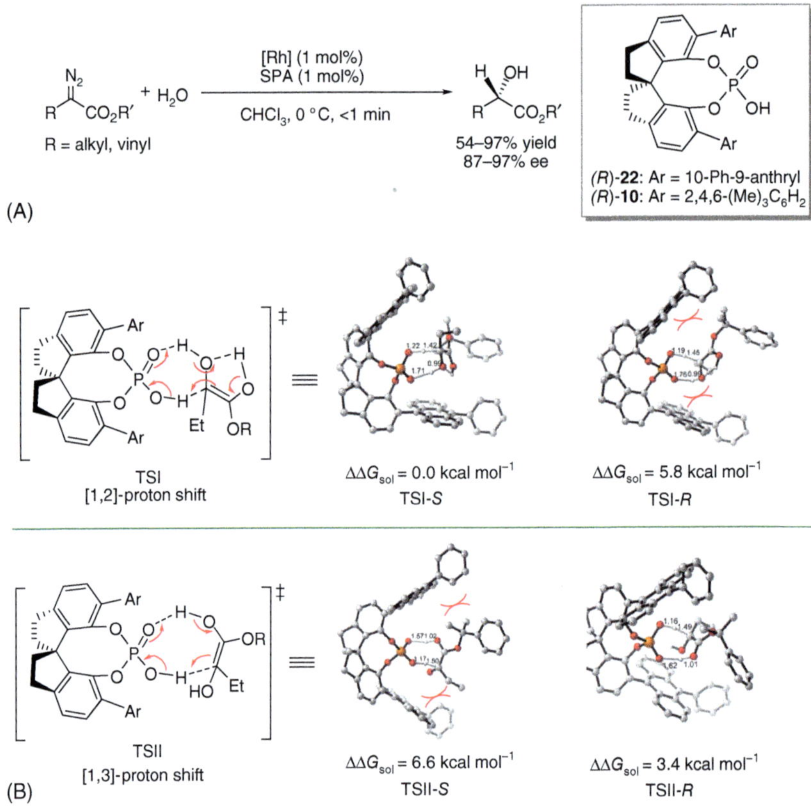

Scheme 3.19 Rh/SPA-catalyzed O–H bond insertion of water.

DFT calculations indicated that phosphoric acid plays as a proton shutter that significantly lows down the transition-state energy barrier of proton transfer, accelerating the O—H bond insertion rate. The chiral proton transfer step of the enol species proceeds through two possible pathways: [1,2]-proton transfer or [1,3]-proton-transfer. DFT calculations illustrated that *S*-pathway of [1,2]-proton transfer is the most energetically favorable, which is consistent with the observed enantioselectivity (Scheme 3.19B).

3.4 S–H Bond Insertion Reactions

Chiral organosulfur compounds have a variety of biological activities and are widely used as pharmaceuticals. Although transition-metal-catalyzed S—H bond insertion is a straightforward method for C—S bond construction, the enantiocontrol of this reaction has long been a challenge [42].

3.4.1 Chiral Metal Catalysts

The first highly enantioselective S—H bond insertion reaction was developed by using Cu/(S_a,*S*,*S*)-**1** catalyst, enabling a facile access to a broad array of

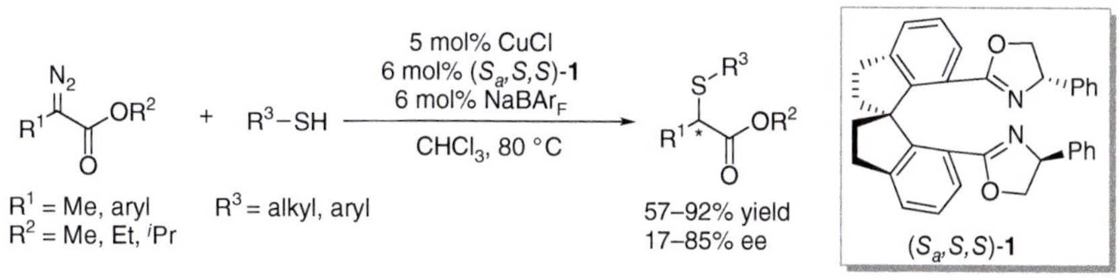

Scheme 3.20 Cu-catalyzed asymmetric S–H bond insertion reactions. Source: Based on Zhang et al. [43].

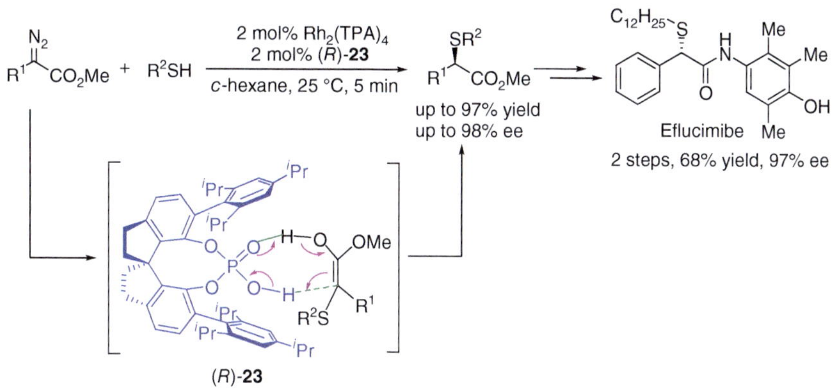

Scheme 3.21 Rh/SPA-catalyzed asymmetric S–H bond insertion reaction.

α-mercaptoesters with moderate-to-good enantioselectivities (up to 85% ee) (Scheme 3.20) [43].

3.4.2 CPTS Catalysts

Zhu, Yu, Zhou, and coworkers [44] found that the CPTS catalyst (*R*)-**23**, combined with $Rh_2(TPA)_4$, exhibited excellent outcomes in the S—H bond insertion reaction (Scheme 3.21). Under very mild and neutral conditions, various α-thioesters were prepared through this reaction in high yields with good-to-high enantioselectivities. (*S*)-Eflucimibe, a medicine for the treatment of atherosclerosis and lipoprotein disorders, was prepared via a two-step reaction in a high yield and high enantioselectivity starting from the S—H bond insertion product.

3.4.3 Enzymes

Fasan and coworkers [45] reported the asymmetric S—H bond insertion of phenyl mercaptan and ethyl α-diazopropionate under the catalysis of engineered variants of sperm whale myoglobin, achieving corresponding insertion product with moderate enantioselectivity (up to 49% ee) (Scheme 3.22A). Hong and coworkers [46] developed a biocatalytic platform based on engineered cytochrome P411/P450 enzymes harbored in *Escherichia coli*. cells. One of the enzymes P450-G8S variant exhibited high catalytic efficiency and good enantioselectivity (up to 86% ee) for the S—H bond insertion of thiophenols and α-diazolactones (Scheme 3.22B).

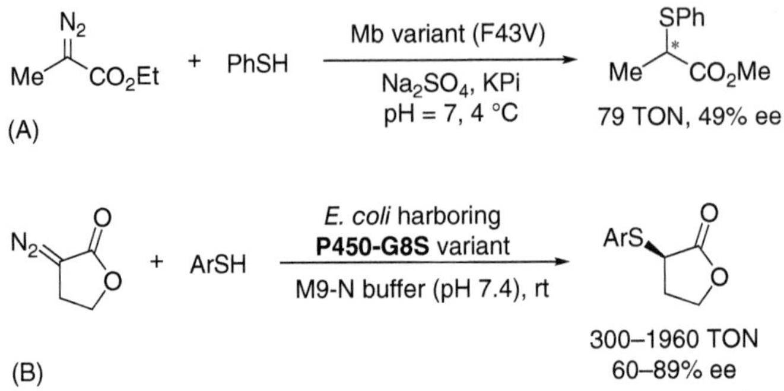

Scheme 3.22 Enzyme-catalyzed asymmetric S–H bond insertion reactions.

3.5 F–H Bond Insertion Reactions

Transition-metal-catalyzed carbene insertion into F—H bond was regarded as a promising protocol for constructing C—F bond. However, compared with other X-H donors (X = N, O, S) with lone pairs, F-H exhibits the weakest nucleophilicity, which results in its lowest reactivity with metal carbenes. Furthermore, the strong toxicity and corrosivity of hydrofluoric acid (HF) impede its direct employment as a F-H donor. Therefore, various F-H surrogates were developed for F-H insertion reactions, including the asymmetric insertions.

Doyle and coworkers [47] reported a Cu/chiral bisoxazoline-catalyzed F—H bond insertion reaction of α-diazocarbonyl compounds. The "F-H" source was generated *in situ* from potassium fluoride (KF) and protic additive HFIP (hexafluoroisopropanol) (Scheme 3.23A). This reaction enabled asymmetric construction C—F/^{18}F bond under mild conditions, affording a variety of α-fluorocarbonyl

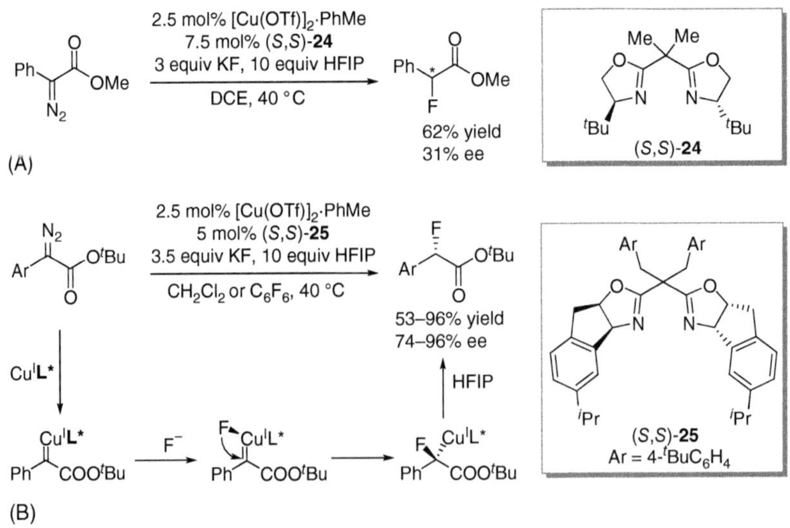

Scheme 3.23 Cu-catalyzed asymmetric F–H bond insertion reactions.

derivatives albeit with moderate enantioselectivity (up to 31% ee). Under similar conditions, Fürstner and coworkers [48] found that indane-based bisoxazoline (*S*,*S*)-**25** combined with $[Cu(OTf)]_2$•(Tol) exhibits high enantioselectivity in the F–H insertion of α-aryl-dizaoesters (Scheme 3.23B). DFT calculations revealed a F^- transfer/protonation mechanism of the F–H insertion process: the F^- nucleophilic attack takes place at the Cu-center rather the carbene C-atom, forming F^- coordinated Cu carbene intermediate, subsequent selective [1,2]-F^- transfer results in a fluoroalkyl copper intermediate, which undergoes a protonation with HFIP to give insertion product with inverse configuration.

3.6 Si–H Bond Insertion Reactions

Transition metal-catalyzed Si—H bond insertion reaction provides an efficient access to organosilanes. The first catalytic enantioselective Si—H bond insertion reaction was reported by Doyle group using a chiral dirhodium catalyst but only with modest enantioselectivity (47% ee) [49]. Great progress has been made recently in this field.

3.6.1 Chiral Rh Catalysts

By using proline-derived tetracarboxylic dirhodium(II) catalysts, Davies et al. [50] and Corey and Ge [51] achieved highly enantioselective Si—H bond insertion of α-alkenyl-diazo esters and α-diazo cyclicketones, respectively (Scheme 3.24A,B).

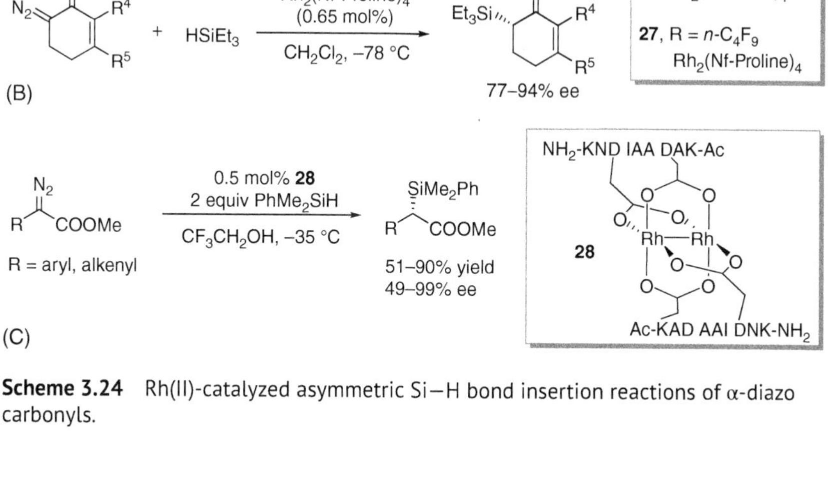

Scheme 3.24 Rh(II)-catalyzed asymmetric Si–H bond insertion reactions of α-diazo carbonyls.

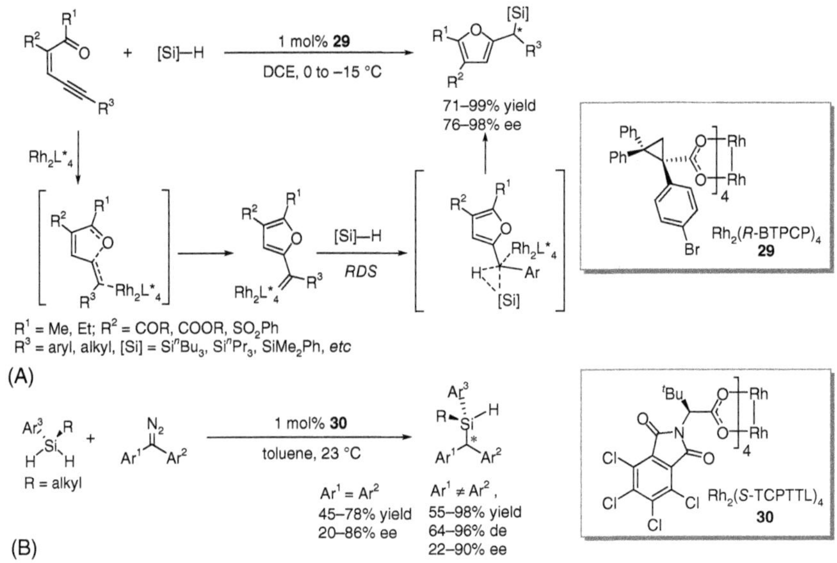

Scheme 3.25 Rh(II)-catalyzed asymmetric Si–H bond insertion reactions of diaryl carbenes.

Ball and Sambasivan [52] developed a series of chiral dirhodium catalysts ligated with peptide sequences with two aspartate side chains for the Si—H bond insertion reaction of α-aryl-diazoesters or α-alkenyl-diazoesters (Scheme 3.24C).

Catalytic enantioselective Si—H bond insertion of diaryl carbenes was developed as a facile access to chiral diarylsilyl methane. By using chiral dirhodium catalysts **29**, Zhu and coworkers [53] realized highly enantioselective Si—H bond insertion reaction of α-furylcarbenes generated *in situ* from carbonyl-ene-ynes (Scheme 3.25A). This reaction is triggered by Rh(II)-catalyzed *5-exo-dig* cyclization that forms α-furylcarbene intermediate, which undergoes Si—H bond insertion to afford the corresponding product. Mechanistic studies demonstrated that the Si—H bond insertion process is probably the rate-determining step. Shaw and coworkers [54] developed a desymmetric Si—H bond insertion reaction of diarylcarbenes under the catalysis of dirhodium catalyst **30** (Scheme 3.25B). When symmetrical diaryl diazo substrates were used, a variety of silicon-stereogenic silanes were obtained with moderate-to-good enantioselectivities, while the use of prochiral diazo substrates resulted in silane products with two stereocenters at carbon and silicon atoms, whose diastereoselectivity and enantioselectivity could be well controlled by introducing an *ortho*-substituent on the phenyl ring or using naphthyl ring.

Zhu, Zhou and coworkers [55] prepared a class of chiral tetraphosphate dirhodium catalysts, which showed high activity and enantioselectivity for the insertion of diarylcarbenes into the Si—H bond of silanes (Scheme 3.26A). Impressively, diazo reagents with different *para*-substituted phenyl rings afforded high enantioselectivities (up to 99% ee), and the log(er) values are linearly related to electronic Hammett parameters, which indicates that the enantioselectivity depends primarily on the electronic characteristics of the carbene substrate. Computational and mechanistic

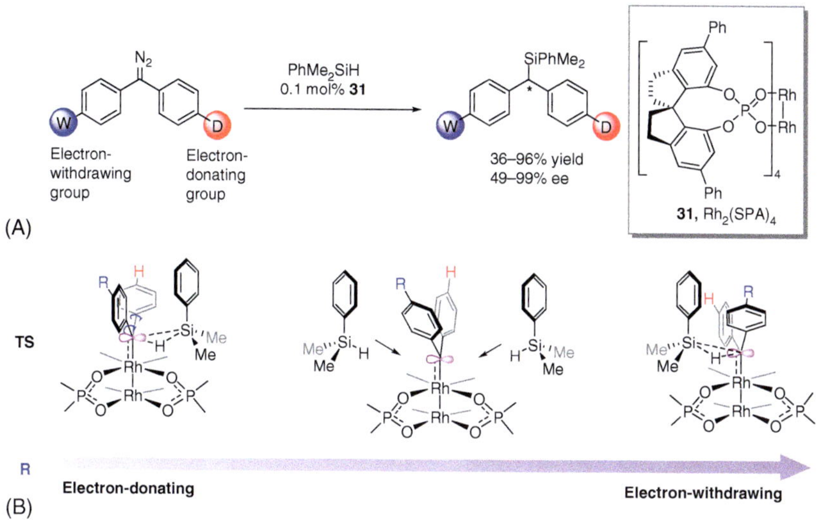

Scheme 3.26 Rh(II)-catalyzed asymmetric Si–H bond insertion reactions of diaryl-carbenes.

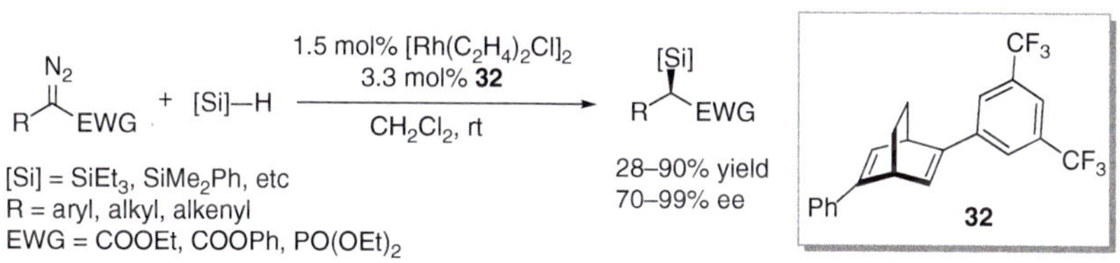

Scheme 3.27 Rh(I)-catalyzed asymmetric Si–H bond insertion reaction.

studies revealed that the electronic differences between the two aryls of the carbene lead to change of enantioselectivity. In the diarylcarbene-rhodium intermediate, the electron-rich phenyl ring adopts a near coplanar orientation with the carbene empty *p* orbital for better conjugation, while the other phenyl ring (especially electron-deficient phenyl ring) rotates to a near orthogonal orientation to minimize the steric repulsions. Controlled by the chiral environment caused by the ligands of catalyst, the Si—H bond of silane approaches the carbene C-center from the side having smaller steric hindrance, forming insertion product (Scheme 3.26B).

Rh(I) complexes can also decompose diazo compounds and catalyze the Si—H bond insertion reactions. Xu and coworkers [56] reported highly enantioselective Si—H bond insertion reactions of α-diazoesters, α-diazophosphonates under the catalysis of Rh(I)-chiral diene complexes (Scheme 3.27).

3.6.2 Chiral Cu Catalysts

Jacobsen and coworkers [57] introduced Cu complexes of chiral diimine ligands in the enantioselective Si—H bond insertion of α-diazophenylacetates and obtained good enantioselectivity (Scheme 3.28A). Similar catalysts were also used

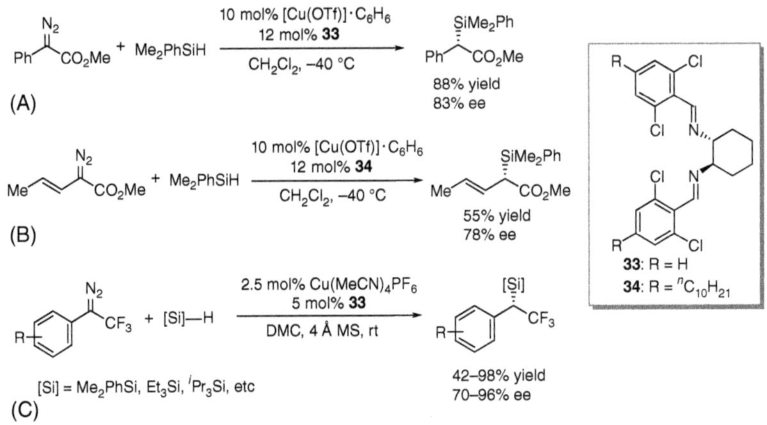

Scheme 3.28 Cu/diamine-catalyzed asymmetric Si–H bond insertion reactions.

for the enantioselective Si—H bond insertions of α-alkenyldiazoesters [58] and α-aryl-diazotrifluoethanes [59] (Scheme 3.28B,C).

Zhou and coworkers [60] developed a series of chiral spirodiimine ligands and found that they are highly effective for Cu-catalyzed asymmetric Si—H bond insertion of α-aryl-diazoesters (Scheme 3.29A). By using Cu/spiro bisoxazoline ligand as the catalyst, Gouverneur and coworkers [61] achieved high level of enantiocontrol in the Si—H bond insertion reaction of α-aryl-diazotrifluoroethanes (Scheme 3.29B).

3.6.3 Other Chiral Metal Catalysts

Katsuki and coworkers [62] reported an Ir(salen)-catalyzed asymmetric Si—H bond insertion of α-diazoacetates and achieved enantioselectivity of up to 99% ee

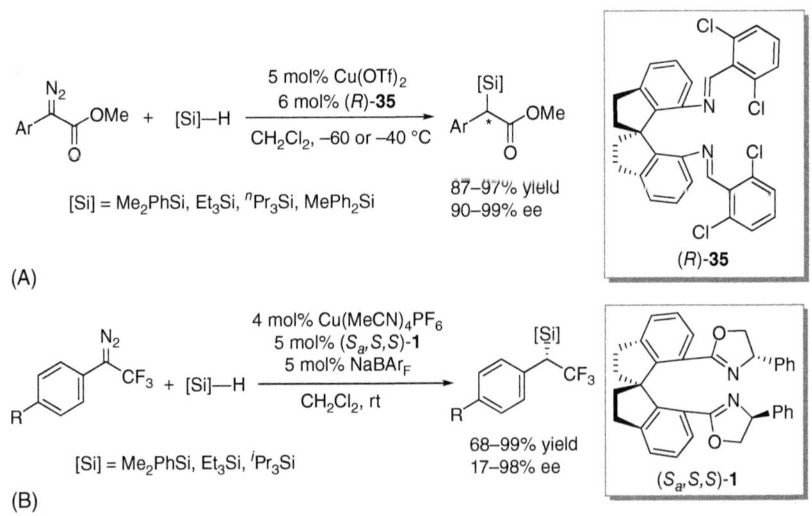

Scheme 3.29 Cu-catalyzed asymmetric Si–H bond insertion reactions using chiral spiro ligands.

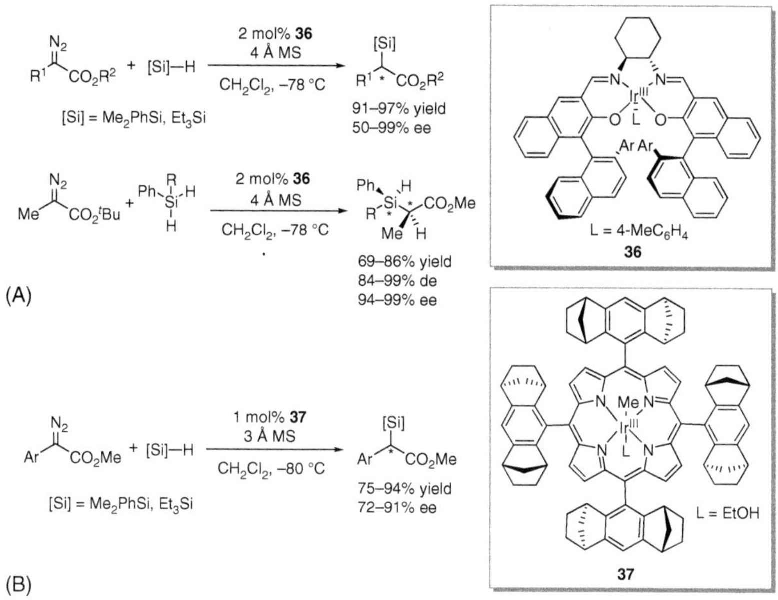

Scheme 3.30 Ir-catalyzed asymmetric Si–H bond insertion reactions.

(Scheme 3.30A). When prochiral dihydrosilanes were used, the reaction proceeded to afford silicon-stereogenic organosilanes with good diastereo- and enantioselectivity. Che and coworkers [63] developed a highly enantioselective Si—H bond insertion reaction by using a chiral Ir(III) complex of a Halterman porphyrin ligand (Scheme 3.30B).

Chanthamath and coworkers [64] developed an Ru-catalyzed enantioselective Si—H bond insertion of trisubstituted and disubstituted silanes (Scheme 3.31A). By using chiral Ru(II) phenyloxazoline catalyst **38**, they achieved high enantioselectivity but poor diastereoselectivity.

Lin and coworkers [65] reported an Fe(II)-catalyzed asymmetric Si—H bond insertion reaction of α-aryl-α-diazoesters by employing a modified chiral spiro bisoxazoline ligand and achieved enantioselectivity of up to 96% ee (Scheme 3.31B).

3.6.4 Enzymes

Arnold and coworkers [66] found that *E. coli* cells harboring wild-type cytochrome *c* from *Rhodothermus marinus* (*Rma* cyt *c*) can promote Si—H bond insertion of α-methyl-diazoesters and dimethylphenylsilane with high enantioselectivity (up to 95% ee) but low efficiency (<40 TTN [total turnover number]). By using directed evolution, various engineered variants of *Rma cyt c* dramatically enhanced the catalytic efficiency (up to >99% ee, 8210 TTN) (Scheme 3.32). This reaction also exhibited good functional group tolerance and could be performed in living systems.

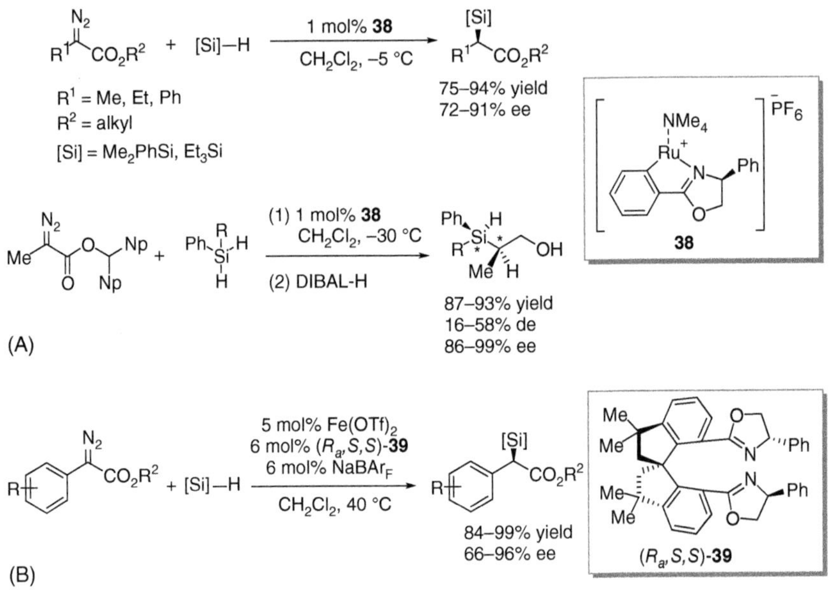

Scheme 3.31 Ru- or Fe-catalyzed asymmetric Si–H bond insertion reactions.

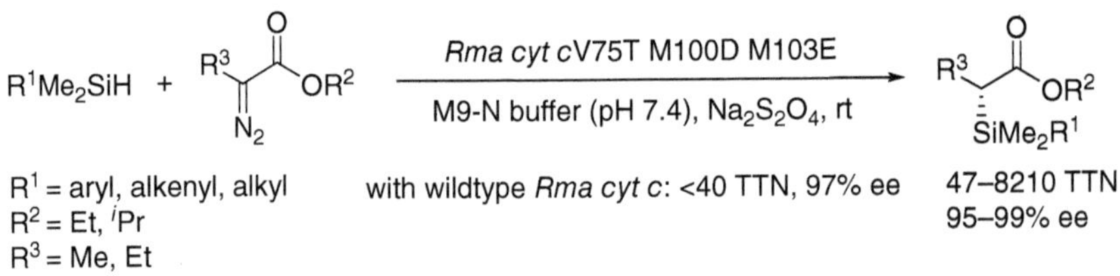

Scheme 3.32 Enzyme-catalyzed asymmetric Si–H bond insertion reaction.

3.7 B–H Bond Insertion Reactions

3.7.1 Chiral Cu Catalysts

The development of methodology for preparation of organoboron compounds, versatile reagents in diverse purpose, has been a highly rewarded topic in organic chemistry. Because the B—H bond of borane is electron-deficient, it is inert to the metal carbene insertion reaction. However, after forming adducts with amine or phosphine, B—H bond insertion of borane becomes possible [67].

Zhu, Zhou and coworkers [68] developed a Cu-catalyzed B—H bond insertion reaction of amine- or phosphine-borane adducts with α-diazo-α-arylacetates (Scheme 3.33A). The asymmetric version of this reaction was also established by using ligand (R_a,*S*,*S*)-**1**, which provides an efficient approach to chiral organoboranes. The mechanism of B—H bond insertion is analogous with Si—H bond insertion: Cu complex catalyzes the decomposition of diazo to form a Cu carbene, which undergoes a concerted B—H bond insertion through a three-member-ring

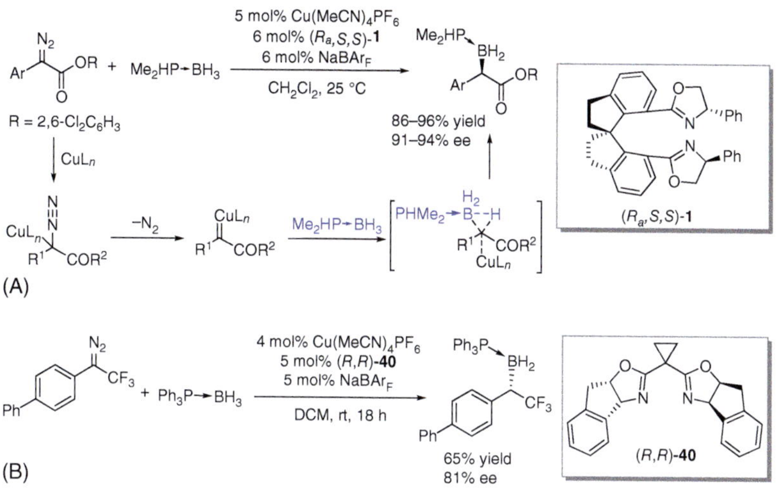

Scheme 3.33 Cu-catalyzed asymmetric B–H bond insertion reactions.

transition state to afford the insertion product. Gouverneur and coworkers [61] reported an example of asymmetric B—H bond insertion of α-CF_3-α-aryl diazomethanes and achieved good enantioselectivity (Scheme 3.33B).

3.7.2 Chiral Rhodium Catalysts

Zhu, Zhou and coworkers [69] developed a highly enantioselective Rh-catalyzed B—H bond insertion reaction using enynones as carbene precursors. It represents the first highly enantioselective heteroatom–hydrogen insertion reaction using enynones as carbene precursors (Scheme 3.34A). Comparing with diazo compounds, the enynones are safe and easily available carbene precursors. The chiral organoboranes prepared through this protocol could be transformed to widely used borates and diaryl methanol compounds without loss of optical purity.

Zhu and coworkers [70] reported an enantioselective Rh-catalyzed B—H bond insertion reaction of unstabilized diazo compounds, which are generated *in situ* from tosylhydrazones (Scheme 3.34B). By using chiral dirhodium catalysts, the protocol exhibited good functional group tolerance and high enantioselectivity. The reaction is the first enantioselective heteroatom–hydrogen insertion reaction to use unstabilized diazo compounds as carbene precursors.

Xu and coworkers [71] reported an asymmetric B—H bond insertion of α-diazo carbonyl compounds with amine-borane adducts by using a Rh(I) catalyst modified with a chiral diene ligand (Scheme 3.34C). Various α-boryl esters and ketones were prepared in high enantioselectivity under mild conditions.

3.7.3 Enzymes

Arnold and coworkers [72] reported that the *E. coli* cells harboring engineered variants of wild-type cytochrome *c* from *Rhodothermus marinus*

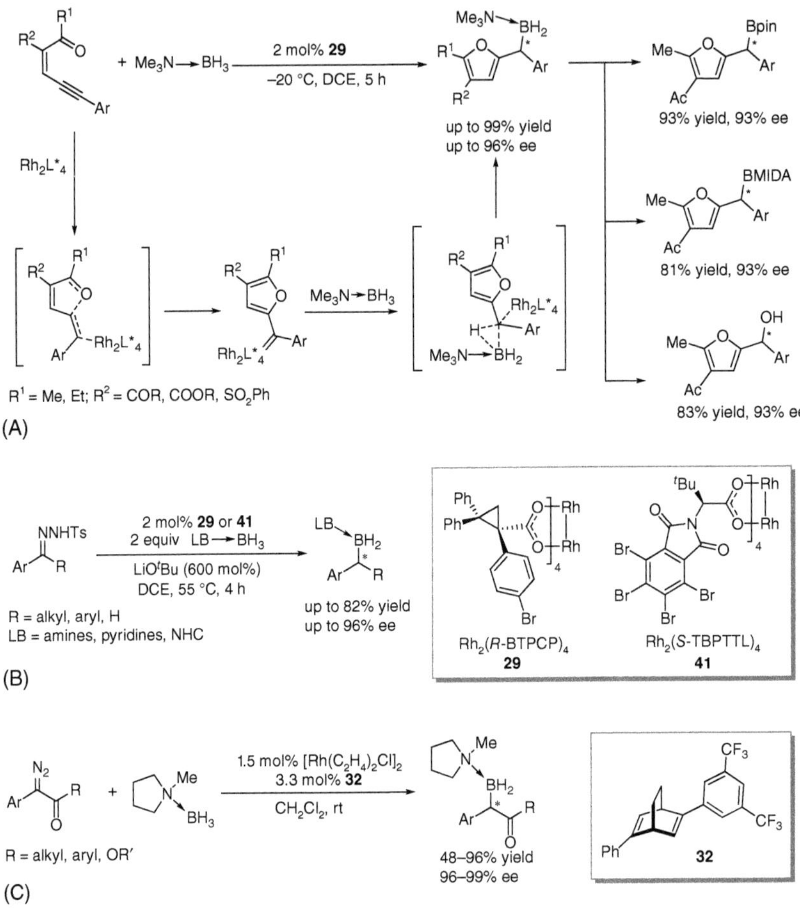

Scheme 3.34 Rh-catalyzed asymmetric B—H bond insertion reactions.

(*Rma* cyt *c*) could promote the B—H bond insertion of α-diazo propionates and (1-diazo-2,2,2-trifluoroethyl)benzene with *N*-heterocarbene (NHC)-borane adducts (Scheme 3.35). The biocatalytic system exhibits high activity (up to 15 300 turnovers and 6100 h^{-1} turnover frequency), as well as selectivity (up to 99 : 1 enantiomeric ratio and 100% chemoselectivity), and is suitable for gram-scale biosynthesis. Either enantiomer of the organoborane products could be obtained through tuning the biocatalyst. This work expands the scope of boron chemistry in living systems.

3.8 Summary and Outlook

Although transition-metal-catalyzed carbene insertion reaction of X—H bonds was known for more than 60 years, its enantiocontrol has been a challenge. The situation was significantly improved in the past 15 years, and the field has become more mature. The development of new chiral catalysts and new catalytic strategies

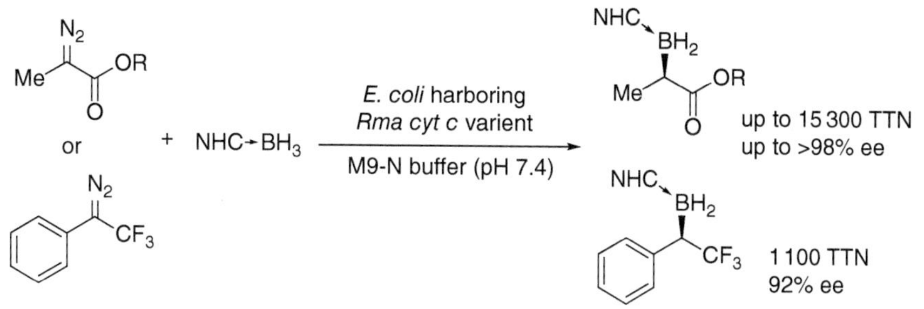

Scheme 3.35 Enzyme-catalyzed asymmetric B—H bond insertion reactions.

is the key to success. Metal complexes modified with chiral spiro bisoxazoline ligands or other nitrogen ligands are proven to be suitable catalysts for various X—H bond insertion reactions. CPTS catalysts, especially chiral SPAs or chiral aminoureas, are also effective for N—H, O—H, and S—H bond insertion reactions. By using the established asymmetric X—H bond insertion reactions, a large number of chiral compounds, such as α-amino acid derivatives, α-amino ketones, α-hydroxy acid derivatives, α-thiol acid derivatives, organosilanes, and organoboranes, have been prepared in high enantioselectivity. Mechanism studies and DFT calculations have revealed more details to the carbene insertions into X—H bonds, especially the enantio-determining hydrogen transfer step, and remarkably promoted the development of new chiral catalysts as well as new asymmetric X—H bond insertion reactions. However, in spite of these achievements, there are still several challenges to be solved for enantioselective X—H bond insertion reactions from the standpoint of synthetic application. For example, the catalytic efficiency of most of the reported X—H bond insertion reactions is not satisfactory, generally requiring more than 1 mol% of the catalyst. Since the carbene precursors suitable for X—H bond insertion reaction are mainly stable diazo compounds, the structural diversity of X—H bond insertion reaction products is limited. Many mechanisms of X—H bond insertion reactions lack reliable experimental evidence, and little is known about the relationship between the structure and function of chiral catalysts. Therefore, the development of new catalysts or catalytic strategies, the extension of carbene precursors and X—H bond donors, and the in-depth understanding of the reaction mechanism will be the research focus of enantioselective X—H bond insertion reactions in the future.

References

1 Yudin, A.K. (2011). *Catalyzed Carbon–Heteroatom Bond Formation*. Weinheim: Wiley-VCH.
2 Doyle, M.P., McKervey, M.A., and Ye, T. (1998). *Modern Catalytic Methods for Organic Synthesis with Diazocompounds*. New York: Wiley.
3 Yates, P. (1952). *J. Am. Chem. Soc.* 74: 5376.

4 Paulissen, R., Hayez, E., Hubert, A.J., and Teyssie, P. (1974). *Tetrahedron Lett.* 15: 607.
5 For leading reviews on enantioselective catalytic X—H bond insertion reactions, see:(a)Moody, C.J. (2007). *Angew. Chem. Int. Ed.* 46: 9148. (b) Zhu, S.-F. and Zhou, Q.-L. (2012). *Acc. Chem. Res.* 45: 1365. (c) Gillingham, D. and Fei, N. (2013). *Chem. Soc. Rev.* 42: 4918. (d) Zhu, S.-F. and Zhou, Q.-L. (2014). *Natl. Sci. Rev.* 1: 580. (e) Ford, A., Miel, H., Ring, A. et al. (2015). *Chem. Rev.* 115: 9981.
6 (a) García, C.F., McKervey, M.A., and Ye, T. (1996). *Chem. Commun.*: 1465. (b) Bachmann, S., Fielenbach, D., and Jørgensen, K.A. (2004). *Org. Biomol. Chem.* 2: 3044.
7 (a) Liu, B., Zhu, S.-F., Zhang, W. et al. (2007). *J. Am. Chem. Soc.* 129: 5834. (b) Zhu, S.-F., Xu, B., Wang, G.-P., and Zhou, Q.-L. (2012). *J. Am. Chem. Soc.* 134: 436. (c) Song, X.-G., Ren, Y.-Y., Zhu, S.-F., and Zhou, Q.-L. (2016). *Adv. Synth. Catal.* 358: 2366.
8 Lee, E.C. and Fu, G.C. (2007). *J. Am. Chem. Soc.* 129: 12066.
9 Hou, Z., Wang, J., He, P. et al. (2010). *Angew. Chem. Int. Ed.* 49: 4763.
10 Shen, H.-Q., Wu, B., Xie, H.-P., and Zhou, Y.-G. (2019). *Org. Lett.* 21: 2712.
11 Zhu, Y., Liu, X.-H., Dong, S.-X. et al. (2014). *Angew. Chem. Int. Ed.* 53: 1636.
12 Arredondo, V., Hiew, S.C., Gutman, E.S. et al. (2017). *Angew. Chem. Int. Ed.* 56: 4156.
13 Xu, X.-F., Zavalij, P.Y., and Doyle, M.P. (2012). *Angew. Chem. Int. Ed.* 51: 9829.
14 Huang, W.-S., Xu, Z., Yang, K.F. et al. (2015). *RSC Adv.* 5: 46455.
15 Steck, V., Carminati, D.M., Johnson, N.R., and Fasan, R. (2020). *ACS Catal.* 10: 10967.
16 Ren, Y.-Y., Zhu, S.-F., and Zhou, Q.-L. (2018). *Org. Biomol. Chem.* 16: 3087.
17 Xu, B., Zhu, S.-F., Xie, X.-L. et al. (2011). *Angew. Chem. Int. Ed.* 50: 11483.
18 Guo, J.-X., Zhou, T., Xu, B. et al. (2016). *Chem. Sci.* 7: 1104.
19 Xu, B., Zhu, S.-F., Zuo, X.-D. et al. (2014). *Angew. Chem. Int. Ed.* 53: 3913.
20 Li, M.-L., Yu, J.-H., Li, Y.-H. et al. (2019). *Science* 366: 990.
21 Bulugahapitiya, P., Landais, Y., Parra-Rapado, L. et al. (1997). *J. Org. Chem.* 62: 1630.
22 Maier, T.C. and Fu, G.C. (2006). *J. Am. Chem. Soc.* 128: 4594.
23 Chen, C., Zhu, S.-F., Liu, B. et al. (2007). *J. Am. Chem. Soc.* 129: 12616.
24 Zhu, S.-F., Chen, W.-Q., Zhang, Q.-Q. et al. (2011). *Synlett*: 919.
25 Zhu, S.-F., Song, X.-G., Li, Y. et al. (2010). *J. Am. Chem. Soc.* 132: 16374.
26 Song, X.-G., Zhu, S.-F., Xie, X.-L., and Zhou, Q.-L. (2013). *Angew. Chem. Int. Ed.* 52: 2555.
27 Osako, T., Panichakul, D., and Uozumi, Y. (2012). *Org. Lett.* 14: 194.
28 Kitagaki, S., Sugisaka, K., and Mukai, C. (2015). *Org. Biomol. Chem.* 13: 4833.
29 Kitagaki, S., Murata, S., Asaoka, K. et al. (2018). *Chem. Pharm. Bull.* 66: 1006.
30 Maux, P.L., Carrié, D., Jéhan, P., and Simonneaux, G. (2016). *Tetrahedron* 72: 4671.
31 Zhu, S.-F., Chen, C., Cai, Y., and Zhou, Q.-L. (2008). *Angew. Chem. Int. Ed.* 47: 932.
32 Zhu, S.-F., Cai, Y., Mao, H.-X. et al. (2010). *Nat. Chem.* 2: 546.

33 Xie, X.-L., Zhu, S.-F., Guo, J.-X. et al. (2014). *Angew. Chem. Int. Ed.* 53: 2978.

34 Shen, H.-Q., Xie, H.-P., Sun, L., and Zhou, Y.-G. (2019). *Organometallics* 38: 3902.

35 Gao, X., Wu, B., Huang, W.-X. et al. (2015). *Angew. Chem. Int. Ed.* 54: 11956.

36 Huang, D.-R., Xu, G.-Y., Peng, S.-Y., and Sun, J.-T. (2017). *Chem. Commun.* 53: 3197.

37 Saito, H., Iwai, R., Uchiyama, T. et al. (2010). *Chem. Pharm. Bull.* 58: 872.

38 Tan, F., Liu, X.-H., Hao, X.-Y. et al. (2016). *ACS Catal.* 6: 6930.

39 Zhang, Y.-L., Yao, Y., He, L. et al. (2017). *Adv. Synth. Catal.* 359: 2754.

40 Li, M.-L., Chen, M.-Q., Xu, B. et al. (2018). *Acta Chim. Sinica* 76: 883.

41 Li, Y., Zhao, Y.-T., Zhou, T. et al. (2020). *J. Am. Chem. Soc.* 142: 10557.

42 (a) Brunner, H., Wutz, K., and Doyle, M.P. (1990). *Monatsh. Chem.* 121: 755. (b) Galardon, E., Roué, S., Maux, P.L., and Simonneaux, G. (1998). *Tetrahedron Lett.* 39: 2333. (c) Zhang, X.-M., Ma, M., and Wang, J.-B. (2003). *Arkivoc* 2: 84.

43 Zhang, Y.-Z., Zhu, S.-F., Cai, Y. et al. (2009). *Chem. Commun.*: 5362.

44 Xu, B., Zhu, S.-F., Zhang, Z.-C. et al. (2014). *Chem. Sci.* 5: 1442.

45 Tyagi, V., Bonn, R.B., and Fasan, R. (2015). *Chem. Sci.* 6: 2488.

46 Chen, K., Zhang, S.-Q., Brandenberg, O.F. et al. (2018). *J. Am. Chem. Soc.* 140: 16402.

47 Gray, E.E., Nielsen, M.K., Choquette, K.A. et al. (2016). *J. Am. Chem. Soc.* 138: 10802.

48 Buchsteiner, M., Martinez-Rodriquez, L., Jerabek, P. et al. (2020). *Chem. Eur. J.* 26: 2509.

49 Buck, R.T., Doyle, M.P., Drysdale, M.J. et al. (1996). *Tetrahedron Lett.* 37: 7631.

50 Davies, H.M.L., Hansen, T., Rutberg, J., and Bruzinski, P.R. (1997). *Tetrahedron Lett.* 38: 1741.

51 Ge, M. and Corey, E. (2006). *Tetrahedron Lett.* 47: 2319.

52 Sambasivan, R. and Ball, Z.T. (2010). *J. Am. Chem. Soc.* 132: 9289.

53 Huang, M.-Y., Yang, J.-M., Zhao, Y.-T., and Zhu, S.-F. (2019). *ACS Catal.* 9: 5353.

54 Jagannathan, J.R., Fettinger, J.C., Shaw, J.T., and Franz, A.K. (2020). *J. Am. Chem. Soc.* 142: 11674.

55 Yang, L.-L., Evans, D., Xu, B. et al. (2020). *J. Am. Chem. Soc.* 142: 12394.

56 Chen, D., Zhu, D.-X., and Xu, M.-H. (2016). *J. Am. Chem. Soc.* 138: 1498.

57 Dakin, L.A., Schaus, S.E., Jacobsen, E.N., and Panek, J.S. (1998). *Tetrahedron Lett.* 39: 8947.

58 Wu, J., Chen, Y., and Panek, J.S. (2010). *Org. Lett.* 12: 2112.

59 Carreras, V., Besnard, C., Gandon, V., and Ollevier, T. (2019). *Org. Lett.* 21: 9094.

60 Zhang, Y.-Z., Zhu, S.-F., Wang, L.-X., and Zhou, Q.-L. (2008). *Angew. Chem. Int. Ed.* 47: 8496.

61 Hyde, S., Veliks, J., Liégault, B. et al. (2016). *Angew. Chem. Int. Ed.* 55: 3785.

62 Yasutomi, Y., Suematsu, H., and Katsuki, T. (2010). *J. Am. Soc. Chem.* 132: 4510.

63 Wang, J.-C., Xu, Z.-J., Guo, Z. et al. (2012). *Chem. Commun.* 48: 4299.

64 Nakagawa, Y., Chanthamath, S., Fujisawa, I. et al. (2017). *Chem. Commun.* 53: 3753.

65 Gu, H.-R., Han, Z., Xie, H.-J., and Lin, X.-F. (2018). *Org. Lett.* 20: 6544.

66 Kan, S.B.J., Lewis, R.D., Chen, K., and Arnold, F.H. (2016). *Science* 354: 1048.

67 (a) Huang, M.-Y. and Zhu, S.-F. (2020). *Chem. J. Chin. Univ.* 41: 1426.
(b) Yang, J.-M., Li, Z.-Q., and Zhu, S.-F. (2017). *Chin. J. Org. Chem.* 37: 2481.

68 Cheng, Q.-Q., Zhu, S.-F., Zhang, Y.-Z. et al. (2013). *J. Am. Chem. Soc.* 135: 14094.

69 Yang, J.-M., Li, Z.-Q., Li, M.-L. et al. (2017). *J. Am. Chem. Soc.* 139: 3784.

70 Pang, Y., He, Q., Li, Z.-Q. et al. (2018). *J. Am. Chem. Soc.* 140: 10663.

71 Chen, D., Zhang, X., Qi, W.-Y. et al. (2015). *J. Am. Chem. Soc.* 137: 5268.

72 Kan, S.B.J., Huang, X., Gumulya, Y. et al. (2017). *Nature* 552: 132.

4

Engineering Enzymes for New-to-Nature Carbene Chemistry

Soumitra V. Athavale[1], Kai Chen[1,2], and Frances H. Arnold[1]

[1] *California Institute of Technology, Division of Chemistry and Chemical Engineering 210-41, Pasadena, CA 91125, USA.*
[2] *University of California, Berkeley, Innovative Genomics Institute, 2151 Berkeley Way, Berkeley, CA 94709, USA*

4.1 Introduction: Biology Inspires Chemistry Inspires Biology

Enzymes – macromolecular catalysts whose properties are modulated by the polypeptide scaffold – are champions of catalytic chemistry. Their highly sophisticated three-dimensional structures offer a profound degree of tunable stereoelectronic control during catalysis. The diversity of protein sequences and cofactor moieties ultimately results in a dazzling array of enzyme-mediated chemistry, ranging from ester hydrolysis and radical methylations to photosynthesis and nitrogen fixation. Indeed, biomimetic schemes of catalysis continue to inspire chemists and have been the source of many extraordinary achievements in organic synthesis [1, 2].

Proteins can also be considered as a source of natural catalysts that can be employed for synthesis. Although individual enzymes are generally specialized for a specific transformation, they may display activity toward similar substrates or reactions following mechanistically related pathways. This is not fundamentally surprising – just as a chiral phosphoric acid optimized for an enantioselective Diels–Alder reaction may also catalyze an unrelated Mannich reaction through the common paradigm of Bronsted acid activation [3, 4] – an enzyme, say hexokinase, may (and does [5]) phosphorylate sugars other than its preferred natural substrate (glucose), albeit with poor efficiency. The appreciation for this critical attribute of enzymes (and catalysts in general) termed “catalytic promiscuity,” is often lost in the notion of enzymes as being reaction and substrate specific [6–8]. An enzyme’s native activity is optimized by natural selection, but its catalytic promiscuity, unless deleterious for the organism, is not under direct selection pressure. Unlike small-molecule catalysts, the enzyme is genetically encoded and can undergo continuous change in small steps through evolution. In some of its dimensions, the evolutionary fitness landscape is smooth and continuous; that is, functionally

Transition Metal-Catalyzed Carbene Transformations, First Edition.
Edited by Jianbo Wang, Chi-Ming Che, and Michael P. Doyle.

distinct variants are linked by a series of single amino acid changes and can be accessed without resorting to low-probability events involving multiple mutations [9]. A promiscuous activity may become advantageous when environmental constraints change, and such novel activity can be improved through mutations under a new selection pressure. Over time, the latent promiscuous activity of an enzyme can be amplified to a point where it becomes predominant – a "new" enzyme is born. With this combination of promiscuous activity and stepwise evolution, nature possesses a winning algorithm to create new catalysts from old ones (and extend the overall repertoire of promiscuous activity even further) – all that is required is a selective environmental pressure that directs such a change.

The catalytic promiscuity of proteins must have been decisive for successful adoption of diverse chemistries during the course of evolution. It follows that every enzyme or its evolutionary progeny potentially possesses promiscuous activity which nature has never utilized, perhaps because the products were not biologically useful and/or the reactants and reaction conditions were biologically inaccessible. Through the gaze of the imaginative synthetic chemist, enzyme reactivity need not be limited to native biological transformations but can perhaps be extended to abiological, new-to-nature reactions by directing the evolution of promiscuous activities.

An effective realization of this paradigm was achieved in 2013 by the demonstration that heme proteins can effect carbene transfer reactions for the cyclopropanation of alkenes (Figures 4.1 and 4.2) [10]. Such a process involving olefins and diazo compounds is unknown in biology. This promiscuous activity was amplified to synthetically useful levels through directed evolution. To begin with, a suite of purified, wild-type proteins was tested for the cyclopropanation of styrene by ethyldiazoacetate (EDA). Among these, horseradish peroxidase (HRP), cytochrome *c* (cyt *c*), myoglobin (Mb), and cytochrome $P450_{BM3}$ (BM3 - *Bacillus megaterium*) all displayed multiple turnovers toward the cyclopropane products. HRP, cyt *c*, and Mb showed negligible enantioinduction and formed the *trans*-cyclopropane with ~90 : 10 diastereoselectivity, which is comparable to the diastereoselectivity induced by free hemin. $P450_{BM3}$, despite forming the cyclopropane products in low yield, catalyzed the reaction with different diasteroselectivity (*cis*:*trans* 37 : 63) and low enantioinduction, suggesting that carbene transfer and selectivity are dictated by the heme cofactor bound in the enzyme active site. Before elaborating on these findings, a detour is warranted – a brief historical summary of carbene chemistry vis-à-vis heme enzymes and metal-porphyrins will assist in providing useful context for further discussion (Figure 4.1).

Heme proteins represent a vast set of structures that enable diverse processes like electron transfer (cytochrome *c*), oxygen transfer (hemoglobin), and myriad oxidative chemistry [11, 12]. The direct, selective oxidation of C–H bonds with molecular oxygen can be considered as the most iconic transformation effected by heme enzymes. This activity has inspired the study of the metalloporphyrin moiety as a scaffold for enabling oxidative transformations – hundreds of synthetic metal-porphyrin complexes have been synthesized as catalytic heme mimics and as models to gain insights into the molecular details of the enzymatic process [13]. The synthesis of substituted porphyrins incorporating varied metals (e.g. Mn, Co, Rh,

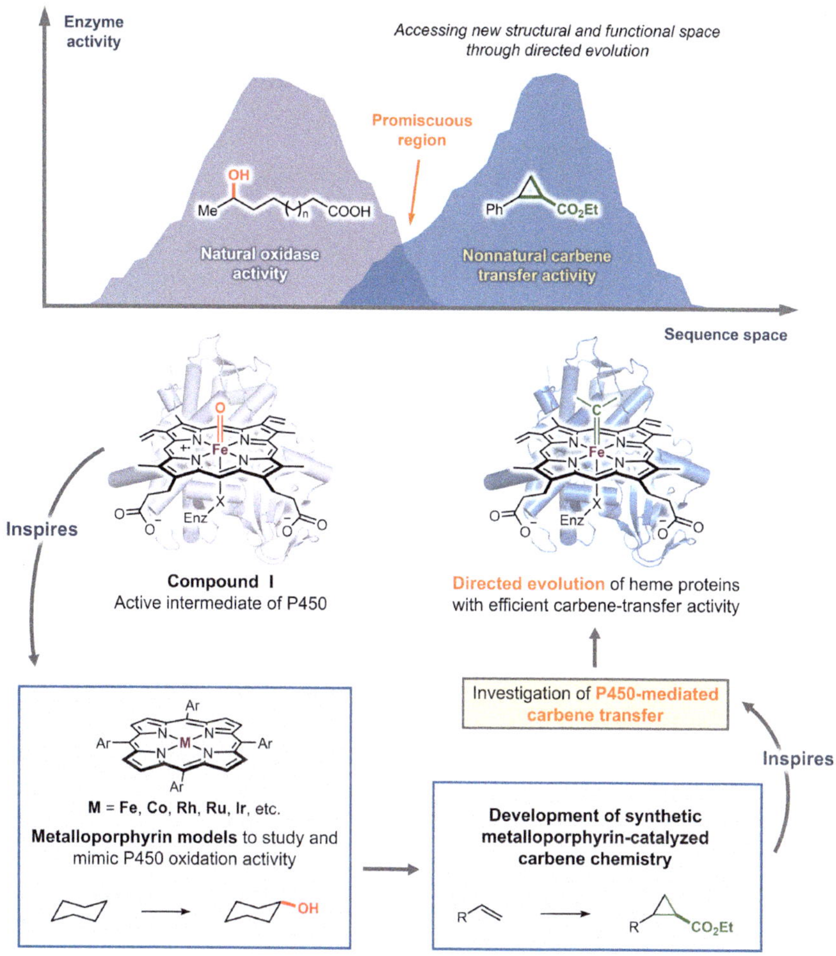

Figure 4.1 Logic behind the discovery of enzymatic carbene transfer. The natural reactivity of P450 enzymes is facilitated by a high-valent iron-oxo intermediate (compound I) in the enzyme active site. In the 1970s, this iconic intermediate inspired the study of metalloporphyrin compounds that could mimic the oxygenase activity and provide mechanistic insights into the active site process. This renaissance in metalloporphyrin chemistry led to the development of carbene (and nitrene) transfer reactions in the coming decades. The resulting small-molecule carbene transfer catalysts inspired investigations into the possibility of heme enzymes effecting analogous processes. Natural heme proteins show low levels of "promiscuous" carbene transfer activity. Using these as starting points, new enzyme lineages having efficient carbene transfer activities can be generated using directed evolution. Source: Based on Coelho et al. [10].

Ru, Os, and Ir) has resulted in remarkable advances in the organometallic chemistry of porphyrins and in the development of catalysts for novel reactions beyond those seen in natural heme enzymes – carbene transfer being one of them. In 1980, Callot reported that the rhodium porphyrin (TPP)RhI catalyzed the *cis*-cyclopropanation of styrene with ethyl diazoacetate [14]. Ru, Os, Rh, Ir, Co, and Fe porphyrins have

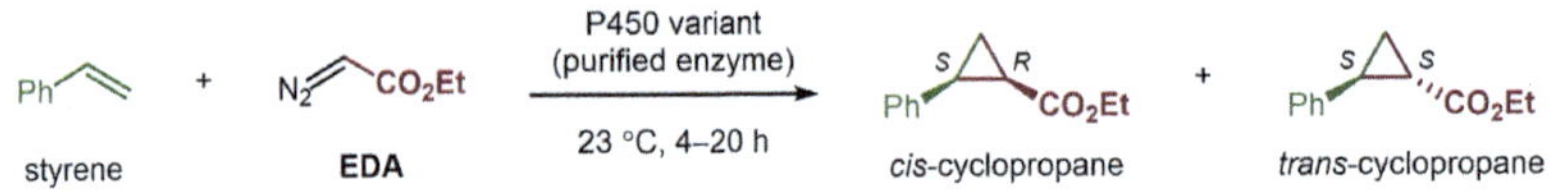

entry	variant	yield (%)*	*cis* : *trans*	% ee	TTN
1	$P450_{BM3}$	1	37 : 63	27, 2	5
2	$P450_{BM3}$-T268A	65	1 : 99	15, 96	323
3	$P450_{BM3}$-CIS	40	71 : 29	94, 91	199
4	$P450_{BM3}$-CIS T438S	59	92 : 8	97, 66	293
5	$P450_{BM3}$-CIS I263A	38	19 : 81	62, 91	190
6**	$P450_{BM3}$-CIS C400S ($P411_{BM3}$-CIS)	55	76 : 24	96, 25	3,700
7**	$P411_{BM3}$-CIS	72	90 : 10	99, 43	67,800

* GC yield for entries 1–5. ** reactions performed with whole-cell catalysts with isolated yields reported; in entry 7, the reaction was performed with neat reagents.

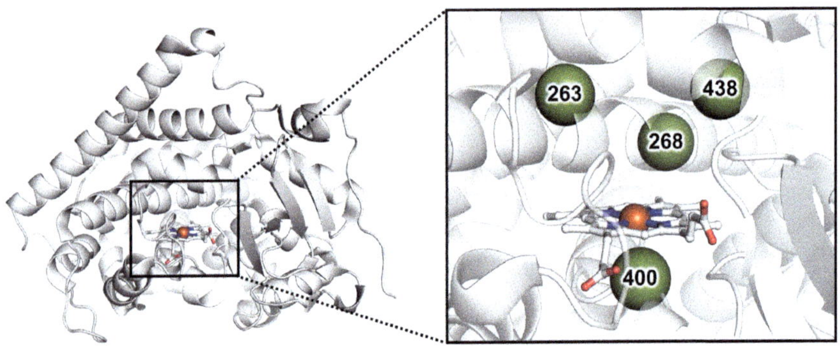

Figure 4.2 Early reports of cyclopropanation activity with engineered P450 and P411 heme enzymes. Screening of P450 variants provided enzymes with enhanced cyclopropanation activity. Note that both diastereoselectivity and enantioselectivity can be tuned. P411s include a substitution of the axial ligand (residue 400) from wild-type cysteine to serine. Some of the important residues in the active site that were targeted for mutation are shown (PDB ID: 4h24). Source: Based on Coelho et al. [10].

since been shown to catalyze efficient cyclopropanation and carbene insertion reactions. Thus, although innovation in metal-porphyrin chemistry originated in the biomimetic study of heme proteins for oxidation transformations, it was soon clear that synthetic metalloporphyrin analogs were chemically competent to catalyze carbene transfer reactions. To bring this arc of innovation a full circle, it was perhaps only natural to enquire if heme proteins themselves are capable of effecting carbene transfer processes, if provided with conducive reaction conditions.

Suggestions that carbene intermediates may be accessed during cytochrome P450 enzyme metabolism were provided as early as 1977 through the remarkable insights of Mansuy and coworkers [15]. While studying the degradation of polyhalogenated methane xenobiotics (e.g. CCl_4, CBr_4, CCl_3F, and $CHCl_3$) in rat liver microsome preparations, based on absorption spectroscopy measurements, the proposal of an active site iron-dihalocarbene intermediate en route to the final degradation product, CO, was made. Under reducing conditions, the initially formed Fe σ-complex was hypothesized to yield the Fe-carbene by further loss of a halide ion. In the following year, this proposal was supported by disclosure of the first crystal structure of a metal-porphyrin-carbene species – the synthetic Fe(TPP)(CCl_2)(H_2O) complex [16]. Mansuy's results, originating in and motivated by P450 enzymology, were foundational in catapulting the development of synthetic metal-porphyrin-carbene chemistry in the coming years [17, 18] (the development of metal-porphyrin nitrene chemistry can be similarly traced to Mansuy's studies with the degradative P450 enzymology of nitrogen compounds and will not be discussed here [18, 19]).

In this context, the backdrop to biocatalytic carbene transfer research may be summarized by the prescient concluding remarks of a 2006 review by Simonneaux and Le Maux [20]:

> *"(i) carbene adducts of metalloporphyrins have been considered as carbon analogs of the porphyrin-iron-oxo species on the basis of similarity in the electronic spectra; (ii) metalloporphyrins, in particular ruthenium porphyrins, are excellent catalysts for carbene transfer reactions; (iii) evidence for cytochrome P450 carbene complexes is now largely recognized. In view of these results, it is tempting to suggest a combination of heme proteins with carbene precursors, such as diazo derivatives to set up a viable organometallic system for asymmetric synthesis. If native heme proteins can be possible candidates for organometallic reactions, one would also expect ruthenium-reconstituted heme proteins to be good candidates for catalyzing asymmetric cyclopropanations… Although many examples have been described an effective system with metalloporphyrins, that utilizes heme proteins as catalysts for carbene transfer, has not yet been forthcoming. However, it is probable that the availability of new artificial metalloproteins will solve this problem in the future."*

Thus, in retrospect, the 2013 finding of carbene-mediated cyclopropanation reactivity in hemeproteins was perhaps unsurprising (indeed, many of the above prophecies, including those about non-iron reconstituted heme proteins have been now fulfilled in principle [21–27] – *cf*. the section on artificial metalloproteins). However, our long-standing experience in the paradigm of directed evolution, especially with P450 enzymes, proved pivotal in demonstrating how this promiscuous reactivity could be amplified in the lab to synthetically useful levels.

4.2 P411-Catalyzed Cyclopropanation

With the detour complete, we return to the discussion of enzymatic cyclopropanation (Figure 4.2) [10, 28]. After obtaining low cyclopropanation activity with a number of heme proteins, our extensive collection of P450 monooxygenase mutants was screened to identify the most active variant as a starting point for further evolution. These mutants were accumulated as a result of years of prior research aimed at manipulating the wild-type fatty acid monooxygenase activity of $P450_{BM3}$. Although wild-type $P450_{BM3}$ is a poor catalyst (1% GC yield, 5 TTN), a single mutation T268A turned on reactivity significantly (65% GC yield, 323 TTN, highly selective for the *trans*-product). Likewise, variant $P450_{BM3}$-CIS (*cis*-selective) which includes T268A among 12 other mutations was used as the starting point for site-saturation mutagenesis at L181, I263, A328, L437, and T438, all of which are active site residues. The *cis*-preference was especially exciting, since small-molecule catalysts were predominantly known to generate *trans*-products. Among these, T438S substantially improved *cis*-cyclopropanation with high enantioselectivity, while I263A promoted a shift to a *trans*-selective catalyst.

The wild-type $P450_{BM3}$ protein [29] is a single-chain, ~120 kD polypeptide made up of two distinct domains, (i) the heme domain, which houses the iron porphyrin, and (ii) the diflavin reductase domain, which provides a binding site for NAD(P)H and facilitates electron transfer to the heme iron center. The heme iron is axially coordinated by a cysteine side chain. In its resting state, the low-spin Fe(III) center is coordinated by a water molecule in the active site. This low-spin configuration (E^o $Fe^{III/II}$ = −430 mV) cannot be reduced by NAD(P)H (E^o = −320 mV). In the native catalytic cycle, substrate binding (typically a long-chain fatty acid) in the active site displaces the water molecule, causing the resting state Fe(III) to transition from a low-spin to a high-spin electronic configuration. This spin shift results in increasing the reduction potential of the iron center (E^o $Fe^{III/II}$ = −290 mV), inducing electron transfer from the NAD(P)H reductase domain and leading to a reduction of the iron to Fe(II). This active ferrous state is the starting point for oxygen binding and further steps in the oxidation cycle. In the cyclopropanation studies, efficient reactions occurred only in the presence of a sufficiently strong external reductant (e.g. dithionite, E^o = −660 mV), suggesting that reduction to heme-Fe(II) was necessary to access the iron-carbenoid. Likewise, the reactions were inhibited by oxygen, likely because of competition with the native oxidation pathway, once the Fe(II) state was reached with dithionite reduction. Presumably in the cyclopropanation system, the high-spin Fe(III) state normally induced by substrate binding is inaccessible because of weak styrene binding (K_m = 5 mM). In any case, the availability and maintenance of the reduced heme-Fe(II) in the P450 active site appears critical for carbenoid formation and efficient catalysis.

The electronic properties of the heme iron can be influenced most directly by substituting the axial, heme-binding ligand at residue 400. Replacement of wild-type cysteine with the weaker coordinating serine drastically affected the behavior of the cyclopropanation enzymes through two favorable changes: (i) it abolished the enzyme's native oxidation activity and (ii) increased the reduction potential of the Fe(III) center (E^o $Fe^{III/II}$ = −293 mV), allowing direct reduction to Fe(II) by endogenous NAD(P)H under biological conditions; an external reductant was no longer required. The serine-ligated enzymes formed a typical Fe(II)–CO complex upon CO exposure, with a signature UV–vis absorption band at 411 nm. Thus, the serine-ligated enzymes are called P411s. $P411_{BM3}$-CIS ($P450_{BM3}$-CIS with a C400S mutation) was a highly efficient cyclopropanation catalyst – gram-scale reaction of styrene and EDA could be performed with TTNs up to 67800 and product yields up to 72% (90 : 10 *cis*:*trans*, 99% ee_{cis}, 43% ee_{trans}). The reactions could be performed either with whole cells (fresh or lyophilized) or cell lysate. In whole-cell reactions, the bacterial cells overexpressing the target enzyme are directly employed as the enzyme source without lysis or extraction (this can be thought of as a heterogenous, immobilized enzyme preparation). Moreover, the enzyme is presented in its preferred, endogenous microenvironment which likely contributes to improved lifetime and activity. Whole-cell catalysis provides superior results, is operationally simple, and is the protocol of choice in carbene transfer reactions.

These studies demonstrated how natural P450s can be converted into cyclopropanation enzymes with just a handful of mutations in the protein structure. As is often the case, the mechanistic/structural basis behind most of the beneficial mutations remains poorly understood. In the P411 enzymes, modification of the axial ligand was pursued with the aim of changing the redox potential of the Fe center and can be considered as a successful example of rational engineering. A change of the axial ligand to histidine (C400H) allowed even higher efficiency of cyclopropanation than the serine analog, along with improved oxygen tolerance (likewise, C400A and C400Y mutations performed comparably to C400S) [30]. It is interesting to note that the axial mutation not only modulates electronics of the heme iron but also contributes to the active site geometry – mutants differing only in the axial ligand show differences in rate as well as enantioselectivity. The precise effect of the other active site mutations is more difficult to rationalize and can rarely be predicted beforehand.

4.3 The Workflow of Directed Evolution

Given a protein with starting activity (fitness) for a reaction, mutational screening/selection allows access to variants within a close sequence space neighborhood of this parent enzyme. Most mutations are deleterious or neutral, but some are beneficial, and variants that differ from the parent by 1–2 amino acid changes but show improved fitness can be identified. With the new, improved variant in hand, the process can be repeated, each round of evolution providing access to a more efficient enzyme while moving through the protein sequence space in small steps. Improvement by small steps is the principle behind directed evolution. Protein catalysts are genetically encoded, can be modified by standard molecular biology techniques, and then screened for the target activity. In practice, the workflow is typically executed as follows (Figure 4.3):

1. A collection of enzymes is screened for target reactivity and the best-performing member(s) is selected as the "parent."
2. Based on available structural and enzymological data, amino acid sites (or regions) expected to be important in the catalytic cycle are selected for mutagenesis. If there is a practical high-throughput screen, mutations can also be made randomly throughout the protein.

3. Mutant libraries made by site-saturation mutagenesis at selected sites or random mutagenesis in a specific region or the whole protein are prepared and screened. This screening is performed in multi-well plates with either whole cells or clarified lysate as the biocatalyst (typically, one 96-well plate is screened for each site-saturation library).
4. Variants which show improved activity are designated as "hits," validated in separate repeat experiments and sequenced to identify the activating mutations. This marks the end of one round of evolution.
5. An improved enzyme becomes the parent for the next round of evolution. The process is repeated until an enzyme with the desired level of activity is obtained or no further improvements are observed, at which point the screening criteria may need to change.

Experimental strategies can differ considerably for each step in this workflow. Various methods for mutagenesis – including but not limited to error-prone PCR, multiple site targeting and recombinational cloning may be utilized. Large mutational libraries can be assembled readily using established molecular biology techniques; however, the assay used for assessing activity and quantifying product formation is generally the throughput-determining step in the workflow and impacts the mutagenesis strategy. Assays utilizing GC(-MS) or LC(-MS) screening continue to be the most generally applicable and provide a throughput that is satisfactory for most directed evolution campaigns (library size: 10^2–10^3). Fluorescence and absorbance assays can enable high throughputs and thus can allow screening larger libraries (10^3–10^5) but are limited in scope. The details of mutagenesis strategies and screening assays will not be discussed here, and the reader is directed to in-depth reviews about the subject [31–33].

4.4 Expanding Cyclopropanation with Diverse Hemeprotein Carbene Transferases

Heme proteins are widespread in nature, and although many of them do not function as enzymes, their catalytic potential as carbene transferases has exploded since the demonstration of engineered $P450_{BM3}$ enzymes for cyclopropanation. Among the many types of carbene transfer reactions, cyclopropanation of alkenes has received the most attention, as summarized in Figure 4.4.

Diverse P450 enzymes were investigated for selective olefin cyclopropanation, where engineering of key residues in these P450s was found crucial. After the discovery of the Ser-ligated P411 enzymes, systematic studies on the axial heme-ligating residue revealed a new variant with histidine as the proximal ligand, BM3-Hstar, which cyclopropanated *N*,*N*-diethyl-2-phenylacrylamide to afford the key precursor of levomilnacipran, with an initial turnover frequency (TOF) of >1000 min^{-1} (Figure 4.5a) [46]. Interestingly, the carbene transfer activity of BM3-Hstar was barely impacted by aerobic conditions and allowed for efficient and selective preparation of an array of precursors to levomilnacipran derivatives. Besides the axial ligand effect, mutagenesis of a highly conserved residue in

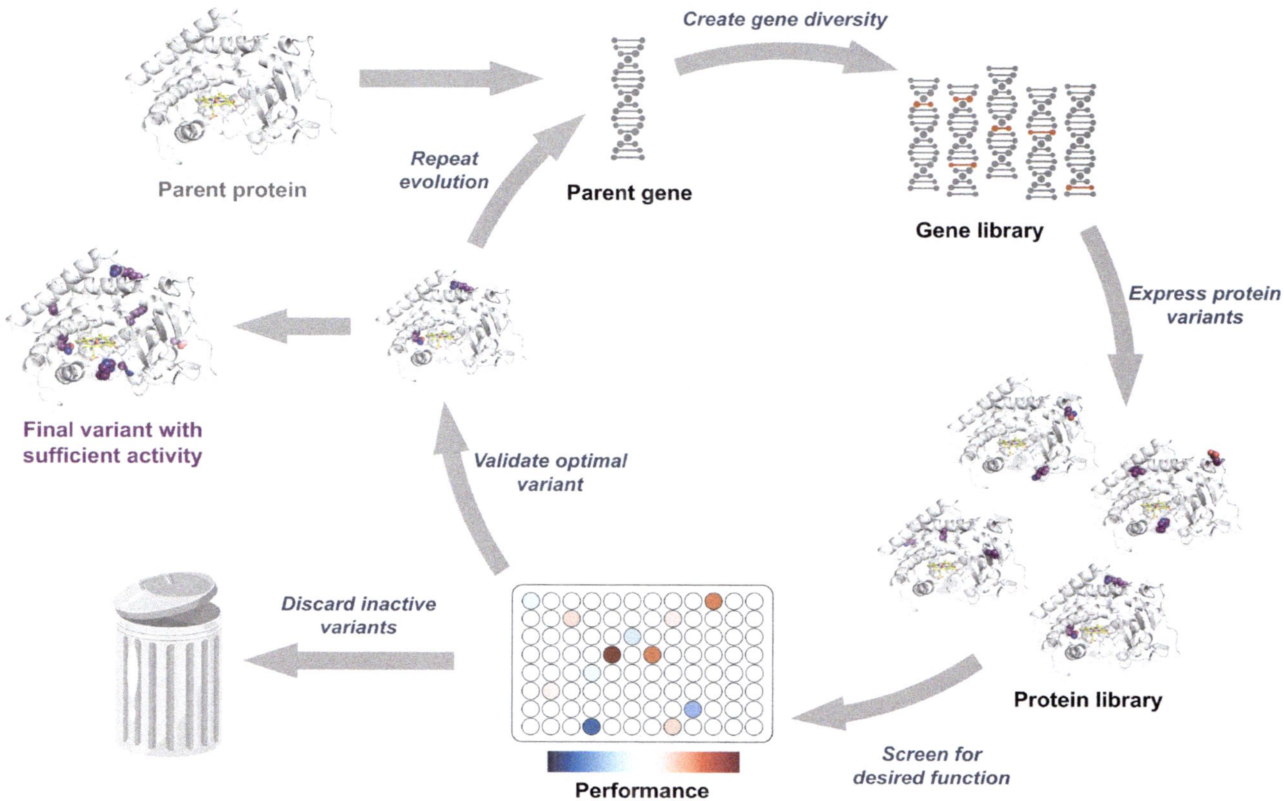

Figure 4.3 The directed evolution workflow. Mutant libraries of the parent enzyme showing starting activity are prepared using molecular biology techniques to modify the gene sequence and express the resulting mutant proteins. These are screened to identify protein variants showing improved performance. The new enzyme is subjected to iterative rounds of mutation and screening to accumulate beneficial mutations.

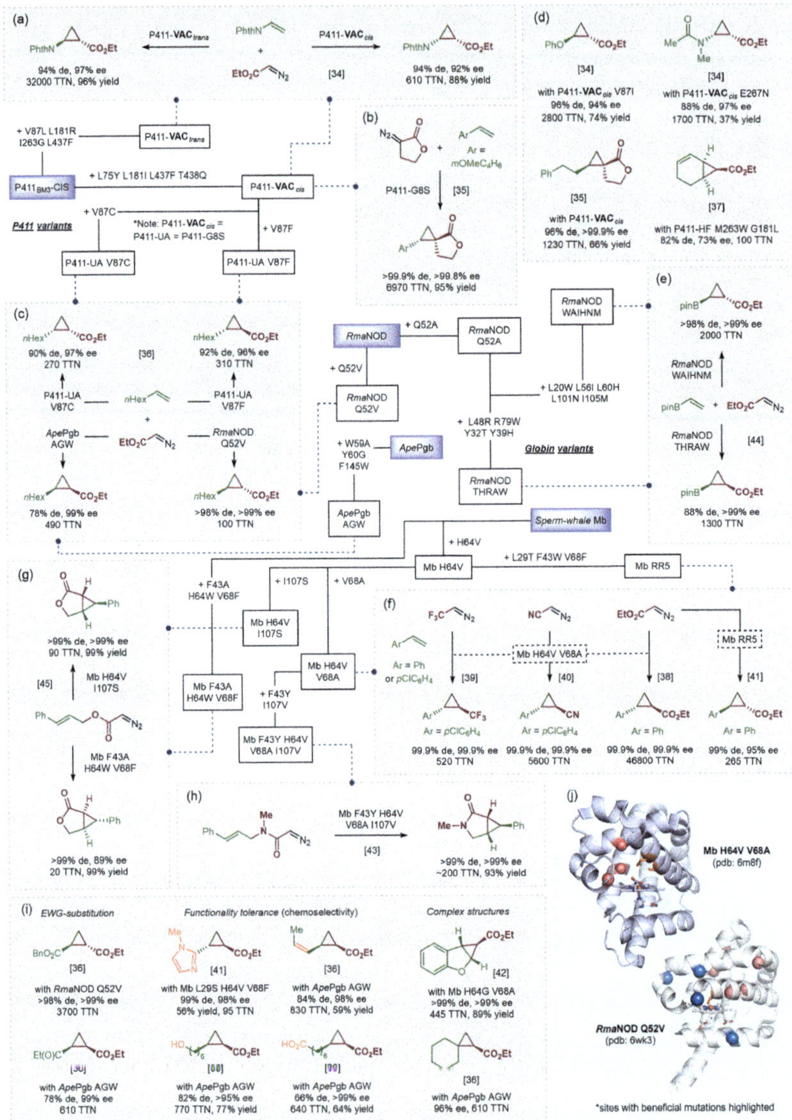

Figure 4.4 Summary of major developments in biocatalytic cyclopropanation. (a) Engineering of P411-CIS for stereodivergent cyclopropanation of heteroatom-substituted olefins. (b) *Spiro*[4.2]lactone synthesis through olefin cyclopropanation using a lactone-based diazo. (c) Cyclopropanation of unactivated substituted olefins using four stereo-complementary heme protein variants. (d) Representative examples of cyclopropanation with engineered P411 carbene transferases. (e) Engineering of *Rma*NOD for stereodivergent cyclopropanation of vinyl boronic ester. (f) Myoglobin-catalyzed styrene cyclopropanation using different diazo reagents. (g) Engineering of Mb for stereodivergent intramolecular cyclopropanation. (h) Cyclopropane-fused lactam synthesis through Mb-catalyzed intramolecular cyclopropanation. (i) Examples of cyclopropanation with engineered globin carbene transferases. (j) Crystal structures of globin variants Mb H64V V68A (pdb: 6m8f) and *Rma*NOD Q52V (pdb: 6wk3).

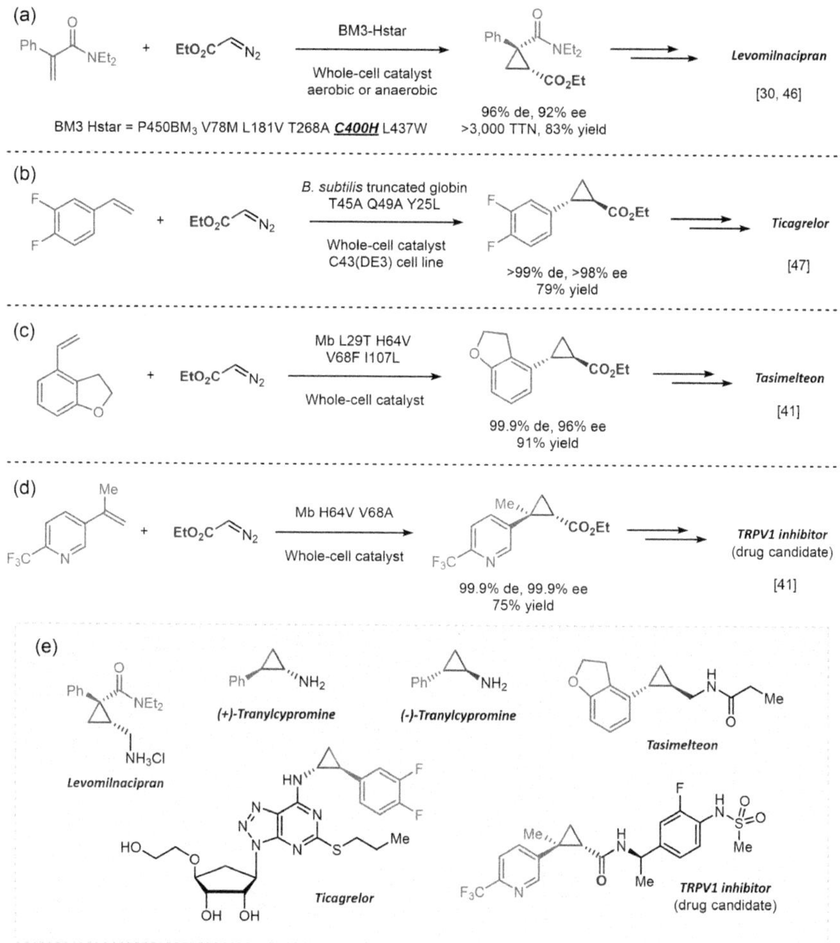

Figure 4.5 Biocatalytic syntheses of cyclopropane-containing intermediates of pharmaceutical compounds, including (a) levomilnaciprmn, (b) ticagrelor, (c) tasimelteon, and (d) TRPV1 inhibitor and (e) structures of corresponding pharmaceutical compounds. Source: (a) Based on Renata et al. [46]. (b) Based on Hernandez et al. [47]. (c) and (d) Based on Bajaj et al. [41].

native P450s, commonly engaged in stabilizing iron-oxo species for natural oxidation chemistries, from threonine to alanine (e.g. T268A in $P450_{BM3}$ and T234A in CPY142), also identified potent P450 mutants with promising diastereoselectivities for cyclopropanation [48]. Additionally, by using decoy molecules, typically peptides based on phenylalanine, the stereocontrol in *cis*-cyclopropane formation catalyzed by a $P450_{BM3}$ variant could be further improved [49].

To expand the scope of enzymatic cyclopropanation, $P411_{BM3}$-CIS, which delivered high activity for carbene transfer in whole cells, served as a suitable template for further engineering of more potent biocatalysts. Brandenberg et al. evolved $P411_{BM3}$-CIS to include heteroatom-substituted alkenes as substrates for cyclopropanation [34]. Only four to five mutations in the enzyme's active site provided P411-$\mathbf{VAC}_{cis}$ and P411-$\mathbf{VAC}_{trans}$ for the efficient and selective synthesis of both *cis*- and *trans*-cyclopropanes bearing various nitrogen, oxygen, and sulfur substitutions (Figure 4.4a,d). The evolved P411-$\mathbf{VAC}_{cis}$ variant also exhibited the extraordinary ability to accommodate other alkene and diazo coupling partners. A structurally different carbene based on a lactone moiety was accepted by P411-$\mathbf{VAC}_{cis}$ for synthesis of *spiro*-lactones with excellent stereocontrol and catalytic efficiency (Figure 4.4b,d) [35]. Beyond styrenyl and heteroatom-substituted alkenes, electronically unactivated or even deactivated olefins were competent substrates for this cyclopropanating enzyme and its variants with a mutation at site 87 (V87C or V87F) (Figure 4.4c) [36]. Evolution of $\mathrm{P411}_{\mathrm{BM3}}$-CIS using a substrate-walking strategy (the progressive evolution of enzymatic activity toward more and more closely related substrates) further afforded new P411 variants able to use internal alkenes for *cis*-cyclopropane formation (Figure 4.4d) [37].

The initial screening of different heme proteins for styrene cyclopropanation indicated the potential of natural heme proteins other than P450s to act as carbene transferases [10]. The Fasan group demonstrated the engineering of sperm whale myoglobin (Mb) for selective cyclopropanation [38]. While the wild-type Mb (180 TTN, 93 : 7 dr) exhibited only slightly improved activity and selectivity toward styrene cyclopropanation using EDA compared to free heme (145 TTN, 87 : 13 dr), replacing its distal histidine ligand with a valine residue (H64V) was found to be beneficial for catalytic efficiency and diastereocontrol. Another mutation in the catalytic pocket, V68A, was highly activating, giving up to 46 800 TTN toward the exclusive formation of (1*S*,2*S*)-cyclopropanes out of four possible stereoisomers (99.9% de and 99.9% ee) (Figure 4.4f). The double mutant, Mb H64V V68A, also utilized other diazo precursors bearing an α-trifluoromethyl or cyano group to furnish the corresponding *trans*-cyclopropanes [39, 40]. Computational analyses pointed to the noncovalent interactions between active site residues and carbene intermediate as well as the styrene substrate in conferring stereoselectivity [50]. Using Mb H64V as the parent, screening a library of 76 variants generated by site-directed mutagenesis at four active site residues (29, 43, 68, and 107) followed by recombination of activating mutations led to a set of Mb mutants (e.g. Mb L29T F43W H64V V68F) with selectivity favoring (1*R*,2*R*)-cyclopropanes (Figure 4.4f) [41]. These engineered Mb variants were then used for the synthesis of key cyclopropane intermediates of drug molecules, including tranylcypromine, tasimelteon, and ticagrelor, with high selectivities (Figure 4.5c,d). An array of cyclopropane-fused ring structures were forged through intermolecular cyclopropanation of benzofurans [42] or intramolecular cyclopropanation [43, 45, 51] using different myoglobin variants (Figure 4.4g–i). These examples show that myoglobin, although natively functioning as an oxygen-binding heme protein, can become a selective carbene transferase. Myoglobins, however, have not yet been shown to switch diastereoselectivity to make the synthetically more difficult *cis*-cyclopropanes.

Nature has generated a diversity of small heme proteins that could be tested for novel cyclopropanation activity. Knight et al. engineered a protoglobin from *Aeropyrum pernix* (*Ape*Pgb) and a nitric oxide dioxygenase from *Rhodothermus marinus* (*Rma*NOD) for cyclopropanation reactions [36]. They generated four enzymes,

*Rma*NOD Q52V and *Ape*Pgb W59A Y60G F145W together with two P411-**VAC**$_{cis}$ variants, to gain access to all four stereoisomers of cyclopropanes selectively and efficiently from electron-deficient and electron-neutral alkenes (Figure 4.4c). This engineered biocatalytic platform accepted a diverse range of alkene substrates and also displayed perfect compatibility with different functionalities, such as free hydroxyl and carboxylic acid groups (Figure 4.4i). *Rma*NOD was subjected to further rounds of evolution via site-saturation mutagenesis and screening to cyclopropanate a vinyl boronic ester substrate (Figure 4.4e) [44]. Two *Rma*NOD variants selectively produced each of the *cis*- and *trans*-cyclopropyl borates; the products could be prepared in multi-gram scale and further converted into more diversified cyclopropane scaffolds through stereoretentive Suzuki cross-coupling. Another study disclosed the engineering of *B. subtilis* group II truncated globin to selectively produce a cyclopropane-containing synthetic intermediate (>99% de, >98% ee, and 79% yield) for ticagrelor (Figure 4.5b) [47].

Analogous to alkene cyclopropanation, carbene transfer to alkynes is expected to deliver cyclopropenes featuring significantly higher strain energy (56 kcal mol^{-1}). A further carbene transfer step would generate a class of even more strained carbocycles, bicyclo[1.1.0]butanes (67 kcal mol^{-1} strain energy) with puckered structures. These strained carbocycles, especially bicyclobutanes, are difficult to assemble chemically. On the other hand, the strain energy makes them useful in applications from strain-release chemistries to stimulus-responsive materials [52, 53]. Cyclopropene synthesis via carbene transfer to alkynes has been demonstrated with different transition metal catalysts based on rhodium [54–56], iridium [57], cobalt [58], and others, but not with any iron catalysts, and only limited examples of constructing bicyclobutanes through such carbene strategies have been reported [52], with no asymmetric versions [59, 60]. Enzymatic carbene transferases would be a powerful addition to the synthetic toolbox for strained carbocycle construction, but the challenges are obvious – the enzymes, as iron catalysts, need to acquire the ability to deliver carbenes to alkynes and also impose precise control on the carbene addition steps.

Chen et al. identified a promiscuous P411 enzyme, P411-**E10**, originally evolved for nitrene transfer reactions [61], which displayed promising activity for bicyclobutane synthesis using terminal aromatic alkynes and EDA while suppressing undesired carbene transfer events such as furan formation [62]. Directed evolution introduced two active site mutations, V78F and S438A, leading to carbene transferase P411-**E10** FA capable of producing the bicyclobutane product as a single stereoisomer (Figure 4.6). This report highlights the enzyme's ability to facilitate carbene transfer to alkynes, stabilize the reactive cyclopropene intermediates, and also exert precise stereocontrol over two carbene transfer steps. Another P411 variant, P411-**P4**, from the same nitrene transferase lineage as P411-**E10**, was found to carry out cyclopropenation using aliphatic alkynes as substrates (260 TTN, 91% ee for substrate 4-phenylbutyne). Starting from P411-**P4**, site-saturation mutagenesis of residue 87 and screening yielded A87F (290 TTN, −94% ee) and A87W (240 TTN, 94% ee) giving opposite enantio-preference. Further evolution of these two mutants improved both catalytic efficiency and selectivity of cyclopropenation,

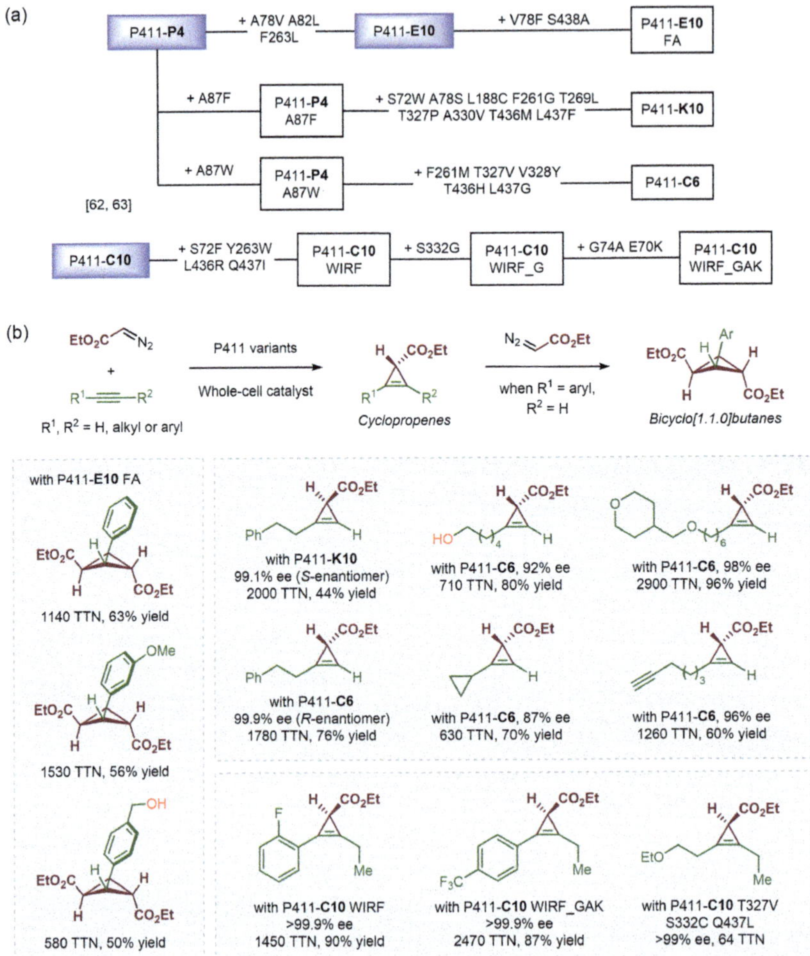

Figure 4.6 Synthesis of cyclopropenes and bicyclobutanes via carbene transfer to alkynes catalyzed by P411 enzymes. (a) Evolutionary trajectories toward P411 enzymes for carbene transfer to alkynes. (b) Examples of cyclopropene and bicyclobutane synthesis with engineered P411 carbene transferases. Source: Based on Chen et al. [62] and Chen et al. [63].

such that either enantiomer could be made with thousands of turnovers and high enantioselectivity (P411-**K10** for *S*-enantiomer, 2000 TTN, 99.1% ee; P411-**C6** for *R*-enantiomer, 1780 TTN, 99.9% ee). Enzymatic reactions using whole-cell catalysts allowed for the convenient preparation of these strained carbocycles on millimole scales.

Carbene transfer to internal rather than terminal alkynes is even more challenging, with only two synthetic systems based on gold/silver [64] or rhodium [65] (co)catalysts reported for such enantioselective cyclopropenation. A variant of P411-**C10**, with 27 amino acid mutations and FAD domain deletion compared to wild-type $P450_{BM3}$, was able to perform cyclopropenation of an internal aromatic alkyne, 1-phenylbutyne, albeit with low activity (55 TTN); however, the perfect enantioselectivity (>99% ee) indicated the enzyme scaffold could bind the alkyne and direct carbene transfer in a well-defined orientation [63]. Introduction of an active site mutation, Y263W, enhanced overall activity by >11-fold; subsequent evolution led to a potent variant with three more beneficial mutations (Q437I, S72F, and L436R, termed P411-**C10** WIRF) giving 2680 total turnovers toward the desired cyclopropene product. Upon further evolution, this biocatalytic system used more diverse substrates: three additional mutations S332G, G74A, and E70K gave P411-**C10** WIRF_GAK, which is capable of cyclopropenating various internal aromatic alkynes with thousands of turnovers and perfect stereoselectivity (all with >99.9% ee, Figure 4.6). Under optimized conditions, preparative-scale reactions readily delivered the cyclopropene products in hundred milligram quantities, and subsequent hydrogenation offered access to tri-substituted all-*cis*-cyclopropanes, which are difficult to synthesize chemically. Screening of variants in the **C10** family identified members that catalyzed enantioselective cyclopropenation of internal aliphatic alkynes, a particularly challenging task. Moreover, variants differing by only a few amino acid substitutions to P411-**C10** were found to switch chemoselectivity to propargylic C–H bond insertion or formal [3+2] cycloaddition, which, once again, speaks to the high tunability of enzyme catalysis in addressing different selectivity problems.

4.5 C–H Functionalization with Carbene Transferases

Just as cyclopropanation is analogous to epoxidation in P450s' native oxidative functions, C–H alkylation is the carbene transfer counterpart to native C–H hydroxylation. Carbene C–H insertion represents a highly sought-after strategy for C–H functionalization for building C–C bonds and has been used extensively in the construction of diverse molecular skeletons as well as late-stage modification of complex molecules [66]. Carbene insertion into C–H bonds, especially asymmetric versions, typically involves the use of transition metal catalysts based on rhodium, iridium, copper, and other metals [67–69]. Iron-based catalysts have shown limited utility for C–H insertion transformations, especially when targeting sp^3-type C–H bonds [70, 71]. The rarity presumably arises from the high activation energy barrier for iron-carbene insertion into a C–H bond. However, it remained plausible that this transformation could take place in a protein active site in which the substrates are properly oriented for the desired C–H insertion processes.

Zhang and coworkers successfully developed an enzymatic platform using P411 variants for selective carbene insertion into $C(sp^3)$–H bonds [72]. Prior to this report, synthetic methods for $C(sp^3)$–H insertion using iron-based catalysts included intramolecular insertion by an iron-pthalocyanine catalyst [73] or a co-catalytic system based on a highly electrophilic iron catalyst with a lithium aluminum alkoxide [74], and intermolecular versions using iron-porphyrins [75] or other electrophilic iron complexes [76]. However, these reactions were not enantioselective and sometimes required substrates in large excess (as solvent), lacked regioselectivity, and/or were associated with significant side reactions. In addition to these catalytic examples, iron-carbenes in the form of $[Cp(CO)_2Fe{=}CHR]^+ X^-$ complexes have been studied for intramolecular C(sp^3)–H insertion since the 1980s [77, 78] and have also been employed in natural product total synthesis [79] but suffer from limitations which include sophisticated preparation and stoichiometric use of the iron-carbene precursors.

The biocatalytic system allowed for efficient carbene C–H insertion, where selectivities could be tuned by directed evolution. P411 variant **P4**-A82L showed trace activity for the insertion reaction of EDA into the benzylic C–H bonds of *p*-methoxybenzyl methyl ether and was utilized as the parent for further directed evolution. Fourteen rounds of evolution yielded variant P411-**CHF** with >140-fold improved activity (2020 TTN) over the parent enzyme (14 TTN) and an enantiomeric excess of 93%. *E. coli* cells harboring P411-**CHF** also functioned smoothly as a whole-cell catalyst at 4 °C, furnishing the desired C–H insertion product (82% yield, 1060 TTN) in millimole scale with even greater enantioselectivity (96% ee) (Figure 4.7a). Preliminary mechanistic study revealed a kinetic isotope effect (KIE) of 5.1 for the enzymatic C–H insertion reaction, indicating C–H bond cleavage is the rate-determining step. The larger KIE value compared to those with rhodium catalysts (1.5–3.2) [80] or other iron catalysts (2.3–4.3) [74, 78] known to undergo a concerted C–H insertion mechanism suggests that the enzymatic system may feature different mechanistic profiles.

Evolved P411-**CHF** could target a variety of C(sp^3)–H bonds, including those at benzylic, allylic, propargylic, and α-amino positions, for carbene insertion. The benefit of using enzyme scaffolds to control selectivity is illustrated by the substrate scope study (Figure 4.7b). For instance, P411-CHF specifically performed carbene insertion into benzylic C–H bonds with substrates bearing terminal olefin or silyl groups, while another P411 variant, P-I263F, carried out cyclopropanation or Si–H insertion exclusively (Figure 4.7c). Using *N*-methyl tetrahydroquinoline as the substrate, regioselectivity between the primary and secondary α-amino C(sp^3)–H bonds, and enantioselectivity of secondary C–H insertion, could be altered by employing different P411 variants. The biocatalytic system was further applied to the chemo-enzymatic synthesis of natural products, (+)-lyngbic acid and both enantiomers of cuspareine.

Continuing effort by Arnold and coworkers further extended the enzymatic platform to include different carbene precursors for C(sp^3)–H functionalization, especially with *N*,*N*-dialkyl aniline substrates. A variant from the lineage of P411-**CHF** was engineered to adopt trifluorodiazoethane as the carbene donor to install a fluoroalkyl group onto α-amino C–H bonds of *N*-aryl pyrrolidines (Figure 4.8a) [81]. Evolved P411-**PFA** accepted a range of pyrrolidine substrates with diverse functionalities on the aromatic rings, giving TTNs up to 4070 and ee's up to >99%, and also utilized pentafluoro diazopropane for C–H insertion with exquisite stereocontrol (>99% ee) (Figure 4.8b). Interestingly, the pentafluoropropylated products were formed with the opposite stereochemistry (*S*-configuration) compared to those trifluoroethylated ones (*R*-configuration), suggesting a potentially different orientation

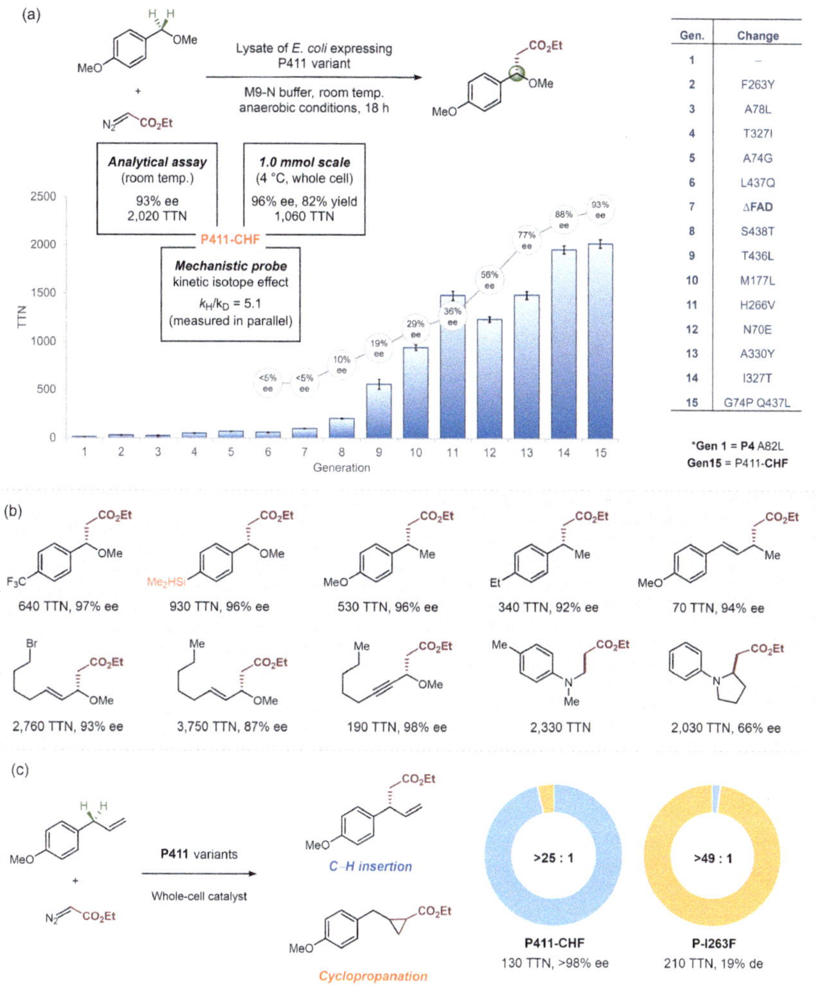

Figure 4.7 Engineering P411 enzymes for carbene insertion into C(sp^3)–H bonds. (a) Directed evolution of P411-**P4** A82L for carbene C(sp^3)–H insertion. (b) Representative examples of carbene C(sp^3)–H insertion catalyzed by P411-**CHF**. (c) The enzyme controls chemoselectivity, choosing between C–H insertion (P411-**CHF**) and cyclopropanation (P-I263F). Source: Based on Zhang et al. [72].

of the bulkier iron-carbene intermediate (the pentafluoroethyl-susbtituted one) in the enzyme's active site, thus effecting the opposite enantio-preference when the prochiral substrates approach the carbene. Screening of various P411 enzymes identified another variant that altered the fluoroalkylation stereochemistry, providing access to both enantiomers of the C–H insertion products. Zhou et al. also described a lactone-based carbene for C–H insertion, forming an α-stereogenic center [82]. A lineage of P411 enzymes, engineered from a single highly promiscuous variant, P411-**C10**, functionalized primary and secondary C(sp^3)–H bonds adjacent to amino groups, offering a route to enantio-enriched β-amino lactones with total

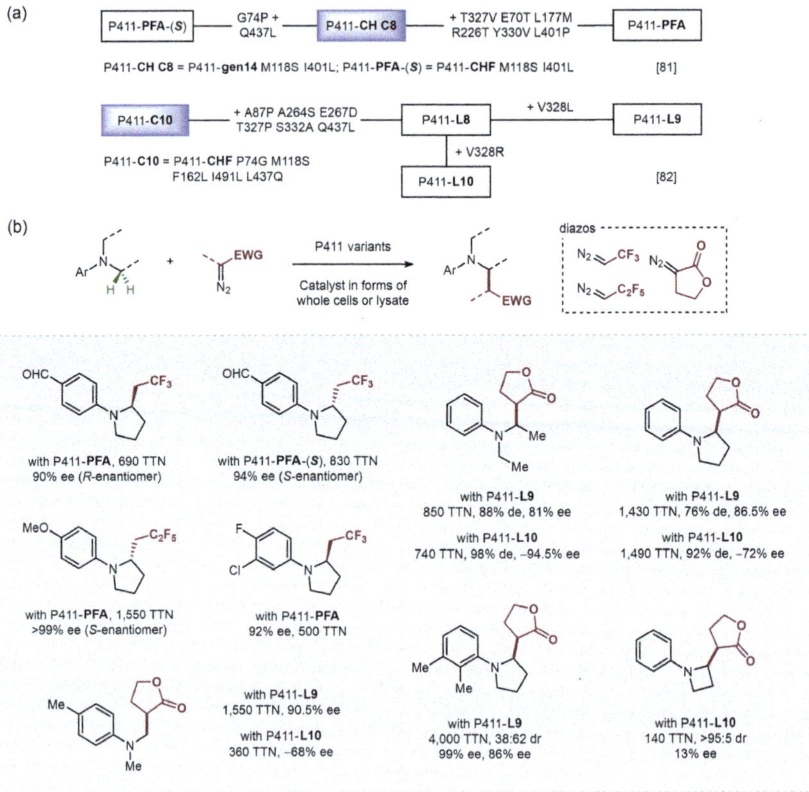

Figure 4.8 Engineering P411 enzymes for carbene insertion into α-amino C(sp^3)–H bonds. (a) Evolutionary trajectories toward P411 variants for carbene C(sp^3)–H insertion. Source: Based on Zhang et al. [81] and Zhou et al. [82]. (b) Representative examples of stereodivergent functionalization of α-amino C(sp^3)–H bonds catalyzed by engineered P411 carbene transferases.

turnovers up to 4000 (Figure 4.8a). It is worth noting that the last two variants in the evolutionary lineage, **L9** and **L10**, differing by a single mutation at site 328 (**L9** with L328 and **L10** with R328), exhibited opposite stereocontrol over the C–H insertion processes, inverting the two contiguous chiral centers simultaneously when targeting secondary C–H bonds for functionalization and leading to opposite enantiomers of the same diastereomers (Figure 4.8b). These examples again showcase the power of directed evolution in tuning enzyme catalysis for divergent selectivities.

Heme protein carbene transferases have also been demonstrated with capacity for C(sp^2)–H alkylation of electron-rich heteroarenes. C_3-Alkylation of indoles through formal carbene C(sp^2)–H insertion is a well-explored reaction with different transition metal catalysts [83–85]. Engineered heme proteins, including myoglobin, peroxidase YfeX and P411s, for indole alkylation provided proof-of-concept demonstrations for enzymatic C(sp^2)–H insertion [86, 87] and provided new enzyme variants for selective alkylation of other (hetero)arenes (Figure 4.9) [37]. Brandenberg and coworkers developed a high-throughput assay based on

UV–vis spectrophotometry to engineer P411 enzymes with improved activity for indole alkylation (Figure 4.9a). The evolved variant, P411-HF, alkylated indole derivatives at C_3 positions using EDA as the carbene precursor with TTNs up to 18 700. A substrate-walking strategy was adopted to evolve selective P411 carbene transferases for more challenging alkylations of indole and pyrrole substrates. With an α-disubstituted carbene derived from ethyl 2-diazopropanoate (Me-EDA), engineering of P411-HF delivered a new variant with 775 TTN and ~50% ee toward the branched alkylation product and displaying ~400-fold increase in catalytic efficiency over the parent enzyme (~2 TTN). Regioselective C(sp^2)–H alkylation of pyrroles at the C_2 or C_3 position, a long-standing challenge in small-molecule catalysis, could be addressed by engineering P411 carbene transferases. P411-HF alkylated *N*-methyl pyrrole with moderate regioselectivity of 3 : 1 favoring C_2. Engineering furnished two P411 variants capable of selectively functionalizing the pyrrole substrate at the C_2 position (50 : 1 selectivity, 485 TTN) or C_3 position (1 : 10 selectivity, 705 TTN) in a regio-divergent fashion (Figure 4.9b).

4.6 Biocatalytic Carbene X–H Insertion

Another class of carbene transfer reactions is insertion into X–H bonds (where X is a heteroatom, including N, S, Si, B, P, and others) to build carbon–heteroatom bonds. The Arnold lab reported enzymatic carbene N–H insertion using engineered P450 enzymes (Figure 4.10a) [88]. Primary and secondary aniline derivatives were favored for N–H insertion using diazo ester or amide as the carbene sources, whereas aliphatic amines, such as morpholine or benzylamine, gave only poor conversions. The Fasan lab reported that myoglobin variants could also accept a variety of aromatic and aliphatic amines for N-alkylation, with high catalytic efficiency (up to 7810 TTN) [90, 91]. Asymmetric N–H insertion using α-disubstituted carbenes is still not quite successful: the only reported enzymatic N–H insertion between α-alkyl-disubstituted diazoacetates and anilines using engineered myoglobins gave poor to moderate enantioselectivities (5–82% ee) [92]. As computational studies disclosed, N–H insertion with heme proteins involves a nucleophilic attack followed by ylide rearrangement and enol tautomerization, where the final proton transfer step forms α-chiral centers [93]. To this end, the enzyme must function not only as an efficient carbene transferase but also must provide a well-defined chiral environment for proton shuttling [94].

Carbene S–H insertion using heme protein carbene transferases was demonstrated by the Fasan laboratory (Figure 4.10b) [89]. Enantioselective S–H insertion was preliminarily studied with Me-EDA, giving 49% ee and 75 TTN under optimized conditions. In this work, the authors proposed a nucleophilic attack mechanism for S–H insertion, similar to that of N–H insertion; Hammett analysis of S–H insertion transformations of (substituted) thiophenols, however, did not establish a simple correlation between reaction rate and substitution effect, which indicates that the mechanism of enzymatic S–H insertion may be more complicated.

Chen et al. subsequently reported the engineering of P450 enzymes for enantioselective S–H insertion with a lactone carbene, giving up to 2190 TTN and 89% ee

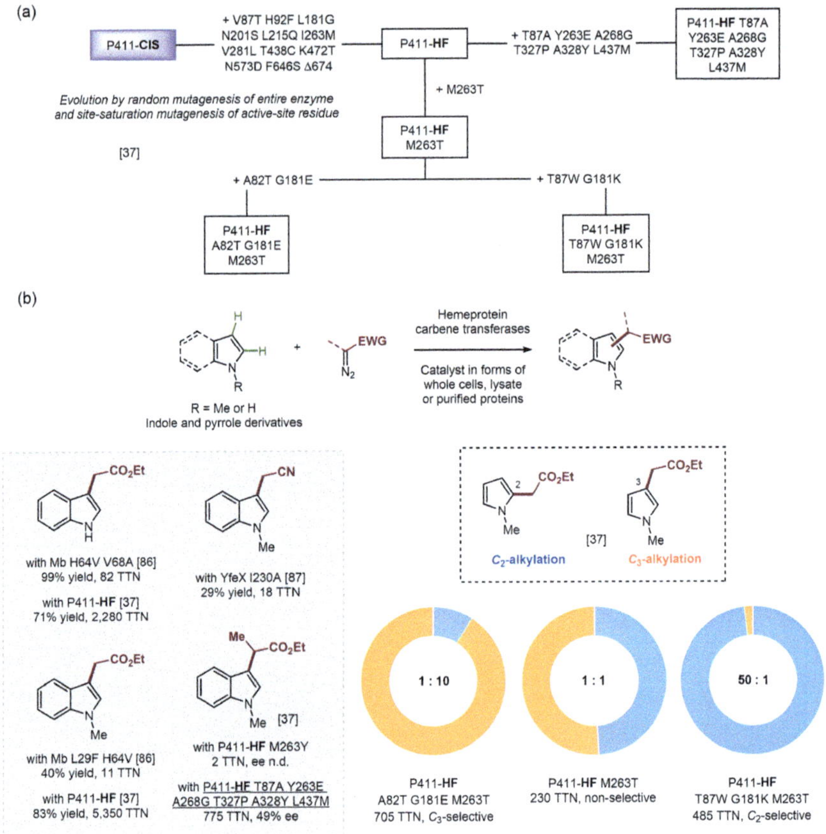

Figure 4.9 Engineered heme protein-catalyzed heteroarene C(sp^2)–H bond alkylation. (a) Evolutionary trajectories toward P411 variants for heteroarene alkylation. (b) Representative examples of indole C_3-alkylation and regio-divergent alkylation of a pyrrole substrate, catalyzed by engineered heme protein carbene transferases. Source: Based on Brandenberg et al. [37].

(Figure 4.10b) [35]. Interestingly, two nearly identical variants, differing only by the heme-ligating residue (serine for P411-G8S and cysteine for P450-G8S), exhibited different activity and selectivity for the desired S–H insertion reactions. Computational study by density functional theory (DFT) uncovered a radical mechanism for S–H insertion, involving a hydrogen atom transfer (HAT) step followed by radical rebound, and also explained how the electronic contribution of the axial ligand drives distinct pathways and leads to different outcomes.

Carbene transferases have been expanded to include carbene insertion into Si–H and B–H bonds. Kan et al. discovered that cytochrome *c* from *Rhodothermus marinus* (*Rma* cyt *c*), whose native function is electron transport, possesses activity for carbene Si–H insertion using Me-EDA and phenyl dimethylsilane as substrates [95]. The wild-type protein catalyzed the desired reaction with excellent enantioselectivity (97% ee) and activity 10-fold higher (44 TTN) than the free heme catalyst. Despite

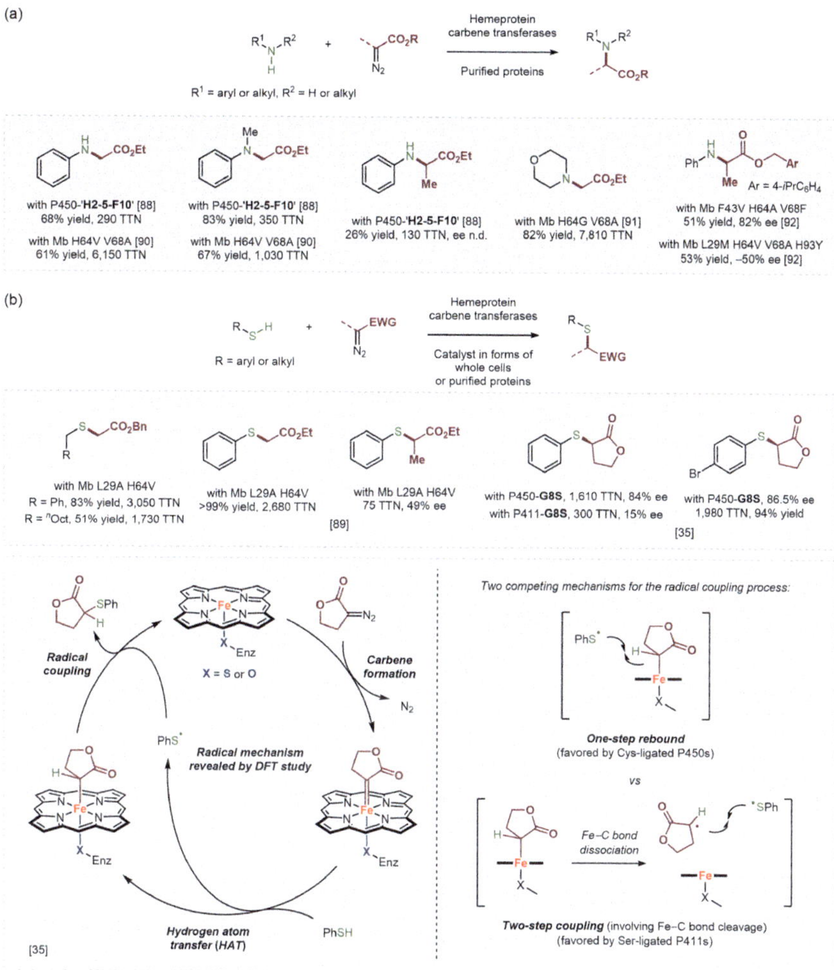

Figure 4.10 Engineered heme protein-catalyzed carbene N/S–H bond insertions. (a) Carbene N–H insertions with heme protein carbene transferases. Source: Based on Wang et al. [88], Sreenilayam et al. [90] and Steck et al. [92]. (b) Carbene S–H insertions with heme protein carbene transferases and a radical mechanism by DFT study. Source: Based on Chen et al. [35] and Tyagi et al. [89].

the fact that *Rma* cyt *c* does not naturally feature an "active site" for catalysis or even substrate binding, it apparently provides a well-defined environment for the desired Si–H insertion reaction to take place with excellent stereocontrol. This latent Si–H insertion activity was improved by directed evolution. Site-saturation mutagenesis targeting the distal heme-ligating residue yielded the M100D mutation, which increased activity 12-fold (550 TTN). Structural analysis showed that this mutation creates an open coordination site on the iron center and promotes formation of an active site for carbene formation and transfer. Further evolution led to the triple mutant *Rma* cyt *c* V75T M100D M103E, which performed carbene Si–H insertion with >1500 TTN and >99% ee (Figure 4.11a,b).

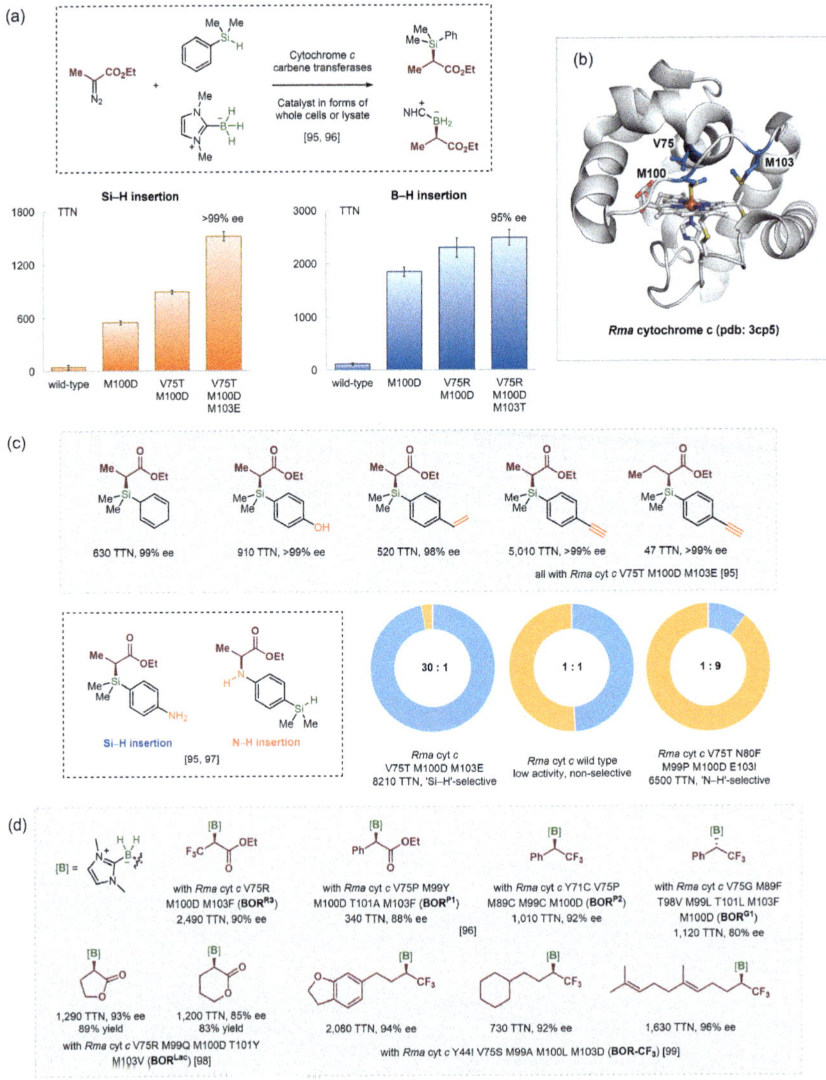

Figure 4.11 Cytochrome *c*-catalyzed C–Si/B bond formation via carbene Si/B–H insertions. (a) Evolution of *Rma* cytochrome *c* for carbene Si/B–H insertions. (b) Crystal structure of *Rma* cytochrome *c* (pdb: 3cp5). (c) Representative examples of carbene Si–H bond insertion catalyzed by engineered cyt *c* and enzyme-controlled chemoselectivity between Si-H insertion and N-H insertion. (d) Representative examples of carbene B–H bond insertion catalyzed by engineered cyt *c* mutants.

The evolved *Rma* cyt *c* accepted a broad range of silanes bearing aryl, vinyl, and/or alkyl substitutions for Si–H insertion, and most of the organosilicon products were obtained cleanly as single enantiomers. The chemo-specificity of this biocatalytic platform was noteworthy: silanes with diverse functionalities such as styrenyl alkene, terminal alkyne, phenol, and aniline, which could engage in other carbene transfer reactions, underwent Si–H insertion almost exclusively (Figure 4.11c). For example, with 4-(dimethylsilyl)aniline, the *Rma* cyt *c* variants in the evolutionary lineage gradually increased their preference for Si–H insertion over N–H insertion, with selectivity going from 1 : 1 in wild-type cyt *c* to 30 : 1 in the triple mutant. The mutations presumably enhanced the orientation and binding of the silane substrate for Si–H insertion. Structural analysis and computational studies also shed light on the accompanying mutational effects [100]: a loop at the active site (T98–E103) was significantly distorted in the crystal structure of the *Rma* cyt *c* triple mutant and appeared highly flexible in the carbene-bound protein structure. As suggested by molecular dynamic (MD) simulations, this loop flexibility is important in endowing an "open" conformation and allowing the silane substrate to approach the carbene intermediate with a specific pro-*R* facial selectivity for Si–H insertion. Assisted by computational modeling, engineering *Rma* cyt *c* to favor N–H insertion over Si–H insertion was also possible: several mutations involved in front loop orientation generated penta-mutant *Rma* cyt *c* V75T N80F M99P M100D E103I, which made the amination product with 9 : 1 selectivity over silylation using 4-(dimethylsilyl)aniline as substrate [97]. These results show that even a small heme protein can be tailored for different selectivities.

Rma cyt *c* also served as a promising biocatalytic template for carbene B–H bond insertion [96]. Wild-type *Rma* cyt *c* was able to catalyze B–H insertion using Me-EDA and an *N*-heterocyclic carbene (NHC)-stabilized borane, giving 120 TTN and 70% ee. Directed evolution enhanced both catalytic efficiency and enantioselectivity and yielded a new triple mutant, *Rma* cyt *c* V75R M100D M103T, which produced a range of organoboron compounds from different boranes when used in whole *E. coli* cells (Figure 4.11a). The bacterial catalyst functioned efficiently with substrates added portion-wise to the reaction system, maintaining both activity and selectivity (up to 15 300 TTN and 92% ee) over time. The *Rma* cyt *c* borylases were readily tuned to accommodate diazo partners structurally distinct from Me-EDA (e.g. α-trifluoromethyl-α-phenyl diazomethane) to synthesize both product enantiomers selectively (Figure 4.11d). Lactone-based organoboron compounds with ring sizes from 5 to 7 were also made using engineered *Rma* cyt *c* variants [98].

Despite the broad scope of carbenes used by cyt *c* borylases, protein engineering was always required to optimize activity and selectivity. Following this initial work, however, Huang et al. reprogrammed *Rma* cyt *c* to accept a spectrum of trifluoromethyl-substituted diazo compounds for B–H insertion [99]. To allow the protein scaffold to tolerate variation on the other alkyl substituent of the carbene, the authors aimed to engineer a borylase that can specifically orient the CF_3 group toward the heme pocket and the α-alkyl chain toward the solvent phase. Residues known to be involved in shaping the protein active site, especially those in the surface loop covering the heme pocket (sites 98–103), were subjected to site-saturation mutagenesis and screening to optimize a B–H insertion reaction between (3-diazo-trifluorobutyl)benzene and an NHC-borane. Variant BOR-CF_3, bearing five active site mutations (Y44I, V75S, M99A, M100L, and M103D) was obtained for the desired borylation with 2460 TTN and 95% ee (Figure 4.11d). Consistent with the hypothesis, BOR-CF_3 proved capable of borylating a variety of diazoalkanes featuring diverse alkyl chains, with 520–2870 turnovers and high enantioselectivities. Computational studies suggested the CF_3 group of the carbene intermediate was positioned in the inner side of the catalytic cavity, while the front loop controlled the approach of the borane substrate. Compared to engineered fluorinases [101] or fluorine-containing precursor-directed biosynthesis of natural product analogs [102], this work represents a complementary biocatalytic strategy to incorporate fluorine-containing motifs into organic molecules.

4.7 Carbene Transfer Reactions with Artificial Metalloproteins

Engineering a natural metalloenzyme usually modifies only the polypeptide scaffold while leaving the metal/cofactor unchanged. New chemistry emerging through this process is ultimately constrained by the reactivity limits of the cofactor. Artificial metalloproteins (ArMs) combining natural protein scaffolds with synthetic metals or organometallic cofactors create new hybrid systems not seen in nature. The history of this approach goes back to the pioneering works of Kaiser, Yamamura, and Whitesides in the 1970s and will not be discussed here [21]. This section will be restricted to important examples of ArMs engineered for carbene chemistry. These can be divided into four categories depending on the degree of protein/cofactor modification: (i) replacement of iron by other transition metals in heme proteins, (ii) introduction of a modified metalloporphyrin in apo heme proteins, (iii) incorporation of a metalloporphyrin in non-heme protein scaffolds, and (iv) combining (non-heme) protein scaffolds with non-porphyrin organometallic moieties.

In 2016, Hartwig and coworkers investigated the carbene chemistry of myoglobin variants with a non-iron metal center [22]. The ArMs were prepared by purifying the heme protein under iron-restrictive conditions (the resulting apo-proteins do not contain any heme cofactor), followed by in vitro reconstitution with the metal-porphyrin of choice. In this way, heme proteins with the iron atom replaced by Co(Cl), Cu, Mn(Cl), Rh, Ir(Me), Ru(CO), or Ag were accessed (the parenthesis entry indicates the axial ligand on the metalloporphyrin). The proteins were tested as enantioselective catalysts for an intramolecular C–H insertion reaction and cyclopropanation of internal alkenes (Figure 4.12a). Among the ArMs screened, the Ir(Me) variants provided the highest TONs for the two reactions (>60 and >20, respectively), while others gave only trace yields. The Fe-bound variants were inactive for these reactions, highlighting the transformational role played by the replaced Ir(Me). Site-directed mutagenesis and screening provided enzymes for the enantiodivergent C–H insertion reaction with good selectivity. In a subsequent study [24], ArMs accessed similarly by Ir(Me) substitution in P450 CYP119 proteins were reported to have superior catalytic attributes for the C–H insertion reaction. Engineering the parent Ir(Me)-CYP enzyme resulted in a variant that catalyzed the reaction with high turnovers, excellent enantioselectivity, and good yields at preparative scales (Figure 4.12b). The kinetic parameters of the engineered variants (k_{cat} = 45.8 min^{-1}, K_M = 0.17 mM, and k_{cat}/K_M = 269 min^{-1} mM^{-1}) were comparable to metabolic enzymes. In another study, the authors reported that these Ir(Me)-CYP119s were also excellent biocatalysts for the enantioselective and diastereoselective cyclopropanation of varied terminal and internal olefins (Figure 4.12c). In a separate report, the group demonstrated a site-selective intermolecular C–H insertion reaction of EDA and phthalan derivatives catalyzed by Ir(Me)-P450-reconstituted proteins (Figure 4.12d) [25].

The Fasan group also described such reactivities with Ir(Me)-substituted myglobins [26]. In this case, the ArMs were accessed by co-expression of a promiscuous heme transporter which facilitated import of exogenously added Ir(Me)-porphyrin in the iron-depleted medium. The ArMs were shown to be competent catalysts for styrene cyclopropanation, aniline N–H insertion, thiophenol S–H insertion, and phthalan C–H insertion reactions. Subsequently, Mn, Co, and Rh-substituted myoglobins were also demonstrated to show activity for X–H insertion and cyclopropanation reactions (Figure 4.13) [27]. While the Mn and Rh-ArMs gave similar reactivity trends as the Fe variants, an intriguing inversion in chemoselectivity was observed with the Co-containing enzymes. For example, only the Co variants showed a preference for cyclopropanation over N–H or Si–H insertion (Figure 4.13a,b). Comparable selectivities were seen for competition between N–H and Si–H insertion (Figure 4.13c), whereas S–H insertion was favored over N–H insertion with the Co variants (Figure 4.13d). The Co-ArMs were also active as cyclopropanation catalysts, although with low stereoselectivities (Figure 4.13e). Such a change in selectivity with just a change in the cofactor metal demonstrates that cofactor replacement can complement conventional reactivity of heme enzymes in surprising ways.

Modification of the porphyrin structure can have a significant effect on metal reactivity, and this motivated efforts to incorporate unnatural porphyrin derivatives into heme proteins (Figure 4.14a). Brustad and coworkers evolved $P450_{BM3}$ variants to selectively bind iron-deuteroporphyrin (Fe-DPIX), a heme derivative lacking the corresponding vinyl substituents [103]. The resulting ArMs were shown to catalyze styrene cyclopropanation, although with low activities. As an extension of this work, a similar incorporation of Ir(Me)-DPIX in the $P450_{BM3}$ scaffold was shown [104]. Fasan and coworkers studied the cyclopropanation activity of a myoglobin mutant with heme substituted by an iron-chlorin e6 cofactor, which carries an additional carboxylate group on the porphyrin. The substitution was found to markedly improve oxygen tolerance: aerobic reactions conducted on a variety of styrenyl olefins (Figure 4.14b) gave high yields and selectivities (the analogous iron-porphyrin variant gave low yields under these conditions) [105]. In another report, the authors accessed iron-2,4-diacetyl deuteropophyrin (Fe-DADP)-substituted myoglobin with a nonnatural, methylhistidine axial ligand [106]. The resulting carbene was hypothesized to be more electrophilic and was predicted to give better reactivity with electron-deficient olefins, a challenging substrate class for cyclopropanation reactions. This ArM indeed showed high activity with electron-deficient terminal olefins; mechanistic investigations suggested that the cofactor substitution results in a stepwise radical pathway, facilitating an expanded substrate scope (Figure 4.14c). Hayashi and coworkers demonstrated that

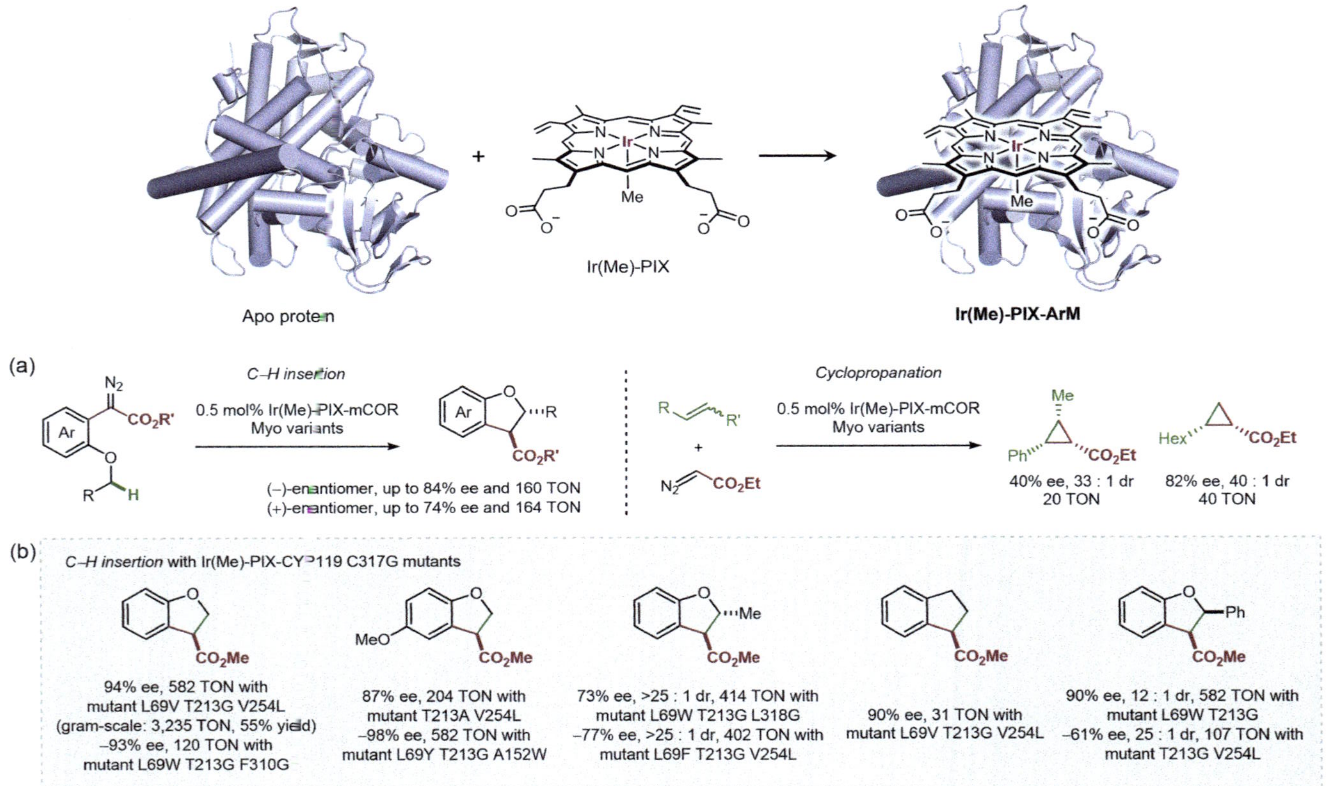

Figure 4.12 Representative examples of carbene transfer reactions catalyzed by Ir(Me)-reconstituted ArMs. (a) Intramolecular C–H insertion and cyclopropanation mediated by an Ir(Me)-myoglobin variant. (b) C–H insertion catalyzed by Ir(Me)-CYP119 variants. (c) Cyclopropanation of terminal and internal olefins. (d) Site-selective C–H insertion into substituted phthalan derivatives. Source: (a) Based on Key et al. [22]. (b) Based on Dydio et al. [23]. (c) Based on Key et al. [24]. (d) Based on Gu et al. [25].

(c)

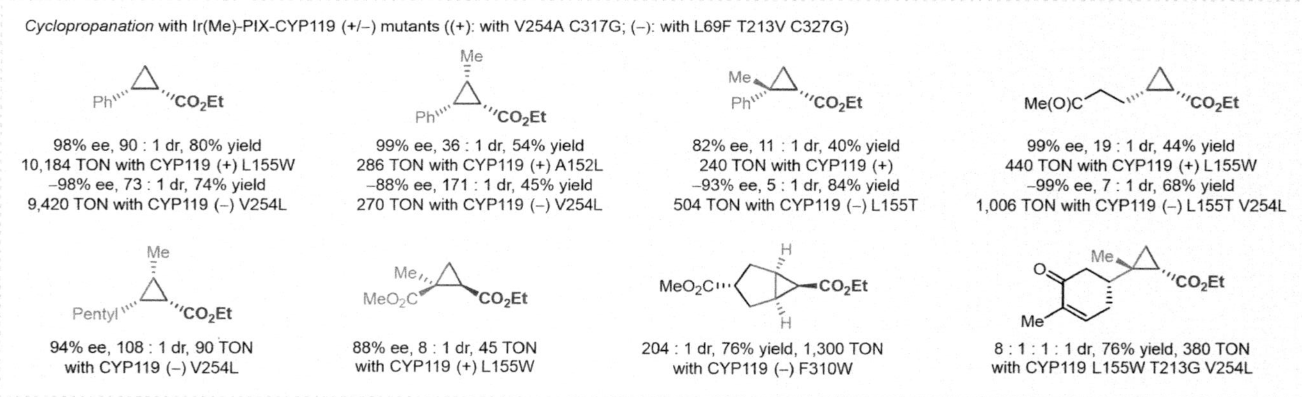

(d)

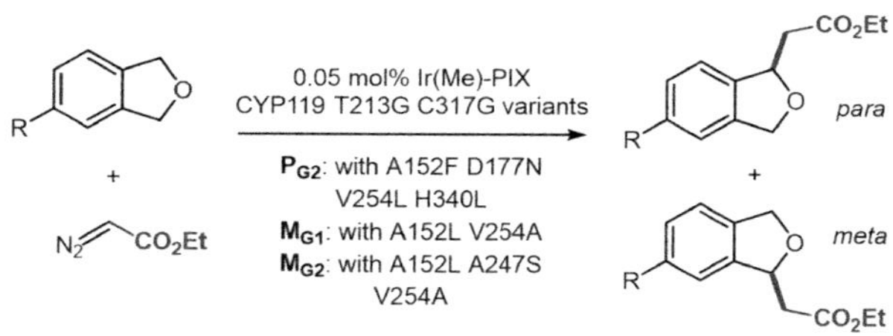

R =	Cl	Br	NO_2	CN
$\mathbf{P_{G2}}$	*p*/*m* 3.5 : 1 2,286 TON 72% yield	*p*/*m* 3.2 : 1 1,160 TON 76% yield	*p*/*m* 5.6 : 1 363 TON	*p*/*m* 9.9 : 1 181 TON
$\mathbf{M_{G2}}$	*p*/*m* 1 : 3.1 1,380 TON	*p*/*m* 1 : 3.4 1,456 TON	*p*/*m* 1 : 2.1 110 TON	*p*/*m* 1 : 1.5 531 TON

R = H: 80% ee, 1,520 TON with $\mathbf{M_{G1}}$ P252S

R = F: *p*/*m* 1 : 1.8, 84% ee (*p*), 94% ee (*m*), 412 TON with $\mathbf{M_{G1}}$ P252S   R = NHAc: *p*/*m* 17.8 : 1, 470 TON with $\mathbf{P_{G2}}$ R = OMe: *p*/*m* 6.4 : 1, 1,760 TON with $\mathbf{P_{G2}}$

Figure 4.12 (*Continued*)

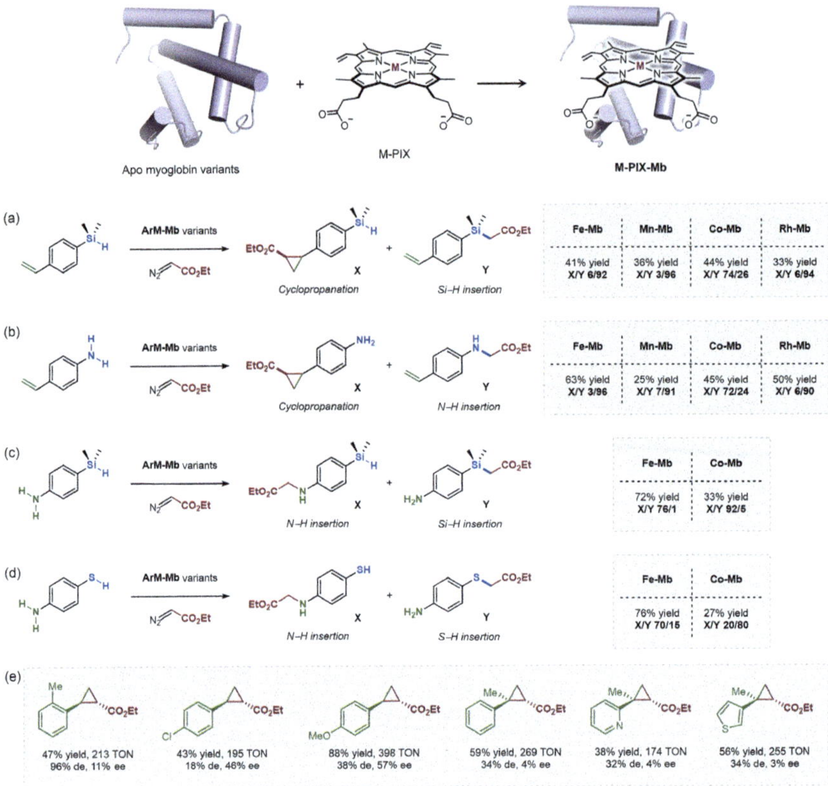

Figure 4.13 Carbene chemistry with metal-substituted myoglobin ArMs. (a–d) Atypical selectivity for X–H insertion reactions observed with Co–Mb variants. (e) Representative examples of cyclopropanation catalyzed by Co–Mb variants. Source: Based on Moore et al. [27].

Fe-porphycene-substituted myoglobin provided a dramatic increase (up to 26-fold, $k_{cat}/K_M = 1.3\ s^{-1}\ mM^{-1}$) in styrene cyclopropanation activity compared to the native Fe-porphyrin variant ($k_{cat}/K_M = 0.05\ s^{-1}\ mM^{-1}$) [107] Kinetic studies attributed the increase to facile carbene formation assisted by the strong field porphycene ligand.

The incorporation of a metalloporphyrin in a non-heme protein is an alternative strategy to access ArMs with potential carbene transfer activity. Roelfes previously utilized the homodimeric LmrR (lactococcal multidrug resistance regulator) protein as a scaffold for enantioselective catalysis by combining it with a Cu^{II} – phenanthroline complex [108]. The promiscuous, hydrophobic binding pocket at the dimer interface of LmrR was also shown to engage the heme cofactor with high affinity (K_D = 38 nM) [109]. The resulting ArM displayed activity for the cyclopropanation of styrene derivatives and improvements in performance were possible by directed evolution (Figure 4.15a). Recently, the Anderson group incorporated heme in a de novo designed helix bundle (termed C45) [111, 112] and used the resulting ArM for carbene transfer reactions (Figure 4.15b) [110]. The C45 scaffold is a small, ~15 kD, thermostable protein that is readily expressed in *E. coli*

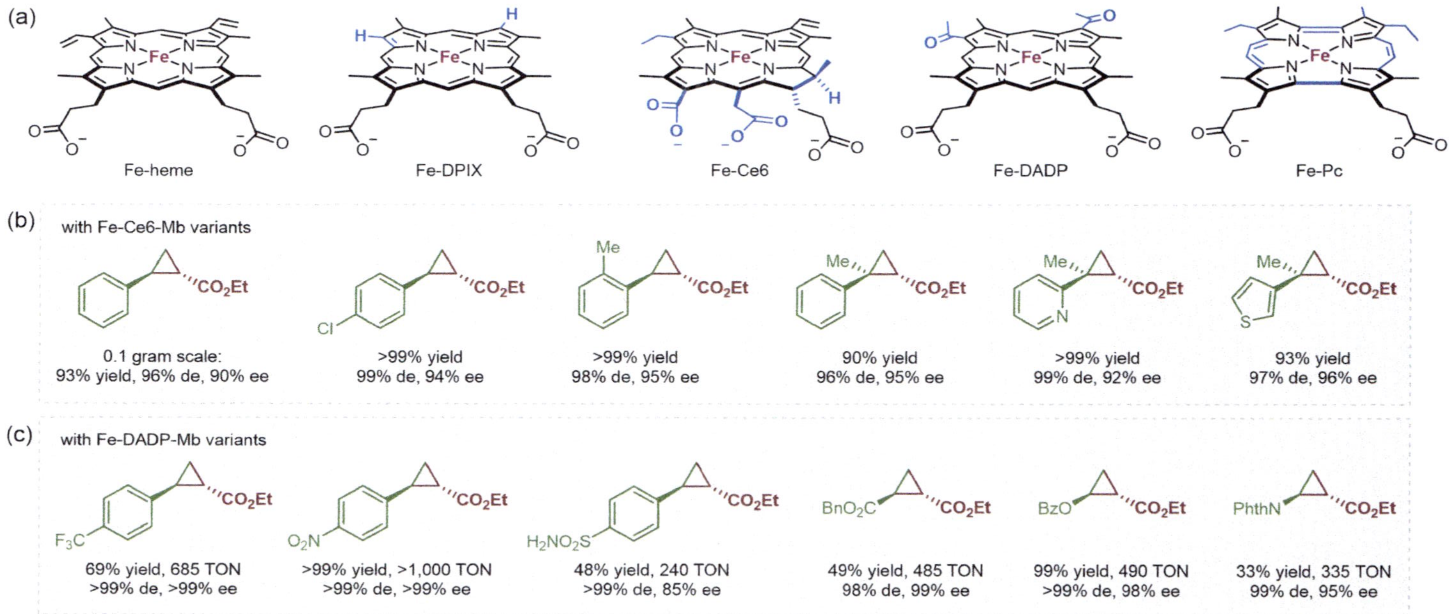

Figure 4.14 Cyclopropanation catalyzed by myoglobins with modified cofactors. (a) Examples of nonnatural porphyrin derivatives utilized in cofactor replacement. Representative examples of cyclopropanation catalyzed by (b). Fe-Ce6- or (c). Fe-PDAP-substituted myoglobin variants.

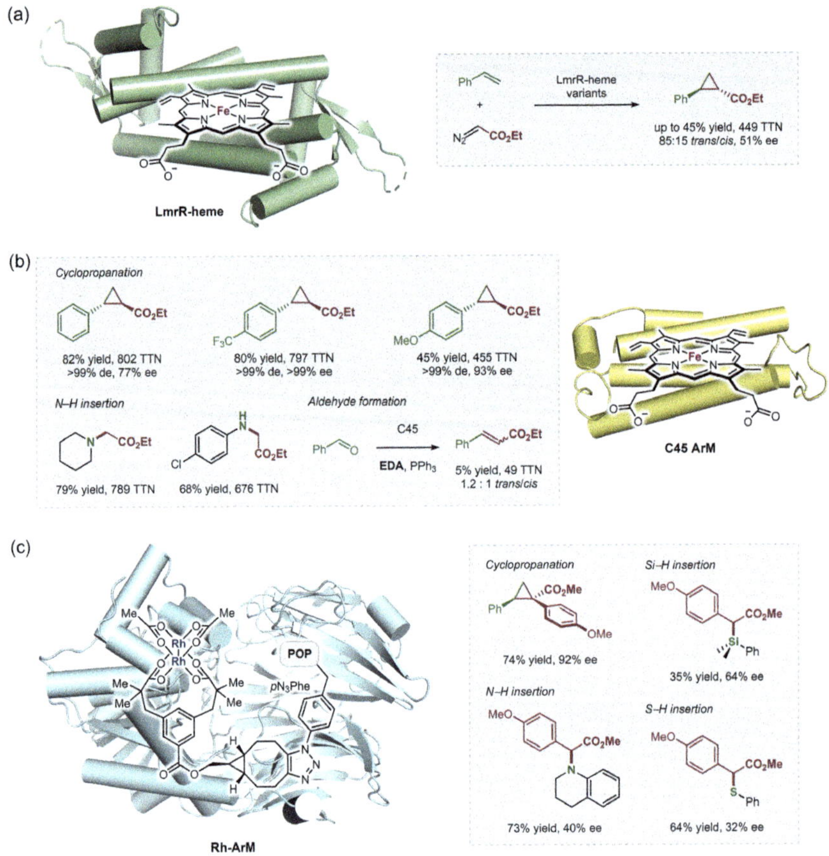

Figure 4.15 Carbene transfer by ArMs resulting from cofactor modification of non-heme protein scaffolds. (a) Dimeric LmrR bound by heme catalyzes cyclopropanation (pdb: 6fuu). Source: Based on Villarino et al. [109]. (b) Carbene transfer reactivity of the *de novo* designed C45 ArM (modeled structure). Source: Based on Stenner et al. [110]. (c) Carbene transfer activity of POP variants covalently modified by a rhodium cofactor (pdb: 5t88). Source: Based on Yang et al. [115].

and can incorporate heme through an axial histidine ligation. This protein was active in the cyclopropanation of styrene derivatives, N–H insertion and a Wittig reaction, and performance could be improved by engineering the protein. The small size and high stability of the scaffold makes this system potentially attractive for further development.

Rhodium-containing organometallic complexes are privileged catalysts for carbene transfer reactions [67], and dirhodium carboxylates have been used to create ArMs by covalently attaching them to a scaffold protein [113]. In 2015, Lewis and coworkers described a prolyl oligopeptidase (POP)-based ArM for asymmetric cyclopropanation [114]. The thermostable POP protein was modified to include the unnatural amino acid *p*-azidophenylalanine (AzF) in its active site for covalent anchoring of dirhodium tetracarboxylate bearing a terminal alkyne. The thermostability of the scaffold made the ArM tolerant to extensive mutagenesis, and variants catalyzing the cyclopropanation of electron-rich and electron-neutral styrenes with modest yields and good enantioselectivity were accessed. A subsequent report [115] described an optimized workflow for the directed evolution of this ArM through random mutagenesis. In addition to the cyclopropanation activity, the POP-Rh ArM was shown to enantioselectively catalyze Si–H, N–H and S–H insertion reactions (Figure 4.15c). Ward and coworkers utilized biotinylated dirhodium tetracarboxylate to noncovalently modify streptavidin (Sav) to make ArMs that cyclopropanate styrene, although with low turnovers and without any stereoselectivity [116].

Advances in ArMs complement and extend the capabilities of carbene transfer seen with natural heme proteins. However, it is unclear if ArMs have any significant advantage in catalyzing reactions that can also be accessed by the directed evolution of natural proteins. Rare earth metals and designed porphyrin derivatives add to cost and expression challenges that can limit synthetic utility.

4.8 Structural Studies of Carbene Intermediates in Heme Proteins

Studies with synthetic metal-porphyrins provide hints into the nature of the iron porphyrin carbene (IPC) intermediates in heme proteins [93, 117–119]. However, extrapolating conclusions from model systems to proteins demands caution, because the polypeptide scaffold and active site residues which must play a crucial role in stabilizing and orienting the IPC for carbene transfer reactions are absent in porphyrin model systems. Arnold and coworkers reported in 2018, the crystal structure of a catalytic IPC intermediate in the active site of a *Rma* cytochrome *c* variant (*Rma* TDE) (Figure 4.16) evolved for carbene insertion into Si–H bonds (Figure 4.11) [100]. A crystal structure of *Rma* TDE at 1.47 Å reveals a pocket on the distal face of the heme. This pocket, absent in the wild-type protein, results from mutation of the axial methionine ligand to aspartate (M100D) and provides a site for carbene formation. The ensuing cavity allows a water molecule (or hydroxide ion) to coordinate the heme iron in the enzyme resting state. Crystals of *Rma* TDE, when soaked with Me-EDA at room temperature under air, reduced with sodium dithionite and flash-frozen in liquid nitrogen, diffracted to 1.29 Å to reveal the active site IPC intermediate (Figure 4.16a). The iron-carbene has nonpolar contacts with five active site side chains (T75, M76, P79, I83, and M89), which appear to restrict the IPC to a single conformation in the predominantly solvent accessible but hydrophobic active site. This crystal structure conformation is recapitulated with MDs simulations, supporting its identity as the solution state catalytic intermediate. The MD studies also indicate the presence of a stabilizing water molecule (Figure 4.16b) that bridges a heme carboxylate with the carbonyl group on the IPC through a hydrogen bond. The active site front loop is flexible and allows substrate access with a pro-*R* facial selectivity, ensuring a highly enantioselective Si–H carbene insertion (Figure 4.16c).

The detailed electronic structure of IPCs is a complex and contested subject [93, 117, 119, 120]. Briefly (and simplistically), classical carbenes can exist in the singlet

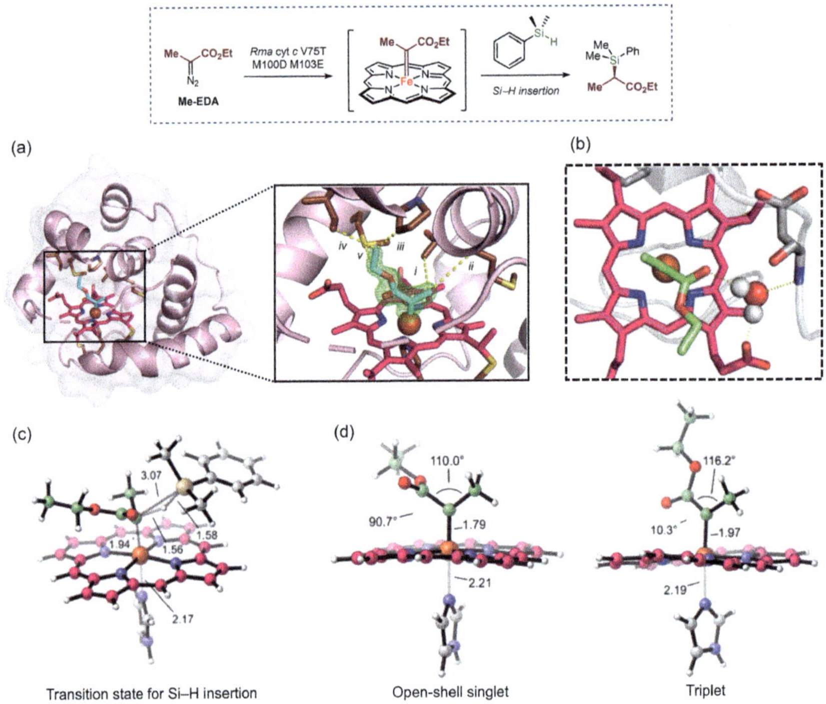

Figure 4.16 Structural studies of IPC intermediate in an enzyme active site [100]. (a) Crystal structure of the iron-carbenoid in the active site of *Rma* TDE. The inset shows side chain contacts (*i*–*v*) with the carbene in the active. (b) MD simulations reveal a bridging water molecule between the carbene carbonyl and the heme. (c) DFT calculated transition state structure of Si–H insertion from the *Re* face. (d) Geometries of the singlet and triplet carbene states as ascertained by DFT/MM calculations. Active site constraints disfavor the triplet state conformation. Source: Lewis et al. [100].

(spin = 0) or triplet (spin = 1) state. The former can be further divided into a closed shell singlet (CSS) with paired electrons in the same orbital and open shell singlet (OSS) with opposite spin electrons in two separate orbitals. Likewise, the triplet state consists of unpaired electrons with parallel spins in separate orbitals. The electronic nature of carbenes is believed to direct their manner of reactivity – OSS and triplet carbenes have a diradical character and react in a stepwise mechanism, whereas the CSS pathway is expected to be concerted (synchronous or asynchronous) and resembles a two-electron mechanism. Spectroscopic and computational studies on IPC model systems suggest that the singlet state is energetically favored, and both the CSS and OSS configurations have been proposed as possible resting states [93, 119]. Despite useful benchmarks, these remain only an approximation of the active site microenvironment. Depending on the active site residues as well as the reaction under consideration, it is possible that either of these electronic states may be stabilized under specific circumstances. In case of *Rma* TDE , UV–vis, electron paramagnetic resonance (EPR), and Mössbauer spectral analysis of the Me-EDA-treated samples were consistent with a singlet state [100]. DFT/MM calculations suggested that the triplet state was disfavored by about 11.7 kcal mol^{-1} relative to the singlet. The significant destabilization of the triplet state was proposed to originate from the conformational restraints imposed in the active site and the presence of the bridging water molecule (Figure 4.16d). In the active site, the carbene ester adopts a conformation parallel to the heme, to minimize steric clashes with the P79 and I83 side chains and is unable to stabilize an unpaired electron in the carbene *p* orbital, thus disfavoring the triplet state. Although a slight preference for the CSS (over the OSS) was seen in the DFT calculations, these states could not be unambiguously distinguished. The study provides a fascinating demonstration of how directed evolution sculpts the *Rma* TDE active site to preferentially stabilize a conformation of the singlet carbene through stereoelectronic restraints and provide complete stereoselectivity during the carbene transfer.

Hilvert and coworkers investigated the IPC intermediate in a myoglobin variant engineered for styrene cyclopropanation with EDA [121]. This ArM (Mb-NMH) contains a nonnatural amino acid side chain – N_δ-methylhistidine – as the axial ligand and could efficiently cyclopropanate styrene with high efficiency (70% yield, > 99% ee) even under aerobic conditions – suggesting that an Fe(III) species was equally competent to effect the carbene transfer. Stopped flow measurements of the reaction between Mb-NMH and EDA revealed an apparent k_{obs} of 5.8 M^{-1} s^{-1} for forming the enzyme-bound IPC, which was stable over hours, could be characterized spectroscopically and which rapidly yielded the cyclopropanated product upon styrene addition. Continuous-wave EPR measurements established the metal to be in an Fe(III) high-spin state which converted to an Fe(III) low-spin configuration upon IPC formation. A crystal structure of the enzyme-bound intermediate revealed an unexpected Fe-C-N(pyrrole) bridging configuration rather than the conventional end-to-end iron-carbene (Figure 4.17a). Such structures had been previously described in synthetic Fe(III) porphyrin complexes and are believed to result in enzyme inactivation. Intriguingly, the bridged Mb-HMN carbenoid did not lead to pyrrole alkylation and was proposed to be the preferred structure in equilibrium with the higher energy, end-to-end carbenoid. DFT studies corroborated this hypothesis – the bridged carbenoid was found to be 14.2 kcal mol^{-1} lower in energy than the end-to-end carbenoid, and while the former was inert toward styrene cyclopropanation (activation energy barrier of 30.7 kcal mol^{-1}), the latter's reaction with styrene was calculated to be barrierless (Figure 4.17b). Thus, the inert, bridged species is thermodynamically favored but is in equilibrium with the reactive, end-to-end carbenoid that is primed for cyclopropanation.

Computational studies and kinetics suggest considerable differences in mechanistic details, depending on the system. For C–H and Si–H insertion, KIEs indicate the concerted insertion step after formation of the singlet Fe(II) carbenoid to be rate-limiting [72, 122]. In cyclopropanation reactions, while carbene formation appears to be rate-determining in some cases [121] and the cycloaddition step maybe limiting in others [50, 119, 123], product release has been proposed to be the slowest step in the case of a porphycene-modified myoglobin [107]. For S–H insertion, the HAT step from the carbenoid appears rate-limiting (studied with DFT calculations,

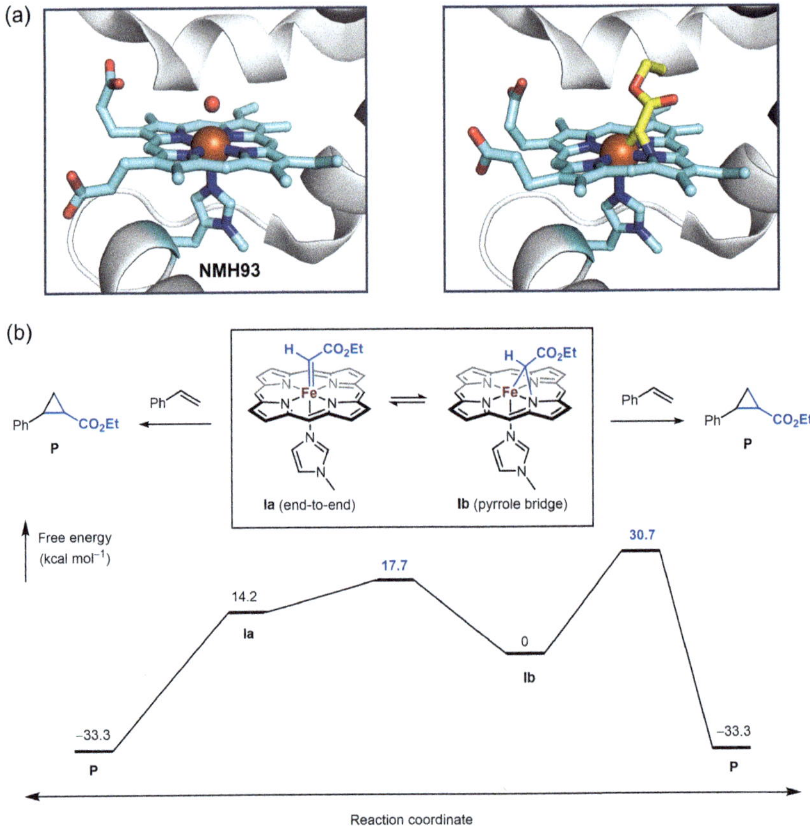

Figure 4.17 Pyrrole-bridged IPC intermediate in Mb-NMH. (a) *From left* – crystal structure of Mb-NMH active site (pdb: 6f17) and the active site with the bridged carbenoid (pdb: 6g5b). The end-to-end species was not observed. (b) DFT calculations indicate that the two species are in equilibrium. The end-to-end carbenoid (**Ia**) rapidly cyclopropanates styrene, whereas the pyrrole bridged (**Ib**) species is inert. Source: Based on Hayashi et al. [121].

Figure 4.10b), with the nature of the axial ligand affecting the stability of the resultant Fe–C species [35]. It seems that different enzyme variants will have their own unique mechanistic intricacies that may be revealed only through a careful combination of structural and computational studies, and unambiguously establishing one pathway over the other remains a challenge. Structural studies present an important foundation to understanding the nature of IPC intermediates in heme proteins, but more investigations are required before declaring generality.

4.9 Summary

Since the first report of olefin cyclopropanation with P450 enzymes in 2013, biocatalytic carbene transfer has expanded rapidly. Today, natural heme proteins and their variants have been evolved to catalyze a wide variety of carbene transfer reactions ranging from cycloadditions to insertions, often with superior selectivities compared to small-molecule catalysts. Artificial metalloenzymes have further contributed to diversifying this reaction portfolio while also providing a challenge for protein design. Initial structural studies have provided an exciting glimpse into how the enzyme active site manages to tame carbene intermediates and control reactivity. As this field has developed, parallel progress, not discussed in the chapter, has also been made in enzymatic nitrene chemistry [124].

Although carbene transferase enzymes have now been demonstrated to show enviable synthetic scope, challenges still remain in their adoption as routine catalysts. Simplified protocols need to be developed to encourage their use in synthetic chemistry settings. Some metalloproteins used in carbene transfer reactions still show only modest expression and stability. Further engineering to improve enzyme lifetime and scale-up potential remains an important challenge. Considerable work remains for these systems to move from the biochemist's bench to process scale settings.

The structural fine-tuning of the active site of an enzyme can be performed with extraordinary sensitivity through directed evolution, which distinguishes them from small-molecule catalysts in a fundamental way. Through steric contacts, hydrophobic effects, hydrogen bonding, electronic changes, and other subtle factors, the active site can stabilize intermediates, enforce enantioselectivity, and guide reaction pathways in directions that may be impossible with simple catalysts. More generally, these discoveries have ushered in the "new-to-nature" era of biocatalysis, with the promise of developing many reactions which enzymes have never catalyzed before.

Acknowledgments

K.C. thanks the Resnick Institute at Caltech for fellowship support. This work is supported by the NSF Division of Molecular and Cellular Biosciences (2016137), the NIH NIGMS (R01GM138740), and the US Army Institute of Collaborative Biotechnologies (W911NF-19-2-0026 and W911NF-19-D0001). The content is solely the responsibility of the authors and does not necessarily represent the official views of the funding agencies.

References

1 Breslow, R. (2009). Biomimetic chemistry: biology as an inspiration. *J. Biol. Chem.* 284 (3): 1337–1342.
2 Swiegers, G.F. (2012). *Bioinspiration and Biomimicry in Chemistry: Reverse-Engineering Nature*. Hoboken, NJ: Wiley.
3 He, L., Bekkaye, M., Retailleau, P., and Masson, G. (2012). Chiral phosphoric acid catalyzed inverse-electron-demand aza-Diels–Alder reaction of isoeugenol derivatives. *Org. Lett.* 14 (12): 3158–3161.

4 Uraguchi, D. and Terada, M. (2004). Chiral brønsted acid-catalyzed direct mannich reactions via electrophilic activation. *J. Am. Chem. Soc.* 126 (17): 5356–5357.

5 Dörr, C., Zaparty, M., Tjaden, B. et al. (2003). The hexokinase of the hyperthermophile *Thermoproteus tenax:* ATP-dependent hexokinases and ADP-dependent glucokinases, two alternatives for glucose phosphorylation in Archaea. *J. Biol. Chem.* 278 (21): 18744–18753.

6 Jensen, R.A. (1976). Enzyme recruitment in evolution of new function. *Annu. Rev. Microbiol.* 30 (1): 409–425.

7 Tawfik, D.S. (2010). Messy biology and the origins of evolutionary innovations. *Nat. Chem. Biol.* 6 (10): 692–696.

8 Khersonsky, O. and Tawfik, D.S. (2010). Enzyme promiscuity: a mechanistic and evolutionary perspective. *Annu. Rev. Biochem.* 79 (1): 471–505.

9 Maynard Smith, J. (1970). Natural selection and the concept of a protein space. *Nature* 225 (5232): 563–564.

10 Coelho, P.S., Brustad, E.M., Kannan, A., and Arnold, F.H. (2013). Olefin cyclopropanation via carbene transfer catalyzed by engineered cytochrome P450 enzymes. *Science* 339 (6117): 307–310.

11 Sykes, A.G. and Mauk, G. (2001). *Advances in Inorganic Chemistry*, vol. 51. San Diego: Academic Press.

12 Poulos, T.L. (2014). Heme enzyme structure and function. *Chem. Rev.* 114 (7): 3919–3962.

13 Kadish, K.M., Smith, K.M., and Guilard, R. (2003). *Bioinorganic and Bioorganic Chemistry*. Amsterdam: Academic Press.

14 Callot, H.J. and Piechocki, C. (1980). Cyclopropanation using rhodium (III) porphyrins: large cis vs trans selectivity. *Tetrahedron Lett.* 21 (36): 3489–3492.

15 Wolf, C.R., Mansuy, D., Nastainczyk, W. et al. (1977). The reduction of polyhalogenated methanes by liver microsomal cytochrome P450. *Mol. Pharmacol.* 13 (4): 698.

16 Mansuy, D., Lange, M., Chottard, J.C. et al. (1978). Dichlorocarbene complexes of iron(II)-porphyrins–crystal and molecular structure of Fe(TPP)(CCl_2)(H_2O). *Angew. Chem. Int. Ed. Engl* 17 (10): 781–782.

17 Montanari, F. and Casella, L. (eds.) (1994). *Metalloporphyrins Catalyzed Oxidations*, Catalysis by Metal Complexes, vol. 17. Dordrecht: Springer Netherlands.

18 Mansuy, D. (2007). A brief history of the contribution of metalloporphyrin models to cytochrome P450 chemistry and oxidation catalysis. *C.R. Chim.* 10 (4–5): 392–413.

19 Mansuy, D., Battioni, P., and Mahy, J.P. (1982). Isolation of an iron-nitrene complex from the dioxygen- and iron porphyrin-dependent oxidation of a hydrazine. *J. Am. Chem. Soc.* 104 (16): 4487–4489.

20 Simonneaux, G. and Le Maux, P. (2006). Carbene complexes of heme proteins and iron porphyrin models. In: *Bioorganometallic Chemistry*, Topics in Organometallic Chemistry, vol. 17 (ed. G. Simonneaux), 83–122. Berlin, Heidelberg: Springer Berlin Heidelberg.

21 Davis, H.J. and Ward, T.R. (2019). Artificial metalloenzymes: challenges and opportunities. *ACS Cent. Sci.* 5 (7): 1120–1136.

22 Key, H.M., Dydio, P., Clark, D.S., and Hartwig, J.F. (2016). Abiological catalysis by artificial haem proteins containing noble metals in place of iron. *Nature* 534 (7608): 534–537.

23 Dydio, P., Key, H.M., Nazarenko, A. et al. (2016). An artificial metalloenzyme with the kinetics of native enzymes. *Science* 354 (6308): 102–106.

24 Key, H.M., Dydio, P., Liu, Z. et al. (2017). Beyond iron: iridium-containing P450 enzymes for selective cyclopropanations of structurally diverse alkenes. *ACS Cent. Sci.* 3 (4): 302–308.

25 Gu, Y., Natoli, S.N., Liu, Z. et al. (2019). Site-selective functionalization of (sp^3)C-H bonds catalyzed by artificial metalloenzymes containing an iridium-porphyrin cofactor. *Angew. Chem. Int. Ed.* 58: 13954–13960.

26 Sreenilayam, G., Moore, E.J., Steck, V., and Fasan, R. (2017). Metal substitution modulates the reactivity and extends the reaction scope of myoglobin carbene transfer catalysts. *Adv. Synth. Catal.* 359 (12): 2076–2089.

27 Moore, E.J., Steck, V., Bajaj, P., and Fasan, R. (2018). Chemoselective cyclopropanation over carbene Y–H insertion catalyzed by an engineered carbene transferase. *J. Org. Chem.* 83 (14): 7480–7490.

28 Coelho, P.S., Wang, Z.J., Ener, M.E. et al. (2013). A serine-substituted P450 catalyzes highly efficient carbene transfer to olefins *in vivo*. *Nat. Chem. Biol.* 9 (8).

29 Ortiz de Montellano, P.R. (2010). Hydrocarbon hydroxylation by cytochrome P450 enzymes. *Chem. Rev.* 110 (2): 932–948.

30 Wang, Z.J., Renata, H., Peck, N.E. et al. (2014). Improved cyclopropanation activity of histidine-ligated cytochrome P450 enables the enantioselective formal synthesis of levomilnacipran. *Angew. Chem. Int. Ed.* 53 (26): 6810–6813.

31 Bershtein, S. and Tawfik, D.S. (2008). Advances in laboratory evolution of enzymes. *Curr. Opin. Chem. Biol.* 12 (2): 151–158.

32 Currin, A., Swainston, N., Day, P.J., and Kell, D.B. (2015). Synthetic biology for the directed evolution of protein biocatalysts: navigating sequence space intelligently. *Chem. Soc. Rev.* 44 (5): 1172–1239.

33 Sun, Z., Wikmark, Y., Bäckvall, J.-E., and Reetz, M.T. (2016). New concepts for increasing the efficiency in directed evolution of stereoselective enzymes. *Chem. Eur. J.* 22 (15): 5046–5054.

34 Brandenberg, O.F., Prier, C.K., Chen, K. et al. (2018). Stereoselective enzymatic synthesis of heteroatom-substituted cyclopropanes. *ACS Catal.* 8 (4): 2629–2634.

35 Chen, K., Zhang, S.-Q., Brandenberg, O.F. et al. (2018). Alternate heme ligation steers activity and selectivity in engineered cytochrome P450-catalyzed carbene-transfer reactions. *J. Am. Chem. Soc.* 140 (48): 16402–16407.

36 Knight, A.M., Kan, S.B.J., Lewis, R.D. et al. (2018). Diverse engineered heme proteins enable stereodivergent cyclopropanation of unactivated alkenes. *ACS Cent. Sci.* 4 (3): 372–377.

37 Brandenberg, O.F., Chen, K., and Arnold, F.H. (2019). Directed evolution of a cytochrome P450 carbene transferase for selective functionalization of cyclic compounds. *J. Am. Chem. Soc.* 141 (22): 8989–8995.

38 Bordeaux, M., Tyagi, V., and Fasan, R. (2015). Highly diastereoselective and enantioselective olefin cyclopropanation using engineered myoglobin-based catalysts. *Angew. Chem. Int. Ed.* 54 (6): 1744–1748.

39 Tinoco, A., Steck, V., Tyagi, V., and Fasan, R. (2017). Highly diastereo- and enantioselective synthesis of trifluoromethyl-substituted cyclopropanes via myoglobin-catalyzed transfer of trifluoromethylcarbene. *J. Am. Chem. Soc.* 139 (15): 5293–5296.

40 Chandgude, A.L. and Fasan, R. (2018). Highly diastereo- and enantioselective synthesis of nitrile-substituted cyclopropanes by myoglobin-mediated carbene transfer catalysis. *Angew. Chem. Int. Ed.* 57 (48): 15852–15856.

41 Bajaj, P., Sreenilayam, G., Tyagi, V., and Fasan, R. (2016). Gram-scale synthesis of chiral cyclopropane-containing drugs and drug precursors with engineered myoglobin catalysts featuring complementary stereoselectivity. *Angew. Chem. Int. Ed.* 55 (52): 16110–16114.

42 Vargas, D.A., Khade, R.L., Zhang, Y., and Fasan, R. (2019). Biocatalytic strategy for highly diastereo- and enantioselective synthesis of 2,3-dihydrobenzofuran-based tricyclic scaffolds. *Angew. Chem. Int. Ed.* 58 (30): 10148–10152.

43 Ren, X., Chandgude, A.L., and Fasan, R. (2020). Highly stereoselective synthesis of fused cyclopropane-γ-lactams via biocatalytic iron-catalyzed intramolecular cyclopropanation. *ACS Catal.* 10 (3): 2308–2313.

44 Wittmann, B.J., Knight, A.M., Hofstra, J.L. et al. (2020). Diversity-oriented enzymatic synthesis of cyclopropane building blocks. *ACS Catal.* 10 (13): 7112–7116.

45 Chandgude, A.L., Ren, X., and Fasan, R. (2019). Stereodivergent intramolecular cyclopropanation enabled by engineered carbene transferases. *J. Am. Chem. Soc.* 141 (23): 9145–9150.

46 Renata, H., Wang, Z.J., Kitto, R.Z., and Arnold, F.H. (2014). P450-catalyzed asymmetric cyclopropanation of electron-deficient olefins under aerobic conditions. *Catal. Sci. Technol.* 4 (10): 3640–3643.

47 Hernandez, K.E., Renata, H., Lewis, R.D. et al. (2016). Highly stereoselective biocatalytic synthesis of key cyclopropane intermediate to ticagrelor. *ACS Catal.* 6 (11): 7810–7813.

48 Gober, J.G., Rydeen, A.E., Gibson-O'Grady, E.J. et al. (2016). Mutating a highly conserved residue in diverse cytochrome P450s facilitates diastereoselective olefin cyclopropanation. *ChemBioChem* 17 (5): 394–397.

49 Suzuki, K., Shisaka, Y., Stanfield, J.K. et al. (2020). Enhanced *cis*- and enantioselective cyclopropanation of styrene catalysed by cytochrome P450BM3 using decoy molecules. *Chem. Commun.* 56 (75): 11026–11029.

50 Tinoco, A., Wei, Y., Bacik, J.-P. et al. (2019). Origin of high stereocontrol in olefin cyclopropanation catalyzed by an engineered carbene transferase. *ACS Catal.* 9 (2): 1514–1524.

51 Ren, X., Liu, N., Chandgude, A.L., and Fasan, R. (2020). An enzymatic platform for the highly enantioselective and stereodivergent construction of cyclopropyl-δ-lactones. *Angew. Chem. Int. Ed.* 59 (48): 21634–21639.

52 Walczak, M.A.A., Krainz, T., and Wipf, P. (2015). Ring-strain-enabled reaction discovery: new heterocycles from bicyclo[1.1.0]butanes. *Acc. Chem. Res.* 48 (4): 1149–1158.

53 Swager, T.M., Dougherty, D.A., and Grubbs, R.H. (1988). Strained rings as a source of unsaturation: polybenzvalene, a new soluble polyacetylene precursor. *J. Am. Chem. Soc.* 110 (9): 2973–2974.

54 Doyle, M.P., Protopopova, M., Muller, P. et al. (1994). Effective uses of dirhodium(II) tetrakis[methyl 2-oxopyrrolidine-5(*R* or *S*)-carboxylate] for highly enantioselective intermolecular cyclopropenation reactions. *J. Am. Chem. Soc.* 116 (19): 8492–8498.

55 Lou, Y., Horikawa, M., Kloster, R.A. et al. (2004). A new chiral Rh(II) catalyst for enantioselective [2 + 1]-cycloaddition. Mechanistic implications and applications. *J. Am. Chem. Soc.* 126 (29): 8916–8918.

56 Briones, J.F., Hansen, J., Hardcastle, K.I. et al. (2010). Highly enantioselective Rh_2(*S*-DOSP)$_4$-catalyzed cyclopropenation of alkynes with styryldiazoacetates. *J. Am. Chem. Soc.* 132 (48): 17211–17215.

57 Uehara, M., Suematsu, H., Yasutomi, Y., and Katsuki, T. (2011). Enantioenriched synthesis of cyclopropenes with a quaternary stereocenter, versatile building blocks. *J. Am. Chem. Soc.* 133 (2): 170–171.

58 Cui, X., Xu, X., Lu, H. et al. (2011). Enantioselective cyclopropenation of alkynes with acceptor/acceptor-substituted diazo reagents via co(II)-based metalloradical catalysis. *J. Am. Chem. Soc.* 133 (10): 3304–3307.

59 Panish, R., Chintala, S.R., Boruta, D.T. et al. (2013). Enantioselective synthesis of cyclobutanes via sequential Rh-catalyzed bicyclobutanation/Cu-catalyzed homoconjugate addition. *J. Am. Chem. Soc.* 135 (25): 9283–9286.

60 Qin, C. and Davies, H.M.L. (2013). Enantioselective synthesis of 2-arylbicyclo[1.1.0]butane carboxylates. *Org. Lett.* 15 (2): 310–313.

61 Prier, C.K., Zhang, R.K., Buller, A.R. et al. (2017). Enantioselective, intermolecular benzylic C–H amination catalysed by an engineered iron-haem enzyme. *Nat. Chem.* 9 (7): 629–634.

62 Chen, K., Huang, X., Kan, S.B.J. et al. (2018). Enzymatic construction of highly strained carbocycles. *Science* 360 (6384): 71–75.

63 Chen, K. and Arnold, F.H. (2020). Engineering cytochrome P450s for enantioselective cyclopropenation of internal alkynes. *J. Am. Chem. Soc.* 142 (15): 6891–6895.

64 Briones, J.F. and Davies, H.M.L. (2012). Gold(I)-catalyzed asymmetric cyclopropenation of internal alkynes. *J. Am. Chem. Soc.* 134 (29): 11916–11919.

65 Zhang, Z., Zheng, M., Xue, X. et al. (2019). Catalytic enantioselective cyclopropenation of internal alkynes: access to difluoromethylated three-membered carbocycles. *Angew. Chem. Int. Ed.* 58 (50): 18191–18196.

66 Doyle, M.P., Duffy, R., Ratnikov, M., and Zhou, L. (2010). Catalytic carbene insertion into C–H bonds. *Chem. Rev.* 110 (2): 704–724.

67 Davies, H.M.L. and Manning, J.R. (2008). Catalytic C–H functionalization by metal carbenoid and nitrenoid insertion. *Nature* 451 (7177): 417–424.

68 Davies, H.M.L. and Liao, K. (2019). Dirhodium tetracarboxylates as catalysts for selective intermolecular C–H functionalization. *Nat. Rev. Chem.* 3 (6): 347–360.

69 Santiago, J.V. and Machado, A.H.L. (2016). Enantioselective carbenoid insertion into C(sp^3)–H bonds. *Beilstein J. Org. Chem.* 12: 882–902.

70 Empel, C., Jana, S., and Koenigs, R.M. (2020). C–H functionalization via iron-catalyzed carbene-transfer reactions. *Molecules* 25 (4): 880.

71 Liu, Y., You, T., Wang, H.-X. et al. (2020). Iron- and cobalt-catalyzed C(sp^3)–H bond functionalization reactions and their application in organic synthesis. *Chem. Soc. Rev.* 49 (15): 5310–5358.

72 Zhang, R.K., Chen, K., Huang, X. et al. (2019). Enzymatic assembly of carbon–carbon bonds via iron-catalysed sp^3C–H functionalization. *Nature* 565 (7737): 67–72.

73 Griffin, J.R., Wendell, C.I., Garwin, J.A., and White, M.C. (2017). Catalytic C(sp^3)–H alkylation via an iron carbene intermediate. *J. Am. Chem. Soc.* 139 (39): 13624–13627.

74 Hernán-Gómez, A., Rodríguez, M., Parella, T., and Costas, M. (2019). Electrophilic iron catalyst paired with a lithium cation enables selective functionalization of non-activated aliphatic C–H bonds via metallocarbene intermediates. *Angew. Chem. Int. Ed.* 58 (39): 13904–13911.

75 Mbuvi, H.M. and Woo, L.K. (2008). Catalytic C–H insertions using iron(III) porphyrin complexes. *Organometallics* 27 (4): 637–645.

76 Cheng, Q.-Q., Yang, J.-M., Xu, H., and Zhu, S.-F. (2017). Iron-catalyzed carbenoid insertion into C(sp^3)–H bonds. *Synlett* 28 (11): 1327–1330.

77 Zhao, S.K., Knors, C., and Helquist, P. (1989). Intramolecular carbon–hydrogen insertion reactions of (cyclopentadienyl)dicarbonyl iron carbene complexes. *J. Am. Chem. Soc.* 111 (22): 8527–8528.

78 Ishii, S., Zhao, S., and Helquist, P. (2000). Stereochemical probes of Intramolecular C–H insertion reactions of iron-carbene complexes. *J. Am. Chem. Soc.* 122 (24): 5897–5898.

79 Zhao, S.K. and Helquist, P. (1990). Short synthesis of (±)-sterpurene. *J. Org. Chem.* 55 (23): 5820–5821.

80 Davies, H.M.L., Hansen, T., and Churchill, M.R. (2000). Catalytic asymmetric C–H activation of alkanes and tetrahydrofuran. *J. Am. Chem. Soc.* 122 (13): 3063–3070.

81 Zhang, J., Huang, X., Zhang, R.K., and Arnold, F.H. (2019). Enantiodivergent α-amino C–H fluoroalkylation catalyzed by engineered cytochrome P450s. *J. Am. Chem. Soc.* 141 (25): 9798–9802.

82 Zhou, A.Z., Chen, K., and Arnold, F.H. (2020). Enzymatic lactone-carbene C–H insertion to build contiguous chiral centers. *ACS Catal.* 10 (10): 5393–5398.

83 DeAngelis, A., Shurtleff, V.W., Dmitrenko, O., and Fox, J.M. (2011). Rhodium(II)-catalyzed enantioselective C–H functionalization of indoles. *J. Am. Chem. Soc.* 133 (6): 1650–1653.

84 Cai, Y., Zhu, S.-F., Wang, G.-P., and Zhou, Q.-L. (2011). Iron-catalyzed C–H fuctionalization of indoles. *Adv. Synth. Catal.* 353 (16): 2939–2944.

85 Gao, X., Wu, B., Yan, Z., and Zhou, Y.-G. (2016). Copper-catalyzed enantioselective C–H functionalization of indoles with an axially chiral bipyridine ligand. *Org. Biomol. Chem.* 14 (35): 8237–8240.

86 Vargas, D.A., Tinoco, A., Tyagi, V., and Fasan, R. (2018). Myoglobin-catalyzed C–H functionalization of unprotected indoles. *Angew. Chem. Int. Ed.* 57 (31): 9911–9915.

87 Hock, K.J., Knorrscheidt, A., Hommelsheim, R. et al. (2019). Tryptamine synthesis by iron porphyrin catalyzed C–H functionalization of indoles with diazoacetonitrile. *Angew. Chem. Int. Ed.* 58 (11): 3630–3634.

88 Wang, Z.J., Peck, N.E., Renata, H., and Arnold, F.H. (2014). Cytochrome P450-catalyzed insertion of carbenoids into N–H bonds. *Chem. Sci.* 5 (2): 598–601.

89 Tyagi, V., Bonn, R.B., and Fasan, R. (2015). Intermolecular carbene S–H insertion catalysed by engineered myoglobin-based catalysts. *Chem. Sci.* 6 (4): 2488–2494.

90 Sreenilayam, G. and Fasan, R. (2015). Myoglobin-catalyzed intermolecular carbene N–H insertion with arylamine substrates. *Chem. Commun.* 51 (8): 1532–1534.

91 Steck, V., Sreenilayam, G., and Fasan, R. (2020). Selective functionalization of aliphatic amines via myoglobin-catalyzed carbene N–H insertion. *Synlett* 31 (03): 224–229.

92 Steck, V., Carminati, D.M., Johnson, N.R., and Fasan, R. (2020). Enantioselective synthesis of chiral amines via biocatalytic carbene N–H insertion. *ACS Catal.* 10 (19): 10967–10977.

93 Sharon, D.A., Mallick, D., Wang, B., and Shaik, S. (2016). Computation sheds insight into iron porphyrin carbenes' electronic structure, formation, and N–H insertion reactivity. *J. Am. Chem. Soc.* 138 (30): 9597–9610.

94 Ren, Y.-Y., Zhu, S.-F., and Zhou, Q.-L. (2018). Chiral proton-transfer shuttle catalysts for carbene insertion reactions. *Org. Biomol. Chem.* 16 (17): 3087–3094.

95 Kan, S.B.J., Lewis, R.D., Chen, K., and Arnold, F.H. (2016). Directed evolution of cytochrome *c* for carbon–silicon bond formation: bringing silicon to life. *Science* 354 (6315): 1048–1051.

96 Kan, S.B.J., Huang, X., Gumulya, Y. et al. (2017). Genetically programmed chiral organoborane synthesis. *Nature* 552 (7683): 132–136.

97 Lewis, R.D. (2019). Evolution and characterization of carbene transferases for cyclopropanation and carbon–silicon bond formation. PhD thesis. Chapter IV: origins of chemoselectivity in cytochrome *c*-catalyzed carbene insertions into silicon–hydrogen bonds.

98 Chen, K., Huang, X., Zhang, S.-Q. et al. (2019). Engineered cytochrome *c*-catalyzed lactone-carbene B–H insertion. *Synlett* 30 (4): 378–382.

99 Huang, X., Garcia-Borràs, M., Miao, K. et al. (2019). Biocatalytic platform for synthesis of chiral α-trifluoromethylated organoborons. *ACS Cent. Sci.* 5 (2): 270–276.

100 Lewis, R.D., Garcia-Borràs, M., Chalkley, M.J. et al. (2018). Catalytic iron-carbene intermediate revealed in a cytochrome *c* carbene transferase. *Proc. Natl. Acad. Sci. USA* 115 (28): 7308–7313.

101 O'Hagan, D. and Deng, H. (2015). Enzymatic fluorination and biotechnological developments of the fluorinase. *Chem. Rev.* 115 (2): 634–649.

102 Walker, M.C. and Chang, M.C.Y. (2014). Natural and engineered biosynthesis of fluorinated natural products. *Chem. Soc. Rev.* 43 (18): 6527–6536.

103 Reynolds, E.W., McHenry, M.W., Cannac, F. et al. (2016). An evolved orthogonal enzyme/cofactor pair. *J. Am. Chem. Soc.* 138 (38): 12451–12458.

104 Reynolds, E.W., Schwochert, T.D., McHenry, M.W. et al. (2017). Orthogonal expression of an artificial metalloenzyme for abiotic catalysis. *ChemBioChem* 18 (24): 2380–2384.

105 Sreenilayam, G., Moore, E.J., Steck, V., and Fasan, R. (2017). Stereoselective olefin cyclopropanation under aerobic conditions with an artificial enzyme incorporating an iron-chlorin e6 cofactor. *ACS Catal.* 7 (11): 7629–7633.

106 Carminati, D.M. and Fasan, R. (2019). Stereoselective cyclopropanation of electron-deficient olefins with a cofactor redesigned carbene transferase featuring radical reactivity. *ACS Catal.* 9 (10): 9683–9697.

107 Oohora, K., Meichin, H., Zhao, L. et al. (2017). Catalytic cyclopropanation by myoglobin reconstituted with iron porphycene: acceleration of catalysis due to rapid formation of the carbene species. *J. Am. Chem. Soc.* 139 (48): 17265–17268.

108 Roelfes, G. (2019). LmrR: a privileged scaffold for artificial metalloenzymes. *Acc. Chem. Res.* 52 (3): 545–556.

109 Villarino, L., Splan, K.E., Reddem, E. et al. (2018). An artificial heme enzyme for cyclopropanation reactions. *Angew. Chem. Int. Ed.* 57 (26): 7785–7789.

110 Stenner, R., Steventon, J.W., Seddon, A., and Anderson, J.L.R.A. (2020). *De novo* peroxidase is also a promiscuous yet stereoselective carbene transferase. *Proc. Natl. Acad. Sci. USA* 117 (3): 1419–1428.

111 Farid, T.A., Kodali, G., Solomon, L.A. et al. (2013). Elementary tetrahelical protein design for diverse oxidoreductase functions. *Nat. Chem. Biol.* 9 (12): 826–833.

112 Watkins, D.W., Jenkins, J.M.X., Grayson, K.J. et al. (2017). Construction and *in vivo* assembly of a catalytically proficient and hyperthermostable *de novo* enzyme. *Nat. Commun.* 8 (1): 358.

113 Lewis, J.C. (2019). Beyond the second coordination sphere: engineering dirhodium artificial metalloenzymes to enable protein control of transition metal catalysis. *Acc. Chem. Res.* 52 (3): 576–584.

114 Srivastava, P., Yang, H., Ellis-Guardiola, K., and Lewis, J.C. (2015). Engineering a dirhodium artificial metalloenzyme for selective olefin cyclopropanation. *Nat. Commun.* 6 (1): 7789.

115 Yang, H., Swartz, A.M., Park, H.J. et al. (2018). Evolving artificial metalloenzymes via random mutagenesis. *Nat. Chem.* 10 (3): 318–324.

116 Zhao, J., Bachmann, D.G., Lenz, M. et al. (2018). An artificial metalloenzyme for carbene transfer based on a biotinylated dirhodium anchored within streptavidin. *Catal. Sci. Technol.* 8 (9): 2294–2298.

117 Khade, R.L., Fan, W., Ling, Y. et al. (2014). Iron porphyrin carbenes as catalytic intermediates: structures, Mössbauer and NMR spectroscopic properties, and bonding. *Angew. Chem. Int. Ed.* 53 (29): 7574–7578.

118 Khade, R.L. and Zhang, Y. (2015). Catalytic and biocatalytic iron porphyrin carbene formation: effects of binding mode, carbene substituent, porphyrin substituent, and protein axial ligand. *J. Am. Chem. Soc.* 137 (24): 7560–7563.

119 Wei, Y., Tinoco, A., Steck, V. et al. (2018). Cyclopropanations via heme carbenes: basic mechanism and effects of carbene substituent, protein axial ligand, and porphyrin substitution. *J. Am. Chem. Soc.* 140 (5): 1649–1662.

120 Chen, B., Rogachev, A.Y., Hrovat, D.A. et al. (2013). How to make the $\sigma^0\pi^2$ singlet the ground state of carbenes. *J. Am. Chem. Soc.* 135 (37): 13954–13964.

121 Hayashi, T., Tinzl, M., Mori, T. et al. (2018). Capture and characterization of a reactive haem–carbenoid complex in an artificial metalloenzyme. *Nat. Catal.* 1 (8): 578–584.

122 Khade, R.L., Chandgude, A.L., Fasan, R., and Zhang, Y. (2019). Mechanistic investigation of biocatalytic heme carbenoid Si–H insertions. *ChemCatChem* 11 (13): 3101–3108.

123 Su, H., Ma, G., and Liu, Y. (2018). Theoretical insights into the mechanism and stereoselectivity of olefin cyclopropanation catalyzed by two engineered cytochrome P450 enzymes. *Inorg. Chem.* 57 (18): 11738–11745.

124 Yang, Y. and Arnold, F.H. (2021). Navigating the unnatural reaction space: directed evolution of heme proteins for selective carbene and nitrene transfer. *Acc. Chem. Res.* 54 (5): 1209–1225.

5

Metal Carbene Cycloaddition Reactions

Kostiantyn O. Marichev[1], *Haifeng Zheng*[2], *and Michael P. Doyle*[2]

[1] *Georgia State University, Department of Chemistry, Atlanta, GA, USA*
[2] *The University of Texas at San Antonio, Department of Chemistry, San Antonio, TX, USA*

5.1 Introduction

Metallocarbenes are classified as electrophilic, free radical, or nucleophilic, and their transformations reflect their metal-centered origin [1]. Electrophilic metallocarbenes, those with electron deficiency at the carbene carbon, have shown remarkable versatility in addition, insertion, and ylide-forming reactions [2]. Control of reactivity and selectivity in their transformations results from modifications in the metal, its ligands, and in the substituents on the carbene carbon that control electron density, as well as in the steric environment through which carbene transfer occurs. Metalloradicals are a more recent discovery whose transformations mirror to some degree those of electrophilic metallocarbenes with broader electronic versatility in addition reactions, but their full potential is still being developed (see Chapter 2). Nucleophilic metal carbenes have increased electron density on the carbenic carbon [3], and their nucleophilic reactivity is evident in metathesis and carbonyl olefination. In this chapter, focus is placed exclusively on electrophilic metal carbenes.

Conjugation increases the stability of metal carbenes and makes possible nucleophilic attack at the vinylogous position, which forms a vinylmetal intermediate that is susceptible to electrophilic attack and the subsequent release of the metal (Scheme 5.1). The overall transformation is *cycloaddition* defined as "a reaction in which two or more unsaturated molecules (or parts of the same molecule) combine with the formation of a cyclic adduct in which there is a net reduction of the bond multiplicity" [4]. The process is stepwise, rather than concerted, but the associated reaction intermediates have not been directly identified experimentally. The ligated metal is associated with the developing cycloaddition product throughout the process. Accordingly, the metal, its ligands, and carbene substituents, as well as the nucleophile, can control whether reaction with the nucleophile takes place at the vinylogous position or the carbene center. These factors also influence the rate for these reactions. In this mechanistic scheme, the initial vinylogous addition

Transition Metal-Catalyzed Carbene Transformations, First Edition.
Edited by Jianbo Wang, Chi-Ming Che, and Michael P. Doyle.

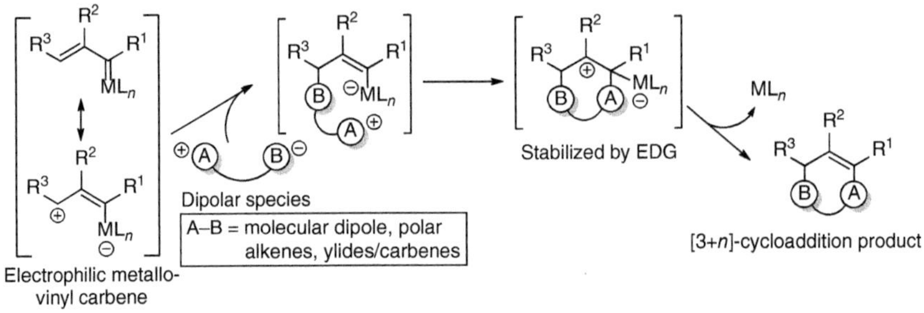

Scheme 5.1 [3+*n*]-Cycloaddition with electrophilic metallo-vinylcarbenes.

can impart stereocontrol, but the close proximity of the ligated metal (ML_n) in the ring-closing step should have a stronger role in catalyst control of stereoselectivity.

Cyclopropenes have long been known to be sources of metallo-vinylcarbenes [5], and their involvements in diverse transformations involving metal carbene intermediates continue to be widely reported [6], especially for cyclopropanation and C—H insertion reactions [7]. Yet, vinyldiazo compounds as metallo-vinylcarbene precursors have elicited greater attention in cycloaddition reactions [8]. Although styryldiazoacetates were the first vinyldiazo compounds to be developed [9], enoldiazo compounds have demonstrated the greatest utility for cycloaddition [10], especially for transformations with high levels of enantiocontrol [11]. However, recently established connectivity between donor–acceptor cyclopropenes (DACPs), formed from enoldiazo compounds by dinitrogen extrusion, and metallo-enolcarbenes (Scheme 5.2) [12] identifies these activated cyclopropenes as effective surrogates for vinyldiazo compounds in cycloaddition reactions. Late transition metal-catalyzed ene-yne-ketone [13, 14] and related cycloisomerization reactions [15–17] are often regarded as furyl- and vinylcarbene alternatives to those formed from furyl- and vinyldiazo compounds, but although these reactions generate metallo-carbenes, their use for intermolecular cycloaddition transformations has yet to be described.

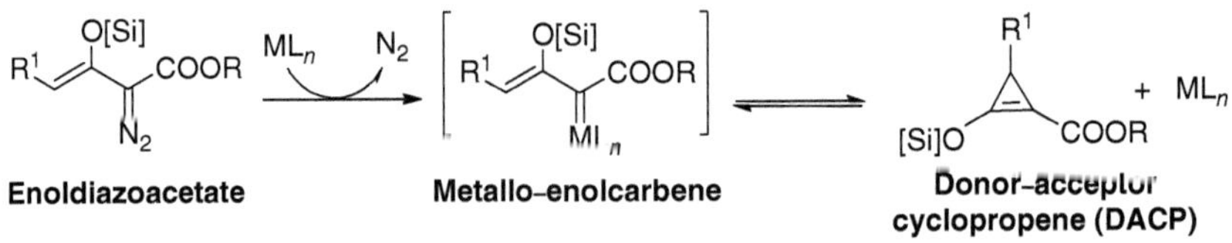

Scheme 5.2 Metallo-vinylcarbene/cyclopropene/vinyldiazo compound connectivity. Source: Based on Dong et al. [12].

Some cycloaddition processes are known to involve metal carbene intermediates but do not follow the mechanistic profile that is outlined in Scheme 5.1. These include the elegantly developed tandem cyclopropanation/Cope rearrangements (formal [4+3]-cycloadditions) of styryl diazoacetates (Scheme 5.3A) [18] and ylide formation/cycloaddition (e.g. Scheme 5.3B) [19] in which the cycloaddition step occurs after the metal carbene reaction and generally without influence by the catalyst. Although such processes can occur in high yields and induce high selectivities, cycloaddition is separate from the metal carbene reaction, and these transformations are not included in this chapter.

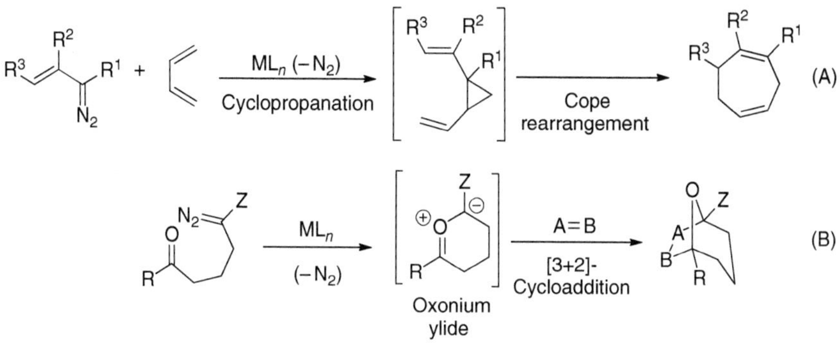

Scheme 5.3 Tandem carbene-cycloaddition reactions. Source: (A) Based on Krüger and Gaich [18]. (B) Based on Hodgson et al. [19].

Metallo-vinylcarbenes are formed catalytically with late transition metals, initially with dirhodium(II) carboxylates [20], but subsequently with silver(I), copper(I), and gold(I) catalysts [10]. Their cycloaddition reactions are normally characterized by high stereocontrol [11]. However, as will become evident in this chapter, cycloaddition reactions involving metallo-vinylcarbene intermediates are poised to be revealed with other transition metals and with different oxidation states of the metals currently employed. Metallo-vinylcarbenes, especially metallo-enolcarbenes, which will be discussed in this chapter, are "masked" 1,3-dipoles with enhanced electrophilic reactivity at the vinylogous carbon, which gives them the capability of sequential addition to dipolarophiles. A variety of dipolarophiles have been used with metallo-enolcarbenes (Scheme 5.4) with the most common being dipolar compounds that undergo [3+3]-cycloaddition or activated alkenes that undergo [3+2]-cycloaddition, and silyl-protected metallo-enolcarbene have been the most successfully applied. Perhaps the most surprising of these transformations is the [3+1]-cycloaddition for which metallo-enolcarbenes cyclize with a carbene or ylide to form a strained four-membered ring.

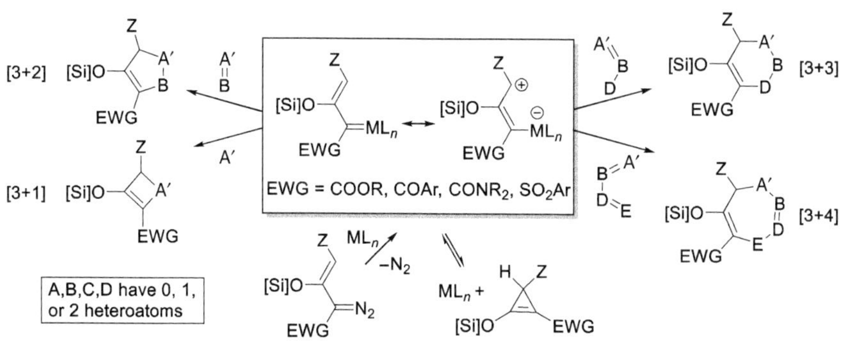

Scheme 5.4 Cycloaddition reactions of metallo-enoldiazo compounds.

5.2 [3+1]-Cycloaddition

The synthesis of four-membered rings is commonly provided by [2+2]-cycloaddition [21–23], which has become a traditional approach for the synthesis of these strained rings. However, until recently, [3+1]-cycloaddition has received scarce attention [24], and there was no reported activity involving metallocarbenes other than ring expansion of cyclopropane- or aziridine-substituted diazo compounds to form four-membered carbo- (**1**) [25] or heterocycles (**2**) [26] (Scheme 5.5). Direct [3+1]-cycloaddition with a vinylcarbene requires its generation in the presence of a one atom surrogate, either a stable carbene/nitrene or an ylide that can serve as a carbene/nitrene donor.

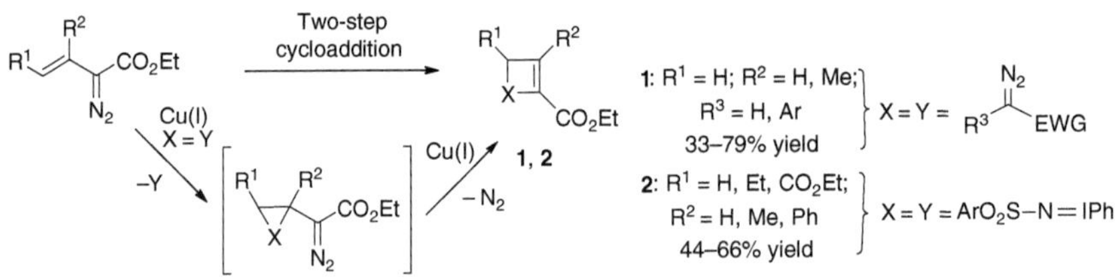

Scheme 5.5 Copper(I)catalyzed [2+1]-cycloaddition/ring expansion.

The first example of a direct [3+1]-cycloaddition with a vinyldiazo compound was reported in 2017 [27]. This transformation is successful with silyl-protected enoldiazoacetates (the three-carbon component) reacting with stable α-acyl sulfur ylides (the one-carbon component) catalyzed by copper(I) salts at room temperature (Scheme 5.6A). The cycloaddition product (**3**) is a 2-siloxy-substituted cyclobutene-1-carboxylate, a compound whose electron-donating and electron-withdrawing substituents across the carbon–carbon double bond identifies it as a donor–acceptor (D—A) cyclobutene. The use of chiral ligands provides enantiocontrol with up to 99% ee. What is especially surprising about this transformation is its catalyst selectivity. No cycloaddition occurs with rhodium(II) acetate, silver(I) salts, or a cationic gold(I) catalyst, but allylpalladium(II) is effective, albeit in low yield; and the nature of the copper(I) salt influences the reaction outcome with CuOTf·$Tol_{1/2}$ providing the highest product yields. Furthermore, simple bisoxazoline (Box) ligands give low enantiocontrol, but modified double-side-armed bisoxazoline (Sabox) ligands are uniquely successful. Without a substituent at the γ- position, enoldiazoacetates give modest enantiocontrol (up to 83% ee) using ligand **L1**, but with alkyl substitution at the γ-position in the (*Z*)-configuration (R^1) enantioselectivity is greatly increased, and diastereoselectivity for the *trans*-isomer is complete.

More recently, a nitrene analogue of the [3+1]-cycloaddition of enoldiazoacetates (the three-atom component) using *N*-aryl-imido sulfur ylides (the one-atom component) to form D—A azetine products **4** was reported (Scheme 5.6B) [28, 29]. Like its D—A cyclobutene counterpart, this chiral copper(I)-catalyzed transformation utilized a chiral Sabox ligand (**L2**) for maximum effect and afforded a broad selection of [3+1]-cycloaddition products in good yields with good-to-excellent enantiocontrol.

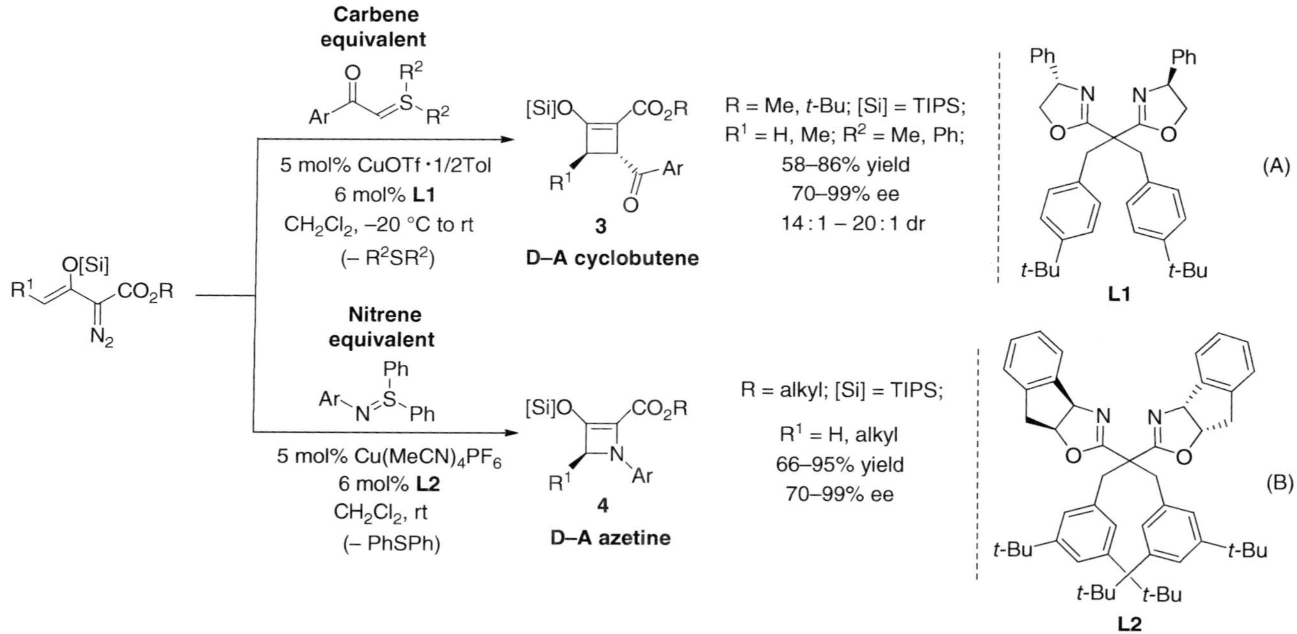

Scheme 5.6 [3+1]-Cycloaddition of enoldiazo compounds for the synthesis of D−A cyclobutenes and azetines.

This cycloaddition reaction occurred with (*Z*)- but not with (*E*)-enoldiazoacetates and, notably, unlike in the formation of **3** from reactions with α-acyl sulfur ylides, the corresponding DACP formed from the enoldiazoacetate (see Scheme 5.2) did not react with *N*-aryl-imido sulfur ylides.

These [3+1]-cycloadditions are reported to occur by nucleophilic addition of the sulfur ylide/imide (the one-atom component) to the vinylogous electrophilic carbon of the Cu(I)-carbene intermediate (the three-atom component). Intramolecular cyclization involves the elimination of sulfide R_2S, aided by electron donation from the silyl ether oxygen, which occurs by apparent S_N2-like backside displacement (Scheme 5.7). Dissociation of the ligated Cu(I) results in the formation of the four-membered ring product and regeneration of the reaction catalyst. The role of DACPs in these reactions differs considerably: they are integral intermediates in reactions with α-acyl sulfur ylides but unreactive with *N*-aryl-imido sulfur ylides [12]. Important applications of strained D—A cyclobutenes **3** and azetines **4** have been realized in synthesis of new chiral glutaric [30] and amino acid [29] derivatives via *retro*-Claisen nucleophilic ring opening demonstrating enormous potential in chiral functionalizations with a broad scope of nucleophiles.

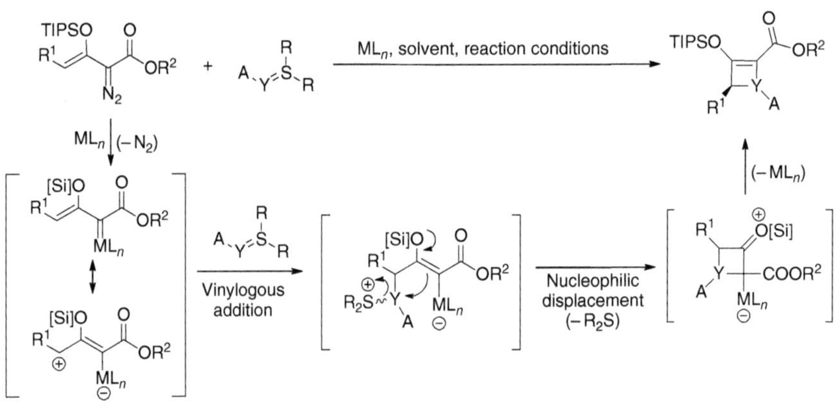

Scheme 5.7 [3+1]-Cycloaddition reaction mechanism.

[3+1]-Cycloaddition is not limited to reactions of enoldiazoacetates with sulfur ylides. A stoichiometric reaction with the stable benzylidene-ruthenium metallocarbene (first-generation Grubbs catalyst **5**) catalyzed by copper(I) triflate proceeded smoothly and afforded cyclobutene **6** in 71% yield (Scheme 5.8) [27]. However, an attempted reaction with classic stable Fischer carbenes did not occur. In [3+1]-cycloaddition transformations, even sulfur ylide reactivities are variable with α-aryl-imido sulfur ylides being reactive with silyl-protected enoldiazoacetates, but α-acyl-imido sulfur ylides are completely unreactive [28, 29]. The applicability of other sulfur ylides, phosphorus ylides, and alternative one-atom donors for [3+1]-cycloaddition is open to future discovery. The success of the catalytic transformations of metallo-enolcarbenes with sulfur ylides and with a stable metal carbene indicates a broad and underexplored generality of [3+1]-cycloaddition reactions.

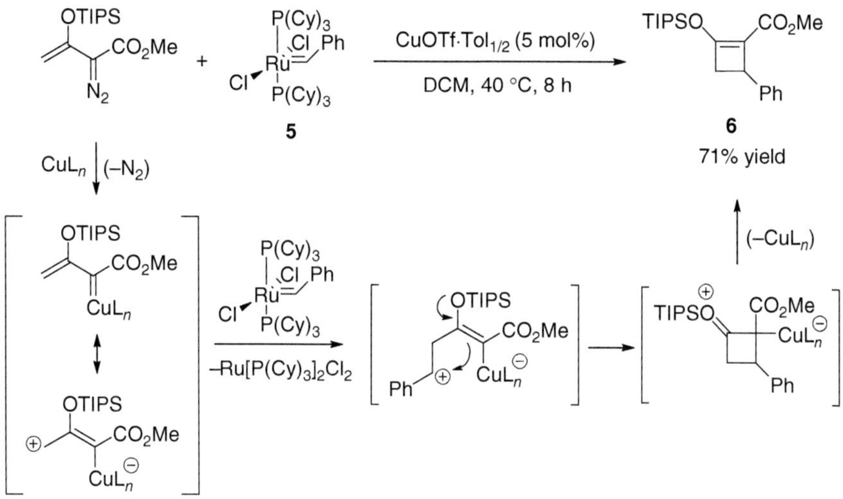

Scheme 5.8 [3+1]-Cycloaddition with a metal carbene and a proposed reaction mechanism.

5.3 [3+2]-Cycloaddition

[3+2]-Cycloaddition of metallo-vinyl carbenes is one of the most straightforward methods for rapid construction of various five-membered carbo- and heterocycles. Imines [31], indoles [32], and polarized alkenes, including enol ethers [33–36], enamides [37], silyl ketene imines [38], and vinyl azides [39], have been applied in this transformation. Thus far, the vast majority of catalysts employed for this transformation have been those of dirhodium(II) carboxylates, and little is known about the applicability of other transition metal compounds. Plausible stepwise reaction pathways for [3+2]-cycloaddition involve initial nucleophilic addition to the carbene center or to the vinylogous position of metallo-vinyl carbenes followed by ring closure with regeneration of the catalyst (see Scheme 5.1).

5.3.1 [3+2]-Cycloaddition with Imines and Indoles

As illustrated in Scheme 5.9, imines and *N*-methylindole nucleophiles are reported to selectively attack the carbene position, rather than the vinylogous position, of rhodium-vinylcarbenes formed from styryldiazoacetates to afford zwitterionic intermediates (**int-1–int-3**). Cyclization occurs prior to or concurrent with dissociation of the rhodium catalysts to form dihydropyrroles **7** [31] and fused indolines **8** and **9** [32] in good-to-excellent yields. In reactions with indoles, the dirhodium(II) catalyst $Rh_2(S\text{-DOSP})_4$ provides excellent enantiocontrol. However, with *N*-methylindole without a methyl group at the 2- or 3-position, a mixture of the two regioisomers was obtained in a 4 : 1 ratio, both with high ee but opposite absolute configurations. The placement of a methyl substituent at the 2- or 3-positions provided single regioisomers, and the cause of the different regioisomers and their opposite absolute configurations was attributed to different vinyl group orientations of the rhodium-bound carbene (**int-2** and **int-3**). The mechanism proposed implies 1,2-dimethylindole reacting with the *s-trans* styrylcarbene and 1,3-dimethylindole reacting with the *s-cis* styrylcarbene. The cause for nucleophilic addition at the carbene versus vinylogous positions of the metal carbene has also been explained as being due to the conformation of the metallo-vinyl carbene in reactions with dirhodium(II) catalysts: vinyl carbenes with a β-substituent, such as those from enoldiazoacetates, preferentially adopt the *s-cis* configuration, whereas vinylcarbenes with a (*Z*)-γ-substituent, such as the styryldiazoacetates, adopt the *s-trans* configuration [40]. Nucleophilic attack of the dipole on the *s-trans* metallo-vinyl carbene is proposed to form an (*E*)-olefin isomer of the addition complex, which closes to a new metal carbene intermediate that undergoes subsequent reactions or, alternatively, does not undergo ring closure. As will be seen, however, metallocarbenes formed from β-substituted enoldiazoacetates undergo cycloaddition reactions that mirror those of styryldiazoacetates, so the actual pathway to their product differences is still open to interpretation.

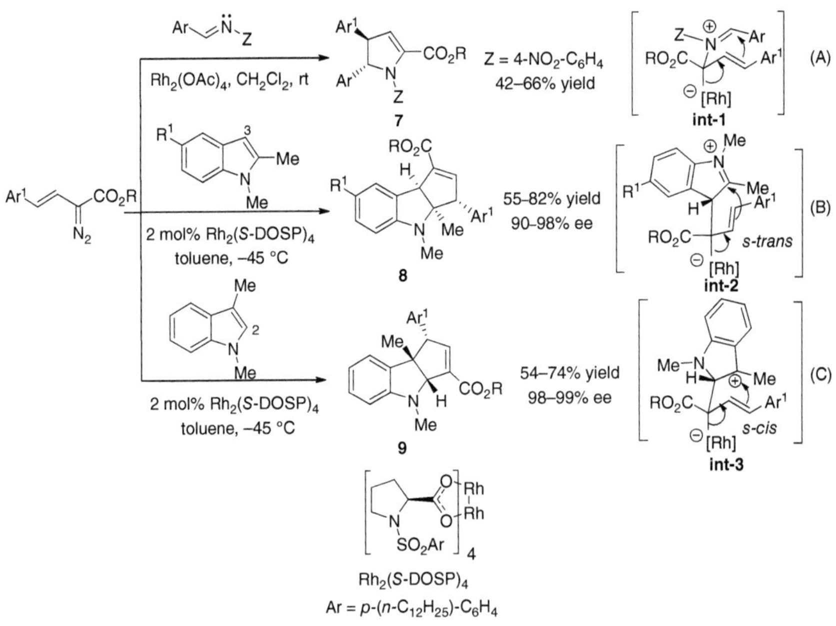

Scheme 5.9 [3+2]-Cycloaddition reactions of imines and indoles with styrydiazoacetates *via* carbene addition/ring closure.

Replacement of the phenyl group in styryldiazoacetate by a methyl group, however, completely changes the transformation with *N*-methylindole from cycloaddition (styryl) to a formal insertion (propenyl) with reaction at the vinylogous position of the metallo-vinyl carbene rather than at the carbene position (Scheme 5.10) [41, 42]. The formation of **10** is consistent with the stepwise pathway to cycloaddition: instead of ring closure, the probable pathway to **10** is initial vinylogous addition followed by proton transfer (Scheme 5.10) with expulsion of the metal. Formation of the

(*Z*)-isomer following proton transfer is attributed to the conformational preference of the *s-trans* vinylcarbene on the dirhodium face [40, 41], but this assumes that proton transfer and catalyst extrusion occur with retention of configuration.

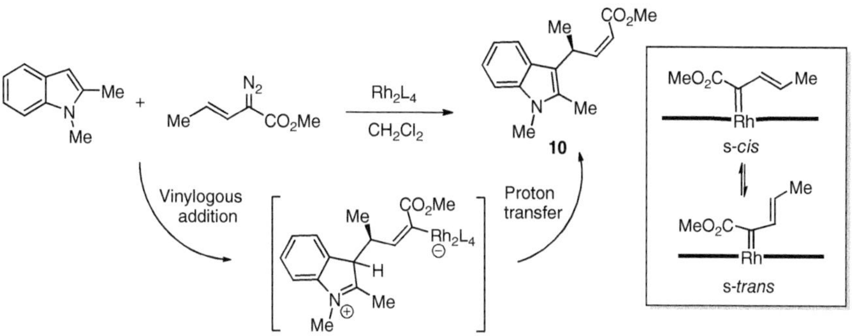

Scheme 5.10 Formal C–H insertion of an indole involving vinylogous addition.

Early in the quest for metallocarbene cycloaddition reactions, alkynyl-Fischer carbenes **11** were treated with variously substituted 2-methylindoles and even with enamines to form corresponding cycloaddition products **12** in good yields and, with the use of (−)-8-phenylmenthyl group as a chiral auxiliary, very high stereoselectivities (Scheme 5.11) [43]. An allenyl intermediate was proposed to account for the remote addition to the 3-position of the indole, cyclizing to a proposed allylic tungsten pentacarbonyl before hydrolysis. Extensions of these studies to vinyl-Fischer-type carbenes have not been reported.

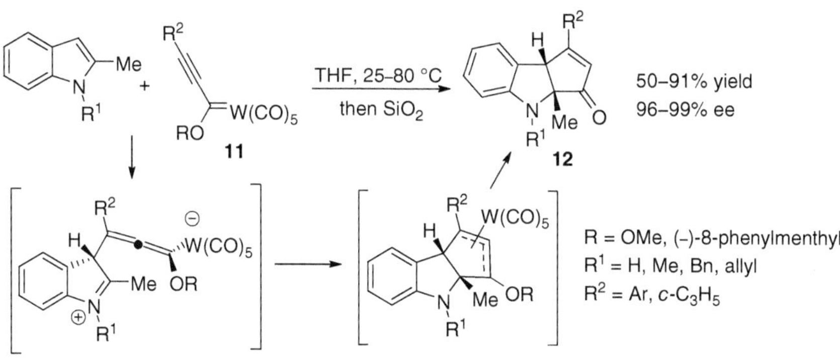

Scheme 5.11 [3+2]-Cycloaddition of an alkynyl-Fischer carbene with 2-methylindoles.

Recently, enoldiazoacetamides were found to undergo exclusive formation of only one cycloaddition regioisomer (**13**) with *N*-methylindole by vinylogous addition of the metallo-enolcarbene (Scheme 5.12A) [44]. Surprisingly, the highest enantioselectivities were obtained using the less sterically encumbered methanesulfonyl analogue of the arylsulfonylprolinate ligated dirhodium(II) carboxylates, $Rh_2(S\text{-}MSP)_4$, and DACPs formed from enoldiazoacetamides were also viable metallo-enolcarbene sources that gave the same outcome as the enoldiazoacetamide reactants. Copper(I) and silver(I) catalysts were ineffective. In contrast, the corresponding enoldiazoacetates reacted with *N*-methylindole to form both [3+2]-cycloaddition regioisomers whose ratio and enantioselectivity for each regioisomer varied with the dirhodium(II) carboxylate catalyst that was employed (Scheme 5.12B) [44]. Chiral phthalimidate-ligated dirhodium(II) carboxylate catalysts favored addition to *N*-methyl indole at its 3-position from the carbenic carbon to give **15** as the major product in high **15:14** ratios, whereas the chiral *N*-arylsulfonylprolinate-ligated dirhodium(II) carboxylates favored reversed regioselectivities with vinylogous addition to afford **14** as the major product. However, enantioselectivities were low.

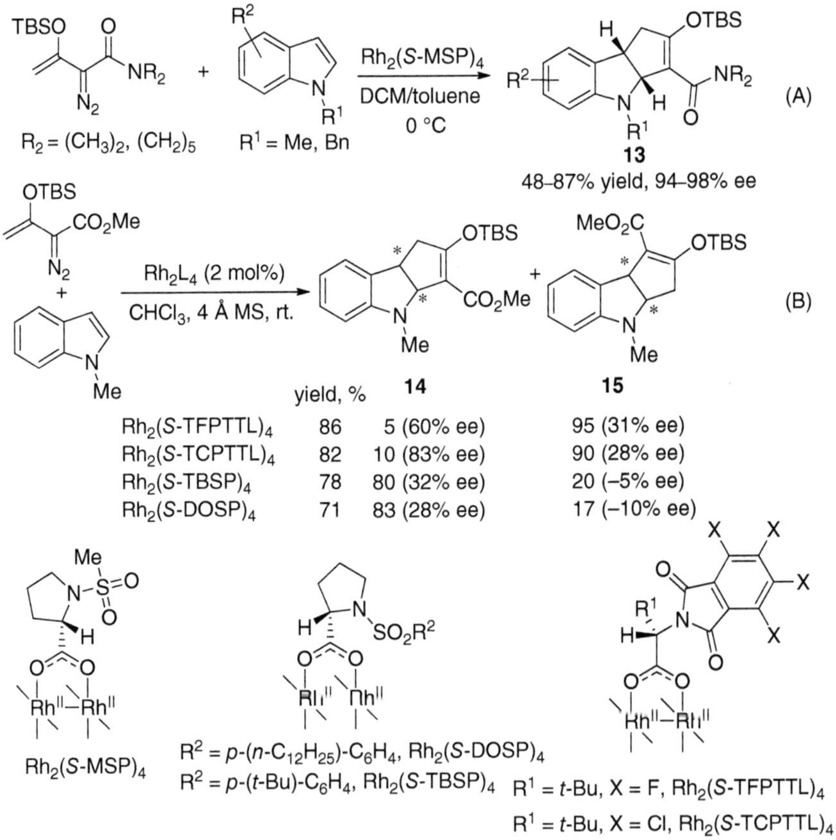

Scheme 5.12 [3+2]-Cycloaddition reactions of enoldiazoacetamides and -acetates with *N*-alkyllindoles.

Reactions at the vinylogous position of metallo-vinylcarbenes may be the most common direction for cycloaddition but, as seen from regioisomer formation in reactions with indoles, the mode of nucleophilic addition at vinylogous versus carbene sites remains open to interpretation. Does initial bond formation occur from the 3-position of the indole to either the vinylogous or carbene position of the metallo-vinylcarbene, or does nucleophilic addition occur from either the 2- or 3-position of the indole to the vinylogous position of the metallo-vinyl carbene?

5.3.2 [3+2]-Cycloaddition with Polarized Alkenes

Enol ethers, enamides, ketene imines, and vinylazides selectively attack the vinylogous position, rather than the carbene position, of β-substituted vinyldiazoacetates (Scheme 5.13) to produce functionalized cyclopentene derivatives in good-to-excellent yields. For enol ether substrates (Scheme 5.13A,B), the use of chiral $Rh_2(S\text{-DOSP})_4$, $Rh_2(S\text{-PTAD})_4$, or binuclear (*R*)-(DTBM)Segphos$(AuCl)_2$ catalysts provides high enantiocontrol [33–35]. Trisubstituted vinyl ethers are suitable

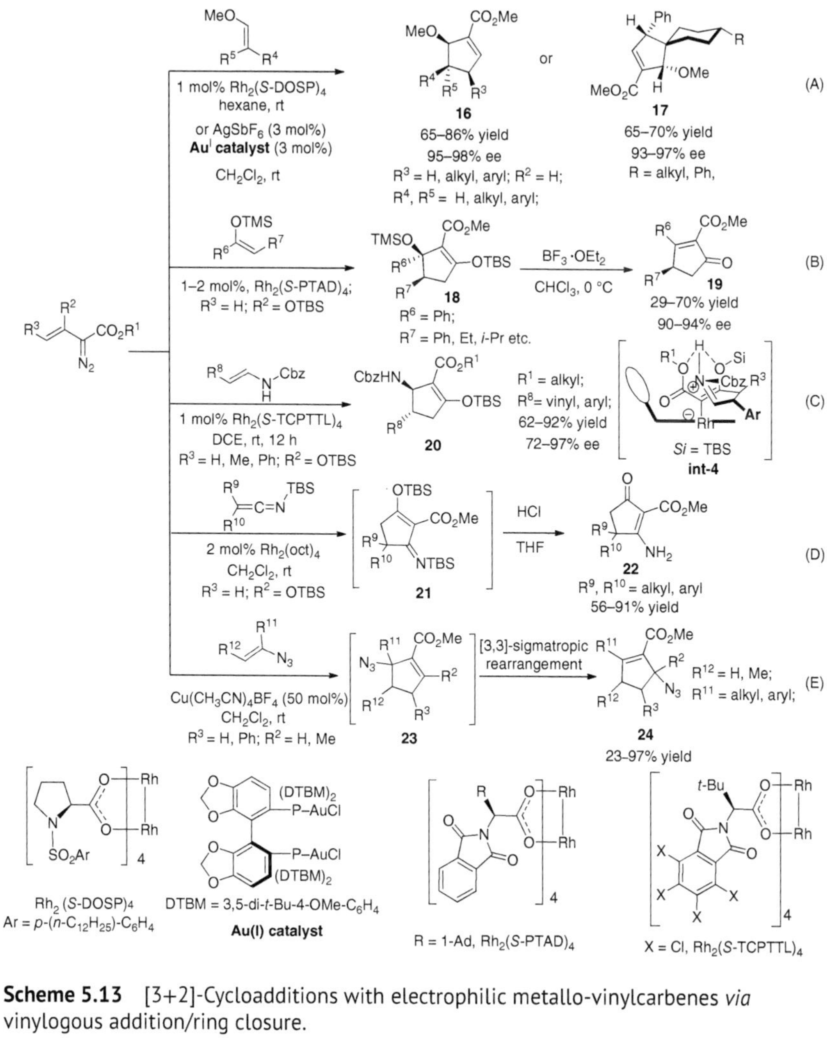

Scheme 5.13 [3+2]-Cycloadditions with electrophilic metallo-vinylcarbenes *via* vinylogous addition/ring closure.

substrates that give products **16** (Scheme 5.13A) bearing quaternary carbon centers in good yields [33, 34]. In addition, dynamic kinetic resolution of 4-substituted-1-(methoxymethylene)cyclohexane could be realized through Au(I)-catalyzed [3+2]-cyclization delivering cyclopentene derivatives **17** (Scheme 5.13A) [35]. The [3+2]-cycloaddition of the metallo-enolcarbenes of enoldiazoacetates with silyl enol ethers forms cyclopentene products **18** [36] that, with $BF_3{\cdot}OEt_2$ for elimination and desilylation, are converted into highly enantioenriched cyclopentenones **19** with no erosion in enantiomeric excess (Scheme 5.13B).

Chiral dirhodium(II) [Rh_2(*S*-TCPTTL)$_4$]-catalyzed [3+2]-cycloaddition of enoldiazoacetates with enecarbamates provides an approach to enantioenriched multifunctionalized cyclopentyl β-amino esters **20** (Scheme 5.13C) [37]. Surprisingly, N—H insertion that is common with amides is not observed [45]. The success of these transformations is attributed to an unprecedented hydrogen-bond association (**int-4**) between the intermediate rhodium-enolcarbene and the *Z*-enecarbamate. A series of cyclopentane β-amino acids was successfully produced from the cycloadducts *via* a deprotection/reduction/hydrogenolysis process.

The rhodium(II) octanoate [Rh_2(oct)$_4$]catalyzed [3+2]-cyclization and acid-promoted deprotection cascade reaction between enoldiazoacetates and silyl ketene imines affords 3-amino-2-cyclopentenone derivatives **21** in moderate-to-high overall yields (Scheme 5.13D) [38]. Dirhodium(II) catalysis is effective at room temperature despite the presence of the basic ketene imine reactant. This methodology represents a general approach for introducing a quaternary carbon onto the 3-amino-2-cyclopentenone skeleton.

Following upon the discovery that reaction of vinyl azides with gold alkenylcarbene species generated from propargylic esters led to buta-1,3-dien-2-yl esters [46], copper(I)-vinylcarbenes were found to undergo [3+2]cycloaddition with vinyl azides to yield 3-cyclopentenyl azides **23**. These allyl azides underwent subsequent [3,3]-sigmatropic rearrangement to form functionalized cyclopentene derivatives **24** in 23–97% yields (Scheme 5.13E) [39]. Ethyl 2-diazobut-3-enoate was effective, but low yields were obtained with diazo compounds having substituents at either the 3- or 4-positions.

5.3.3 [3+2]-Cycloaddition with Nitrones

Interestingly, the reaction of 2-diazo-3-butenoate with nitrones resulted in an unusual tricyclic derivative **25** (Scheme 5.14) that was formed with extensive rearrangement and appeared to be produced in a cascade reaction [47]. A reaction mechanism was proposed that begins by [3+2]-cycloaddtion of the electrophilic dirhodium(II)-vinylcarbene with the nitrone to produce intermediate **int-5**. Subsequently, an intramolecular Buchner reaction generated the norcaradiene **int-6**, which after a 6π-electrocyclic ring expansion followed by N—O cleavage and [1,7]-oxygen migration formed the observed product **25**. This mechanistic pathway was confirmed in a subsequent computational study [48]. The overall transformation was general for *N,C*-diarylnitrones, but was vastly different from

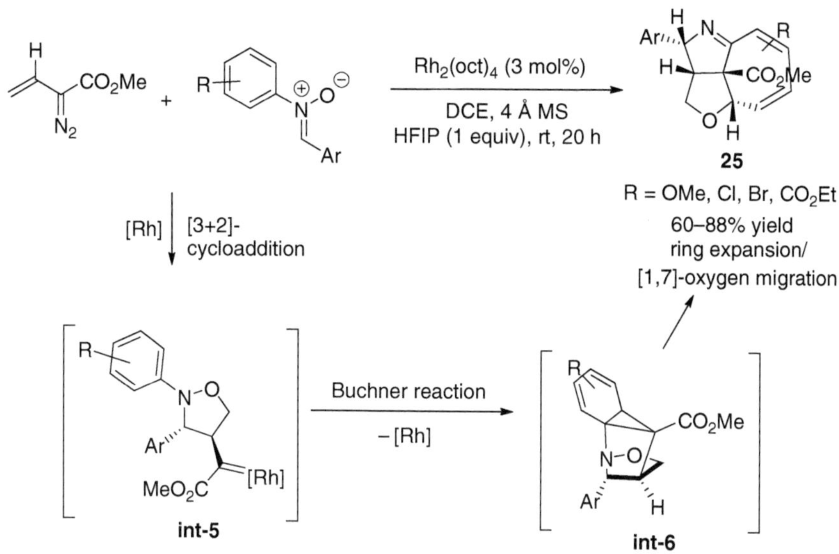

Scheme 5.14 [3+2]-Cycloaddition of electrophilic dirhodium(II)-vinylcarbenes with nitrones and subsequent cyclopropanation/rearrangement.

the [3+3]-cycloaddition reactions between nitrones and enoldiazo compounds that are discussed in Section 5.4.1.

5.3.4 Divergent Behavior of Catalysts

Different catalysts used for cycloaddition can direct the same reactants to different products. This was made evident in reactions of styryldiazoacetates with imines, catalyzed by rhodium acetate (reported in Scheme 5.9A) and by copper(II) triflate (Scheme 5.15A) in the same solvent at the same temperature. Rhodium acetate catalyzes the formation of the metallo-vinyl carbene that produced an ylide with the imine that cyclizes to dihydropyrroles **7**, whereas the more Lewis acidic copper(II) triflate forms a complex with the imine, activating it for addition to the diazo carbon that in cyclization with loss of dinitrogen forms **26** [31]. But perhaps the most divergent behavior of catalysts is described for the reactions of enoldiazoamides with nitrosoarenes catalyzed by two "paddlewheel" dirhodium compounds (Scheme 5.15B) [49]. Rhodium(II) octanoate catalyzes formal [3+2]-cycloaddition of enoldiazoacetamide *via* dinitrogen extrusion to the DACP to produce multifunctionalized 5-isoxazolones **28**, whereas reactions that occur with the less Lewis acidic γ-caprolactamate-ligated dirhodium catalyst direct reaction to 1,3-oxazin-4-ones **27** that also occurs with extensive structural rearrangement of the reactants. Copper(I) and palladium(II) catalysts produce mixtures of the two products, but in low-to-moderate yields. High enantioselectivity accompanies the formation of 5-isoxazolones **28a** from the γ-ethyl-enoldiazoacetamide using the $Rh_2(S\text{-}PTTL)_4$ catalyst (Scheme 5.15C). These are just two of a growing number of transformations of diazo compounds whose product is determined by the catalyst that is employed.

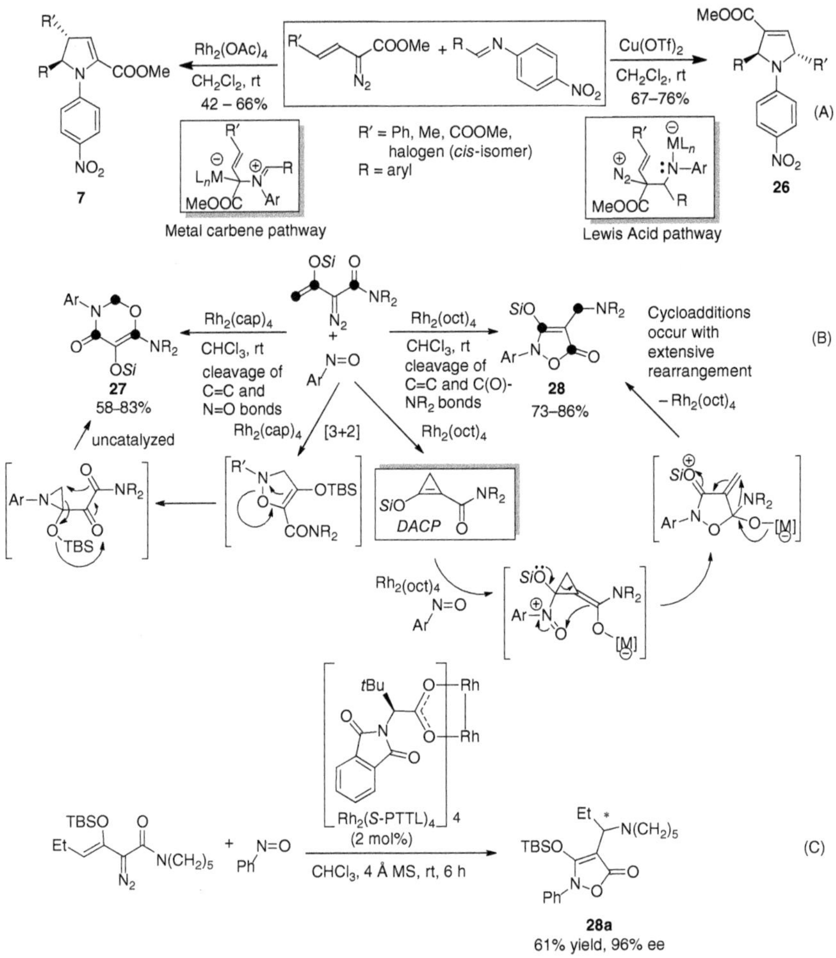

Scheme 5.15 Catalytic divergent cycloaddition reactions of vinyldiazoacetates with imines and enoldiazoacetamides with nitrosoarenes.

5.4 [3+3]-Cycloaddition of Enoldiazo Compounds

5.4.1 [3+3]-Cycloaddition with Nitrones

Nitrones are among the most widely used dipoles for cycloaddition reactions [50]. Easily formed from imines, stable, and commercially available, it is no surprise that the first demonstration of highly enantioselective catalytic [3+3]-cycloaddition reactions with vinyldiazo compounds originated with nitrones. With the sterically small phthalimide-alanine ligated $Rh_2(S\text{-PTA})_4$ catalyst, trisubstituted 3,6-dihydro-2*H*-1,2-oxazines **29** (EWG = CO_2Me) were obtained in moderate-to-high yields and enantioselectivities (Scheme 5.16A) [51]. After the demonstration that [3+3]-cycloaddition using dirhodium(II) catalysts was easily achieved, focus was placed on the suitability of other ligated metals as catalysts. In the search for enhanced selectivities and yields, enoldiazoacetamides (EWG = NR_2) with nitrones were reported to undergo copper(I)-catalyzed asymmetric [3+3]-cycloaddition with higher yields and, with Box ligand **L3**, exceptional enantioselectivities (Scheme 5.16B) [52]. Extension to enoldiazoketones (EWG = COR) in [3+3]-cycloaddition with nitrones was also successful using chiral Cu(I) catalysis, although optimization of a Sabox ligand was necessary (Scheme 5.16C) to achieve high enantiocontrol [53]. Sabox ligand **L4** was optimal for the [3+3]-cycloaddition reactions with nitrones, not only for enoldiazoketones but also for enoldiazo-acetates and -acetamides. However, chiral dirhodium(II) catalysts were not suitable for the [3+3]-cycloaddition of enoldiazoketones with nitrones due to a competing self-reaction of the enoldiazoketone to form 2,5-disubstituted furans [54]. In spite of this, when both sets of catalysts could be used in cycloaddition reactions of enoldiazoacetates, copper(I) with (*R*)-Sabox ligands allowed access to the opposite enantiomer of 3,6-dihydro-2*H*-1,2-oxazines **31** than was produced by chiral dirhodium(II) with (*S*)-carboxylate ligands (Scheme 5.16C versus 5.16A).

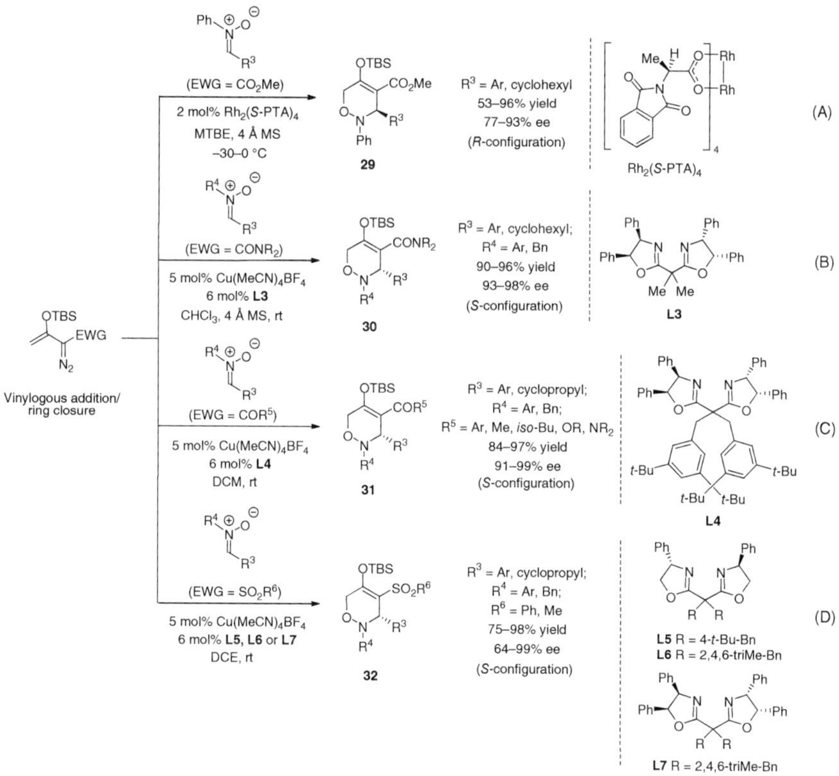

Scheme 5.16 Dirhodium(II)- and copper(I)-catalyzed asymmetric [3+3]-cycloaddition of enoldiazo compounds with nitrones.

Enoldiazosulfones were also suitable for [3+3]-cycloaddition with nitrones using chiral copper(I) catalysis with optimized Sabox ligands (**L5**–**L7**) to afford

3,6-dihydro-2*H*-1,2-oxazines **32** with EWG = SO_2R (Scheme 5.16D) [55]. Use of chiral dirhodium(II) catalysts did not result in the production of [3+3]-cycloaddition products with enoldiazosulfones; instead, [3+2]-cycloaddition occurred between nitrones with the D—A cyclopropene formed directly from the diazo compound, indicating the anticipated enhanced stability of sulfonyl-substituted DACPs. That the sulfone DACP was not formed or was a minor product in the copper-catalyzed reactions suggested that [3+3]-cycloaddition occurs at a much faster rate than does DACP formation (Scheme 5.17).

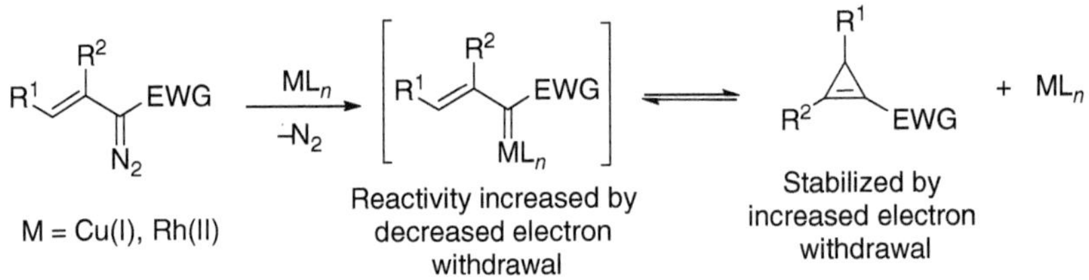

Scheme 5.17 Stability and reactivity of metalloene-carbenes versus DACPs.

All examples of [3+3]-cycloaddition with nitrones described in Scheme 5.16 utilize enoldiazo compounds without substituents at the 4-position. 4-Alkyl- or 4-aryl-substituted enoldiazo compounds require special attention not only because they can exist as (*Z*)- and (*E*)-geometrical isomers that have different reactivity and stereoselectivity but, also, because they can provide a steric barrier to cycloaddition. Different reactivities and selectivities of reactions with (*Z*)- and (*E*)-isomers, as well as with their corresponding DACPs, are reported in [3+*n*]-cycloaddition reactions [12]. The copper(I)-catalyzed [3+3]-cycloaddition reaction of (*Z*)-4-alkyl-substituted enoldiazoacetates occurs through a (*Z*)-copper(I) vinylcarbene that is formed directly from (*Z*)-4-alkyl-substituted enoldiazoacetate or from the DACP – the resting state of the carbene (Scheme 5.2) [12, 53]. (*E*)-4-Alkyl-substituted enoldiazoacetates form the corresponding (*E*)-copper(I) vinylcarbene and, then, the DACP after elimination of Cu(I); however, only the (*Z*)-copper(I) vinylcarbene, and not its (*E*)-configured isomer, is formed from the DACP. The outcome of [3+3]-cycloadditions with nitrones demonstrates that there is no difference in reaction yields and stereoselectivities when either (*Z*)- or (*E*)-enoldiazoacetates, their mixtures, or the DACPs are used. This outcome highlights the central role that DACPs, formed by dinitrogen extrusion, have in cycloaddition process [56]. Ligand optimization reveals that Box ligand **L8** is optimal for the copper(I)-catalyzed [3+3]-cycloaddition of 4-substituted enoldiazoacetates with nitrones (Scheme 5.18) to afford highly substituted 3,6-dihydro-2*H*-1,2-oxazines **33** in high yields and stereocontrol [56].

Diastereoselectivity in cycloaddition reactions with 4-substituted enoldiazoacetates is exclusively *cis*. Conceptually, enantiocontrol occurs in the initial vinylogous addition to prochiral carbon-4, even though the chiral catalyst is not in close proximity to the developing chiral center, and diastereocontrol is set in the ring-closing step. In contrast, enantiocontrol occurs during the ring-closing step

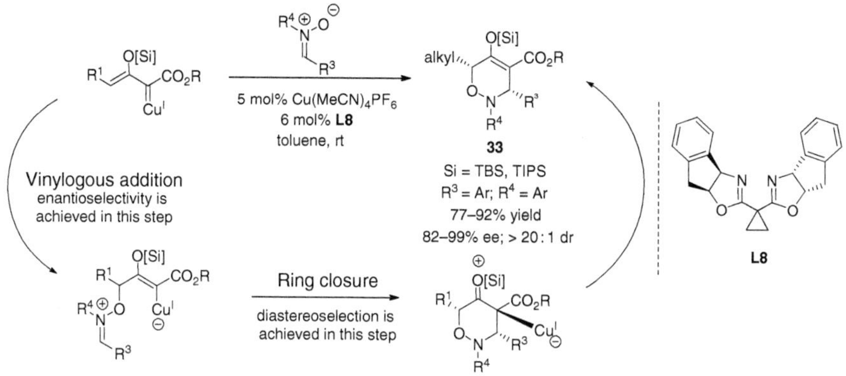

Scheme 5.18 Copper(I)-catalyzed asymmetric [3+3]-cycloaddition of 4-substituted enoldiazoacetates with nitrones.

in reactions with enoldiazoacetates without a substituent at carbon-4, which is a more likely scenario because of the proximity of the developing chiral center to the chiral catalyst. This suggests, but awaits computational verification, that the initial vinylogous addition is reversible, and the observed enantioselectivity is actually dependent on the compatibility of the initial addition with ring closing.

(*Z*)-4-Phenyl-substituted enoldiazoacetates did not undergo cycloaddition with nitrones under standard conditions in copper(I) or dirhodium(II)-catalyzed reactions. However, a two-step silver(I)-catalyzed pathway provided an alternative process (Scheme 5.19) [57]. DACP **34** was generated *in situ* using dirhodium(II) acetate and its chiral silver(I)-catalyzed (using classic Box ligand **L9**) [3+3]-cycloaddition with nitrones afforded 3,6-dihydro-2*H*-1,2-oxazines **35** with the silyl group being removed by treatment with TBAF.

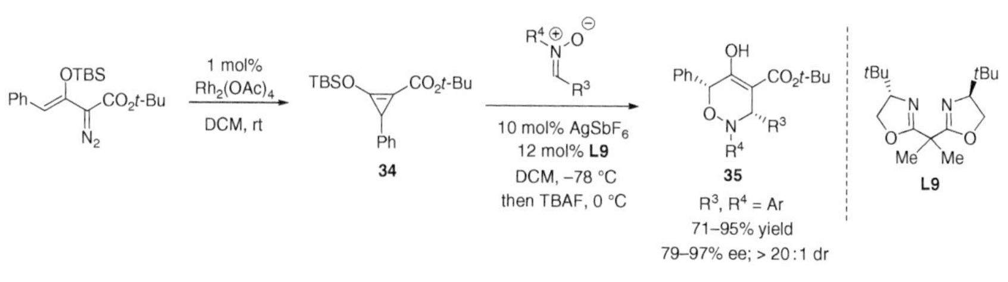

Scheme 5.19 Silver(I)-catalyzed asymmetric formal [3+3]-cycloaddition of DACP **34** with nitrones.

5.4.2 [3+3]-Cycloaddition with Pyridinium Ylides and Hydrazones

Asymmetric [3+3]-cycloaddition reactions with pyridinium ylides have been less explored [58–62] than those with nitrones. The challenge with this dipole is the ring-closing step of the cycloaddition reaction, which requires dearomatization of the aromatic ring. Pioneering efforts on dearomatizing [3+3]-cycloaddition of enoldiazoacetates with acylimino-pyridinium or -quinolinium ylides were reported for the synthesis of chiral bi- or tricyclic diazines **36** (EWG = CO_2R) using chiral dirhodium(II) [Rh_2(*S*-PTTL)$_4$ or Rh_2(*S*-PTAD)$_4$] catalysis in moderate-to-high yields and stereocontrol (Scheme 5.20A) [63]. However, in a recent report, chiral copper(I) catalysis using Sabox ligand **L10** was established as superior to Rh(II) catalysis in product yields and enantiocontrol, including derivatives with EWG = COR, CO_2Me, and $CONR_2$ (Scheme 5.20B) [53]. Copper(I) catalysis produces the enantiomers of bicyclic diazines **37** formed with chiral Rh(II) catalysts having phthalimide-amino acid ligands.

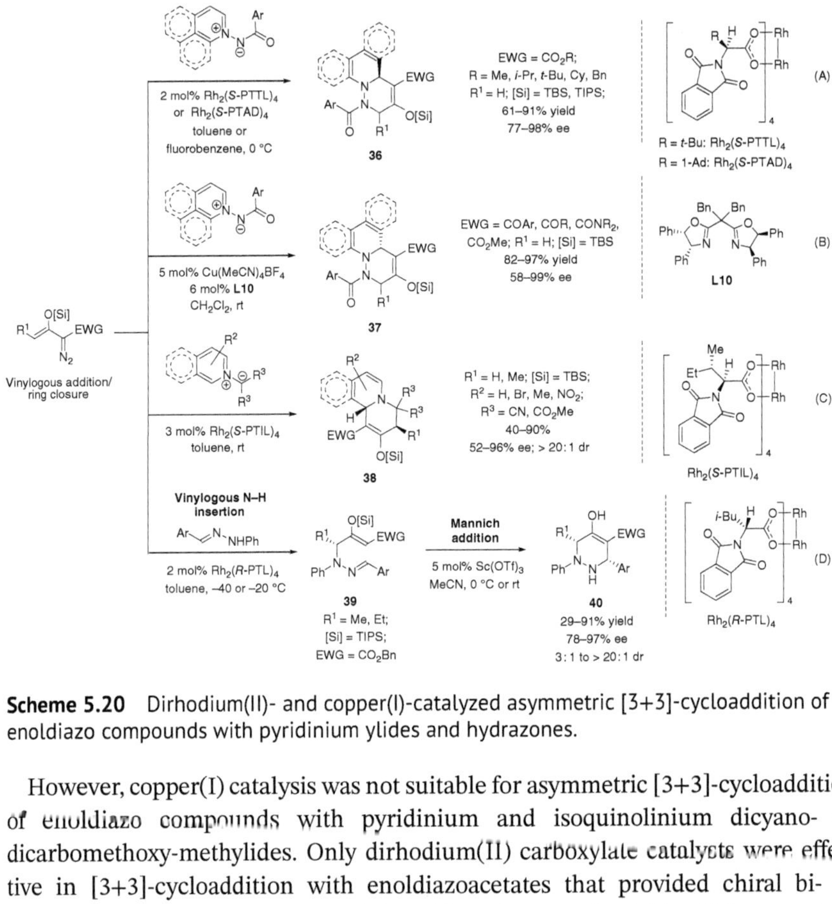

Scheme 5.20 Dirhodium(II)- and copper(I)-catalyzed asymmetric [3+3]-cycloaddition of enoldiazo compounds with pyridinium ylides and hydrazones.

However, copper(I) catalysis was not suitable for asymmetric [3+3]-cycloaddition of enoldiazo compounds with pyridinium and isoquinolinium dicyano- or dicarbomethoxy-methylides. Only dirhodium(II) carboxylate catalysts were effective in [3+3]-cycloaddition with enoldiazoacetates that provided chiral bi- or tricyclic azines **38** in moderate-to-high yields and stereoselectivities (Scheme 5.20C) [64]. High yields and stereocontrol were achieved in [3+3]-cycloaddition only when at least 3 mol% of Rh_2(*S*-PTIL)$_4$ was used. Lowering catalyst loading resulted in the predominant formation of products of a competing [3+2]-cycloaddition with the corresponding DACP, and this observation led to the discovery of DACPs as the carbene resting state.

The cycloaddition of phenylhydrazones with enoldiazoacetates is a special case because the product of the initial vinylogous nucleophilic addition to the metallo-enolcarbene does not undergo the traditional ring closure (Scheme 5.20D) [65]. Instead, the initial step in the reaction occurs by N—H insertion, probably by ammonium ylide formation followed by proton induced release of the catalyst to form hydrazone derivative **39**. Use of chiral Rh_2(*R*-PTL)$_4$ provides the asymmetric vinylogous N—H insertion product that in a separate step, catalyzed by $Sc(OTf)_3$, forms chiral diazines **40** (EWG = CO_2Bn) by Mannich addition/cyclization and silyl group removal.

5.4.3 Diastereoselective [3+3]-Cycloaddition with Achiral Catalysts

Certain [3+3]-cycloaddition reactions of enoldiazo compounds occur with high diastereocontrol without the use of chiral catalysts, and, when reacted with a substrate that possesses a chiral center, enantiomerically enriched cycloaddition products are formed. In these cases the dipole reacting with the metallo-vinylcarbene has a substituent at a tetrahedral carbon that enforces diastereocontrol and preserves enantioselectivity.

The application of azomethine imines is generally limited to asymmetric metal-catalyzed [3+2]-cycloaddition reactions [66]. However, two reports describe highly diastereoselective syntheses of bicyclic pyrazolines via [3+3]-cycloaddition reactions of azomethine imines with trimethylenemethane (palladium catalysis) [67] and propargyl esters (gold catalysis) [68], and one report has documented the [3+3]-cycloaddition of enoldiazoacetates (EWG = CO_2Me) with azomethine imines using $Rh_2(OAc)_4$ catalysis to form bicyclic pyrazolines **41** (Scheme 5.21A) [69]. Confirmed by computational studies of this reaction [70], high diastereocontrol for the *cis*-stereoisomer of **41** originates from steric and electronic effects that reduce the repulsion between the phenyl group and the OTBS group in the dirhodium-vinylcarcene complex and in the transition state for cycloaddition. Computationally, the reaction process occurs in two steps: vinylogous addition (C—N bond formation) and intramolecular ring closure (C—C bond formation), with the second step being rate-limiting.

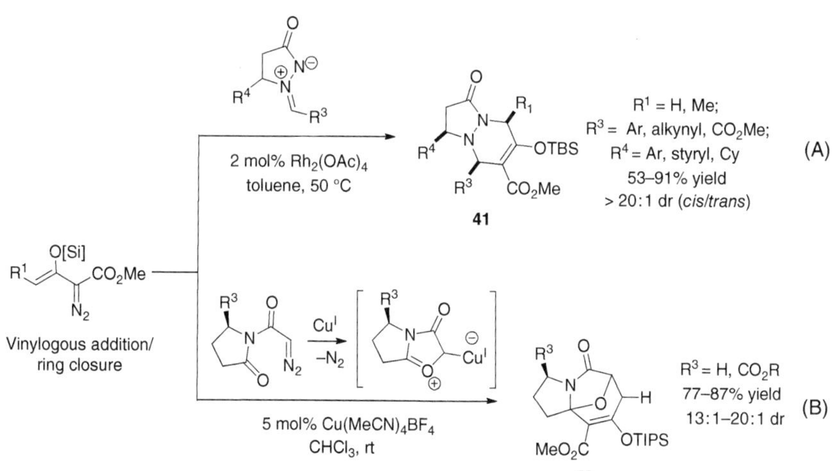

Scheme 5.21 Examples of diastereoselective [3+3]-cycloaddition of enoldiazo compounds.

An example of catalytic discrimination between two diazo compounds is the regio-divergent copper(I)- and rhodium(II)-catalyzed [3+3]- versus [3+2]-cycloadditions of enoldiazoacetates with α-diazocarboximides [71]. Tricyclic azepines **42** are obtained from [3+3]-cycloaddition of enoldiazoacetates with cyclic carbonyl ylides (generated from α-diazocarboximides) using a copper(I) catalyst in high yields and with good-to-excellent diastereoselectivities (Scheme 5.21B) [71].

5.4.4 [3+3]-Cycloaddition with Diaziridines

A novel chiral copper(I)-catalyzed (with optimal Box ligand **L11**) formal [3+3]-cycloaddition of (*Z*)-4-substituted enoldiazoacetates with symmetric diaziridines is reported to occur with desymmetrization of the reactant diaziridines and N—N bond cleavage to afford 1,5-diazabicyclo[*n*.3.1]non-2-ene derivatives **43** in moderate-to-high yields and stereoselectivities (Scheme 5.22) [72]. Interestingly, both the (*Z*)-isomers of 4-substituted enoldiazoacetates and their corresponding DACPs showed high reactivity and stereoselectivity in this asymmetric [3+3]-cycloaddition, while the (*E*)-isomer reacted slowly and with low stereoselectivity. In contrast, [3+3]-cycloaddition of 4-substituted enoldiazoacetates with nitrones

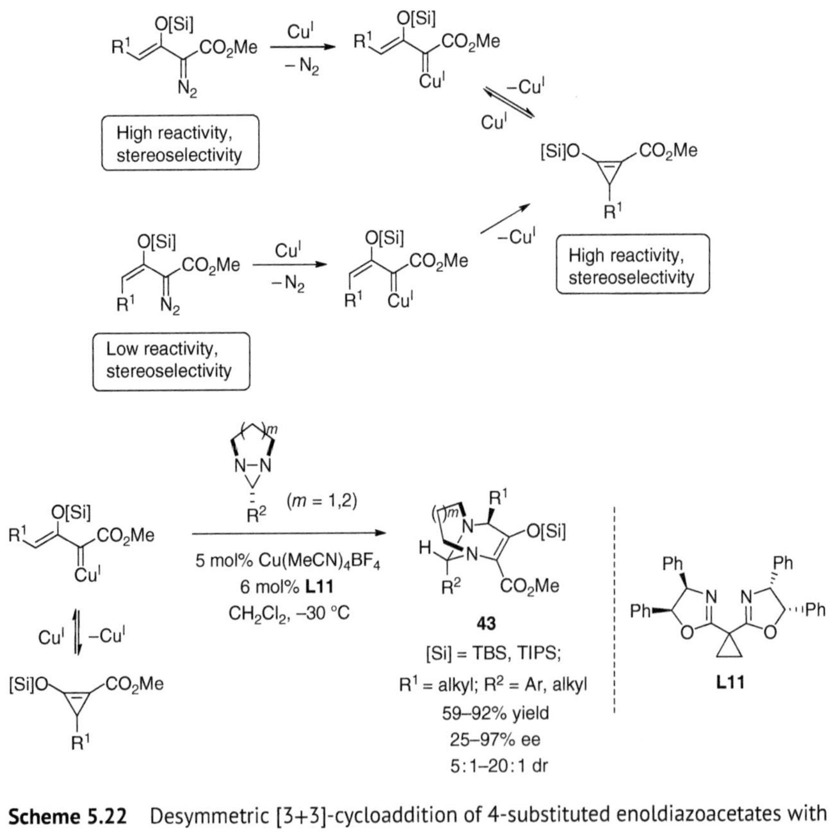

Scheme 5.22 Desymmetric [3+3]-cycloaddition of 4-substituted enoldiazoacetates with diaziridines.

demonstrated no influence of geometrical isomers (*Z*- or *E*-) to the outcome of the cycloaddition in terms of yields and stereocontrol [12, 56], indicating that the DACP intervened to convert the (*E*)-metallo-enolcarbene into its (*Z*)-counterpart prior to cycloaddition.

5.4.5 [3+3]-Cycloaddition with Donor–Acceptor Cyclopropanes and Oxiranes

Vinyldiazo compounds undergo Lewis-acid-catalyzed cycloaddition at their carbon–carbon double bond with strained ring compounds to form products with an exocyclic diazo functionality that are capable of ring expansion through dinitrogen extrusion. An example of this process was already seen in Scheme 5.5 with its formal [3+1]-cycloaddition. An analogous process is possible for formal [3+3]-cycloaddition. For example, silyl-protected enoldiazoacetates readily undergo Lewis acid–catalyzed [3+2]-cycloaddition with highly reactive D—A cyclopropanes or oxiranes under mild conditions to produce diazo-functionalized five-membered carbo- and heterocycles **44** with high diastereocontrol (Scheme 5.23) [73]. Their subsequent ring expansion that involves a metal carbene intermediate occurs under dirhodium(II) catalysis to form six-membered carbo- and heterocycles **45**. A robust synthesis of **45** was also performed via a one-pot formal [3+3]-cycloaddition (involves [2+3]-cycloaddition and ring expansion steps) of enoldiazoacetates with D—A cyclopropanes or oxiranes by sequential addition of Lewis acid [$Yb(OTf)_3$] and dirhodium(II) [$Rh_2(cap)_4$] catalysts.

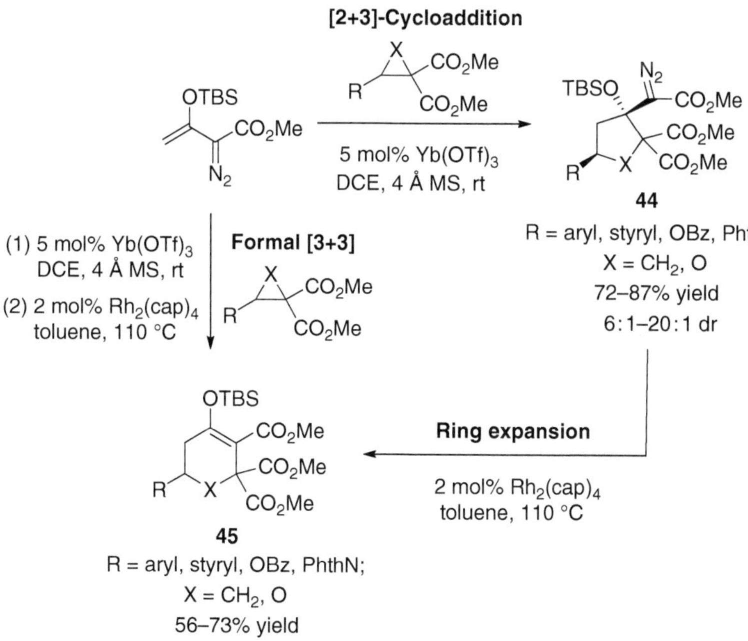

Scheme 5.23 Lewis acid/rhodium-catalyzed formal [3+3]-cycloaddition of an enoldiazoacetate.

The significance of the siloxy group in these metallo-enolcarbene cycloaddition reactions is evident in their ability to generate cycloaddition products compared with, for example, metallo-vinylcarbenes from styryl- and related vinyldiazoacetates, which undergo [3+2]-cycloaddition **46** [74] but not [3+3]-cycloaddition **29** [51]. Scheme 5.24 provides one example of this difference from their comparative reactions with nitrones [74]. Although conformational influences have been used to explain this difference, the relative inability of metallocarbenes in which the siloxy group is replaced by phenyl or methyl to produce cycloaddition products in even modest yields and stereocontrol by dirhodium(II) catalysis (Scheme 5.24B, unpublished results) confirms the importance of electron donation by oxygen in facilitating cycloaddition. Obviously, other strong electron-donating groups in place of the siloxy group (e.g. OR, OAc, halide) in these vinyldiazo compounds should facilitate cycloaddition reactions with the same or similar yields and selectivities, but these compounds have not yet been reported.

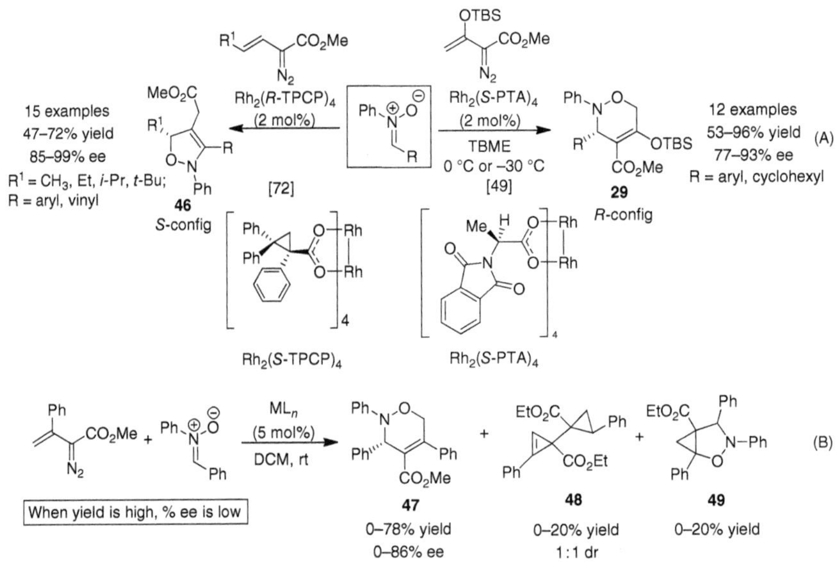

Scheme 5.24 The siloxy group of vinyldiazo compounds provides a unique pathway for [3+n]-cycloaddition.

5.5 [3+4]-Cycloaddition

Because of the unfavorable kinetic and/or thermodynamic factors that typically impede medium-sized ring formation, [3+4]-cycloaddition reactions of electrophilic metallo-vinyl carbenes through zwitterionic intermediates are less established. So far, only two examples have been described, and the use of vinyl carbenes without β-silyloxy substituents has a negative influence on product yields and selectivities. As illustrated in Scheme 5.25, dienes and hexahydro-*s*-triazines react with enolvinylcarbenes *via* vinylogous addition/ring closure to give seven-membered carbocycles **50** and *N*-containing heterocycles **51** in good yields with excellent diastereoselectivities [75]. The use of chiral dirhodium(II) carboxylate catalyst Rh_2(*S*-BTPCP)$_4$ in reactions with dienes gave [3+4]-cycloaddition products **50** in moderate yields with 82–99% ee (Scheme 5.25A) [75]. The *s*-triazines reacted directly, rather than as formaldimine precursors. Gold-enolcarbenes, generated from either enoldiazoacetates or an enoldiazoacetamide with Au[^{t}BuXPhos]Cl, furnished [3+4]-cycloadducts in low-to-moderate yields (Scheme 5.25B). However, gold-vinylcarbenes formed from β-unsubstituted vinyldiazoacetates participated only as one-carbon components (the carbene carbon) to form five-membered ring products **52** in a formal [4+1]-cycloaddition (Scheme 5.25C) [76].

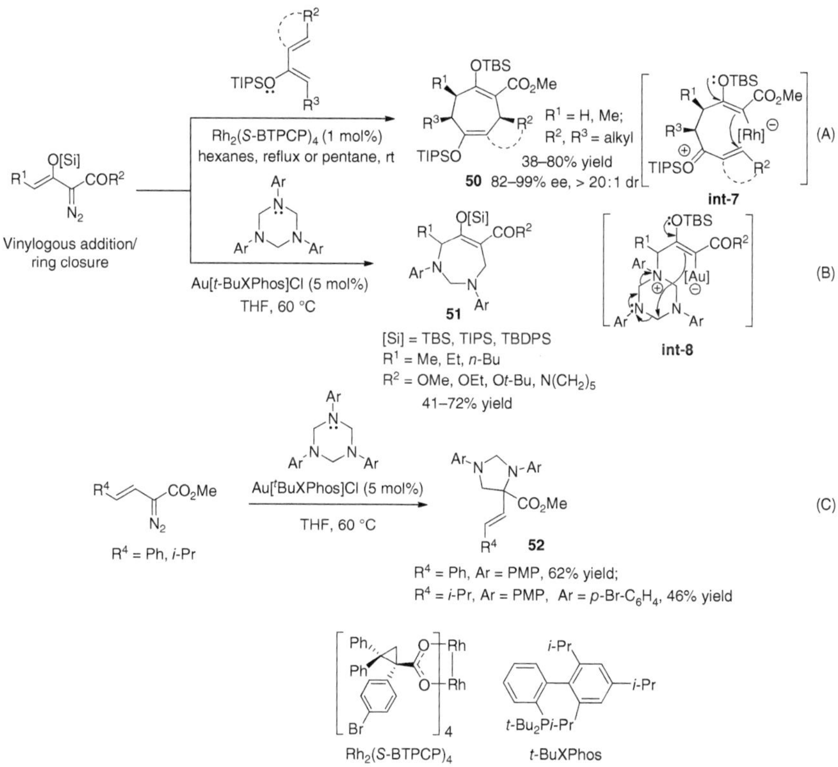

Scheme 5.25 [3+4]-Cycloadditions with electrophilic metallo-vinylcarbenes *via* vinylogous addition/ring closure.

5.6 [3+5]-Cycloaddition

[3+5]-Cycloaddition involving metallo-vinylcarbenes is rare. In the only recorded example thus far, stable pyridinium zwitterions selectively attack the vinylogous position of dirhodium(II)-enolcarbenes to form remote zwitterionic intermediates (**int-9**) that undergo fast intramolecular ring closure to diazocine derivatives **53** in 43–73% yields (Scheme 5.26) [77]. The asymmetric version of this process was reported to occur with high enantioselectivity (**54**, 63% yield and 90% ee) using the chiral dirhodium(II) carboxylate catalyst Rh_2(*S*-PTAD)$_4$. This initial result demonstrates the possibility for remote asymmetric control *via* chiral dirhodium(II) catalysis.

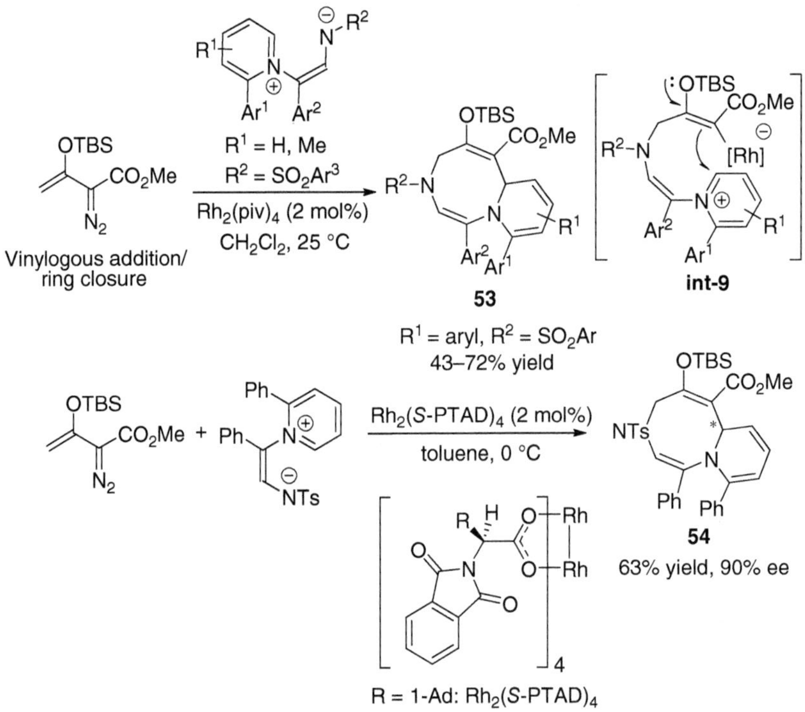

Scheme 5.26 [3+5]-Cycloaddition of pyridinium zwitterions with electrophilic rhodium(II)-vinylcarbenes *via* vinylogous addition/ring closure.

5.7 Summary

Catalytic cycloaddition of vinylcarbenes is a rapidly growing general methodology for the synthesis of carbo- and heterocyclic compounds. As is evident in this chapter, compounds with ring sizes from five to eight have been prepared *via* cycloaddition and in most cases with high yields and stereocontrol. Access to vinylcarbenes has expanded from vinyldiazo compounds to DACPs that undergo reactions at or below room temperature. Easily accessible β-silyloxy-α-diazo compounds appear to have the greatest generality for cycloaddition and, in contrast to styryldiazo compounds, they are stable to intramolecular pyrazole formation. Catalyst development has been central to these achievements, and those of copper and dirhodium(II) have the broadest applications. The use of chiral bisoxazoline ligands, formed from natural amino acid derivatives, for copper(I), evolving from simple Box to more structurally encumbered Sabox, have afforded very high stereoselectivities. Somewhat lower enantioselectivities, but of the opposite enantiomer, are produced by chiral dirhodium(II) carboxylate ligands that are also formed from natural amino acid derivatives.

There is much more to be accomplished in metallo-vinylcarbene cycloaddition reactions. The scope of vinyldiazo compounds that are suitable for cycloaddition awaits further evaluation, as does the extent to which cyclopropenes or other vinylcarbene generators can be employed. Catalysts appropriate for cycloaddition are any that could be used to achieve cyclopropanation with the caveat that competing reactions may be more favorable. This chapter reports many dipolar species that have been used for cycloaddition, but their number is far from complete, and this is particularly true for [3+1]- and those beyond [3+3]-cycloaddition. Adding to these conjugate extensions of vinylcarbenes that would be dienylcarbenes, the list of structural opportunities is very long.

References

1 Moss, R.A. and Doyle, M.P. (eds.) (2014). *Contemporary Carbene Chemistry*. New York, NY: Wiley.

2 Doyle, M.P., McKervey, M.A., and Ye, T. (1998). *Modern Catalytic Methods for Organic Synthesis with Diazo Compounds*. New York, NY: Wiley.

3 Won, J., Jung, H., Mane, M.V. et al. (2019). *Adv. Inorg. Chem.* 73: 385–443.

4 IUPAC (1997). *Compendium of Chemical Terminology* (the "Gold Book"). Compiled by A. D. McNaught and A. Wilkinson., 2e. Oxford: Blackwell Scientific Publications Online version (2019-) created by S. J. Chalk. ISBN 0-9678550-9-8. https://doi.org/10.1351/goldbook.

5 Müller, P., Pautex, N., Doyle, M.P., and Bagheri, V. (1990). Rh(II) catalyzed isomerizations of cyclopropenes. Evidence for Rh(II)-complexed vinylcarbene intermediates. *Helv. Chim. Acta* 73: 1233–1241.

6 Li, P., Zhang, X., and Shi, M. (2020). Recent developments in cyclopropene chemistry. *Chem. Commun.* 56: 5457–5471.

7 Archambeau, A., Miege, F., Meyer, C., and Cossy, J. (2015). Intramolecular cyclopropanation and C-H insertion reactions with metal carbenoids generated from cyclopropenes. *Acc. Chem. Res.* 48: 1021–1031.

8 Cheng, Q.-Q., Yu, Y., Yedoyan, J., and Doyle, M.P. (2018). Vinyldiazo reagents and metal catalysts: a versatile toolkit for heterocycle and carbocycle construction. *ChemCatChem* 10: 488–496.

9 Davies, H.M.L., Clark, D.M., and Smith, T.K. (1985). [3+4] Cycloaddition reactions of vinyl carbenoids with furans. *Tetrahedron Lett.* 26: 5659–5662.

10 Cheng, Q.-Q., Deng, Y., Lankelma, M., and Doyle, M.P. (2017). Cycloaddition reactions of enoldiazo compounds. *Chem. Soc. Rev.* 46: 5425–5443.

11 Marichev, K.O. and Doyle, M.P. (2019). Catalytic asymmetric cycloaddition reactions of enoldiazo compounds. *Org. Biomol. Chem.* 17: 4183–4195.

12 Dong, K., Marichev, K.O., and Doyle, M.P. (2019). The role of donor-acceptor cyclopropenes in metal carbene reactions. Conversion of *E*-substituted enoldiazoacetates to *Z*-substituted metallo-enolcarbenes. *Organometallics* 38: 4043–4050.

13 Koji, M., Sakae, U., and Kouichi, O. (2005). Transition metal-catalyzed reactions using alkynes as precursors of carbene and vinylidene complexes. *Chem. Lett.* 34: 1068–1073.

14 Chen, L., Chen, K., and Zhu, S. (2018). Transition-metal-catalyzed intramolecular nucleophilic addition of carbonyl groups to alkynes. *Chem* 4: 1208–1262.

15 Rautenstrauch, V. (1984). 2-Cyclopentenones from 1-ethynyl-2-propenyl acetates. *J. Org. Chem.* 49: 950–952.

16 Wang, J., Cao, X., Lv, S. et al. (2017). Synthesis and structures of gold and copper carbene intermediates in catalytic amination of alkynes. *Nat. Commun.* 8: 14625.

17 Tang, X.-T., Yang, F., Zhang, T.-T. et al. (2020). Recent progress in *N*-heterocyclic carbene gold-catalyzed reactions of alkynes involving oxidation/amination/cycloaddition. *Catalysts* 10: 350.

18 Krüger, S. and Gaich, T. (2014). Recent applications of the divinylcyclopropane–cycloheptadiene rearrangement in organic synthesis. *Beilstein J. Org. Chem.* 10: 163–193.

19 Hodgson, D.M., Brückl, T., Glen, R. et al. (2004). Catalytic enantioselective intermolecular cycloadditions of 2-diazo-3,6-diketoester-derived carbonyl ylides with alkene dipolarophiles. *Proc. Natl. Acad. Sci. U.S.A.* 101: 5450–5454.

20 Xu, X. and Doyle, M.P. (2014). The [3+3]-cycloaddition alternative for heterocycle syntheses: catalytically generated metalloenolcarbenes as dipolar adducts. *Acc. Chem. Res.* 47: 1396–1405.

21 Alcaide, B., Almendros, P., and Aragoncillo, C. (2010). Exploiting [2+2] cycloaddition chemistry: achievements with allenes. *Chem. Soc. Rev.* 39: 783–816.

22 Xu, Y., Conner, M.L., and Brown, M.K. (2015). Cyclobutane and cyclobutene synthesis: catalytic enantioselective [2+2] cycloadditions. *Angew. Chem. Int. Ed.* 54: 11918–11928.

23 Poplata, S., Tröster, A., Zou, Y.-Q., and Bach, T. (2016). Recent advances in the synthesis of cyclobutanes by olefin [2+2] photocycloaddition reactions. *Chem. Rev.* 116: 9748–9815.

24 Li, C., Jiang, K., Ouyang, Q. et al. (2016). [3+1]- and [3+2]-cycloadditions of azaoxyallyl cations and sulfur ylides. *Org. Lett.* 18: 2738–2741.

25 Barluenga, J., Riesgo, L., López, L.A. et al. (2009). Discrimination of diazo compounds toward carbenoids: copper(I)-catalyzed synthesis of substituted cyclobutenes. *Angew. Chem. Int. Ed.* 48: 7569–7572.

26 Barluenga, J., Riesgo, L., Lonzi, G. et al. (2012). Copper(I)-catalyzed [3+1] cycloaddition of alkenyldiazoacetates and iminoiodinanes: easy access to substituted 2-azetines. *Chem. Eur. J.* 18: 9221–9224.

27 Deng, Y., Massey, L.A., Zavalij, P.Y., and Doyle, M.P. (2017). Catalytic asymmetric [3+1]-cycloaddition reaction of ylides with electrophilic metallo-enolcarbene intermediates. *Angew. Chem. Int. Ed.* 56: 7479–7483.

28 Marichev, K.O., Wang, K., Dong, K. et al. (2019). Synthesis of chiral tetrasubstituted azetidines from donor-acceptor azetines via asymmetric copper(I)-catalyzed imido-ylide [3+1]-cycloaddition with metallo-enolcarbenes. *Angew. Chem. Int. Ed.* 58: 16188–16192.

29 Marichev, K.O., Dong, K., Massey, L.A. et al. (2019). Chiral donor–acceptor azetines as powerful reactants for synthesis of amino acid derivatives. *Nat. Commun.* 10: 5328.

30 Wang, R., Marichev, K.O., Dong, K. et al. (2020). Chiral 3-acylglutaric acid derivatives from strain-induced nucleophilic retro-claisen ring-opening reactions. *J. Org. Chem.* 85: 9475–9490.

31 Doyle, M.P., Yan, M., Hu, W.H., and Gronenberg, L.S. (2003). Highly selective catalyst-directed pathways to dihydropyrroles from vinyldiazoacetates and imines. *J. Am. Chem. Soc.* 125: 4692–4693.

32 Lian, Y. and Davies, H.M.L. (2010). Rhodium-catalyzed [3+2] annulation of indoles. *J. Am. Chem. Soc.* 132: 440–441.

33 Davies, H.M.L., Kong, N., and Churchill, M.R. (1998). Asymmetric synthesis of cyclopentenes by [3+2] annulations between vinylcarbenoids and vinyl ethers. *J. Org. Chem.* 63: 6586–6589.

34 Davies, H.M.L., Xiang, B.P., Kong, N., and Stafford, D.G. (2001). Catalytic asymmetric synthesis of highly functionalized cyclopentenes by a [3+2] cycloaddition. *J. Am. Chem. Soc.* 123: 7461–7462.

35 Briones, J.F. and Davies, H.M.L. (2013). Enantioselective gold(I)-catalyzed vinylogous [3+2] cycloaddition between vinyldiazoacetates and enol ethers. *J. Am. Chem. Soc.* 135: 13314–13317.

36 Smith, A.G. and Davies, H.M.L. (2012). Rhodium-catalyzed enantioselective vinylogous addition of enol ethers to vinyldiazoacetates. *J. Am. Chem. Soc.* 134: 18241–18244.

37 Deng, Y.M., Yglesias, M.V., Arman, H., and Doyle, M.P. (2016). [3+2] Cycloaddition of enecarbamates with electrophilic metalloenolcarbene intermediates. *Angew. Chem. Int. Ed.* 55: 10108–10112.

38 Xu, X., Leszczynski, J.S., Mason, S.M. et al. (2014). Expedient access to substituted 3-amino-2-cyclopentenones by dirhodium-catalyzed [3+2]-annulation of silylated ketene imines and enoldiazoacetates. *Chem. Commun.* 50: 2462–2464.

39 López, E. and López, L.A. (2017). Synthesis of functionalized cyclopentene derivatives from vinyldiazo compounds and vinylazides through sequential copper-promoted [3+2] cycloaddition/azide rearrangement. *Angew. Chem. Int. Ed.* 56: 5121–5124.

40 Zhang, B. and Davies, H.M.L. (2020). Rhodium-catalyzed enantioselective [4+2] cycloadditions of vinylcarbenes with dienes. *Angew. Chem. Int. Ed.* 59: 4937–4941.

41 Lian, Y.J. and Davies, H.M.L. (2010). Rhodium carbenoid approach for introduction of 4-substituted (Z)-pent-2-enoates into sterically encumbered pyrroles and indoles. *Org. Lett.* 12: 924–927.

42 Lian, Y.J. and Davies, H.M.L. (2012). Rh_2(S-biTISP)$_2$-catalyzed asymmetric functionalization of indoles and pyrroles with vinylcarbenoids. *Org. Lett.* 14: 1934–1937.

43 Barluenga, J., Tudela, E., Ballesteros, A., and Tomás, M. (2009). Asymmetric C_2-C_3 cyclopentannulation of the indole ring. *J. Am. Chem. Soc.* 131: 2096–2097.
44 Jing, C., Cheng, Q.-Q., Deng, Y. et al. (2016). Highly regio- and enantioselective formal [3 + 2]-annulation of indoles with electrophilic enol carbene intermediates. *Org. Lett.* 18: 4550–4553.
45 Ford, A., Miel, H., Ring, A. et al. (2015). Modern organic synthesis with α-diazocarbonyl compounds. *Chem. Rev.* 115: 9981–10080.
46 Wagh, S.B. and Liu, R.-S. (2015). Gold-catalyzed reactions of propargylic esters with vinylazides for the synthesis of *Z*- or *E*-configured buta-1,3-dien-2-yl esters. *Chem. Commun.* 51: 15462–15464.
47 Wang, X.C., Abrahams, Q.M., Zavalij, P.Y., and Doyle, M.P. (2012). Highly regio- and stereoselective dirhodium vinylcarbene induced nitrone cycloaddition with subsequent cascade carbenoid aromatic cycloaddition/N-O cleavage and rearrangement. *Angew. Chem. Int. Ed.* 51: 5907–5910.
48 Yang, X., Xu, P., and Xue, Y. (2014). Mechanism and regioselectivity of the cycloaddition between nitrone and dirhodium vinylcarbene catalyzed by $Rh_2(O_2CH)_4$: a computational study. *Theor. Chem. Acc.* 133: 1549.
49 Cheng, Q.-Q., Lankelma, M., Wherritt, D. et al. (2017). Divergent rhodium-catalyzed cyclization reactions of enoldiazoacetamides with nitrosoarenes. *J. Am. Chem. Soc.* 139: 9839–9842.
50 Anderson, L.L. (2016). Diverse applications of nitrones for the synthesis of heterocyclic compounds. *Asian J. Org. Chem.* 5: 9–30.
51 Wang, X., Xu, X., Zavalij, P.Y., and Doyle, M.P. (2011). Asymmeric formal [3+3]-cycloaddition reactions of nitrones with electrophilic vinylcarbene intermediates. *J. Am. Chem. Soc.* 133: 16402–16405.
52 Cheng, Q.-Q., Yedoyan, J., Arman, H., and Doyle, M.P. (2016). Copper-catalyzed divergent addition reactions of enoldiazoacetamides with nitrones. *J. Am. Chem. Soc.* 138: 44–47.
53 Marichev, K.O., Adly, F.G., Carranco, A.M. et al. (2018). Catalyst choice for highly enantioselective [3+3]-cycloaddition of enoldiazocarbonyl compounds. *ACS Catal.* 8: 10392–10400.
54 Marichev, K.O., Wang, Y., Carranco, A.M. et al. (2018). Rhodium(II)-catalysed generation of cycloprop-1-en-1-yl ketones and their rearrangement to 5-aryl-2-siloxyfurans. *Chem. Commun.* 54: 9513–9516.
55 Adly, F.G., Marichev, K.O., Jensen, J.A. et al. (2019). Enoldiazosulfones for highly enantioselective [3 + 3]-cycloaddition with nitrones catalyzed by copper(I) with chiral BOX ligands. *Org. Lett.* 21: 40–44.
56 Dong, K., Xu, X., and Doyle, M.P. (2020). Copper(I)-catalyzed highly enantioselective [3+3]-cycloaddition of γ-alkyl enoldiazoacetates with nitrones. *Org. Chem. Front.* 7: 1653–1657.
57 Xu, X., Zavalij, P.J., and Doyle, M.P. (2013). A donor-acceptor cyclopropene as a dipole source for a silver(I) catalyzed asymmetric catalytic [3+3]-cycloaddition with nitrones. *Chem. Commun.* 49: 10287–10289.
58 Chrzanowska, M. and Rozwadowska, M.D. (2004). Asymmetric synthesis of isoquinoline alkaloids. *Chem. Rev.* 104: 3341–3370.

59 Perreault, C., Goudreau, S.R., Zimmer, L.E., and Charette, A.B. (2008). Cycloadditions of aromatic azomethine imines with 1,1-cyclopropane diesters. *Org. Lett.* 10: 689–692.

60 Zhang, O., Yang, L., and Tong, X. (2010). 2-(Acetoxymethyl)buta-2,3-dienoate, a versatile 1,4-biselectrophile for phosphine-catalyzed (4+n) annulations with 1,n-bisnucleophiles (n = 1, 2). *J. Am. Chem. Soc.* 132: 2550–2551.

61 Zhou, Y.-Y., Li, J., Ling, L. et al. (2013). Highly enantioselective [3+3] cycloaddition of aromatic azomethine imines with cyclopropanes directed by π–π stacking interactions. *Angew. Chem. Int. Ed.* 52: 1452–1456.

62 Guo, C., Fleige, M., Janssen-Müller, D. et al. (2015). Switchable selectivity in an NHC-catalysed dearomatizing annulation reaction. *Nat. Chem.* 7: 842–847.

63 Xu, X., Zavalij, P.Y., and Doyle, M.P. (2013). Highly enantioselective dearomatizing formal [3+3]-cycloaddition reactions of *N*-acyliminopyridinium ylides with electrophilic enolcarbene intermediates. *Angew. Chem. Int. Ed.* 52: 12664–12668.

64 Xu, X., Zavalij, P.Y., and Doyle, M.P. (2013). Catalytic asymmetric syntheses of quinolizidines by dirhodium-catalyzed dearomatization of isoquinolinium/pyridinium methylides – the role of catalyst and carbene source. *J. Am. Chem. Soc.* 135: 12439–12447.

65 Xu, X., Zavalij, P.Y., and Doyle, M.P. (2012). Synthesis of tetrahydropyridazines by a metal-carbene directed highly enantioselective vinylogous N-H insertion/Lewis acid catalyzed diastereoselective Mannich addition. *Angew. Chem. Int. Ed.* 51: 9829–9833.

66 Požgan, F., Mamari, H.A., Grošelj, U. et al. (2018). Synthesis of non-racemic pyrazolines and pyrazolidines by [3+2] cycloadditions of azomethine imines. *Molecules* 23: 3.

67 Shintani, R. and Hayashi, T. (2006). Palladium-catalyzed [3+3] cycloaddition of trimethylenemethane with azomethine imines. *J. Am. Chem. Soc.* 128: 6330–6331.

68 Shapiro, N.D., Shi, Y., and Toste, F.D. (2009). Gold-catalyzed [3+3]-annulation of azomethine imines with propargyl esters. *J. Am. Chem. Soc.* 131: 11654–11655.

69 Qian, Y., Zavalij, P.J., Hu, W., and Doyle, M.P. (2013). Bicyclic pyrazolidinone derivatives from diastereoselective catalytic [3+3]-cycloaddition reactions of enoldiazoacetates with azomethine imines. *Org. Lett.* 15: 1564–1567.

70 Zhang, Y., Yang, Y., Zhao, J., and Xue, Y. (2018). Mechanism and diastereoselectivity of [3+3] cycloaddition between enol diazoacetate and azomethine imine catalyzed by dirhodium tetracarboxylate: a theoretical study. *Eur. J. Org. Chem.*: 3086–3094.

71 Deng, Y., Massey, L.A., Rodriguez Núñez, Y.A. et al. (2017). Catalytic divergent [3+3]- and [3+2]-cycloaddition by discrimination between diazo compounds. *Angew. Chem. Int. Ed.* 56: 12292–12296.

72 Zheng, H. and Doyle, M.P. (2019). Catalytic desymmetric cycloaddition of diaziridines with metalloenolcarbenes. The role of donor-acceptor cyclopropenes. *Angew. Chem. Int. Ed.* 58: 12502–12506.

73 Cheng, Q.-Q., Qian, Y., Zavalij, P.Y., and Doyle, M.P. (2015). Lewis acid/rhodium-catalyzed formal [3+3]-cycloaddition of enoldiazoacetates with donor–acceptor cyclopropanes. *Org. Lett.* 17: 3568–3571.

74 Qin, C. and Davies, H.M.L. (2013). $Rh_2(R\text{-}TPCP)_4$-catalyzed enantioselective [3+2]-cycloaddition between nitrones and vinyldiazoacetates. *J. Am. Chem. Soc.* 135: 14516–14519.

75 Guzmán, P.E., Lian, Y.J., and Davies, H.M.L. (2014). Reversal of the regiochemistry in the rhodium-catalyzed [4+3] cycloaddition between vinyldiazoacetates and dienes. *Angew. Chem. Int. Ed.* 53: 13083–13087.

76 Zhu, C., Xu, G., and Sun, J. (2016). Gold-catalyzed formal [4+1]/[4+3] cycloadditions of diazo esters with triazines. *Angew. Chem. Int. Ed.* 55: 11867–11871.

77 Lee, D.J., Ko, D., and Yoo, E.J. (2015). Rhodium(II)-catalyzed cycloaddition reactions of non-classical 1,5-dipoles for the formation of eight-membered heterocycles. *Angew. Chem. Int. Ed.* 54: 13715–13718.

6

Metal-Catalyzed Decarbenations by Retro-Cyclopropanation

Mauro Mato[1,2] *and Antonio M. Echavarren*[1,2]

[1] *Institute of Chemical Research of Catalonia (ICIQ), Av. Països Catalans 16, 43007 Tarragona, Spain*
[2] *Universitat Rovira i Virgili, Departament de Química Analítica i Química Orgànica, 43007 Tarragona, Spain*

6.1 Introduction

Metal carbenes are one of the most studied organometallic intermediates in organic synthesis due to the unique and diverse reactivity patters that they display [1, 2]. However, the access to these very reactive species often has to be unlocked by the use of highly energetic precursors, such as diazo compounds [3]. Accordingly, great efforts have been made on the discovery and development of new carbene precursors over the years [4]. In this regard, *metal-catalyzed decarbenations* via formal retro-cyclopropanation processes have emerged as a powerful, versatile, and safe alternative for the generation of non-acceptor metal carbenes [5].

6.2 Reactivity and Generation of Metal Carbenes

Analogously to singlet free carbenes, the benchmark reaction displayed by metal carbenes is the cyclopropanation of alkenes [1]. Besides, the Buchner ring expansion [6] (first reported in 1885) [7] is a special type of cyclopropanation, in which a carbene reacts with an aromatic ring giving rise to a norcaradiene, which is in equilibrium with the corresponding cycloheptatriene through a thermally allowed disrotatory electrocyclic ring opening (Scheme 6.1) [8].

Besides (2+1) cyclopropanations, metal carbenes can engage in higher formal cycloadditions ($m+n$) [9], and take part in insertion [10], or coupling reactions [11], among other transformations [1]. Due to this rich and unique reactivity, the controlled catalytic generation of metal carbenes is considered one of the cornerstone challenges of modern organometallic chemistry, and a large number of approaches have been explored in this regard.

Transition Metal-Catalyzed Carbene Transformations, First Edition.
Edited by Jianbo Wang, Chi-Ming Che, and Michael P. Doyle.

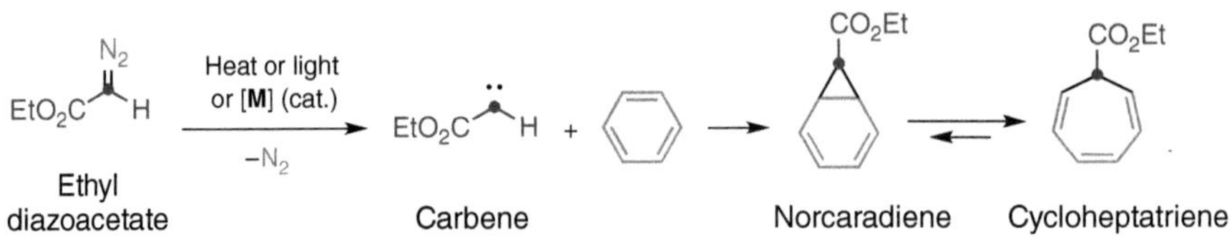

Scheme 6.1 The Buchner reaction of ethyl diazoacetate with benzene and the norcaradiene–cycloheptatriene equilibrium.

6.2.1 Decomposition of Diazo Compounds

The first and most studied way of generating carbenes is the decomposition of diazo compounds through release of a nitrogen molecule, which can be promoted by heat, light, or a metal catalyst [3]. This process was first reported by Buchner and Curtius in 1885 [7], and since then it has evolved to be the most widespread strategy for the generation of carbene intermediates, finding numerous applications in organic synthesis [12, 13].

6.2.2 Alternative Methods for the Generation of Metal Carbenes

Diazo compounds bearing a stabilizing (electron-withdrawing) group are generally employed for the generation of carbenes, since they are easier to handle. However, this limits this methodology to the use of acceptor carbenes: non-acceptor diazo compounds are usually challenging to prepare, inherently unstable, explosive in pure form, cannot be stored for long periods of time [14], and are prone to dimerization [15]. For these reasons, many groups have recently turned their attention to the development of versatile, practical, and safe donor carbene precursors [16]. In this regard, substrates such as enynes [17], diynes [18], propargyl carboxylates or ethers [19], cyclopropenes [20], *N*-tosylhydrazones [14], and alkynes under oxidative conditions [21] have been explored as precursors of carbene intermediates under metal catalysis [22].

Over the last decade, 7-substituted 1,3,5-cycloheptatrienes (and other derivatives) have emerged not only as a safe alternative for the generation of donor carbenes (through aromative decarbenations by retro-Buchner reactions), but also as a powerful incorporation to the reaction discovery toolbox and for the rapid construction of molecular complexity [5].

6.2.3 Decarbenation Reactions: General Process and Definition

A *decarbenation* is a process involving the fragmentation of a molecular entity to release a carbene fragment, upon generation of a stable molecule as the driving force.

In the context of this review, the term *metal-catalyzed decarbenation* is employed to define any process in which a metal carbene is generated catalytically, by cleavage of two C—C sigma bonds, upon release of an unsaturated organic fragment (namely an aromatic unit or an alkene), through a formal *retro-Buchner* or *retro-cyclopropanation* reaction. The driving force of these decarbenation reactions is the opening of a strained three-membered ring while releasing a much more stable alkene or aromatic unit, such as benzene, mesitylene, or naphthalene. The most studied decarbenation process by retro-cyclopropanation is the metal-catalyzed aromative decarbenation by retro-Buchner reaction of 7-substituted 1,3,5-cycloheptatrienes, through cleavage of their norcaradiene tautomer, which generates a metal carbene upon release of an aromatic molecule (Scheme 6.2) [5].

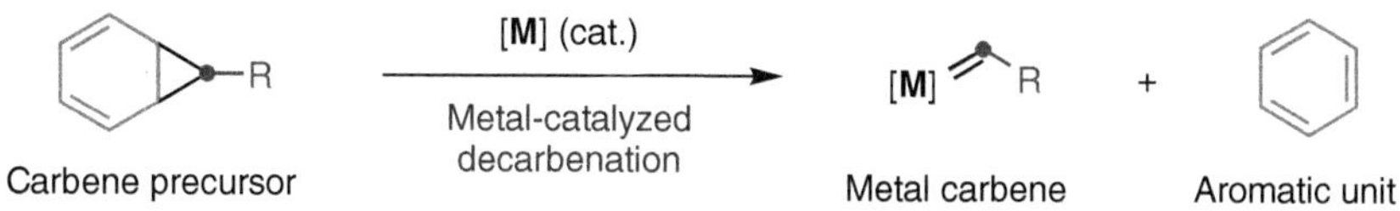

Scheme 6.2 Metal-catalyzed aromative decarbenation by retro-Buchner reaction. Source: Based on Mato et al. [5].

Some other processes leading to carbene intermediates, such as the decomposition of diazo compounds or alkene metathesis cycles [23], would also fall under the wide definition of *decarbenation* reactions. However, since they involve the cleavage of double C—C or C—heteroatom bonds (instead of sigma C—C bonds via retro-cyclopropanation), they are out of the focus of this chapter.

6.3 Retro-Cyclopropanation Reactions: A Historical Walkthrough

From the 1970s, almost 100 years after the discovery of the Buchner ring expansion, some isolated examples of the opposite process were reported. However, the possibility of using the formally released carbene fragments in a synthetically useful manner was not clear at the time. It was not until homogeneous gold(I) catalysis came into play that decarbenation reactions emerged as a powerful tool in organic synthesis [5].

6.3.1 Early Observations

To the best of our knowledge, the first example of a metal-catalyzed retro-cyclopropanation process was reported by Volger and coworkers in 1973 [24]. This transformation involved the Rh(I)-catalyzed formal extrusion of a methylene carbene unit from hexamethylnorcaradiene **2**, to produce hexamethylbenzene quantitatively (Scheme 6.3). The authors observed the trapping of the corresponding methylene carbene with cyclohexene to a certain extent, but the concept did not evolve into any synthetic methodology at the time.

In 1976, Gassman and Johnson showed that, in the presence of the highly electrophilic $PhWCl_3$—$EtAlCl_2$ pair, simple ethylcyclopropane (**3**) reacted to give a 7 : 1 mixture of 1-butene and ethylene, resulting from a formal retro-cyclopropanation or, as it was originally termed, a *retrocarbene addition* [25]. Subsequently, they found that if the same decarbenation process was carried out in the presence of an external alkene such as ethyl acrylate (**4**), the released methylene fragment could be trapped

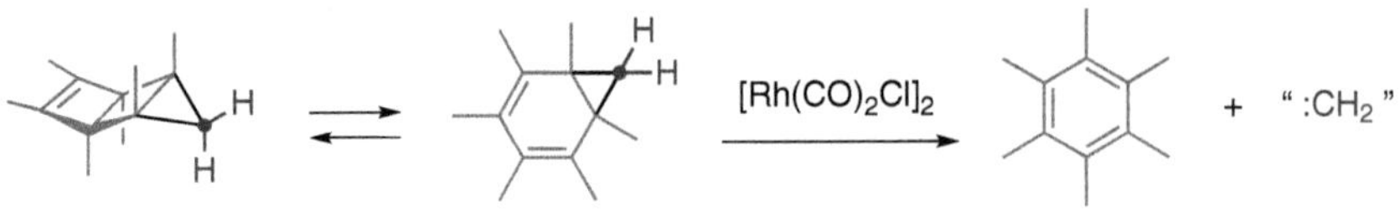

1 **2**

Scheme 6.3 First metal-catalyzed retro-cyclopropanation. Source: Based on Volger et al. [24].

to give cyclopropyl ester **6**, resulting in an overall cyclopropane–alkene-metathesis process (Scheme 6.4) [26].

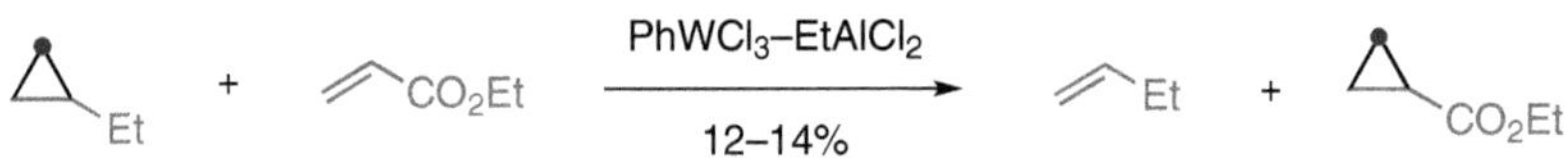

3 **4** **5** **6**

Scheme 6.4 First cyclopropane–alkene cross-metathesis. Source: Based on Gassman and Johnson [25].

Without the aid of a metal catalyst, Dvoretzky and coworkers showed that phenanthrene derivatives **7** could undergo a retro-cyclopropanation process under photochemical conditions, releasing phenanthrene [27] (Scheme 6.5). This type of process was further studied by other groups [28], both experimentally and computationally [29].

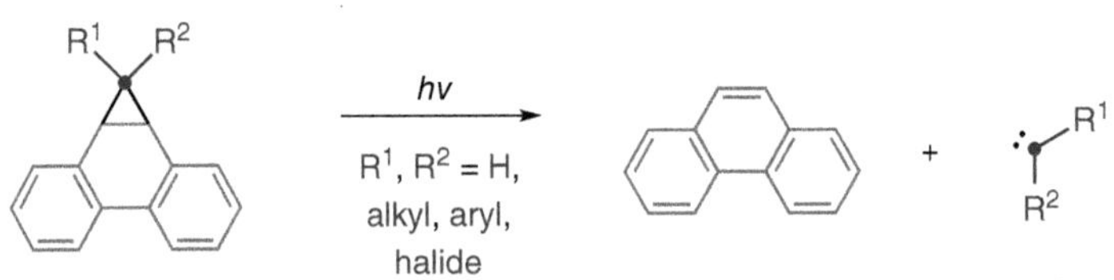

7 **8** **9**

Scheme 6.5 Retro-cyclopropanation of phenanthrene derivatives under photochemical conditions. Source: Based on Richardson et al. [27].

Further examples of this type of reactivity were isolatedly reported throughout the years. Thus, it was found that $Pd(OAc)_2$ promoted the formation of trace amounts of retro-cyclopropanation products from ethyl 1,3,5-cycloheptatriene-7-carboxylate [30]. Interestingly, simple 1,3,5-cycloheptatriene was transformed into benzene at 600 K on the surface of W(100)-(5 × 1)-C metal carbide [31]. Highly strained biscyclobutanes were also shown to be cleaved under nickel [32], and rhodium catalysis [33]. Echegoyen, Diedrich, Martín, and coworkers showed that methanofullerene derivatives can also undergo a retro-cyclopropanation under both chemical [34] and electrochemical conditions [35].

6.3.2 Decarbenation Reactions from Gas Phase to Solution

Recently, the group of Chen studied in detail the generation and detection of gold(I)–carbene complexes by gas-phase decarbenation, upon collision-induced dissociation (CID) of gold(I) phosphonium yilides such as **10**. This procedure allowed the detection of highly energetic intermediates **11**, which display the typical reactivity of metal carbenes: cyclopropanation of alkenes. Furthermore, the resulting cyclopropanes were found to undergo retro-cyclopropanation under the same conditions. Thus, both an alternative carbene fragment **13**, and the formal carbene–alkene metathesis product **14** were detected, pointing strongly toward the reversibility of the cyclopropanation steps (Scheme 6.6) [36].

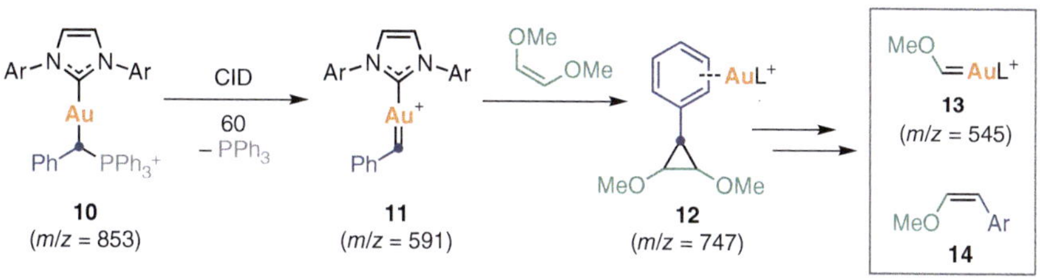

Scheme 6.6 Gas-phase generation, detection, and reactivity of gold(I) carbenes. Source: Based on Fedorov et al. [36].

The same group applied the knowledge gathered during their study in gas phase for the design of a carbene precursor in solution. They found that reaction of sulfone-imidazolium gold(I) carbenoids **15** with styrenes at 120 °C affords *cis*-1,2-diarylcyclopropanes **16** (Scheme 6.7). Using stoichiometric amounts of an external imidazolium salt containing the aryl carbene fragment precursor, the same transformation could be performed catalytically with a gold(I) complex **15** [37].

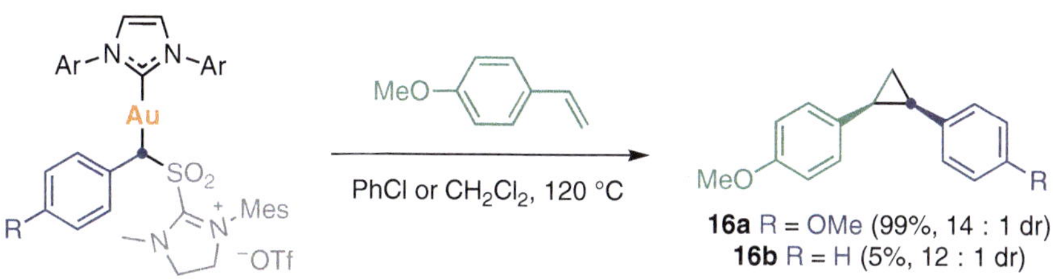

Scheme 6.7 Generation of aryl carbenes in solution from sulfone-imidazolium carbenoids. Source: Based on Ringger and Chen [37].

6.3.3 The Discovery of the Gold(I)-Catalyzed Retro-Buchner Reaction

In 2010, within the context of the gold(I)-catalyzed cycloisomerization of aryl-linked 1,6-enynes **17**, the generation of carbenes by retro-cyclopropanation of electron-rich benzo-fused norcaradienes **19** was observed (Scheme 6.8). In the presence of cationic gold(I) complexes, enyne **17** gives rise to a mixture of naphthalene **20** and biscyclopropane **22**. Naphthalene **20** was proposed to arise from the retro-cyclopropanation of intermediate **19**, and the resulting free aryl gold(I) carbene **21** would be trapped by another unit of dihydronaphthatlene **19**, giving biscyclopropane **22**. The mechanistic proposal was supported by the independent synthesis of gold-free **19**: treatment of this dihydro-1*H*-cyclopropa[*a*]naphthalene with a cationic gold(I) complex led to a similar distribution of products [38]. Years later, a similar strategy was employed to generate cyclopropyl gold(I) carbenes from 1-cyclopropyl-1,6-enynes, which evolved into 1,3-dienes [39].

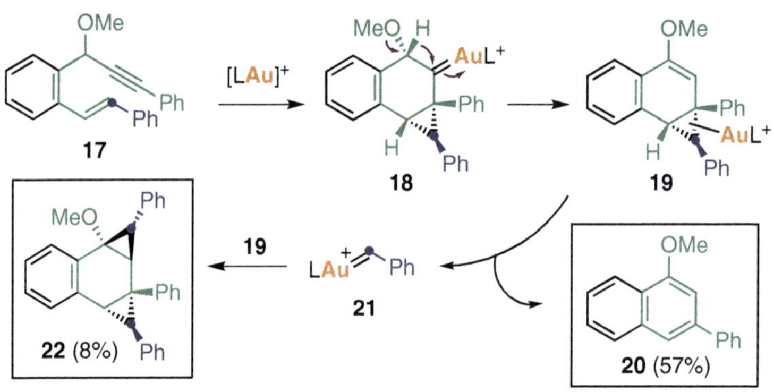

Scheme 6.8 Gold(I)-catalyzed annulation/fragmentation of 1,6-enynes.

The structural resemblance of carbene precursor **19** to 7-aryl norcaradienes (**23**) paved the way for the discovery and implementation of 7-substituted cycloheptatrienes (**24**) as practical alternatives for the generation of gold(I) carbenes. Thus, it was found that the reaction of **24** with (*E*)-stilbene in the presence of a cationic gold(I) complex at 120 °C affords 1,2,3-triphenylcyclopropane **25**, upon release of benzene (Scheme 6.9) [40].

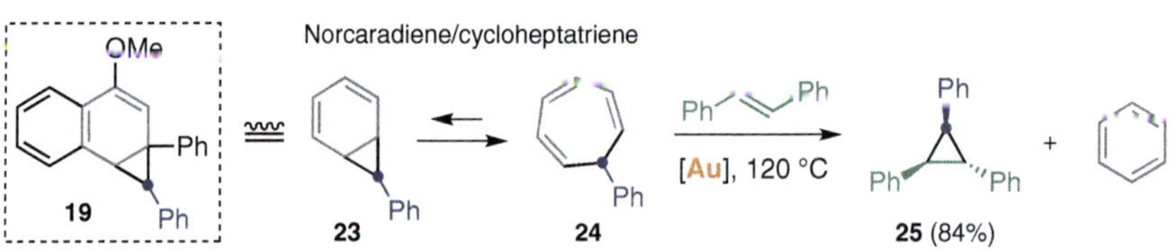

Scheme 6.9 Gold(I)-catalyzed decarbenation or retro-Buchner reaction of cycloheptatrienes, followed by cyclopropanation. Source: Based on Solorio-Alvarado et al. [40].

In 2015, Hashmi and coworkers reported the reactivity of hydroxydiynes **26** in the presence of gold catalysts, obtaining naphthalenes such as **28**. The mechanistic proposal involved the intermediacy of benzo-fused norcaradiene **27**, which would undergo a retro-cyclopropanation process, resulting in the thermodynamically downhill release of ethylene, which was detected by means of mass spectrometry (Scheme 6.10) [41].

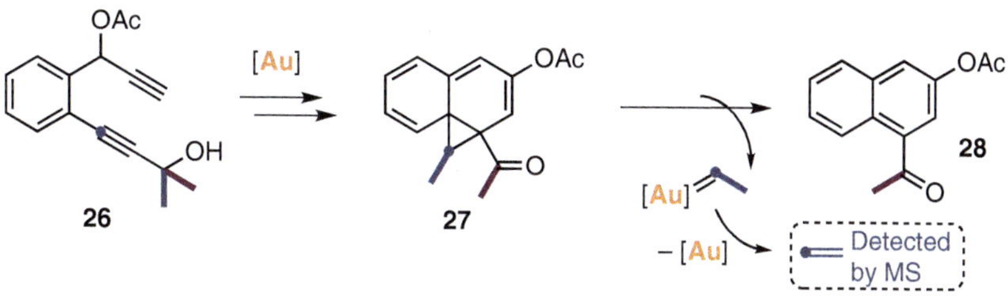

Scheme 6.10 Gold(I)-catalyzed naphthalene synthesis through ethylene release. Source: Based on Lauterbach et al. [41].

6.4 Metal-Catalyzed Aromative-Decarbenation Reactions: A Mechanistic Analysis

Although the generation and reactivity of metal carbenes from diazo compounds has been studied extensively in the past [42, 43], the mechanism of metal-catalyzed decarbenations based on retro-cyclopropanation processes has remained unexplored until recent years. Today, there is enough experimental and theoretical data to form a clear mechanistic picture.

6.4.1 Basic Mechanistic Picture

Overall, the mechanistic proposal for metal-catalyzed decarbenations via retro-cyclopropanation or retro-Buchner reaction is analogous to that for the decomposition of diazo compounds (Scheme 6.11). First, the norcaradiene tautomer reacts with the electrophilic metal complex, in a process in which the first C—C bond of the three-membered ring is cleaved. This is often the rate-limiting step of the process (with barriers in the range of 15–25 kcal mol^{-1}), and leads to a Wheland-type carbenoid intermediate (Scheme 6.11, gray box, top). The latter evolves smoothly to generate a metal carbene, upon cleavage of the second C—C bond, while releasing an aromatic molecule [44].

Similarly, nucleophilic diazo compounds react with electrophilic metal centers, forming carbenoid adducts [43] which evolve to metal carbenes upon downhill release of molecular nitrogen (Scheme 6.11, bottom).

6.4.2 Alternative Generation of the Same Carbenes from Carbenoids

All metal-catalyzed retro-cyclopropanations reported to date have to be carried out at temperatures above 0 °C. This makes the detection or characterization of the highly energetic carbene intermediates a significant challenge, which has not met with

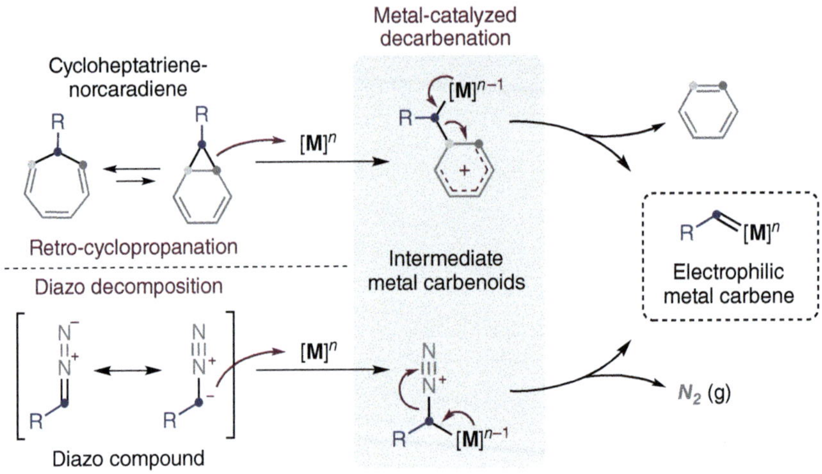

Scheme 6.11 Overall mechanistic picture for decarbenation reactions. Analogy between retro-cyclopropanations and diazo decomposition.

success thus far. However, alternative methods have been explored in order to access the very same reactive species [45, 46]. One of these methods is the use of stable metal carbenoids as precursors [47].

Metal carbenoids are organometallic species of the type $(X)(M)C(R^1)(R^2)$ (Scheme 6.12, left, **29**), in which an sp^3 carbon atom C (for which R^1 and R^2 are H or organic substituents) is bound both to a metal M and to a leaving group X. These species can evolve or be in equilibrium with the corresponding carbenes $(M)=CR^1R^2$ (**30**) upon release of the leaving group X and display similar reactivity than the latter, such as cyclopropanating alkenes [48]. For instance, gold(I) chloromethyl carbenoid **31** allows the methylene cyclopropanation of alkenes, upon activation by chloride abstraction [47].

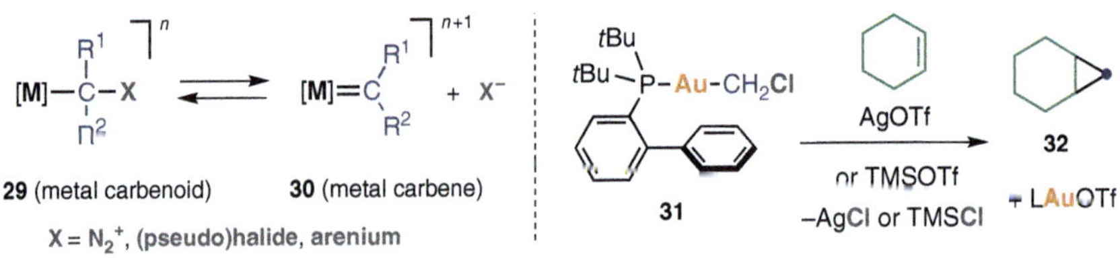

Scheme 6.12 Chemical relationship between metal carbenes and carbenoids.

Both the proposed intermediates for the decomposition of diazo compounds or for the retro-Buchner reaction can be considered metal carbenoids (Scheme 6.11, gray box), but they are so reactive that very rarely can be detected [43]. On the other hand, chloromethyl aryl carbenoids such as **33** are relatively inert precursors, which upon treatment with $GaCl_3$ as chloride abstractor, evolve to the same type of aryl carbenes (**34**) generated in the gold(I)-catalyzed retro-Buchner reaction (Scheme 6.13) [49]. These intermediates were characterized by NMR at −90 °C, and in the presence of an alkene, they undergo cyclopropanations smoothly at the same temperature. These cyclopropanations taking place at −90 °C further support the hypothesis of the decarbenation being the rate-limiting step in retro-Buchner/cyclopropanation sequences [44].

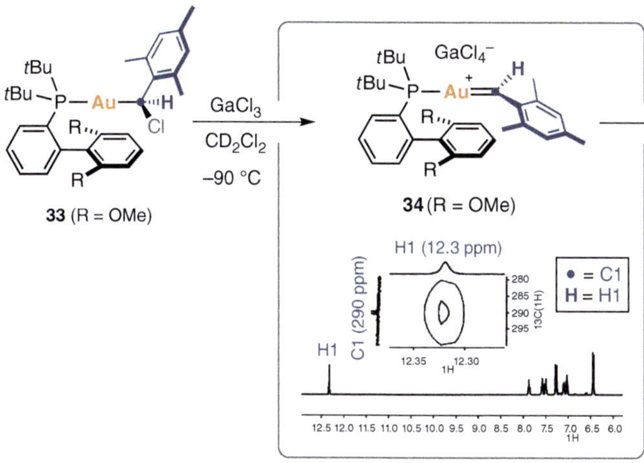

Scheme 6.13 Generation and characterization of gold carbenes via gold carbenoids. Source: Based on García-Morales et al. [49].

This strategy was also employed by the groups of Fürstner to generate other types of metal carbenes under mild conditions [50].

6.4.3 Theoretical Studies on the Mechanism of the Retro-Buchner Reaction

The first close look at the mechanism of metal-catalyzed decarbenations by retro-cyclopropanation was reported in 2014, in the context of the intramolecular reactivity of *ortho*-substituted 7-aryl-1,3,5-cycloheptatrienes under gold catalysis. The decarbenation of **I** was analyzed theoretically by means of DFT calculations, using trimethylphosphine gold(I) as model catalyst (Scheme 6.14) [44].

Several gold(I)-coordinated species were analyzed, being cyclopropane-coordinated **III** the productive one in the decarbenation process. The first C—C bond in **III** is cleaved through $\mathbf{TS_{III\text{-}IV}}$, giving Wheland carbenoid-type intermediate **IV**. This step is the turnover-limiting step (TLS) of the entire process, with an activation barrier of 23.3 kcal mol^{-1} from the most stable species **II**. This energy is consistent with a catalytic process which takes place at 100–120 °C, as observed experimentally. Intermediate **IV** is a shallow minimum that evolves smoothly through $\mathbf{TS_{IV\text{-}V}}$ to give benzene-coordinated gold(I) carbene **V**. Even though the entire process is moderately endothermic, subsequent reactions, with alkenes for instance, are highly exothermic and often take place through transitions states with small activation barriers.

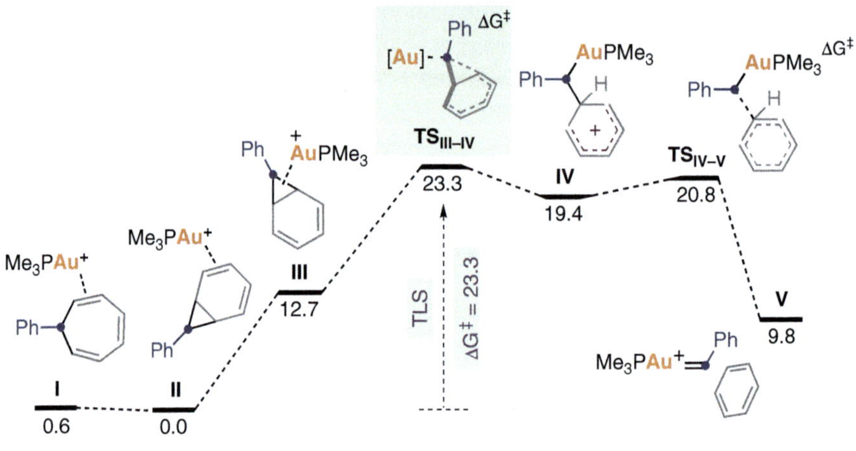

Scheme 6.14 Energy profile for the decarbenation of 7-phenyl-1,3,5-cycloheptatriene. Source: Based on Wang et al. [44].

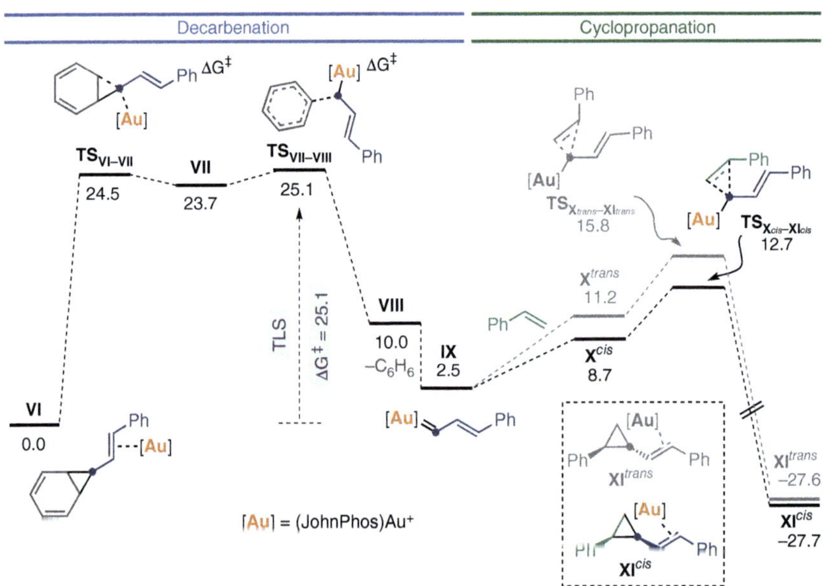

Scheme 6.15 Energy profile for the decarbenation of 7-styryl-1,3,5-cycloheptatriene and the stereoselective *cis*-vinylcyclopropanation of styrene. Source: Based on Herlé et al. [51].

Another computational study was carried out three years later on a gold(I)-catalyzed *cis*-vinylcyclopropantion from 7-alkenyl-1,3,5-cycloheptatrienes (Scheme 6.15) [51]. A similar scenario was observed: a stepwise cleavage of the two C—C bonds of gold(I)-coordinated norcaradiene **VI**. Although this process does not occur in a concerted manner, the potential energy surface linking $\mathbf{TS_{VI\text{-}VII}}$ and $\mathbf{TS_{VI\text{-}VII}}$ is rather shallow and, in this case, the highest energy point corresponds to the cleavage of the second C—C bond in Wheland intermediate **VII**. The activation energy of this step is 25.1 kcal mol^{-1}, consistent with the temperature (c. 75 °C) required for the catalytic cycle to proceed at a reasonable rate. Again, the slightly endothermic formation of gold(I) carbene **IX** is followed by reaction with styrene (and related activated alkenes) through lower-energy barriers to give preferentially *cis*-vinylcyclopropane $\mathbf{XI}^{cis}$ by a concerted cyclopropanation pathway. A complete stereochemical model was also developed, accounting for the excellent *cis*/*trans* selectivity observed in these reactions: even though cyclopropanes $\mathbf{XI}^{cis}$ and $\mathbf{XI}^{trans}$ have almost identical energy, ratios of >20 : 1 *cis*/*trans* are observed experimentally. This diastereoselectivity may arise from the attractive noncovalent π–π interactions between the two aromatic fragments, which are present in $\mathbf{TS}_{\mathbf{X}cis\text{-}\mathbf{XI}cis}$, but absent in $\mathbf{TS}_{\mathbf{X}trans\text{-}\mathbf{XI}trans}$ ($\Delta\Delta G^{\ddagger} = 3.1$ kcal mol^{-1}).

Certain vinylcyclopropane products, for which their *trans* isomer is thermodynamically more stable, undergo a *cis*-to-*trans* isomerization in the presence of the same gold(I) catalyst, observable at 75 °C over several days. This is the result of an equilibration process by the rotation of the cationic intermediates which would arise from a reversible opening of gold-coordinated vinylcyclopropanes **XI** [51]. The *cis*-to-*trans* isomerization of cyclopropanes was also observed by the group of Fiksdahl in the presence of Au(III) complexes [52].

6.4.4 Second-Generation Cycloheptatrienes: Low Temperature and Other Metals

At this point in time, two of the main limitations of the retro-Buchner reaction were the requirement of high temperatures and the exclusive access to gold(I)–carbene intermediates. This was circumvented by the development of a new generation of more reactive cycloheptatrienes.

The working hypothesis for the new design was aiming for the stabilization of the transitions states and intermediates leading to the formation of metal carbenes **V** or **IX**. In particular, transitions states leading to Wheland intermediates such as **IV** or **VII** would be stabilized by the introduction of electron-donating substituents in the cycloheptatriene ring, lowering the activation barrier of the proposed rate-limiting step of the decarbenation process. Methyl substituents were found to be the ideal electron-donating elements, since more coordinating groups such as MeO- could potentially lead to more stable resting states, translating into higher overall activation barriers (Scheme 6.16) [53].

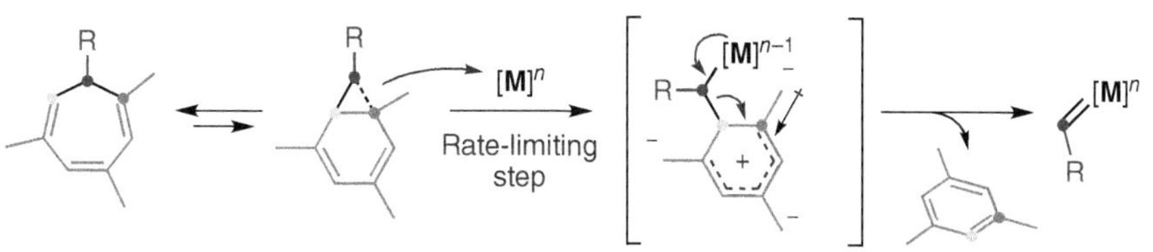

Scheme 6.16 Working hypothesis for the design of more reactive cycloheptatrienes.

Regarding the synthesis of these carbene precursors, simple 7-substituted 1,3,5-cycloheptatrienes **38** can be prepared by reaction of nucleophiles (especially organometallic reagents) with commercially available tropylium tetrafluoroborate (**37**) (Scheme 6.17, left) [54]. A powerful alternative for the synthesis of 7-alkenyl-1,3,5-cycloheptatrienes is the Julia–Kocienski olefination with sulfones such as **39** [51]. This strategy allowed for the gram-scale preparation of a variety of more reactive trimethyl-subtituted cycloheptatrienes (**40**) in excellent yields (Scheme 6.17, right) [53].

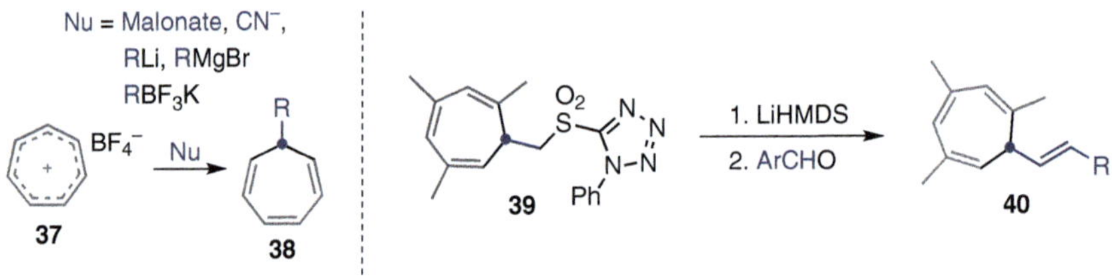

Scheme 6.17 Strategies for the synthesis of 7-substituted cycloheptatrienes. Source: Based on Mato et al. [53].

These more reactive methylated carbene precursors **40** indeed unlocked an improved protocol for the gold-catalyzed decarbenation/cyclopropanation sequence, allowing it to proceed at room temperature and more selectively. More importantly, the same reaction could be, for the first time, carried out under catalysis of other metals: zinc(II) [53] and rhodium(II) (Scheme 6.18, see Section 6.5.1.2) [55].

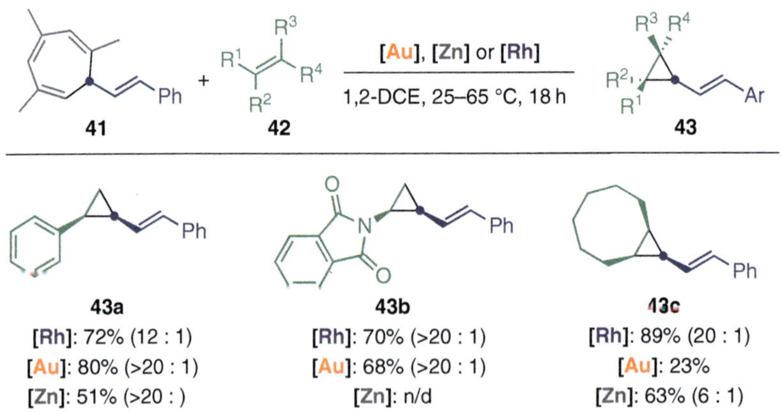

Scheme 6.18 [(JohnPhos)Au(MeCN)]SbF_6 (25 °C), $ZnBr_2$ (65 °C) or Rh_2TFA_4 (25–60 °C) as catalysts for the decarbenation/cyclopropanation sequence. Source: Based on Mato and Echavarren [55].

Zinc(II) salts such as $ZnBr_2$ emerge as an inexpensive alternative to traditional cationic gold(I) complexes. On the other hand, analogous non-acceptor rhodium(II) carbenes display a richer reactivity, allowing the development of new synthetic methodologies previously inaccessible using gold (see Section 6.5) [55, 56].

All in all, the design and development of more reactive cycloheptatrienes allowed the extension of the retro-Buchner reaction as a general method for accessing metal carbenes, while also proceeding under much milder conditions.

6.4.5 Mechanism of the Rh(II)-Catalyzed Aromative Decarbenation

The mechanism of the rhodium(II)-catalyzed decarbenation of 1,3,5-trimethyl-7-styryl-1,3,5-cycloheptatriene **41** was studied computationally in the context of the alkenylcyclopropanation of enol ethers (Scheme 6.19) [56]. For this transformation, lower energies were found for the key intermediates and transition states, in accordance with the established hypothesis, corresponding to a catalytic process that can take place at 25 °C, as observed experimentally.

The decarbenation process starts with norcaradiene intermediate **XIV**, which is in a c. 1:1 equilibrium with cycloheptatriene **XIII**. Although both C1—C2 and C1—C3 bonds could potentially be cleaved first, it was found that the electron-donating methyl group in C2 favors significantly ($\Delta\Delta G^{\ddagger} = 6.4\,\text{kcal mol}^{-1}$) the cleavage of C1–C2. This leads to Wheland-type carbenoid **XV**, which is significantly lower in energy than for the analogous gold(I)-decarbenation processes (Schemes 6.14 and 6.15), mainly as a consequence of stabilization of the positive charge by the three methyl groups. Accordingly, alternative intermediate **XV***, with only two methyl groups stabilizing the positive charge, is significantly higher in energy ($\Delta\Delta G^{\ddagger} = 7.0\,\text{kcal mol}^{-1}$). The cleavage of the first C—C bond is the TLS of the process, with an activation barrier of 20.3 kcal mol^{-1}, allowing for the retro-Buchner reaction to take place at room temperature, even starting from a considerably stable resting state, **XII**. Wheland intermediate **XV** evolves through $\mathbf{TS_{XV\text{-}XVI}}$ to release rhodium carbene **XVI**, resulting in an overall energetically neutral process after a mesitylene–enol ether ligand exchange. Then, this intermediate evolves through a highly exothermic stepwise *cis*-cyclopropanation process (see Section 6.5.1.2) [56].

6.5 Synthetic Methodologies and Applications

Besides awakening academic interest, decarbenation reactions via retro-cyclopropanation processes are a safe and attractive alternative for the controlled generation of metal carbenes. Specifically, non-acceptor metal carbenes can be easily accessed with this strategy, as a practical complement to the classical decomposition of acceptor diazo compounds. Not only that, but these processes led to the discovering of new synthetic methodologies, which found applications in the assembly of synthetically complex structures, natural products, or derivatives of other bioactive compounds [5].

6.5.1 Cyclopropanation Reactions

The benchmark reaction of metal carbenes is the cyclopropanation of alkenes. Thus, several methodologies to assemble a wide variety of cyclopropanes were developed using metal-catalyzed retro-Buchner/cyclopropanation sequences.

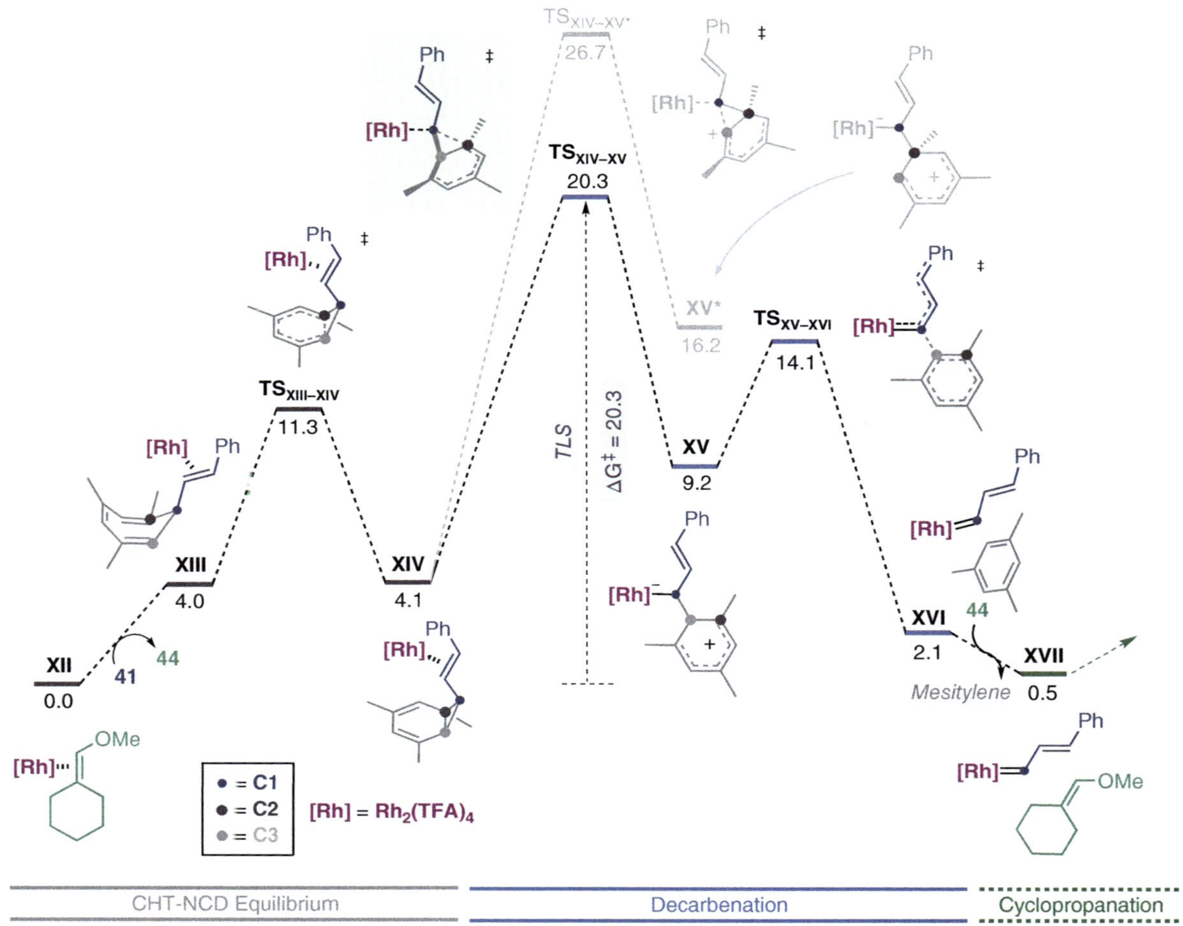

Scheme 6.19 Energy profile for the R┐(II)-catalyzed decarbenation of 1,3,5-trimethyl-7-styryl-1,3,5-cycloheptatriene. Source: Based on Mato et al. [56].

6.5.1.1 Aryl Cyclopropanations

One year after the initial discovery of the generation of aryl gold(I) carbenes from intermediate benzo-fused norcaradienes (see Section 6.3.3) [38], 7-aryl-1,3,5-cycloheptatrienes **45** were described as general precursors for the synthesis of diaryl- or triaryl-substituted cyclopropanes **47** (Scheme 6.20) [40]. These compounds were assembled by trapping of the corresponding aryl carbenes with styrene derivatives, upon release of benzene, under gold(I) catalysis. A variety of styrenes and related alkenes could be employed, and also aryl carbene fragments with different electronic properties, generally giving the corresponding cyclopropane as the *cis* or *endo* diastereoisomer. However, while electron-poor or electron-neutral aryl substitutents followed this trend, the presence of electron-donating groups in the aryl rings led to mixtures of diastereoisomers (**47e**), or even in some cases to the *trans* product as the preferred one (**47h**). Presumably, formation of *trans* cyclopropanes is a result of the high temperature (120 °C) necessary for the reaction to proceed.

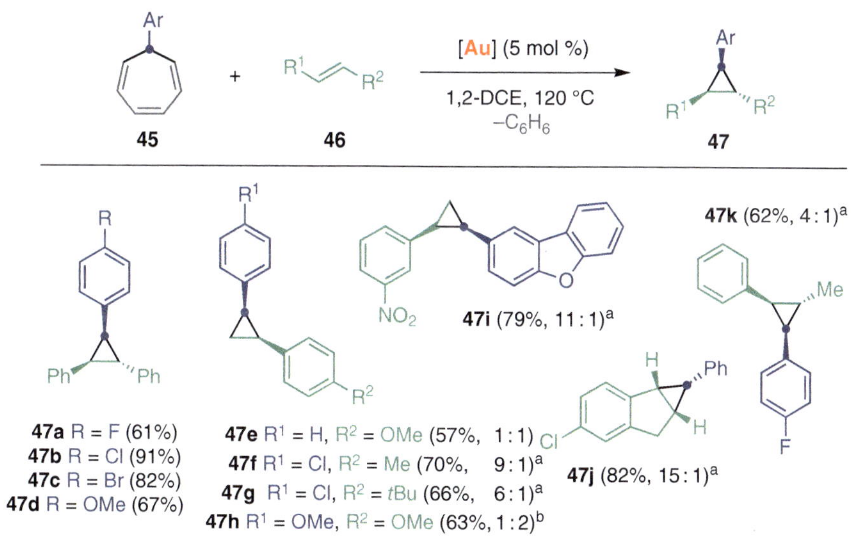

Scheme 6.20 Gold-catalyzed aryl cyclopropanation by decarbenation of cycloheptatrienes. [Au] = [(JohnPhos)Au(MeCN)]SbF_6. [a]The *cis* or *endo* diastereoisomer was obtained as major product. [b]The *trans* or *exo* diastereoisomer was obtained as major product. Source: Based on Solorio-Alvarado et al. [40].

The same process could be carried out intramolecularly with substrate **48**, affording **49** as the *exo* diastereoisomer exclusively, in quantitative yield (Scheme 6.21) [40].

An unusual case of cyclopropanation was observed when heating cycloheptatriene **50** in the presence of a gold(I) catalyst (Scheme 6.22) [44]. Intermediate gold(I) carbene **51** displays a geometry which facilitates a Buchner-type cyclopropanation giving **52**, which undergoes a electrocyclic ring opening to give **53**.

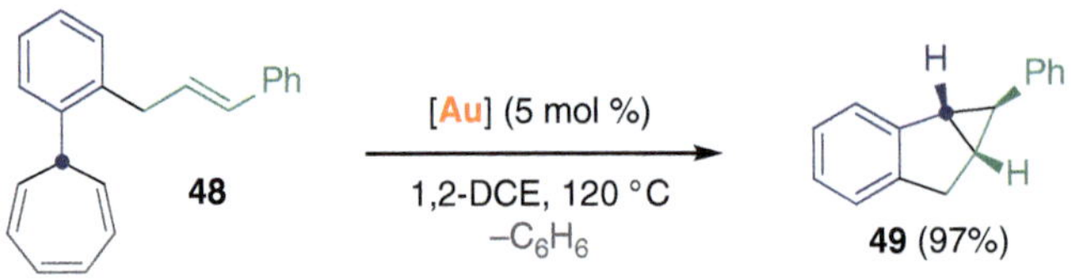

Scheme 6.21 Decarbenation and intramolecular arylcyclopropanation sequence. [Au] = [(*t*BuXPhos)Au(MeCN)]SbF_6. Source: Based on Wang et al. [44].

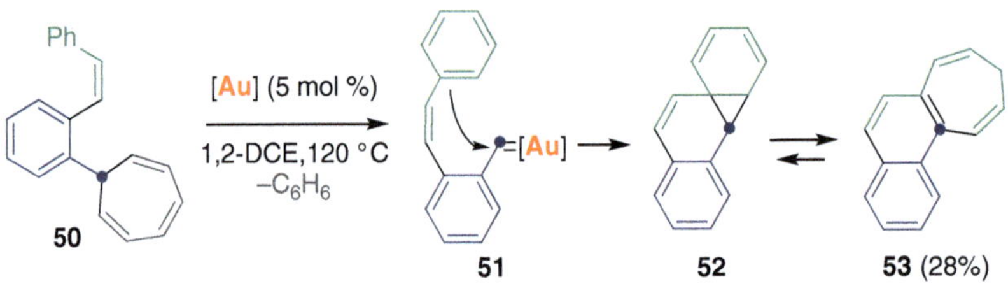

Scheme 6.22 Retro-Buchner decarbenation/Buchner ring expansion sequence [Au] = [(JohnPhos)Au(MeCN)]SbF_6.

This results in a rare retro-Buchner/Buchner reaction sequence, which arises from the favorable geometric arrangement in **51**, since Buchner-type processes are not reactions usually observed with donor gold(I) carbenes. For instance, typical retro-Buchner/cyclopropanation sequences can be carried out in common aromatic solvents, such as toluene, giving no trace of the corresponding Buchner ring expansion product.

6.5.1.2 Alkenyl Cyclopropanations

Among all cyclopropanes, vinylcyclopropanes are of great interest due to their rich downstream chemistry [57], particularly for taking part in a variety of formal cycloaddition processes [58]. Furthermore, vinylcyclopropanes can be found in drugs or natural bioactive molecules [59]. However, their synthesis is often challenging: the most obvious retro-synthetic approach would involve the selective mono-cyclopropanation of a 1,3-diene. In this regard, 7-alkenyl cycloheptatrienes can be employed as vinyl or styryl carbene precursors, which could be used to cyclopropanate styrenes and related activated alkenes (Scheme 6.23). This has been established as a straightforward approach to the stereoselective assembly of *cis*-alkenyl cyclopropanes. An in-depth mechanistic study of the reaction was performed, which is presented in Section 6.2.3 (Scheme 6.15) [51].

An improved protocol for the vinylcyclopropanation of alkenes was later developed using more reactive 1,3,5-trimethyl-7-styryl-1,3,5-cycloheptatrienes (see Section 6.4.4). These new vinyl carbene precursors allowed the generation of not only gold(I) carbenes at room temperature, but also zinc(II) [53] and rhodium(II) analogs, through the release of mesitylene instead of benzene (Scheme 6.24) [55]. While gold(I) shines on the cyclopropanation of styrenes or related activated

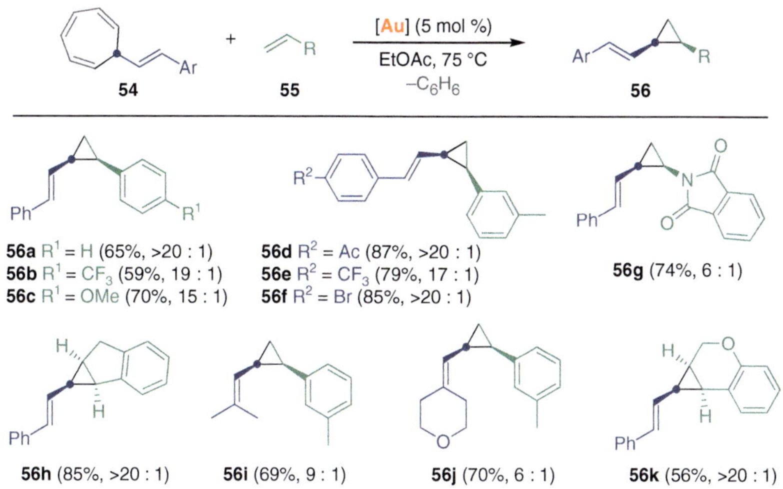

Scheme 6.23 Gold(I)-catalyzed *cis*-alkenylcyclopropanation with cycloheptatrienes. [Au] = [(JohnPhos)Au(MeCN)]SbF_6.

alkenes, zinc(II) performs better with non-activated alkenes, being also a cheaper alternative. The rhodium(II)-based system discovered later performs better overall than the other two. These mild protocols could be employed for obtaining vinylcyclopropane derivatives of bioactive molecules such as **59d** and **59e**. Furthermore, cyclopropane **59f** was transformed in two more steps into the natural fatty acid lactobacillic acid (**59h**, Scheme 6.24, bottom).

More recently, this cyclopropanation strategy was further extended to the *cis*-styryl cyclopropanation of highly electron-rich enol ethers (Scheme 6.25, left) [56]. This led to the stereoselective assembly of a variety of spirocyclic cyclopropyl ethers (**62**). Furthermore, while the cyclopropanation step can be performed under both gold(I) and rhodium(II) catalysis, these cyclopropyl ethers can be opened to give trienes **63** upon heating in the presence of [$Rh_2(TFA)_4$]. This allowed the development of a one-pot procedure in which cyclopropanes are both assembled and disassembled through the release of mesitylene and methanol, respectively, to construct a variety of all-*E* conjugated trienes (Scheme 6.25, right).

6.5.1.3 Reactions with Furans

Furans were found to react with alkynes under gold(I) catalysis to give phenols [60]. This reaction was proposed to take place through the formation of a cyclopropyl gold carbene, which would open and then close back in a process driven by aromatization [61, 62]. Similarly, gold(I) carbenes react with furans triggering ring-opening processes which can afford α,β,γ,δ-unsaturated carbonyl compounds. These carbenes could be generated from 1,6-enynes or propargyl acetates, but also from 7-aryl-1,3,5-cycloheptatrienes via decarbenation. The resulting carbenes were proposed to react with furans to give cyclopropane intermediates **65a** (or open cationic intermediates such as **65b**), which would readily undergo ring opening

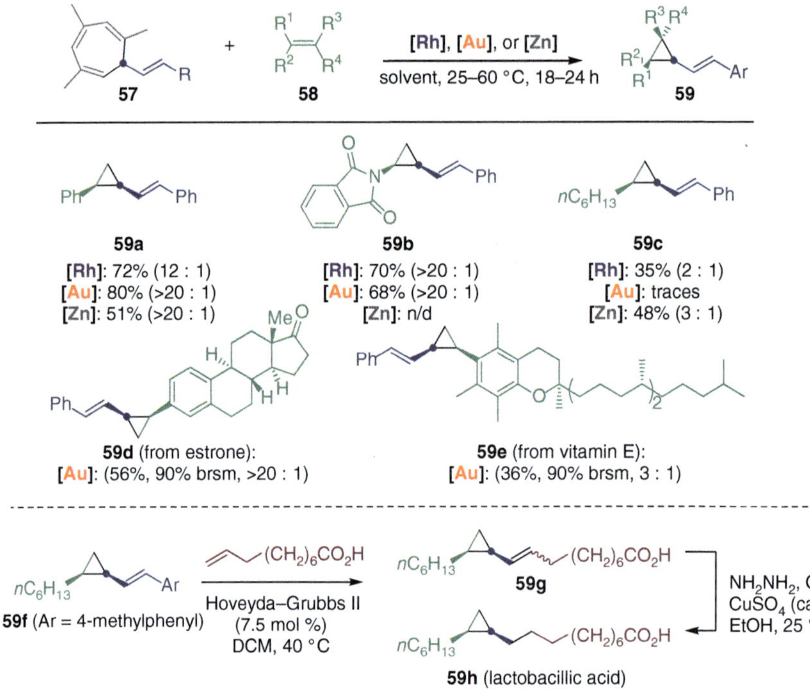

Scheme 6.24 Rhodium(II)-, gold(I)- or zinc(II)-catalyzed cyclopropanation (top). [Rh] = [$Rh_2(TFA)_4$]; [Au] = [(JohnPhos)Au(MeCN)]SbF_6]; [Zn] = $ZnBr_2$. Total synthesis of lactobacillic acid (bottom).

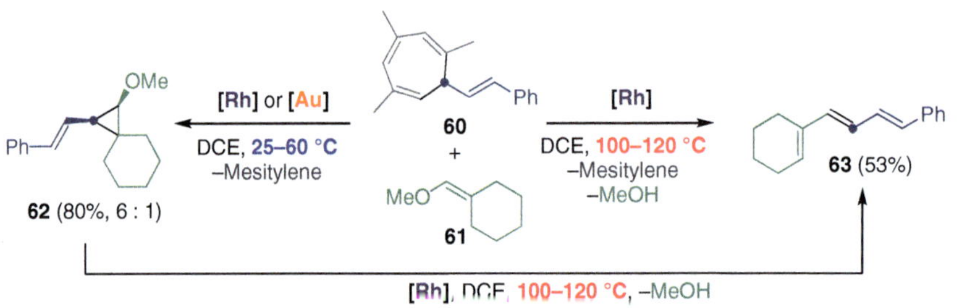

Scheme 6.25 Cyclopropanation of enol ethers (left) and rhodium(II)-catalyzed cyclopropyl ether opening to give all-*E* trienes (right). [Rh] = [$Rh_2(TFA)_4$]; [Au] = [(JohnPhos)Au(MeCN)]SbF_6.

(Scheme 6.26). Depending on the substitution pattern of the furan (R^1 and R^2), some of the products were found to react further giving polycyclic compounds [63].

6.5.2 Higher Formal Cycloadditions

Besides (2+1) cycloadditions in cyclopropanations, this type of donor carbenes can also engage in higher (*m*+*n*) cycloaddition processes. This led to the development of methods for the rapid generation of molecular complexity in a single reaction flask.

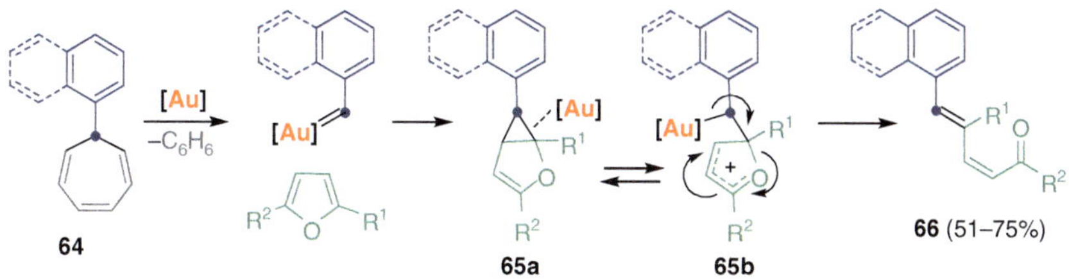

Scheme 6.26 Reaction of aryl gold(I) carbenes with furans through the formation of cyclopropane intermediates.

6.5.2.1 (4+1) Cycloadditions

Aryl gold(I) carbenes generated by retro-Buchner reaction of precursors such as **67** undergo formal (4+1)-cycloaddition processes with methylenecyclopropanes as **68**. The reaction starts by a gold(I)-catalyzed methylenecyclopropane–cyclobutene rearrangement. Then, the resulting cyclobutene reacts with the aryl carbene through a formal cyclopropanation/ring expansion sequence delivering cyclopentene **70**. This mechanistic proposal is strongly backed up by the development of another methodology in which cyclobutenes **69** were used directly instead of methylenecyclopropanes **68**, leading to the same type of products (Scheme 6.27) [64]. The entire process can be seen as a formal insertion of an aryl carbene into a double C—C bond of a cyclobutene.

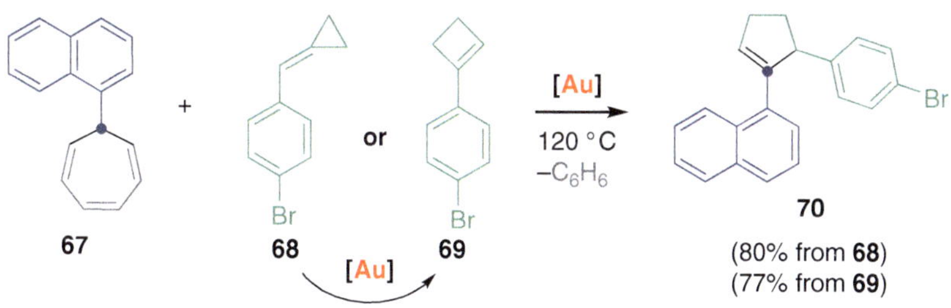

Scheme 6.27 (4+1) Cycloaddition of aryl gold(I) carbenes with methylenecyclopropanes or cyclobutanes to give cyclopentenes. Source: Based on Wang et al. [64].

The versatility of this approach was further demonstrated by carrying out a gold(I)-catalyzed one-pot [2+2] cycloaddition [65]/(4+1) cycloaddition using an alkene (**71**), an alkyne (**72**) and 7-naphthyl-1,3,5-cycloheptatriene **67**. Overall, the carbene carbon gets formally inserted into the triple C—C bond of arylacetylene **72** (Scheme 6.28) [66].

6.5.2.2 (3+2) Cycloadditions

Both aryl (**74**) and styryl (**75**) gold(I) carbenes generated by decarbenation of 7-substituted 1,3,5-cycloheptatrienes engage in formal (3+2) cycloaddition reactions with allenes to give densely substituted indenes **77** and cyclopentadienes **78**, respectively (Scheme 6.29). Both reactions take place by nucleophilic attack of the central carbon of the allene to the corresponding electrophilic aryl or styryl

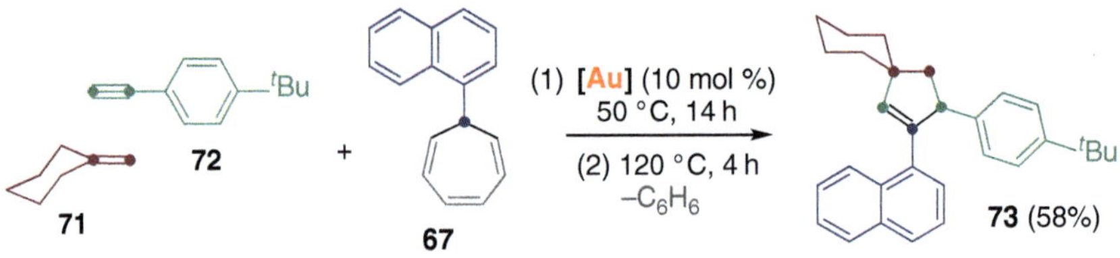

Scheme 6.28 Gold(I)-catalyzed [2+2] cycloaddition/decarbenation/(4+1) cycloaddition sequence. Source: Based on de Orbe and Echavarren [66].

gold(I) carbene. Then, Friedel–Crafts-type cyclization and further double-bond isomerization affords the corresponding products [67].

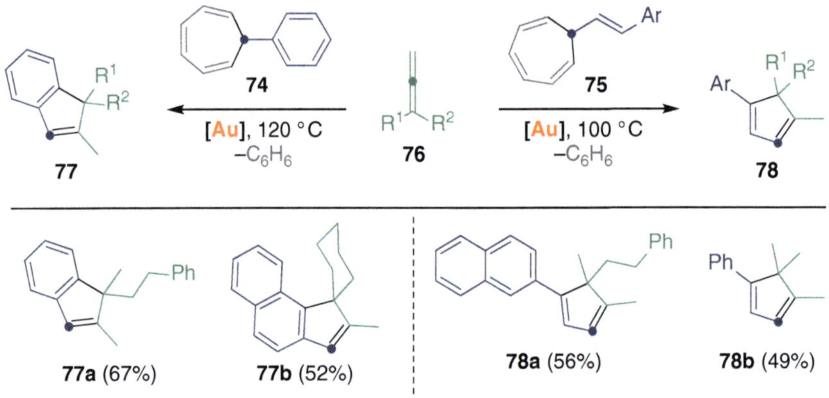

Scheme 6.29 (3+2) Cycloaddition of aryl (left) and styryl (right) gold(I) carbenes with allenes. Source: Based on Yin et al. [67].

Both (3+2) cycloadditions were applied in the context of natural product synthesis. Thus, selective hydrogenation of cyclopentadiene **78c** with Wilkinson's catalyst afforded natural product laurokamurene B (**79**, Scheme 6.30) [67].

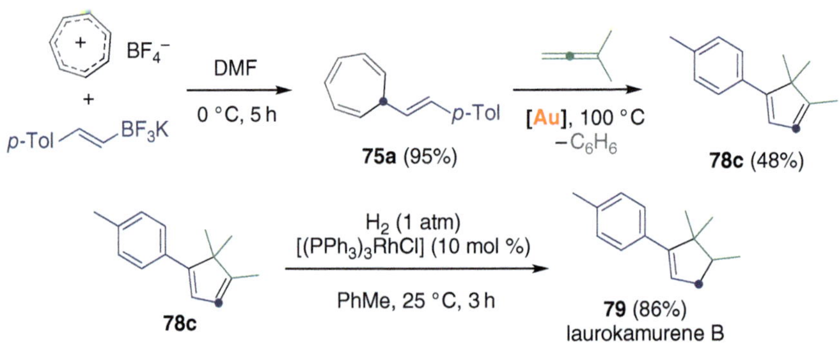

Scheme 6.30 Three-step total synthesis of (±)-laurokamurene B. Source: Based on Yin et al. [67].

The same methodology allowed the assembly (and separate isolation) of both *cis* and *trans* isomers of **80**, which could be further elaborated to afford the carbon skeletons of the cycloaurenones (**81**) and the dysiherbols (**82**) (Scheme 6.31) [67].

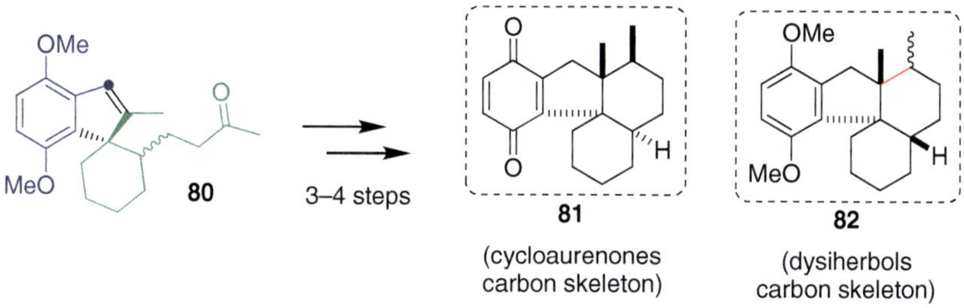

Scheme 6.31 Assembly of the carbon skeleton of the cycloaurenones and dysiherbols.

6.5.2.3 (4+3) Cycloadditions

Vinyl rhodium(II) carbenes generated by decarbenation of cycloheptatrienes **83** were found to react cleanly with 1,3-dienes to afford 1,4-cycloheptadienes **86** through a cyclopropanation/Cope rearrangement sequence. In some cases, depending on the substitution pattern of the 1,3-diene, the corresponding divinylcyclopropane **85** could be isolated. Thermal treatment (>140 °C) of such intermediates (**85a**) allowed obtaining the corresponding 1,4-cycloheptadienes. A mechanistic investigation allowed establishing a model for the Cope rearrangement, in which the outcome of the transformations at lower temperatures could be predicted depending on the calculated barriers (see activation energies in Scheme 6.32). Finally, this methodology was applied for the one-step total synthesis of dictyopterene C' (**86d**), a natural pheromone [68].

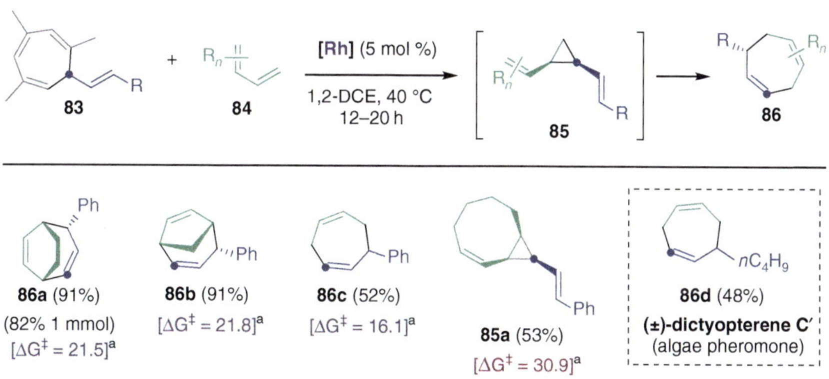

Scheme 6.32 Reactivity of non-acceptor vinyl rhodium(II) carbenes with 1,3-dienes. [a]Calculated activation energy (kcal mol^{-1} at 25 °C) for the free Cope rearrangement of the corresponding dinvinylcyclopropanes **85**.

6.5.3 Intramolecular Friedel–Crafts Reactivity

Aryl gold(I) carbenes bearing aryl or alkenyl *ortho* substitutents take part in intramolecular Friedel–Crafts-type reactions, leading to the formation of fluorenes and indenes, respectively (Scheme 6.33). Following this principle, a range of these products could be efficiently prepared. More complex polyaromatic structures could also be assembled starting from substrates bearing more than one 7-aryl-1,3,5-cycloheptatriene unit. Furthermore, a detailed mechanistic study of both the decarbenation process and the fate of the thereby generated aryl cabrenes was carried out using DFT calculations (see Section 6.2.3) [44].

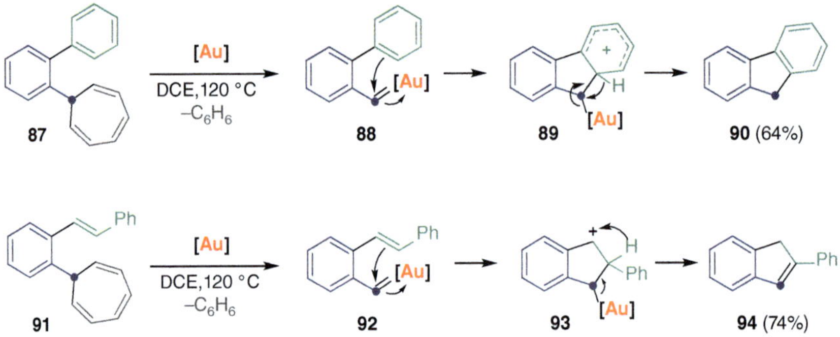

Scheme 6.33 Intramolecular reactions of *ortho*-substituted aryl gold(I) carbenes.

6.5.4 Insertion Reactions

In general, donor metal carbenes are less reactive toward C–H or X–H insertion than their acceptor or donor–acceptor counterparts. Furthermore, the fact that cycloheptatrienes are often used as carbene precursors (which can potentially be cyclopropanated in a competitive manner) makes the optimization of insertion processes by this method a significant challenge. Nevertheless, several insertion reactions have been recently developed based on metal-catalyzed decarbenations via retro-cyclopropanation.

6.5.4.1 C–H Insertion

The challenge of C–H insertion reactions is illustrated by early attempts to develop such transformations with these carbene intermediates. Thus, reaction of **95** in the presence of cationic gold(I) complexes only led to very poor yields (less than 10%) of **97**, even when the insertion process would take place intramolecularly into a benzylic position (Scheme 6.34). The major product observed (>50% yield) was the one of cyclopropanation of the norcaradiene tautomer of **95** by carbene **96** [40].

Further work showed that, in the absence of external nucleophiles, gold(I) carbene **99** reacts intramolecularly, presumably through a Friedel–Crafts-type mechanism, to give **100**. In the presence of 1,3-dicarbonyl compound **101**, **99** undergoes a formal C–H insertion process to afford benzylated product **102** (Scheme 6.35) [44].

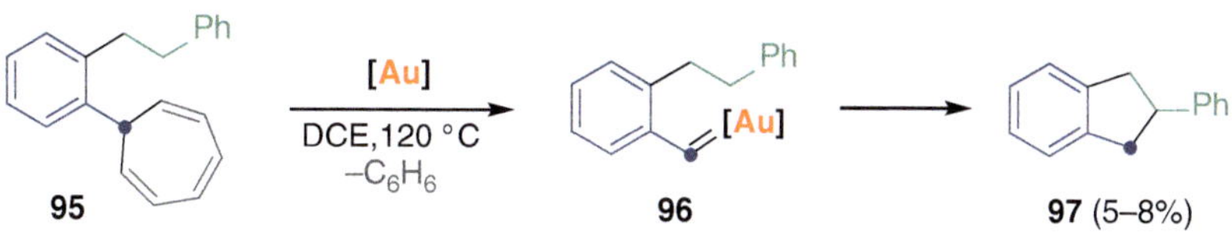

Scheme 6.34 Preliminary studies on the intramolecular C–H insertion of aryl gold(I) carbenes.

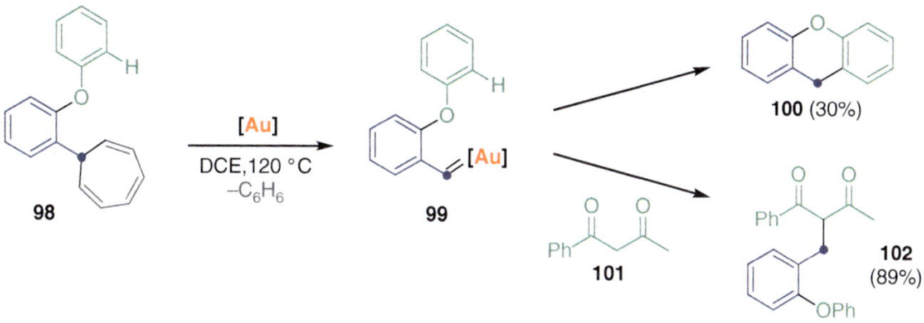

Scheme 6.35 Intra- and intermolecular reactivity of aryl gold(I) carbenes. Source: Based on Wang et al. [44].

Through careful substrate design, the first general intramolecular C–H insertion process via decarbenation of *ortho*-substituted cycloheptatrienes was developed (Scheme 6.35) [69]. Introducing a tertiary carbon as *ortho*-substituent (such as a *tert*-butyl group) in the aryl moiety leads to an ideal geometrical disposition of the C–H element attached to said tertiary carbon, favoring the insertion process in contrast to the example in Scheme 6.34 (Scheme 6.36). This allowed the efficient preparation of a variety of indenes **104**. Furthermore, through activation of a well-characterized 2-(*tert*-butyl)aryl gold(I) carbenoid, the same type of carbene intermediate was generated in a controlled manner. This species evolved analogously through intramolecular C–H insertion, leading to the quantitative formation of the corresponding indene product.

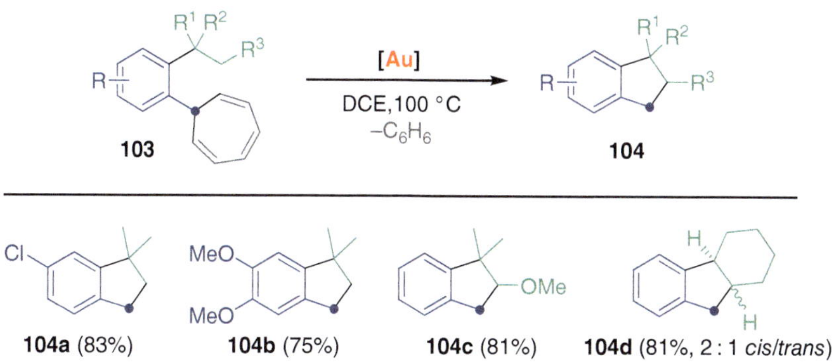

Scheme 6.36 Intramolecular C–H insertion of aryl gold(I) carbenes.

Similarly, carbene precursors such as **105** allowed to carry out a gold(I)-catalyzed formal C–H insertion/elimination sequence, in which dihydronaphthalenes were assembled in a regioselective manner (Scheme 6.37) [69].

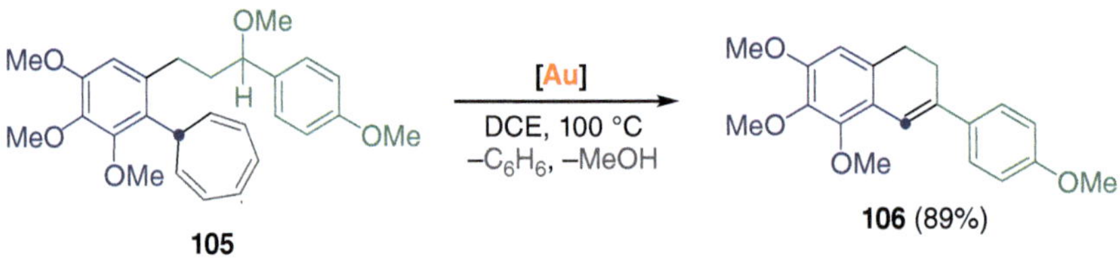

Scheme 6.37 Synthesis of dihydronaphthalenes by gold(I)-catalyzed decarbenation and formal insertion/elimination. Source: Based on Yin et al. [69].

6.5.4.2 Si–H Insertion

In spite of the fact that general intermolecular C–H insertion processes with carbenes generated by retro-Buchner reaction have not been reported so far, it was shown that more reactive Si–H bonds undergo insertion reactions with alkenyl donor rhodium(II) carbenes. This allowed the preparation of a wide variety of allyl silanes **109** in good yields, through a concerted insertion process (Scheme 6.38) [55]. A similar reaction was not observed with cationic gold(I) complexes as catalysts.

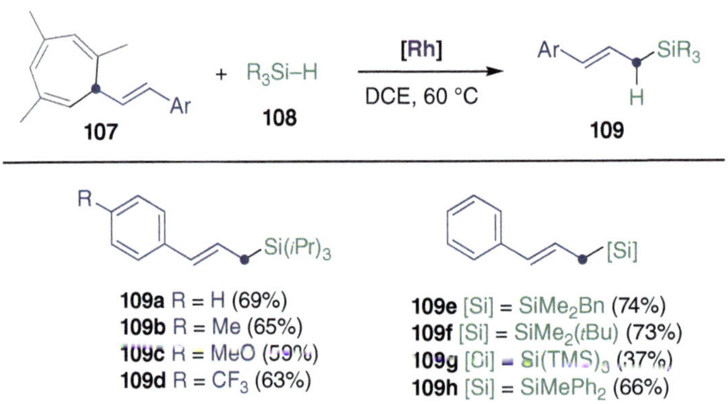

Scheme 6.38 Intermolecular Si–H insertion of donor rhodium(II) carbenes. Source: Based on Mato and Echavarren [55].

6.5.5 Oxidation Reactions

Vinyl carbene intermediates generated by decarbenation of cycloheptatrienes **112** can be trapped by a mild oxidant, such as diphenylsulfoxide, to afford α,β-unsaturated aldehydes (Scheme 6.39) [55]. This process allowed the development of a two-step vinylogation of aldehydes: 7-alkenyl cycloheptatrienes **112** are obtained from aldehydes **110** by Julia–Kocienski olefination with sulfone-lithium salt **111**. After an oxidative decarbenation catalyzed by $[Rh_2(TFA)_4]$, vinylogous aldehydes **113** are assembled. This two-step process can be repeated iteratively, growing chains of all-*E* conjugated polyenes. This strategy was applied to the total synthesis of two natural polyenes, navenones B and C, through three iterations from the corresponding benzaldehydes.

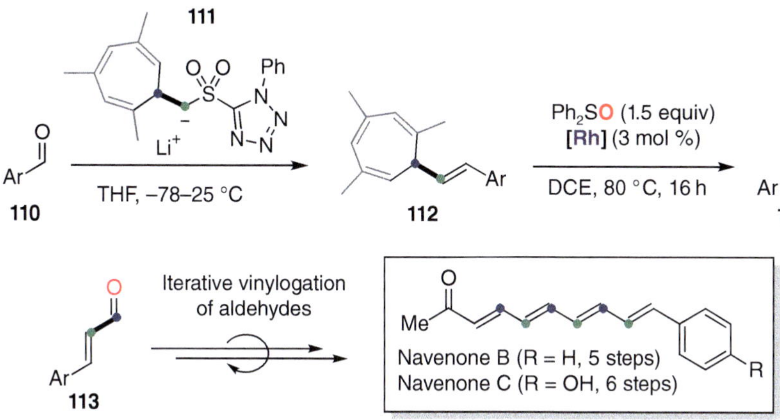

Scheme 6.39 Iterative vinylogation of aldehydes by oxidative decarbenation. Source: Based on Mato and Echavarren [55].

6.5.6 Alternative Precursors

Besides 7-substituted 1,3,5-cycloheptatrienes, other systems were designed and employed to carry out metal-catalyzed decarbenations by retro-cyclopropanation. Originally, this process had been observed to take place from electron-rich benzo-fused norcaradienes [38]. Accordingly, different naphthalene (**114**) and phenanthrene (**115**) derivatives were established as alternative gold(I) carbene precursors through decarbenation by retro-cyclopropanation of cyclopropanes. These carbenes can also be trapped by alkenes to afford a variety of non-acceptor cyclopropanes, resulting in an overall cyclopropane–alkene metathesis process (Scheme 6.40) [70].

6.5.7 Decarbenations Based on the Release of Alkenes

Besides the early observations reported by Gassman [25], several groups have developed methodologies based on formal retro-cyclopropanations which release alkenes instead of aromatic units, in spite of the considerable weaker driving force. Most of these transformations rely on the entropically favored release of ethylene gas. Thus, Takai and coworkers reported in 2018 that a molybdenum-based catalytic system promotes the retro-cyclopropanation of substrates such as **118** to afford

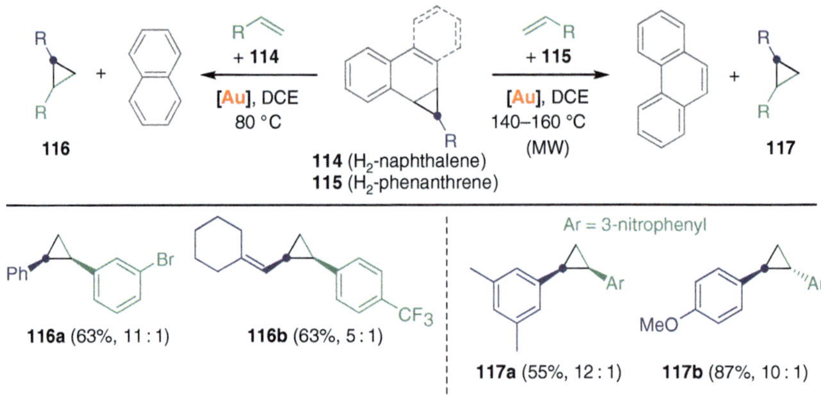

Scheme 6.40 Gold(I)-catalyzed decarbenation of persistent cyclopropanes. Source: Based on Mato et al. [70].

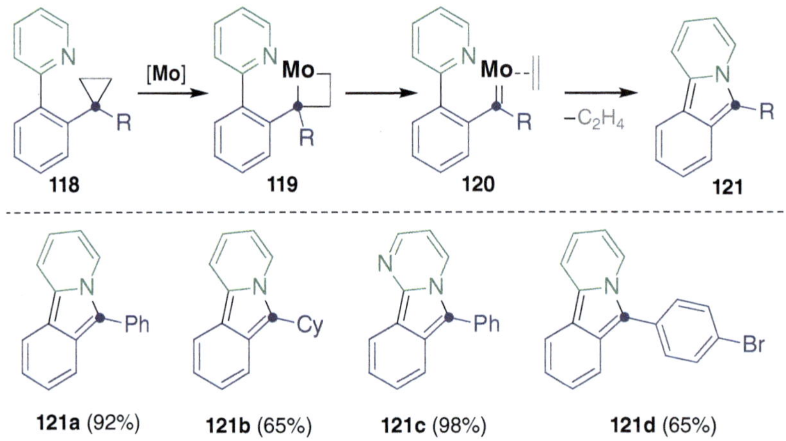

Scheme 6.41 Molybdenum-catalyzed decarbenation through ethylene release. [Mo] = Mo(benzene)$_2$ (20 mol%) and 9,10-phenanthrenequinone (20 mol%), 160 °C. Source: Based on Asako et al. [71].

pyrido[2,1-*a*]isoindoles **121** in excellent yields, through the release of ethylene (Scheme 6.41) [71].

A similar release of ethylene from cyclopropanes was observed in the context of the rhodium(I)-catalyzed cycloisomerization of allene-allenylcyclopropanes **122** [72] (Scheme 6.42A), or alkylidene cyclopropanes **124** [73] (Scheme 6.42B). The latter were proposed to form metal vinylidenes, which then react intermolecularly with diynes.

In contrast to retro-Buchner reactions which release aromatic fragments (see Section 6.2.2), by release of ethylene, not only short-lived reactive intermediates could be generated. An interesting example was disclosed by Mascareñas, Esteruelas, and coworkers in 2010 in which rather stable metal vinylidenes were generated

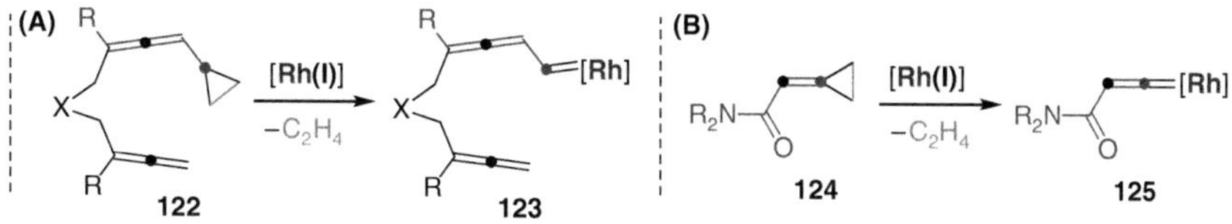

Scheme 6.42 Retro-cyclopropanation of allenyl- or alkylidene cyclopropanes. Source: Based on Yoshizaki et al. [73].

by retro-cyclopropanation of alkylidene cyclopropanes (Scheme 6.39) [55]. The corresponding osmium vinylidene complexes could be characterized by X-ray diffraction, in which the released ethylene fragment is coordinated to the metal center (Scheme 6.43).

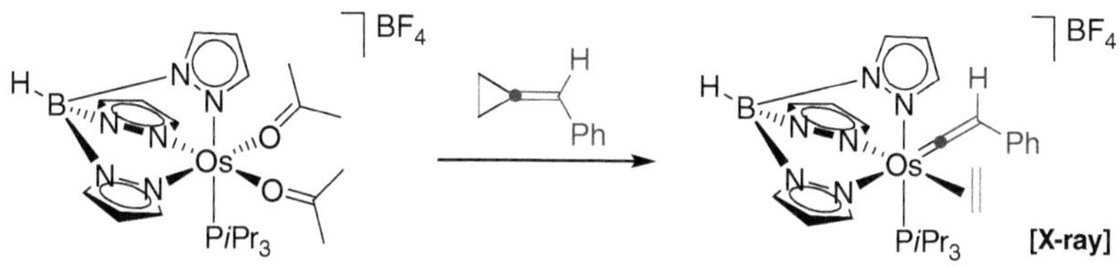

Scheme 6.43 Osmium vinylidenes by retro-cyclopropanation.

6.6 General Outlook and Concluding Remarks

Among all methods available to generate metal carbene intermediates, decarbenation reactions via retro-cyclopropanation have reached an interesting compromise between reactivity, versatility, generality, safety, and relative simplicity of the corresponding precursors. Different families of 7-substituted cycloheptatrienes and analogous substrates have been developed over the past 10 years, emerging not only as complementary alternatives to the use of diazo compounds, but also as great tools for reaction discovery. Using these precursors, non-acceptor carbenes of different metals can be generated catalitically. These reactive intermediates can cyclopropanate alkenes, engage in higher cycloadditions or in insertion processes, and undergo intramolecular reactivity. This led to the development of new synthetic methodologies which allow for the rapid construction of complex molecular architectures of interest, such as drug-like compounds or natural products [5].

Despite all the advances in the field, these strategies still face significant challenges, which will hopefully be overcome in the future. For instance, several types of 7-substituents in 1,3,5-cycloheptatrienes do not undergo decarbenation, such as acceptor groups (typical functionalities in diazo compounds), or alkynes (which lead to products of 6-*endo*-dig cyclization) [74]. As mentioned in the text, intermolecular C–H insertions with carbenes generated by retro-cyclopropanation reaction have not been reported [47]. Similarly, the detection and characterization of metal carbene intermediates generated by retro-Buchner reaction is a significant challenge, which has not met with success thus far [49]. Furthermore, the possibility of engaging these carbenes in alkene cross-metathesis cycles has not been explored yet. Finally, the development of enantioselective transformations involving the transfer of non-acceptor metal carbenes is a challenging and underexplored field. In this regard, the use of 7-substituted cycloheptatrienes as non-acceptor carbene precursors could be an attractive starting point [53].

References

1 Dötz, K.H. (ed.) (2004). *Metal Carbenes in Organic Synthesis*. Berlin, Heidelberg: Springer-Verlag.
2 de Frémont, P., Marion, N., and Nolan, S.P. (2009). *Coord. Chem. Rev.* 253: 862.
3 Ford, A., Miel, H., Ring, A. et al. (2015). *Chem. Rev.* 115: 9981.
4 Jia, M. and Ma, S. (2016). *Angew. Chem. Int. Ed.* 55: 9134.
5 Mato, M., García-Morales, C., and Echavarren, A.M. (2019). *ChemCatChem* 11: 53.
6 Dave, V. and Warnhoff, E.W. (1970). *Org. React.* 18: 217.
7 Buchner, E. and Curtius, T. (1885). *Ber. Dtsch. Chem. Ges.* 18: 2377.
8 McNamara, O.A. and Maguire, A.R. (2011). *Tetrahedron* 67: 9.
9 Frühauf, H.-W. (1997). *Chem. Rev.* 97: 523.
10 Doyle, M.P., Duffy, R., Ratnikov, M., and Zhou, L. (2010). *Chem. Rev.* 110: 704.
11 Xia, Y., Qiu, D., and Wang, J. (2017). *Chem. Rev.* 117: 13810.
12 Davies, H.M.L. and Beckwith, R.E.J. (2003). *Chem. Rev.* 103: 2861.
13 Doyle, M.P. and Forbes, D.C. (1998). *Chem. Rev.* 98: 911.
14 Fulton, J.R., Aggarwal, V.K., and de Vicente, J. (2005). *Eur. J. Org. Chem.*: 1489.
15 Doyle, M.P., Griffin, J.H., Bagheri, V., and Dorow, R.L. (1984). *Organometallics* 3: 53.
16 Zhu, D., Chen, L., Fan, H. et al. (2020). *Chem. Soc. Rev.* 49: 908.
17 Dorel, R. and Echavarren, A.M. (2015). *Chem. Rev.* 115: 9028.
18 Hashmi, A.S.K., Braun, I., Nösel, P. et al. (2012). *Angew. Chem. Int. Ed.* 51: 4456.
19 Shiroodi, R.K. and Gevorgyan, V. (2013). *Chem. Soc. Rev.* 42: 4991.
20 Miege, F., Meyer, C., and Cossy, J. (2011). *Beilstein J. Org. Chem.* 7: 717.
21 Ye, L.-W., Zhu, X.-Q., Sahani, R.L. et al. (2020). *Chem. Rev.* 121: 9039.
22 Mato, M., Franchino, A., García-Morales, C., and Echavarren, A.M. (2021) *Chem. Rev.* 121: 8613.
23 Ogba, O.M., Warner, N.C., O'Leary, D.J., and Grubbs, R.H. (2018). *Chem. Soc. Rev.* 47: 4510.
24 Volger, H.C., Hogeveen, H., and Roobeek, C.F. (1973). *Recl. Trav. Chim. Pays-Bas* 92: 1223.
25 Gassman, P.G. and Johnson, T.H. (1976). *J. Am. Chem. Soc.* 98: 6057.
26 Gassman, P.G. and Johnson, T.H. (1976). *J. Am. Chem. Soc.* 98: 6058.
27 Richardson, D.B., Durrett, L.R., Martin, J.M. et al. (1965). *J. Am. Chem. Soc.* 87: 2763.

28 Nigam, M., Platz, M.S., Showalter, B.M. et al. (1998). *J. Am. Chem. Soc.* 120: 8055.
29 Moore, K.A., Vidaurri-Martinez, J.S., and Thamattoor, D.M. (2012). *J. Am. Chem. Soc.* 134: 20037.
30 Saito, K., Kozaki, M., and Takahashi, K. (1993). *Chem. Pharm. Bull.* 41: 2187.
31 Pearlstine, K.A. and Friend, C.M. (1985). *J. Am. Chem. Soc.* 107: 5898.
32 Takaya, H., Suzuki, T., Kumagai, Y. et al. (1981). *J. Org. Chem.* 46: 2854.
33 Walczak, M.A.A. and Wipf, P. (2008). *J. Am. Chem. Soc.* 130: 6924.
34 Herranz, M.A., Rivera, J.A., Alvarado, R.J. et al. (2001). *J. Supramol. Chem.* 1: 299.
35 Herranz, M.A., Beulen, M.W.J., Rivera, J.A. et al. (2002). *J. Mater. Chem.* 12: 2048.
36 Fedorov, A., Moret, M.-E., and Chen, P. (2008). *J. Am. Chem. Soc.* 130: 8880.
37 Ringger, D.H. and Chen, P. (2013). *Angew. Chem. Int. Ed.* 52: 4686.
38 Solorio-Alvarado, C.R. and Echavarren, A.M. (2010). *J. Am. Chem. Soc.* 132: 11881.
39 de Orbe, M.E., Amenós, L., Kirillova, M.S. et al. (2017). *J. Am. Chem. Soc.* 139: 10302.
40 Solorio-Alvarado, C.R., Wang, Y., and Echavarren, A.M. (2011). *J. Am. Chem. Soc.* 133: 11952.
41 Lauterbach, T., Higuchi, T., Hussong, M.W. et al. (2015). *Adv. Synth. Catal.* 357: 775.
42 Caballero, A. and Pérez, P.J. (2017). *Chem. Eur. J.* 23: 14389.
43 Maxwell, J.L., Brown, K.C., Bartley, D.W., and Kodadek, T. (1992). *Science* 256: 1544.
44 Wang, Y., McGonigal, P.R., Herlé, B. et al. (2014). *J. Am. Chem. Soc.* 136: 801.
45 Harris, R.J. and Widenhoefer, R.A. (2016). *Chem. Soc. Rev.* 45: 4533.
46 Werlé, C., Goddard, R., and Fürstner, A. (2015). *Angew. Chem. Int. Ed.* 54: 15452–15456.
47 Sarria Toro, J.M., García-Morales, C., Raducan, M. et al. (2017). *Angew. Chem. Int. Ed.* 56: 1859.
48 Wang, Y., Muratore, M.E., and Echavarren, A.M. (2015). *Chem. Eur. J.* 21: 7332.
49 García-Morales, C., Pei, X.-L., Sarria Toro, J.M., and Echavarren, A.M. (2019). *Angew. Chem. Int. Ed.* 58: 3957.
50 Tskhovrebov, A.G., Lingnau, J.B., and Fürstner, A. (2019). *Angew. Chem. Int. Ed.* 131: 8926.
51 Herlé, B., Holstein, P.M., and Echavarren, A.M. (2017). *ACS Catal.* 7: 3668.
52 Reiersølmoen, A.C., Østrem, E., and Fiksdahl, A. (2018). *Eur. J. Org. Chem.*: 3317.
53 Mato, M., Herlé, B., and Echavarren, A.M. (2018). *Org. Lett.* 14: 4341.
54 Conrow, K. (1963). *Org. Synth.* 43: 101.
55 Mato, M. and Echavarren, A.M. (2019). *Angew. Chem. Int. Ed.* 58: 2088.
56 Mato, M., García-Morales, C., and Echavarren, A.M. (2020). *ACS Catal.* 10: 3564.
57 Kulinkovich, O.G. (2015). *Cyclopropanes in Organic Synthesis*. Wiley.
58 Rubin, M., Rubina, M., and Gevorgyan, V. (2007). *Chem. Rev.* 107: 3117.

59 Chen, D.Y.K., Pouwer, R.H., and Richard, J.-A. (2012). *Chem. Soc. Rev.* 41: 4631.
60 Hashmi, A.S.K., Frost, T.M., and Bats, J.W. (2000). *J. Am. Chem. Soc.* 122: 11553.
61 Martín-Matute, B., Nevado, C., Cárdenas, D.J., and Echavarren, A.M. (2003). *J. Am. Chem. Soc.* 125: 5757.
62 Huguet, N., Leboeuf, D., and Echavarren, A.M. (2013). *Chem. Eur. J.* 19: 6581.
63 Leboeuf, D., Gaydou, M., Wang, Y., and Echavarren, A.M. (2014). *Org. Chem. Front.* 1: 759.
64 Wang, Y., Muratore, M.E., Rhong, Z., and Echavarren, A.M. (2014). *Angew. Chem. Int. Ed.* 53: 14022.
65 López-Carrillo, V. and Echavarren, A.M. (2010). *J. Am. Chem. Soc.* 132: 9292.
66 de Orbe, M.E. and Echavarren, A.M. (2018). *Eur. J. Org. Chem.*: 2740.
67 Yin, X., Mato, M., and Echavarren, A.M. (2017). *Angew. Chem. Int. Ed.* 56: 14591.
68 Armengol-Relats, H., Mato, M., and Echavarren, A.M. (2021). *Angew. Chem. Int. Ed.* 60: 1916–1922.
69 Yin, X., Zuccarello, G., García-Morales, C., and Echavarren, A.M. (2019). *Chem. Eur. J.* 25: 9485.
70 Mato, M., Martín-Torres, I., Herlé, B., and Echavarren, A.M. (2019). *Org. Biomol. Chem.* 17: 4216.
71 Asako, S., Kobashi, T., and Takai, K. (2018). *J. Am. Chem. Soc.* 140: 15425.
72 Kawamura, T., Kawaguchi, Y., Sugikubo, K. et al. (2015). *Eur. J. Org. Chem.*: 719.
73 Yoshizaki, S., Shibata, Y., and Tanaka, K. (2017). *Angew. Chem. Int. Ed.* 56: 3590.
74 McGonigal, P.R., de León, C., Wang, Y. et al. (2012). *Angew. Chem. Int. Ed.* 51: 13093.

7

Gold-Catalyzed Oxidation of Alkynes by *N*-Oxides or Sulfoxides

Kaylaa Gutman, Tianyou Li, and Liming Zhang

University of California, Department of Chemistry and Biochemistry, Santa Barbara, CA 93106, USA

7.1 Introduction: Gold-Activated Alkynes Attacked by Nucleophilic Oxidants

π-Acidic cationic Au(I) complexes or Au(III) salts are versatile and potent soft Lewis acids and can effectively activate alkynes toward attack by nucleophiles. With sulfoxides and *N*-oxides as the nucleophilic oxidants, the reactions can lead to ultimate alkyne oxidation. As shown in Scheme 7.1, the initial attack of the gold-activated C—C triple bond by a generic nucleophilic oxidant Z^+—O^- forms the adduct **A**, which can undergo heterolysis of its weak O—Z bond to generate the α-oxo gold carbene **B**. As gold is the most electronegative metal on the Pauling's scale, this carbene intermediate is remarkably electrophilic and can react with a range of nucleophiles to afford various functional products. Traditionally, α-diazo carbonyl compounds have been used to access this type of metal carbene; however, the α-diazo carbonyl compounds are hazardous and potentially explosive, and their preparation requires energetic reagents and necessitates additional steps [1, 2]. The process outlined in Scheme 7.1 permits alkyne compounds to function as surrogates for α-diazo carbonyl compounds in the presence of nucleophilic oxidants and a gold catalyst, thereby offering valuable synthetic advantages in the realm of gold chemistry. However, the α-oxo gold carbene intermediate may be bypassed in an alternative pathway, where the initial adduct **A** is directly converted to the same product. As a result, **A** can be considered as a carbenoid. Density function theory (DFT) calculations and mechanistic studies have been employed to elucidate whether the reaction follows a carbene or carbenoid reaction pathway. In this chapter, this class of oxidative gold catalysis is organized first by the employed oxidants and then by the types of nucleophiles trapping the carbene or carbenoid intermediates. The discussions are focused on outlining the versatility and variety of this approach. For more comprehensive coverage of this topic, please refer to the recent review [3].

Transition Metal-Catalyzed Carbene Transformations, First Edition.
Edited by Jianbo Wang, Chi-Ming Che, and Michael P. Doyle.

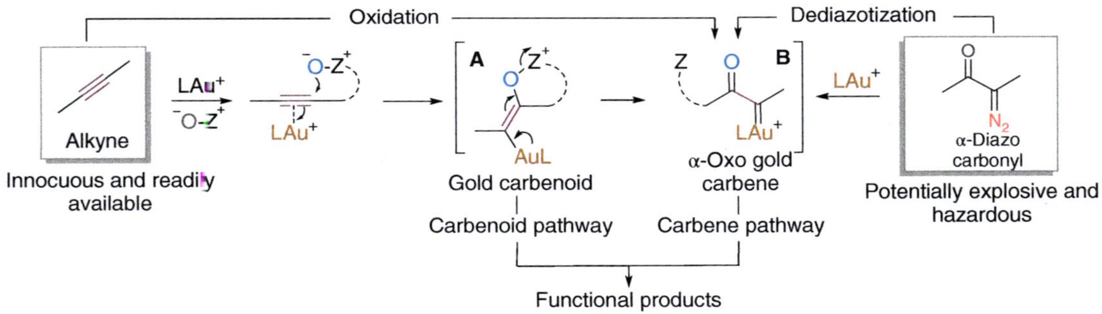

Scheme 7.1 The use of alkynes as surrogates for α-diazo ketones.

7.2 Sulfoxides as Nucleophilic Oxidants

Sulfoxides, though being mild oxidants, are the first class of nucleophilic oxidants employed in this type of oxidative gold catalysis. Toste and Shapiro [4] and L. Zhang and G. Li [5] reported separately in 2007 that the tethered sulfoxide of (but-3-yn-1-ylsulfinyl)benzene **1** can oxidize its terminal C—C triple bond in the presence of a cationic Au(I) catalyst or an Au(III) salt, affording 2,3-dihydrobenzo[*b*]thiepin-4(5*H*)-one (**2**) in >90% yield in both cases (Scheme 7.2). The initially proposed reaction mechanism by both research groups entails an intramolecular nucleophilic attack of the gold-activated C—C triple bond by the sulfoxide oxygen in a 5-*exo-dig* cyclization, which is followed by the heterolytic fragmentation of the weak S—O bond of the adduct **C** to form the α-oxo gold carbene **D**, which in turn undergoes the Friedel–Crafts-type cyclization to form the product.

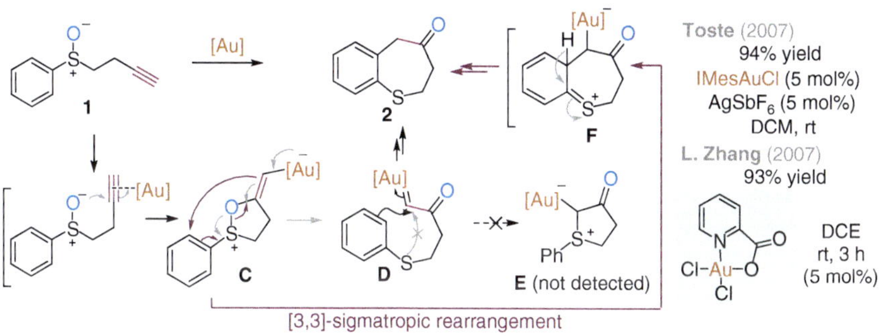

Scheme 7.2 Sulfoxides as oxidant: early studies and a carbenoid pathway. Source: Based on Shapiro et al. [4] and Li et al. [5].

An early DFT study supported this carbene pathway; however, the alternative carbenoid pathway was not considered [6]. In the initially proposed carbene mechanism, the formation of the five-membered sulfonium salt **E**, the result of nucleophilic cyclization of the tethered sulfur to the electrophilic gold carbene center, should readily occur and would likely hinder the observed reactivity, yet appears nonexistent or inconsequential. The subsequent experimental work and DFT calculations by L. Zhang and Y. Li concluded that the cyclized intermediate **F** is formed directly from the gold carbenoid **C** via a [3,3]-sigmatropic rearrangement [7].

A perhaps definitive demonstration of the carbene pathway in the presence of a sulfoxide oxidant is the work reported by Davies [8]. As shown in Scheme 7.3, the proposed mechanism involves trapping the gold carbene moiety of **G** by sulfur to generate the sulfonium salt **H**, which is consistent with the mechanistic consideration related to **E** in Scheme 7.2. The sulfur ylide **I**, formed upon gold dissociation, can undergo a 2,3-sigmatropic rearrangement to afford the observed 2-allyldihydrothiophen-3-one products. There appears to be no readily available carbenoid alternative for this reaction. Of note, $PtCl_2$ can promote this reaction as well, albeit being less effective than the Au(III) catalyst.

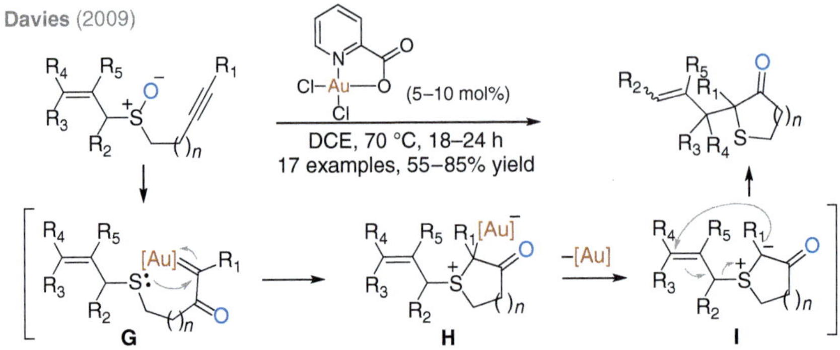

Scheme 7.3 Generation of sulfur ylides through gold-catalyzed oxidation of alkynes by tethered sulfoxides. Source: Modified from Davies and Albrecht [8].

In addition to tethered sulfoxides, external sulfoxides (primarily aryl sulfoxides) are also suitable for oxidizing gold-activated alkynes. Ujaque and Asensio reported the gold-catalyzed intermolecular oxoarylation of terminal alkynes in 2009 (Scheme 7.4) [9]. The exclusive functionalization at the *ortho*-position of the sulfoxide **3** is consistent with a 3,3-sigmatropic rearrangement of the initial adduct **J** in a carbenoid pathway and does not support the alternative carbene pathway involving a Friedel–Crafts reaction by an α-oxo gold carbene intermediate. This conclusion is further supported by DFT studies and inspired the subsequent mechanistic work on intramolecular variants by L. Zhang and Y. Li shown in Scheme 7.2 [7]. A related reaction was reported later by Davies and Grainger using dibenzo[*b,d*]thiophene 5-oxide as the oxidant [10].

Diphenyl sulfoxide was subsequently shown to be the preferred external sulfur-based oxidant by R.-S. Liu in 2010. In this work, cyclopropylalkynes are converted to cyclobutenyl ketones **4** in moderate to excellent yields (Scheme 7.5 [11]. High temperature (100 °C) and a large excess of diphenyl sulfoxide (5 equiv) were necessary to achieve good yields and full substrate conversions. The mechanistic studies revealed a carbenoid pathway, where the ring expansion is concurrent with the expulsion of diphenyl sulfide, as outlined in the structure **L**.

Additional studies employing diphenyl sulfoxide as the oxidant led to alkyne double oxidation to form 1,2-dicarbonyl products [12] and oxidative cyclization to deliver 2-acyl-1-naphthols [13].

7.3 *N*-Oxides as Nucleophilic Oxidants

In comparison to sulfoxides, *N*-oxides are in general stronger nucleophiles, which can be correlated from their stronger Brønsted basicity, and can be stronger oxidants. The use of *N*-oxides as nucleophilic oxidants has gained significant popularity and become dominant in gold-catalyzed alkyne oxidation reactions. Among the *N*-oxides, aniline- or tertiary amine-derived ones are only employed intramolecularly. Their participation in oxidative gold catalysis mostly follows carbenoid

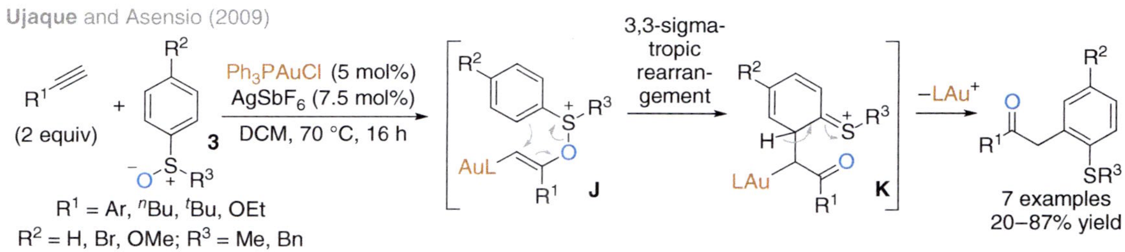

Scheme 7.4 Gold-catalyzed intermolecular oxoarylation of alkynes. Source: Modified from Cuenca et al. [9].

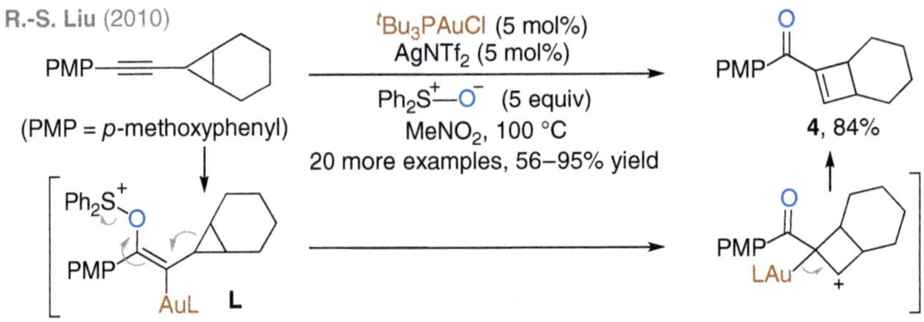

Scheme 7.5 Gold-catalyzed oxidative ring expansion of cyclopropylalkynes by using diphenyl sulfoxide as the oxidant. Source: Modified from Li et al. [11].

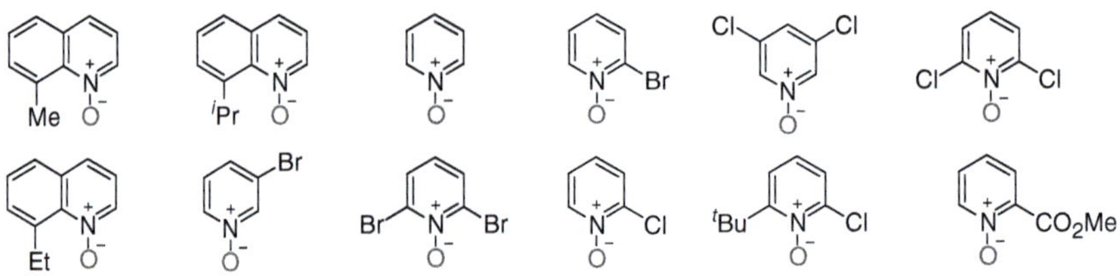

Figure 7.1 N-heteroarene *N*-oxides employed as the external oxidants in oxidative gold catalysis.

mechanistic pathways. On the other hand, *N*-oxides of *N*-heteroarenes, particularly pyridines and quinolines are the overwhelming choices of external nucleophilic oxidants. They readily participate in intermolecular alkyne oxidations, which have conceptually and in many instances practically achieved the replacement of hazardous diazo carbonyl compounds with innocuous and easily accessible alkynes. A list of *N*-heteroarene *N*-oxides that have been successfully employed in multiple instances in oxidative gold catalysis is shown in Figure 7.1. Among them, 8-alkylquinoline *N*-oxides including the methyl, ethyl, and isopropyl variants are the most versatile and often preferred. Of the pyridine *N*-oxides, the electron-deficient and/or sterically hindered ones are preferred. In the early studies using sterically unhindered pyridine *N*-oxides, stoichiometric amounts of an acid additive (typically MsOH) are added to reactions in order to neutralize the pyridine byproducts, which are basic and can impede or inhibit gold catalysis via metal binding. With sterically hindered 8-alkylquinoline *N*-oxides and 2-substituted or 2,6-disubstituted pyridine *N*-oxides, no acid additives are needed since the *N*-arene byproducts are too sterically hindered to bind to gold effectively. In addition, nitrones can be effective internal oxidants and in some cases serve as valuable external oxidants but are not included in the discussion. Hashmi and coworkers [14] conducted kinetic studies of several oxidative gold catalysis reactions by using an extensive collection of *N*-heteroarene *N*-oxides, which revealed that 2,5-dibromopyridine *N*-oxide, 2-trifluoromethylpyridine *N*-oxide, and 8-methylquinoline *N*-oxide are highly effective, and recommended that they should be screened during condition optimizations.

7.3.1 Reactions of Carbene/Carbenoid Intermediates with Oxygen-Based Nucleophiles

In 2010, L. Zhang reported the first examples of intermolecular generation of the proposed α-oxo gold carbenes [15]. As exemplified in Scheme 7.3, 7.5, 7.6A, dichloropyridine *N*-oxide is employed as the nucleophilic oxidant, and a cyclohexane-fused homopropargylic alcohol is converted into the strained *trans*-fused dihydrofuranone **5** in 88% yield. The proposed mechanism entails the oxidative generation of the α-oxo gold carbene **M**, which is followed by an intramolecular carbene O–H insertion. Due to the highly electrophilic nature of the gold carbene center, it is likely, as demonstrated later [17], that the insertion is not concerted and may involve the gold *C*-enolate **N** and/or gold *O*-enolate **O** intermediates prior to protodeauration to furnish the final product. In this early work, stoichiometric amounts of MsOH were requisite to neutralize the 3,5-dichloropyridine byproduct generated during the reaction.

This oxidative strategy was later successfully applied in the synthesis of strained oxetan-3-ones [18]. As shown in Scheme 7.6B, secondary propargylic alcohols or tertiary propargylic alcohols bearing an electron-withdrawing alkoxycarbonyl group terminating the C—C triple bond proceed well under different optimized conditions. Due to the strong acidity of Tf_2NH employed to neutralize the pyridine byproducts, the ester group in the cases of tertiary propargylic alcohol substrates is essential to avoid propargylic cation formation. This method provides a straightforward synthesis of enantiomerically enriched oxetan-3-ones from chiral propargylic alcohol substrates, as evident in the formation of the cyclobutanone **7** (81% ee) from the propargylic alcohol **6** (80% ee). While a carbenoid pathway for the formation of the five-membered ring product in Scheme 7.6A could not be ruled out, this construction of strained four-membered rings most likely involves highly reactive α-oxo gold carbene intermediates.

In 2019, W. Hu and X. Xu reported that the *O*-enolate **O** could be trapped selectively over protodeauration in the form of an aldol interception by isatin (Scheme 7.6C) [16]. The high diastereoselectivity of this step is explained by the assembly shown in the intermediate **P** which involves hydrogen bonding between isatin oxygens and the oxonium proton. In this case, trifluoroacetic acid is employed to neutralize the 2-methyl-5-bromopyridine byproduct.

The α-oxo gold carbene intermediates can be trapped by ethereal oxygen in the gold-catalyzed oxidations of homopropargylic ethers [19], 2-ethynylanisoles [20], and 5-methoxypent-3-en-1-ynes [21]. In the latter two studies, subsequent demethylations afford oxacyclic ketone products.

A combination of trapping the α-oxo gold carbene by an allyl ether and subsequent allyl migration was reported by Z. Yang and Y. Tang in 2013 [22]. As outlined in Scheme 7.7A, the author proposed three reaction pathways for the transformation of the allyl oxonium intermediate **Q** based on the observed reaction intermediates and initial mechanistic studies. Among them, Path A involves a concerted [2,3]-sigmatropic rearrangement and was evoked at the same time by J. Tae and coworkers in a similar study under a different set of reaction conditions

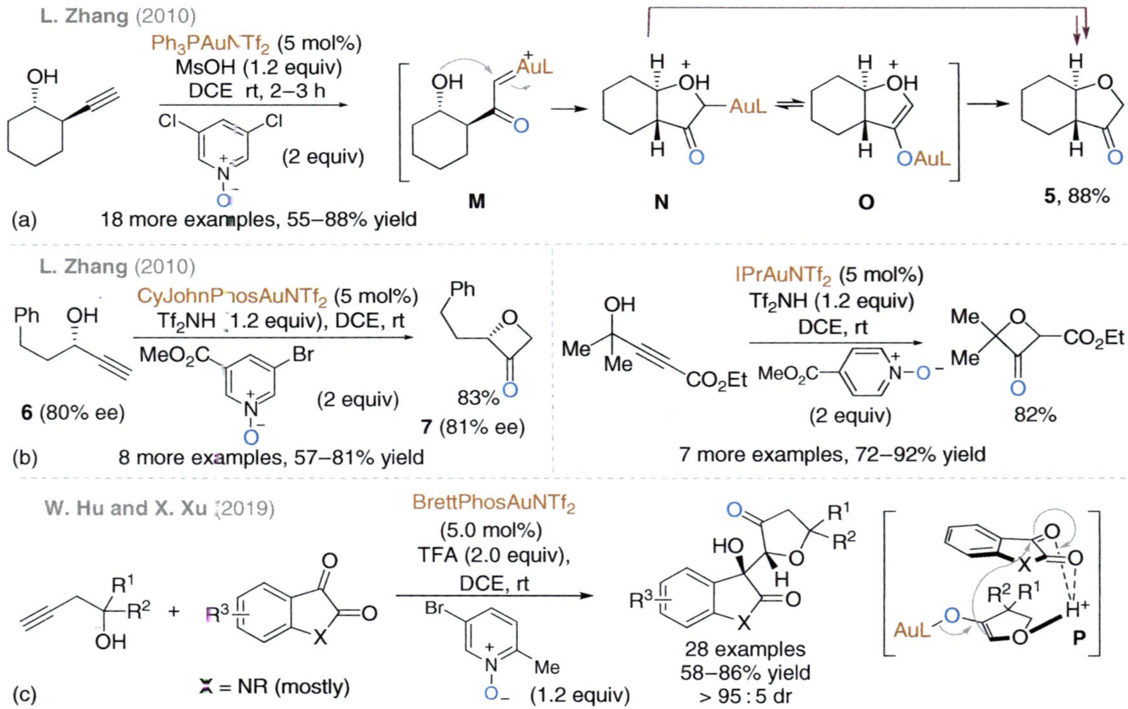

Scheme 7.6 Intramolecular O–H insertion of α-oxo gold carbenes. Source: (C) Modified from Cai et al. [16].

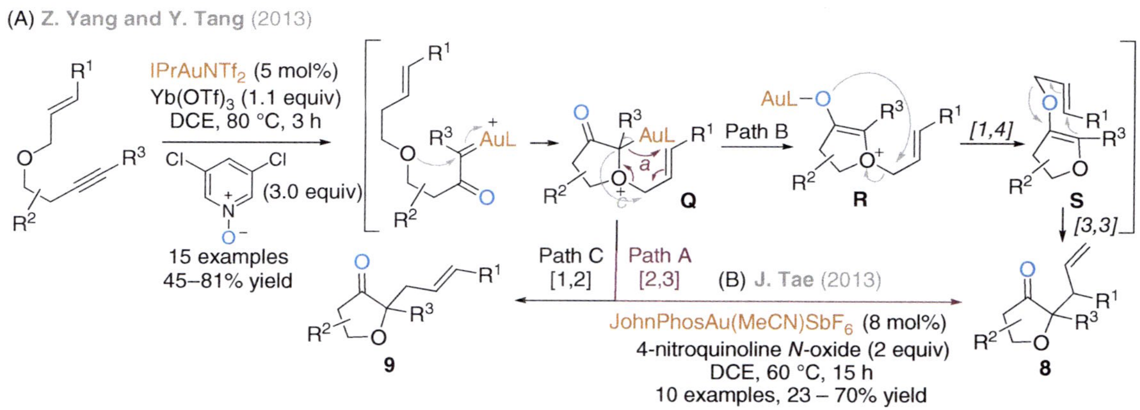

Scheme 7.7 Rearrangements of allyl oxonium ions generated from α-oxo gold carbenes. Source: (A) Modified from Shan et al. [23]. (B) Modified from Tae et al. [24].

[24]. The synthetic utility of this chemistry was demonstrated in the total synthesis of (±)-kumausallene [24].

With a carboxyl group α to an oxidatively generated carbene center, its 1,2-migration ensues, resulting in the formation of an α-carboxyenone. This method of carbene trapping was first reported by K. Ohe in 2012. As shown in Scheme 7.8A [25], in this rare case employing a tethered pyridine *N*-oxide, the α-oxo gold carbene **T**, which is generated upon intramolecular alkyne oxidation, is attacked by the carboxyl carbonyl oxygen during the migration, which delivers the α-carboxyenone **10** [25]. This product can be further converted to the indolizinone product **11** upon cyclization and benzoyl migration. In 2013, L. Zhang and coworkers reported gold-catalyzed oxidation of propargylic carboxylates, which also features a 1,2-carboxy migration of the gold carbene intermediate **V** (Scheme 7.8B) [26]. Notably, this process is favored over the typically facile 1,2-C–H insertion with IPr as the ligand. Unlike the work reported by K. Ohe, the gold carbene intermediate was generated using an external oxidant, i.e. 8-methylquinoline *N*-oxide. This intermolecular oxidation is highly regioselective as the enone **14** was barely detected. Even better selectivities of 1,2-acetoxy migration over the 1,2-C–H insertion by the carbene intermediate can be achieved by using the *P,N*-bidentate ligand Mor-DalPhos or a *P,S*-variant. In 2014, Y. Liu reported a related 1,2-carbonate migration approach to accessing enones featuring an α-carbonate group [27].

Despite the advances in their intramolecular trappings, the intermolecular trapping of α-oxo gold carbenes with stoichiometric amounts of trapping reagents remained challenging until the introduction of bidentate ligands into the field by L. Zhang in 2012 [28]. The highly electrophilic nature of the α-oxo gold carbene is evident in the work shown in Scheme 7.15 (*vide infra*) and often results in intractable mixtures in the absence of rapid intramolecular trapping or further oxidation by the nucleophilic oxidant to deliver 1,2-dicarbonyl products [30]. As shown in Scheme 7.9, the work realized the Au(I)-catalyzed synthesis of 2,4-disubstituted oxazoles via a dehydrative [3+2] annulation between the α-oxo gold carbene moiety and a primary carboxamide. The annulation entails an intermolecular attack of the amide carbonyl oxygen, yielding the protonated imidate **X** and followed by protodeauration and dehydrative cyclization. It was reported that various monodentate ligands are completely ineffective and only the *P,N*-bidentate ones such as Mor-DalPhos and Me-DalPhos can facilitate this reaction with moderate to excellent yields. To rationalize this finding, the author proposed that the formation of a *tris*-coordinated gold carbene complex **W** should lead to the increased electron density at the gold metal in comparison to the *bis*-coordinated counterpart. Consequently, the carbene center becomes less electron-deficient due to the increased back donation from the metal center and hence exhibits attenuated electrophilicity. Further evidence was provided by DFT calculations, which revealed that **W** is 5.4 kcal mol^{-1} more stable than its *bis*-coordinated counterpart.

Later in 2013, L. Zhang reported a related intermolecular trapping of the α-oxo gold carbene with carboxylic acids in slight excess (i.e. 1.2 equiv) in the presence of a modified *P,N*-bidentate ligand (Scheme 7.10A) [31]. The reaction affords α-carboxyl ketone products in mostly excellent yields. Around the same time, W. He and J. Xiang

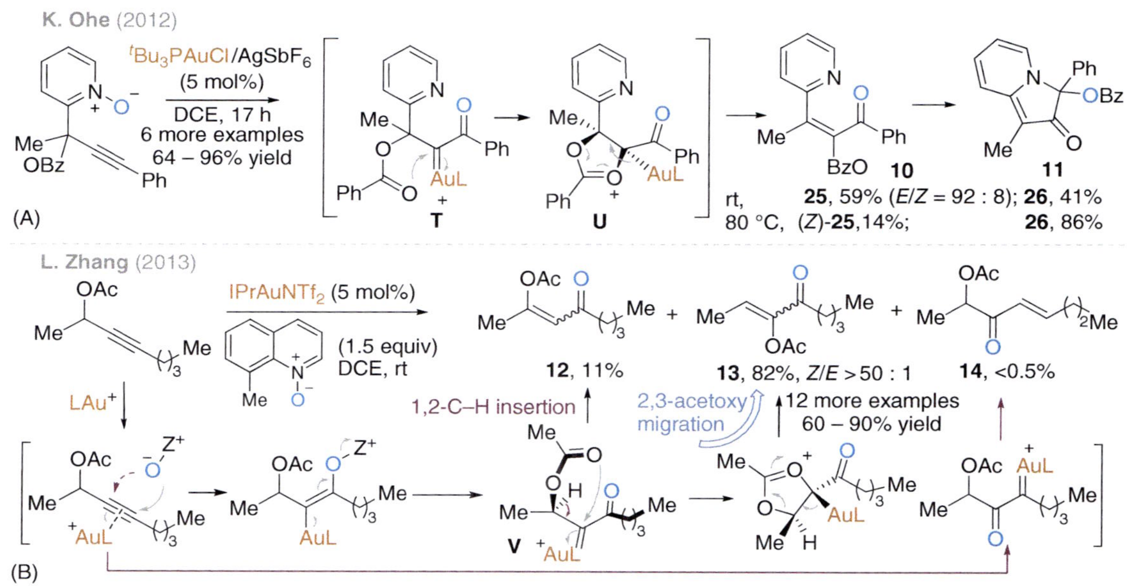

Scheme 7.8 Intramolecular trapping by carbonyl group leading to a 1,2-carboxy migration. Source: (A) Modified from Murai et al. [25]. (B) Modified from Ji et al. [26].

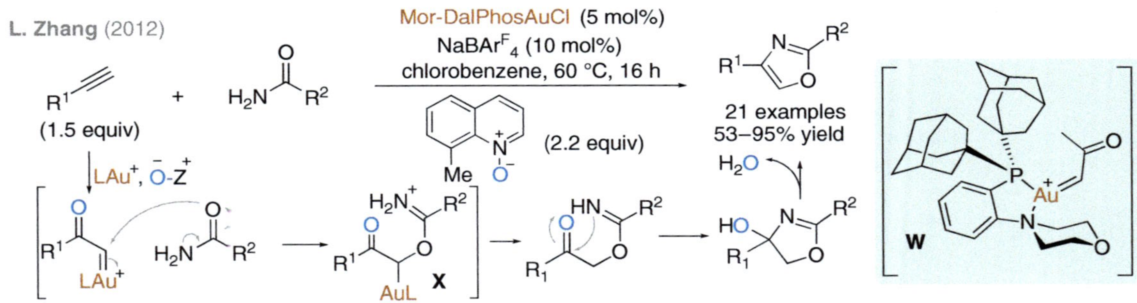

Scheme 7.9 Intermolecular trapping of the α-oxo gold carbene intermediate with a primary amide.

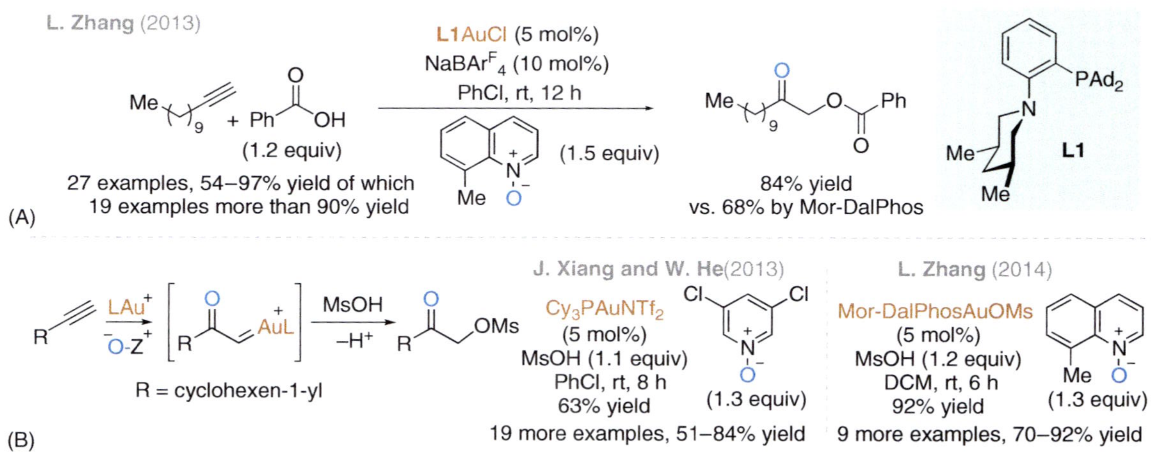

Scheme 7.10 Intermolecular trapping of the α-oxo gold carbene intermediate with acids. Source: (A) Modified from Ji et al. [31]. (B) Modified from He et al. [32].

reported the same trapping using the monodentate PCy_3 as the ligand [33]. However, the carboxylic acids are mostly used as the solvent in the reaction, there needs the addition of stoichiometric MsOH, and the reaction yield tends to be lower.

In 2013, W. He and J. Xiang extended their carbene trapping approach to MsOH, which affords α-mesyloxy ketones (Scheme 7.10B) [32]. In this case, MsOH is not used as the solvent but in slight excess. In 2014, L. Zhang reported a generally more efficient synthesis of the same products by using Mor-DalPhos as the ligand and mesylate as the counter anion to the cationic gold catalyst [34]. Additionally, they combined this reaction with the one-pot synthesis of 2,4-disubstituted thiazoles by the subsequent addition of a thioamide and mild heating. This oxidative approach to methanesulfonyloxymethyl ketones was employed by T. Luo in 2016 as a key step in the total synthesis of (±)-vinblastine [35].

In addition, the gold carbene intermediate can be trapped by the nucleophilic oxidant to afford substituted glyoxals in double oxidation of terminal alkynes [30].

7.3.2 Reactions of Carbene/Carbenoid Intermediates with Nitrogen-Based Nucleophiles

In 2011, L. Zhang described the intermolecular trapping of α-oxo gold carbene intermediates with nitriles, offering rapid access to 2,5-disubstituted oxazoles (Scheme 7.11) [36]. To achieve efficient trapping, nitriles are typically used as the solvent. In the cases of expensive liquid nitriles, 3 equiv are used under neat conditions to achieve moderate yields. Mechanistically, the nitrilium intermediate **Y**, which is generated from trapping the anticipated carbene with a nitrile, undergoes cyclization and deaurative aromatization to yield the final product. The overall reaction is an [2+2+1] annulation between the C—C triple bond, a nitrile, and the *N*-oxide oxygen. The alkyne scope of this chemistry was later expanded to include indole-3-ylacetylenes [37], ynoates [38], and alkynylamides [39]. Cyanamides were also shown as a suitable nitrile source [40].

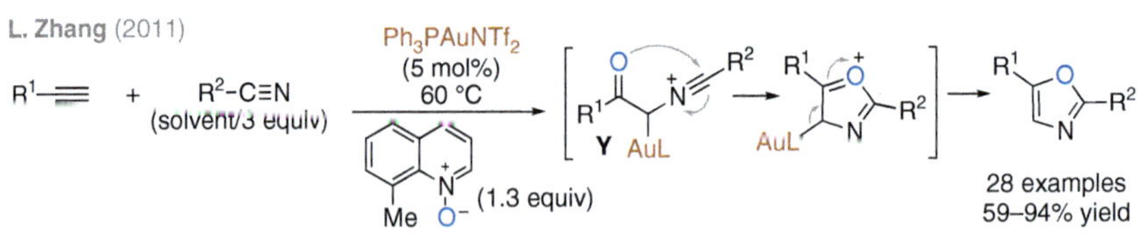

Scheme 7.11 Intermolecular reaction of α-oxo gold carbenes with nitriles. Source: Modified from He et al. [36].

In 2014, R.-S. Liu reported an intramolecular variant of carbene trapping by nitriles, which allows rapid access to *N*-(pyridine-2-yl)-ε-lactams (Scheme 7.12) [41]. The proposed mechanism entails the formation of a seven-membered nitrilium intermediate upon the trapping and a subsequent nucleophilic attack at the nitrilium carbon by the pyridine *N*-oxide to form the intermediate **AA**, which then undergoes a Nazarov-type cyclization to generate the tricyclic dihydropyridine intermediate **AB**. This species can then undergo rearomatization of the pyridine ring via elimination and protodeauration to produce the product. This reaction is applicable to arylacetylenes as well as cycloalkenylacetylenes.

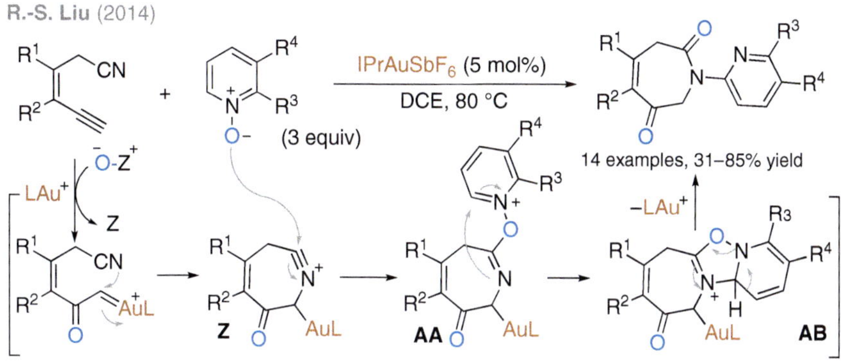

Scheme 7.12 Intramolecular reaction of α-oxo gold carbene with nitrile. Source: Modified from Karad and Liu [41].

In 2011, L. Zhang reported a one-step synthesis of azetidin-3-ones from propargylic sulfonamides (Scheme 7.13A) [42]. This reaction is related to the oxetan-3-one chemistry depicted in Scheme 7.6B, but differs by the trapping nucleophile, which is a much bulkier sulfonamide. As such, the optimized conditions employ 2,6-dibromopyridine as the oxidant and BrettPhos as the ligand. Due to the bulkiness of the oxidant, no acid additive is required in this chemistry. By combining with the chiral ^{t}butylsulfinamide chemistry [17], this reaction permits access to azetidin-3-ones with excellent ee. In 2014, employing a similar strategy but with different optimized conditions, Ye described the synthesis of chiral pyrrolidin-3-ones **15** [43, 44].

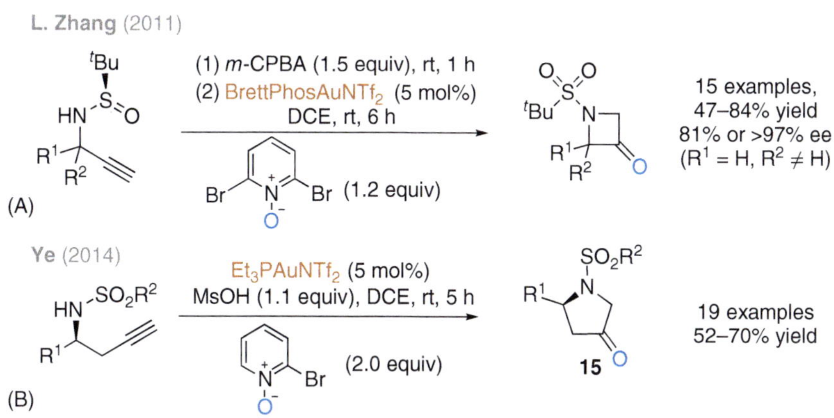

Scheme 7.13 Intramolecular reaction of α-oxo gold carbene with sulfonamide. Source: (A) Modified from Ye et al. [42]. (B) Modified from Ye 2014.

In 2013, R.-S. Liu reported that the α-oxo gold carbene intermediate **AC**, generated from a terminal dienyne substrate, can be trapped by the quinoline byproduct to initiate an intramolecular [3+2] cycloaddition between a remote electron-deficient alkene and the quinolinium ylide to deliver a tetracyclic dihydroquinoline product (Scheme 7.14) [45]. Interestingly, the R^3 group in the oxidant and the added quinoline can be H or a sterically hindered isopropyl group. Despite the quinoline being the reaction byproduct, its addition to the reaction increases product yield. A related carbene trapping by pyridine byproduct was reported by Toste in 2014 [46].

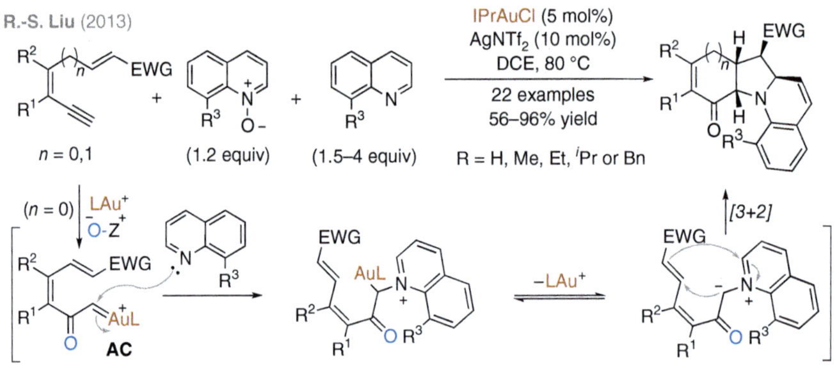

Scheme 7.14 Trapping α-oxo gold carbene by quinoline. Source: Modified from Huple et al. [45].

7.3.3 Reactions of Carbene/Carbenoid Intermediates with Other Heteronucleophiles

In 2012, J. Xiang and L. Zhang reported that α-oxo gold carbenes generated from terminal alkynes are highly electrophilic and capable of abstracting halogens from halogenated solvents, such as 1,2-dichloroethane and 1,2-dibromoethane (Scheme 7.15) [29]. With 8-methylquinoline *N*-oxide as the oxidant and PPh_3AuNTf_2 as the catalyst, α-halogenated ketones are formed in moderate

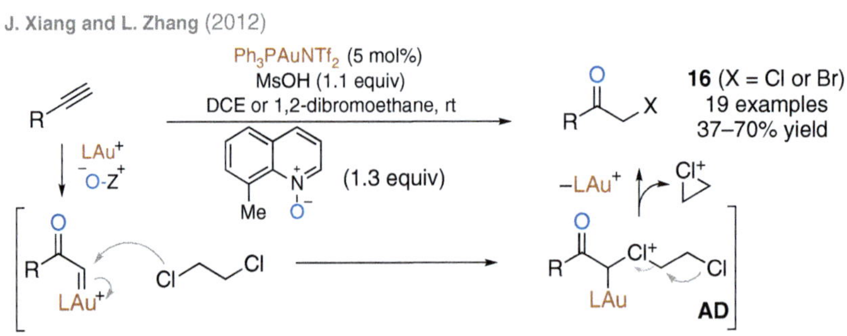

Scheme 7.15 Halogen abstraction by an α-oxo gold carbene intermediate. Source: Modified from He et al. [29].

yields. In the proposed mechanism, which uses dichloroethane as the solvent, the α-oxo gold carbene is trapped by the alkyl chloride to form the chloronium species **AD**. The subsequent loss of ethylene chloronium and protodeauration afford the α-chloro ketone **16**. This chemistry elucidates the difficulty in trapping α-oxo gold carbene intermediates with stoichiometric amounts of nucleophiles due to their high electrophilicity while offering early support of their intermediacy in this type of oxidative gold catalysis.

In 2014, L. Zhang [47] and Davies [48] simultaneously reported that the α-oxo gold carbene intermediate can be trapped by an allyl sulfide to generate the sulfonium salt **AE** and then likely the sulfur ylide **AF**. A subsequent [2,3]-sigmatropic rearrangement of either **AE** or **AF** affords an α-thio-γ,δ-unsaturated ketone as the product (Scheme 7.16). L. Zhang discovered that the use of bidentate ligands such as Mor-DalPhos and the *P,S*-bidentate phosphine ligand **17** lead to good reaction yields, while the conventional ligands such as Ph_3P and BrettPhos result in <5% yield. Notably, this chemistry displays a broad scope of terminal alkyne substrates, with Y being alkyl, alkenyl or aryl groups, and substitutions at the allyl sulfide are also allowed. Davies found the α-oxo gold carbenes generated from aryl-terminated ynamides can be likewise trapped to afford imides bearing sulfur-substituted quaternary carbon centers. The ligand used is IPr. The fact that a bidentate ligand is not needed for this trapping can be attributed to the less reactive nature of the donor–acceptor-type gold carbene intermediate, in comparison to the ketone gold carbene generated in L. Zhang's work. An intramolecular version of this chemistry was later reported [49].

In 2019, R.-S. Liu reported the 1,2-arylthio migration in the oxidation of thioalkynes (Scheme 7.17). An arylthioketene intermediate **AH** is generated in the reaction and subsequently trapped by an alcohol to afford the corresponding α-arylthio ester product **18**. Their DFT calculations exclude the formation of an α-oxo gold carbene intermediate and favor the carbenoid pathway, where the initial adduct **AG**, undergoes a direct 1,2-migration of the arylthio group to generate the ketene **AH**.

B. Xu reported that the putative α-oxo gold carbene generated via oxidative gold catalysis could react with pyridine-HF to afford α-fluoro ketones [50] and α-fluorinated β-ketoesters [51]. In 2018, Zhu and Zhou reported the α-oxo gold carbene **AI** inserts into the B—H bond of trimethylamine-borane (Scheme 7.18) [52]. Sterically hindered $Me_4{}^tBuXPhos$ is the optimal ligand and slows the detrimental reduction of gold(I) by the borane complex. The insertion is believed to be a concerted process. Their kinetic isotope effect (KIE) studies indicate that B–H insertion is a fast step and the expulsion of 8-isopropylquinoline should be the rate-determining step in the catalytic cycle.

7.3.4 Friedel–Crafts Reactions of Carbene/Carbenoid Intermediates with Arenes

Gold-catalyzed alkyne oxidation leading to the Friedel–Crafts-type alkylation was reported by L. Zhang in 2009, in which tetrahydrobenz[*b*]azepin-4-ones are

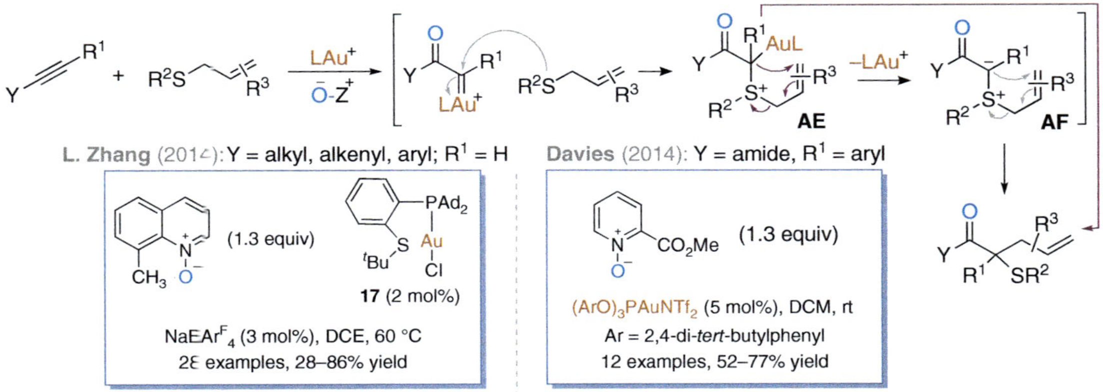

Scheme 7.16 Reaction of α-oxo gold carbene with allylic sulfide. Source: Based on Wu et al. [34] and Santos and Davies [48].

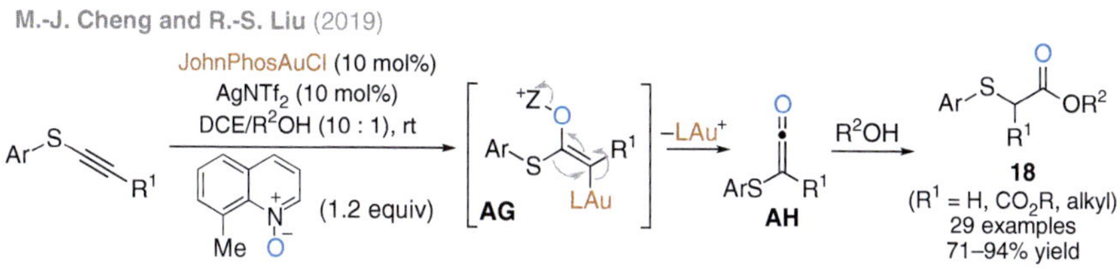

Scheme 7.17 Gold-catalyzed oxidation of thioalkynes forming arylthioketenes.

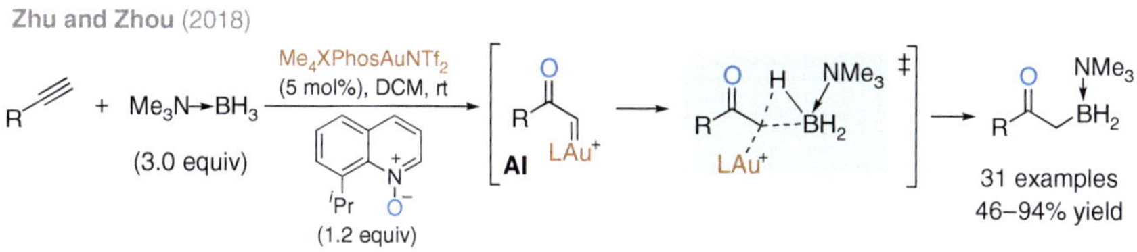

Scheme 7.18 Insertion into the B−H bond by an α-oxo gold carbene. Source: Modified from Yang et al. [52].

prepared in a one-pot synthesis from tertiary aniline substrates (Scheme 7.19) [53]. This two-step, one-pot reaction commences with the oxidation of the aniline substrate by *m*-CPBA to the corresponding amine *N*-oxide, which then undergoes gold-catalyzed intramolecular alkyne oxidation. The initially proposed mechanism evokes the formation of an α-oxo gold carbene intermediate, which could subsequently cyclize to the aromatic ring to afford the benzene-fused heterocycle, as illustrated in path A. However, in light of the revised mechanism of the related sulfoxide chemistry (Scheme 7.2), it is most likely that this chemistry follows a similar [3,3]-sigmatropic rearrangement (i.e. path B, a carbenoid pathway). Interestingly, if the substrate features an electron-withdrawing group (EWG)-terminated alkyne, it undergoes this reaction without any gold catalyst. This result is consistent with path B. This chemistry provides rapid access to benzene-fused azepinones.

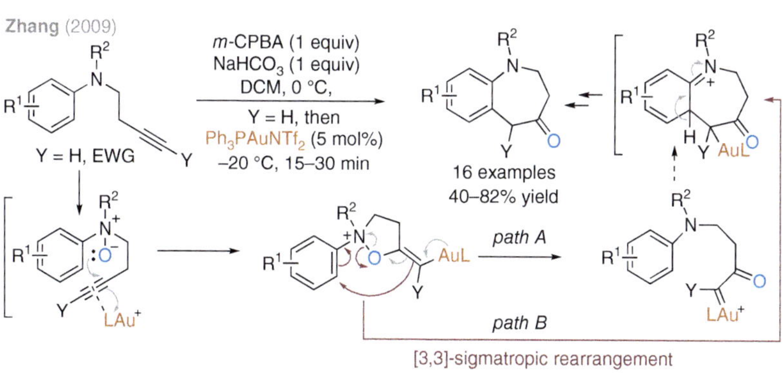

Scheme 7.19 Gold-catalyzed cycloisomerization of alkynylaniline *N*-oxides. Source: Modified from Cui et al. [53].

In 2012, L. Zhang reported the synthesis of chroman-3-ones through the electrophilic cyclization of α-oxo gold carbenes generated from propargyl aryl ethers in the presence of an external pyridine *N*-oxide [54]. As shown in Scheme 7.20A, the reaction follows the general trend of the Friedel–Crafts reactions, with electron-deficient aryl groups generally giving lower yields. The sterically hindered $Me_4{}^t$BuPhos is the optimal ligand. Shortly after, Gagosz reported the synthesis of indan-2-ones using propynyl arenes as substrates (Scheme 7.20B) [55]. The authors found that the Au(I)-complex derived from the electron-poor biarylphosphonite ligand **L2** exhibits higher catalytic activity than the more electron-donating ^{t}BuXPhos and proposed a mechanism in between the carbene and the carbenoid pathways.

In addition to the terminal aliphatic alkyne substrates discussed above, ynamides, featuring an electronically activated C—C triple bond, have been successfully employed in this reaction manifold [51, 56–59], affording various functionalized aromatic products. In a rare intermolecular implementation, Ye reported that the use of water as the solvent does not affect the generation of α-oxo gold carbenes from ynamides and moreover, largely suppresses the undesired further oxidation at the carbene center (Scheme 7.21) [59]. The reaction features a broad scope, and C3-alkylated indoles are formed in mostly good yields.

Electron-deficient C—C triple bonds bearing a carbonyl substituent (e.g. ynones and propiolamides) have also been successfully employed in this type of oxidative gold catalysis. As shown in Scheme 7.22, the gold carbene intermediates (e.g. **AJ** and **AK**) are acceptor/acceptor-type of carbenes and hence are expected to be highly reactive. J. Zhang reported the generation of such species from *N*-phenylpropiolamides and their subsequent Friedel–Crafts cyclization to afford 3-acyloxindoles (Scheme 7.22A) [60]. K. Ji reported a related reaction where an aryl enynone or biphenyl ynone was used as the α-oxo gold carbene precursor (Scheme 7.22B) [61]. In this reaction, the intermediacy of **AK** was corroborated by the observation of its cyclopropanation of a tethered alkene.

The intermediates generated upon the carbene Friedel–Crafts cyclization can participate in subsequent transformations, which ultimately lead to polycyclic products [58–64].

7.3.5 Reactions of Carbene/Carbenoid Intermediates with Alkenes

In the gold-catalyzed cycloisomerizations of 1,5-enynes [65], a cyclopropyl gold carbene intermediate, e.g. **AL**, is generated. As shown in Scheme 7.23A, in the presence of a nucleophilic oxidant, this intermediate could be oxidized to deliver the cyclopropyl ketone product **19** (*path A*). However, with the advent of facile gold-catalyzed intermolecular alkyne oxidation, alternative pathways where alkyne oxidation precedes cyclopropanation have been proposed in the literature, that is, the carbene pathway (i.e. *path B*) and the carbenoid pathway (i.e. *path C*).

In 2011, R.-S. Liu reported the first examples of oxidative cycloisomerization of 1,5-enynes [66]. As shown in Scheme 7.23B, benzene-fused 1,5-enynes or 1,3,5-dienynes **20** are converted into bicyclo[3.1.0]hex-3-en-2-ones **21**. The reaction tolerates a range of benzene substituents, R^1/R^2 can be alkyl, aryl, and in one case

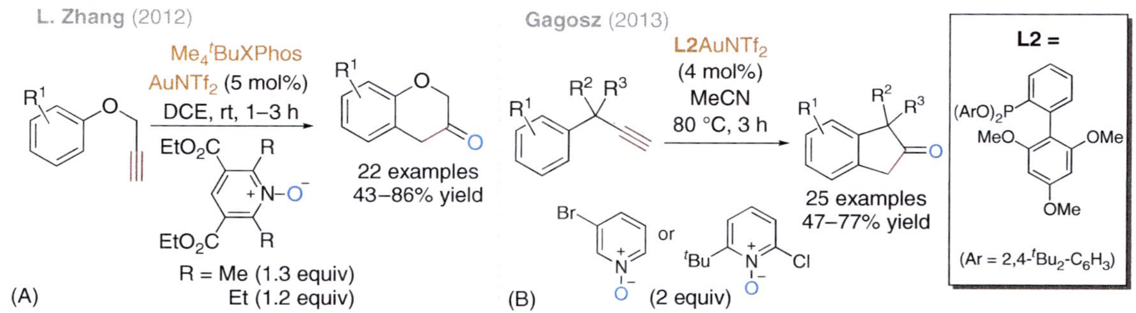

Scheme 7.20 The Friedel–Crafts reactions of α-oxo gold carbenes generated from terminal alkynes. Source: (B) Modified from Henrion et al. [55].

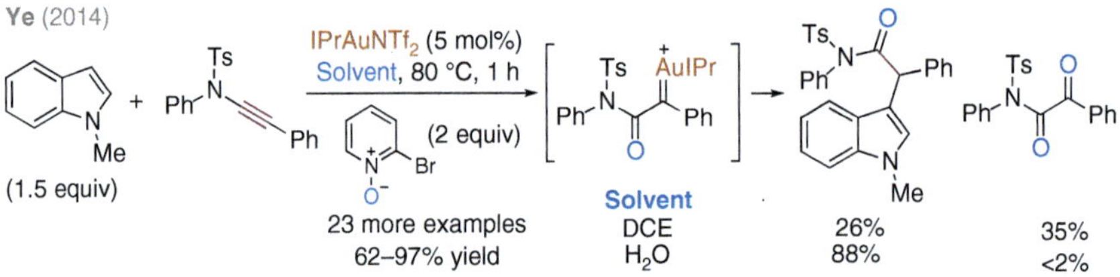

Scheme 7.21 In-water generation of α-oxo gold carbene and its reaction with indoles. Source: Modified from Li et al. [59].

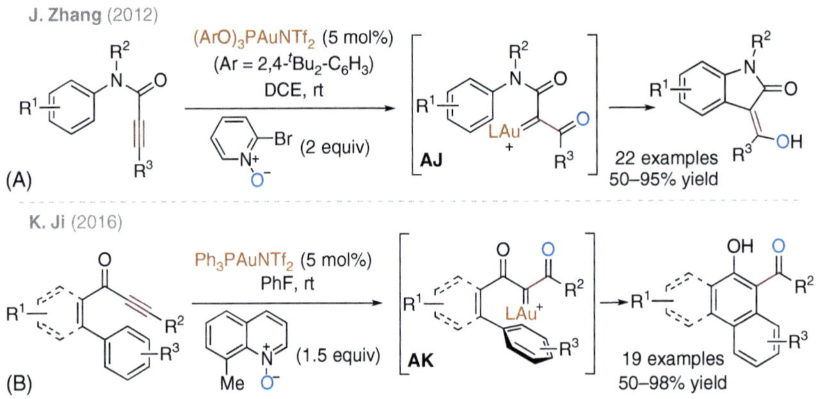

Scheme 7.22 The Friedel–Crafts reaction of 1,3-dioxo-2-gold carbenes. Source: (A) Modified from Qian and Zhang [60]. (B) Modified from Ji et al. [61].

an electron-withdrawing CO_2Et, and R^4 can be methyl or hydrogen. It was proposed that the reaction follows the carbene path (i.e. *path B*). However, the tolerance of an internal alkyne substrate, i.e. R^4 = Me, indicates that *path C* might be plausible as well, as α-oxo gold carbenes possessing α′-hydrogens typically undergo fast 1,2-C–H insertion, leading to enone products. In 2015, L. Zhang developed an enantioselective version of this chemistry [69]. By using the designed chiral ligand shown in the structure **AM**, the reaction exhibits in general higher enantioselectivities (up to 94% ee) with substrates featuring a *trans* electron-deficient C—C double bond than with those possessing an electron-neutral C—C double bond.

In 2013, C.-Y. Li reported another oxidative 1,5-enyne cyclopropanation by using substrates featuring an electronically activated ynamide moiety (Scheme 7.23D) [68]. With MsOH (1.2 equiv) as the pyridine scavenger, the parent pyridine *N*-oxide can be employed as a suitable oxidant. The inferior results from a corresponding diazo substrate rule out a carbene pathway. The reaction mechanism was proposed to follow *path A* or the carbenoid pathway (i.e. *path C*), although the latter was preferred.

By using readily accessible *t*-butyldimethylsilyl (TBS)-protected 1,5-enyn-3-ol substrates, L. Zhang realized efficient oxidative cyclopropanation under gold catalysis in 2014 (Scheme 7.23C) [67]. 8-Methylquinoline *N*-oxide is the optimal oxidant. Ligand screening revealed that **L1**, previously developed in the carboxylic acid-trapping chemistry (see Scheme 7.10A), performed substantially better than Mor-DalPhos

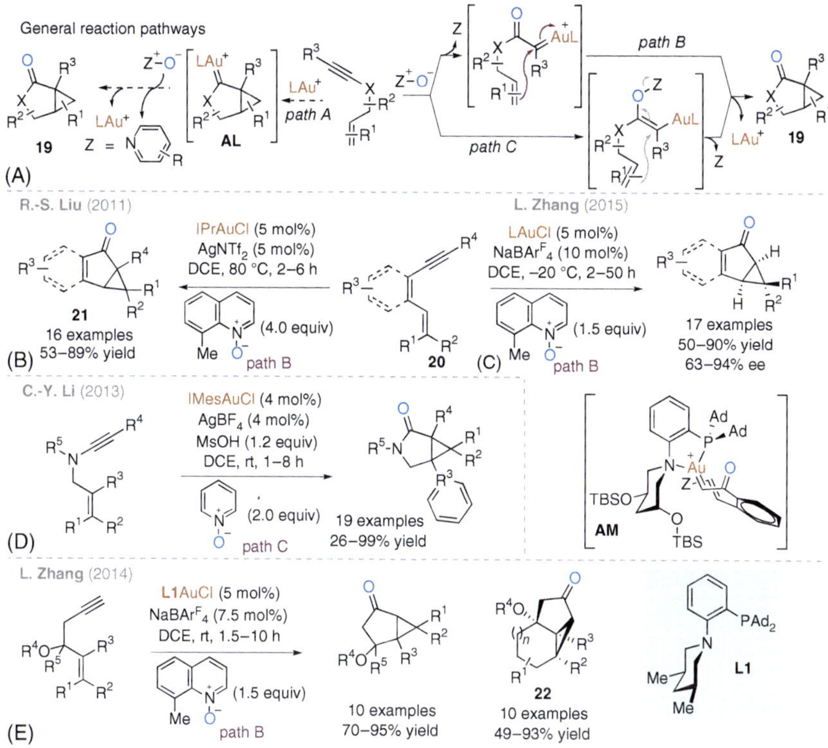

Scheme 7.23 Gold-catalyzed oxidative cyclization of 1,5-enynes. Source: (B) Modified from Liu et al. [66]. (C) Modified from Ji and Zhang [67]. (D) Modified from Wang et al. [68].

and other conventionally employed phosphine and *N*-heterocyclic carbene (NHC) ligands (e.g. 92% versus <35% yield). When this reaction is applied to cycloalkene substrates, angularly fused tricyclic cyclopropanes, e.g. **22**, are formed in moderate to excellent yields.

When a C—C double bond is too close to the gold carbene center for cyclopropanation to occur, the two react via cationic pathways. In the work published by R.-S. Liu in 2011 (Scheme 7.24A) [66], the gold carbene center is generated regioselectively due to the electronically biased nature of the ynamide C—C triple bond and cyclizes to the C—C double bond as an electrophile to generate the carbocation **AN**, which would then undergo hydride migration and deauration to afford the indenecarboxamide **23**. Scheme 7.24B outlines another study by R.-S. Liu [70], where the carbene cyclization is followed by the migration of the alkene *cis*-substituent, i.e. R^2. This is explained, with the help of DFT calculations, by that gold in the carbocation **AO** facilitates the migration of its neighboring *anti*- substituent R^2 via hyperconjugation.

The gold-catalyzed oxidative cyclizations of 1,6-enynes can generate either 5-*exo* carbonyl or 6-*endo* carbonyl products. The 5-*exo* mode was reported by J. Zhang in 2011 in the conversion of the 1,6-enynones or enynamides **24** to the bicyclic cyclopropanones **25** (Scheme 7.25A) [71]. In this study, the combination of 4-acetylpyridine *N*-oxide and MsOH was initially employed, but it was found later

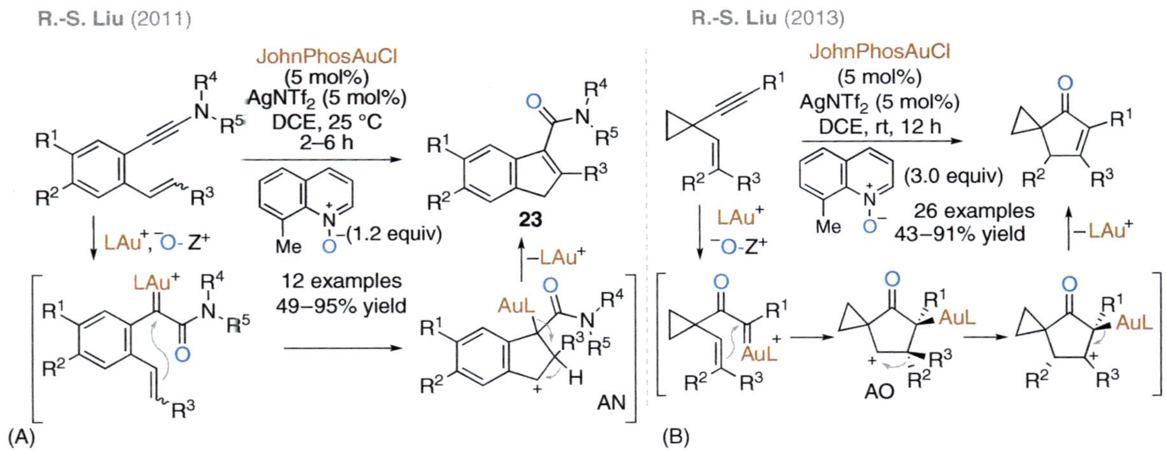

Scheme 7.24 Gold-catalyzed oxidative cyclization of enynes. Source: (A) Modified from Vasu et al. [66]. (B) Modified from Ghorpade et al. [70].

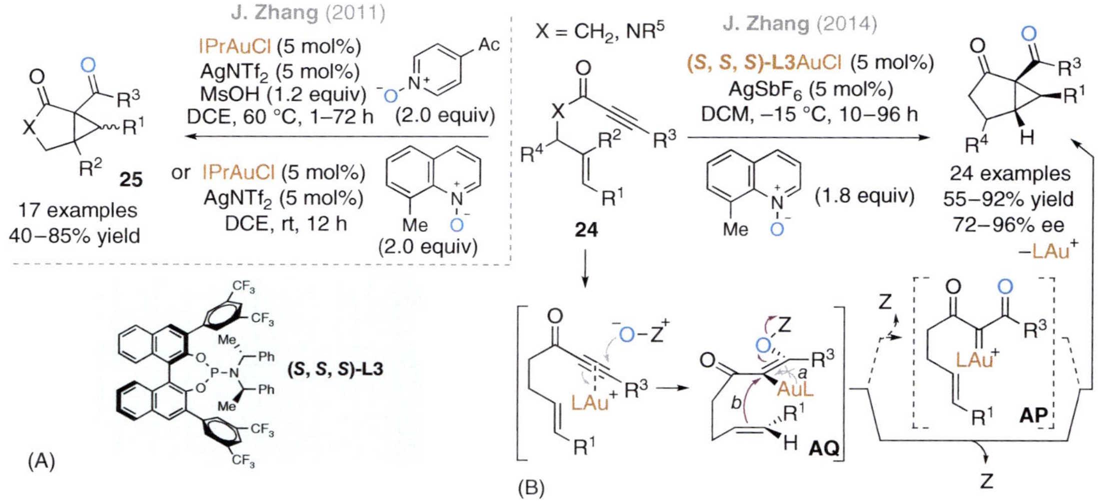

Scheme 7.25 Gold-catalyzed oxidative cyclization of 1,6-enyne. Source: (A) Modified from Qian and Zhang [71]. (B) Modified from Qian et al. [72].

8-methylquinoline *N*-oxide without acid additives results in faster reactions and in some cases better yields. The reaction works with an amide or a ketone linkage between the two π systems, but not with the ester counterpart. The regiochemistry of the oxidation is dictated by the electron-withdrawing carbonyl group. In 2014, the same group published an asymmetric version of this reaction (Scheme 7.25B) [72]. Using the chiral binol phosphoramidite (i.e. ***(S,S,S)*-L3**) as the ligand, the gold catalysis exhibits good to excellent enantioselectivities. The mechanistic studies indicate that an α-oxo gold carbene species **AP** is likely not involved and the reaction instead follows a carbenoid pathway, as outlined by **AQ**. The scope study mostly employed ketone substrates, and the one amide (X = NMe, R^3 = PMP, and $R^1 = R^2$ = H) resulted in 84% ee.

1,6-enyn-5-sulfoxides were later shown to undergo similar oxidative cyclopropanation [73]. The 6-*endo* mode of the oxidative cycloisomerization of 1,6-enynes has also been documented [66, 74].

The oxidatively generated α-oxo gold carbene can cyclopropanate a benzene ring, thereby accessing the Büchner reaction manifold. In 2018, K. Ji and L. Zhang reported such a reaction en route to dihydropyranone-fused cycloheptatrienes (Scheme 7.26) [75]. Mechanistically, the norcaradiene intermediate **26** is formed via benzene cyclopropanation by the expected α-oxo gold carbene. Subsequent treatment of the reaction mixture with Et_3N drives the equilibrium between **26** and the ring-expanded cycloheptatriene structure **27** toward the observed product by isomerization of one of the C—C double bonds of **27**. When a substrate featuring an all-carbon backbone was tested, a mixture of the Büchner product and the Friedel–Crafts product was formed.

7.3.6 Reactions of Carbene/Carbenoid Intermediates with C—C Triple Bonds

Carbenes or carbenoids generated from single alkyne oxidation of 1,5-diynes or 1,6-diynes can undergo subsequent reactions with the remaining C—C triple bond. Several mechanistic scenarios have been documented. As shown in Scheme 7.27A, a plausible carbene metathesis with the pendant C—C triple bond would directly deliver the vinyl carbene intermediate **AR** (path A), which could be alternatively generated via ring opening of the cyclopropene intermediate **AS** (path B); on the other hand, a nucleophilic attack by the tethered alkyne would generate a reactive β-gold vinyl cation intermediate **AT** (path C).

Since 2013, Hashmi has described several gold(I)-catalyzed oxidative transformations of 1,2-dialkynylbenzenes (Scheme 7.27B). It is proposed that the alkenyl carbene **AU** is generated via either path A or path B and can undergo different subsequent transformations depending on its substitution patterns. For example, with a *t*-butyl substituent at the carbene center, methyl migration affords the indenone **28** as the final product (X = halide, R^2 = Me) [76]. When the X group of **AR** is a carbonyl group as in the carbene **AV**, it can undergo carbonyl cyclization to form the indenofuranone **29** product [78]. Notably, the author indicated that the vinyl cation pathway, i.e. path C, is also possible. It is noteworthy that these studies all involve the

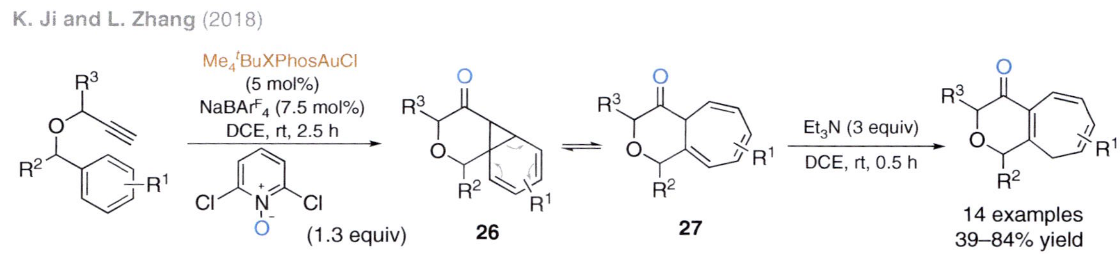

Scheme 7.26 Oxidative gold(I)-catalyzed cyclopropanation of benzene rings. Source: Modified from Ji and Zhang [75].

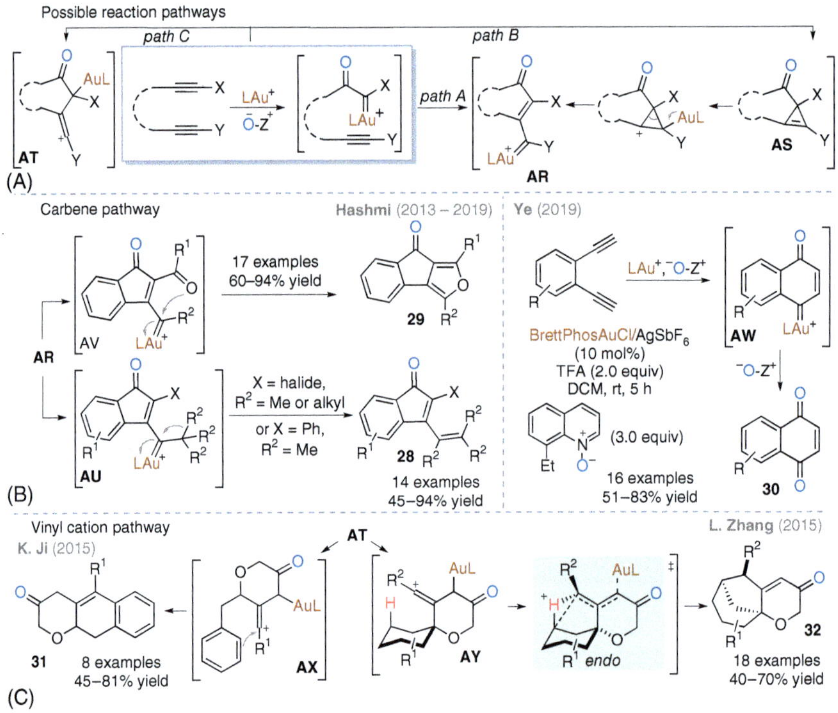

Scheme 7.27 Reaction of carbene/carbenoid intermediates with C–C triple bonds. Source: (B) Based on Nösel et al. [76], Wang et al. [77], Rode et al. [78], Shu et al. [79]. (C) Based on Ji et al. [80] and Zheng et al. [81].

exocyclic vinyl carbene intermediate **AR**. In 2019, Ye proposed that the endocyclic vinyl carbenoid intermediate **AW** is generated from 1,2-diethynylbenzenes en route to the naphthoquinone product **30** [79].

In 2015, K. Ji described the gold(I)-catalyzed oxidative cyclization of non-arene-fused 1,6-diynes and proposed a vinyl cation reaction pathway (i.e. path C) [80]. For example, the β-gold vinyl cation intermediate **AX** generated in the reaction can undergo Friedel–Crafts cyclization to form the chromen-3(4*H*)-ones **31** (Scheme 7.27C). In the same year, L. Zhang reported that the related vinyl cation intermediate **AY** undergoes a concerted C–H insertion process, leading to the formation of tricyclic 2*H*-pyran-3(6*H*)-ones **32** [81]. The observed diastereoselectivities are consistent with a vinyl cation pathway.

7.3.7 1,2-C–C and 1,2-C–H Insertions of Carbene/Carbenoid Intermediates

Among the most valuable transformations of metal carbenes are the 1,2-C–C/C–H insertions, which lead to the formation of C—C double bonds. With α-oxo gold carbenes being highly electrophilic, the 1,2-C–H insertion is facile and can constitute the main carbene decomposition pathway if an α′-C—H bond is present.

The absence of the corresponding alkene products is often a mechanistic indication against the formation of a discrete α-oxo gold carbene species, although the contrary does not rule out a carbenoid pathway.

In 2010, L. Zhang [82] reported the oxidation of internal alkynes to α,β-unsaturated ketones (Scheme 7.28A). For the first time, 8-alkylquinoline *N*-oxides were shown to be highly effective external oxidants and the reaction does not require a stoichiometric acid additive. Similarly, 2-bromopyridine *N*-oxide was found to be a suitable oxidant without an acid additive. The regioselectivity challenge associated with electronically unbiased dialkylalkynes can be largely addressed by using bulky 8-isopropylquinoline *N*-oxide as the oxidant, IPr as the ligand, and by running the reaction at 0 °C or −20 °C. For example, with R^1 = Me and R^2 = *n*-hexyl, this oxidation can achieve a selectivity of 11 : 1 favoring the delivery of the oxidant at the less hindered alkyne end by differentiating a methyl over a linear primary alkyl group. For electronically biased internal alkynes including arylalkynes, enynes, and ynamides, a more acidic bulky phosphite ligand is employed to maximize the bias. As such, the selectivity is routinely above 10 : 1 and the major products have the carbonyl group α to the π-system or the amide moiety.

This enone-forming reaction has been successfully employed in the total synthesis of citrinadin alkaloids. As shown in Scheme 7.28B, Martin [84] and Wood [83] separately reported the conversion of **33** to **34** in the presence of an array of other functional groups. Notably, 1.1 equiv of $Ph_3PAuNTf_2$ was used in both instances due to the binding and deactivation of the gold catalyst by the basic tertiary amine moiety.

This transformation of internal alkynes to α,β-unsaturated carbonyl compounds has been further explored with functionalized substrates including ynamides [85], propargylic ethers [86], enynes [87], and 1-alkynyl triazenes [88], leading to regioselective oxidations.

The 1,2-C–C insertions by α-oxo gold carbenes are often assisted by the release of ring strain and/or the stabilization of carbocation intermediates by an α′ heteroatom. When the heteroatom is oxygen from a hydroxyl group, the transformation resembles a semi-pinacol rearrangement. In 2012, Hashmi reported the gold-catalyzed oxidation of the tertiary propargyl alcohols **35** to afford 1,3-diketone products **36** (Scheme 7.29A) [89]. The reaction regioselectivity, i.e. oxygen delivery distal to the hydroxyl group, is likely due to the inductive effect of the alcohol oxygen and is augmented by steric hindrance at the proximal C(sp). S. Li and Y. Liu reported related alkyne oxidation using tertiary alcohols prepared from quinones [90]. As shown in Scheme 7.29B, the 1,2-alkenyl group migration to the carbene center enables a facile synthesis of various functionalized tropones and their analogs.

In 2013, Y. Liu reported the synthesis of dihydro-γ-carbolines using oxidative gold catalysis [91]. As shown in Scheme 7.30A, the reaction features generally good yields and is compatible with a series of aryl and alkyl substituents on the alkyne and the imine. A subsequent DFT study by Ariafard favors a carbenoid reaction route involving a facile 1,2-aryl migration of the alkyne/*N*-oxide adduct **AZ** to give the β-ketoaldehyde intermediate **BA**, which is then trapped by the imine starting material and subsequently undergoes cyclization to furnish the final product. In

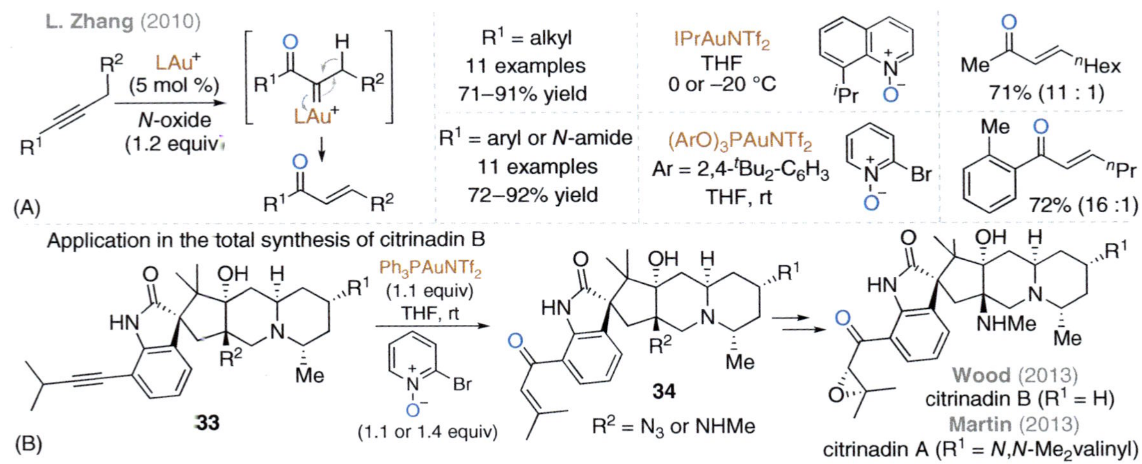

Scheme 7.28 1,2-C–H insertions with oxidatively generated gold carbenes: early studies and application in total synthesis. Source: (A) Modified from Zhang 2010. (B) Based on Kong et al. [83] and Bian et al. [84].

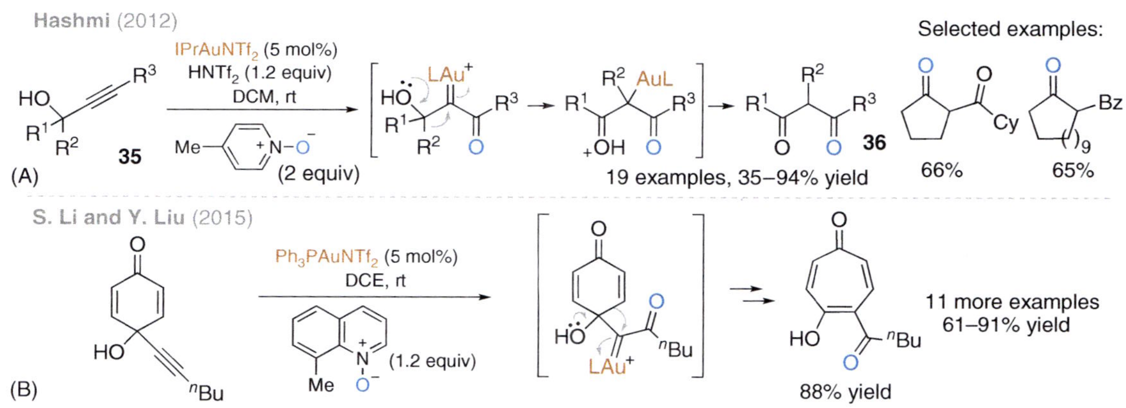

Scheme 7.29 Gold-catalyzed oxidation of tertiary propargylic alcohols. Source: Modified from Hashmi et al. [89]. (B) Modified from Zhao et al. [90].

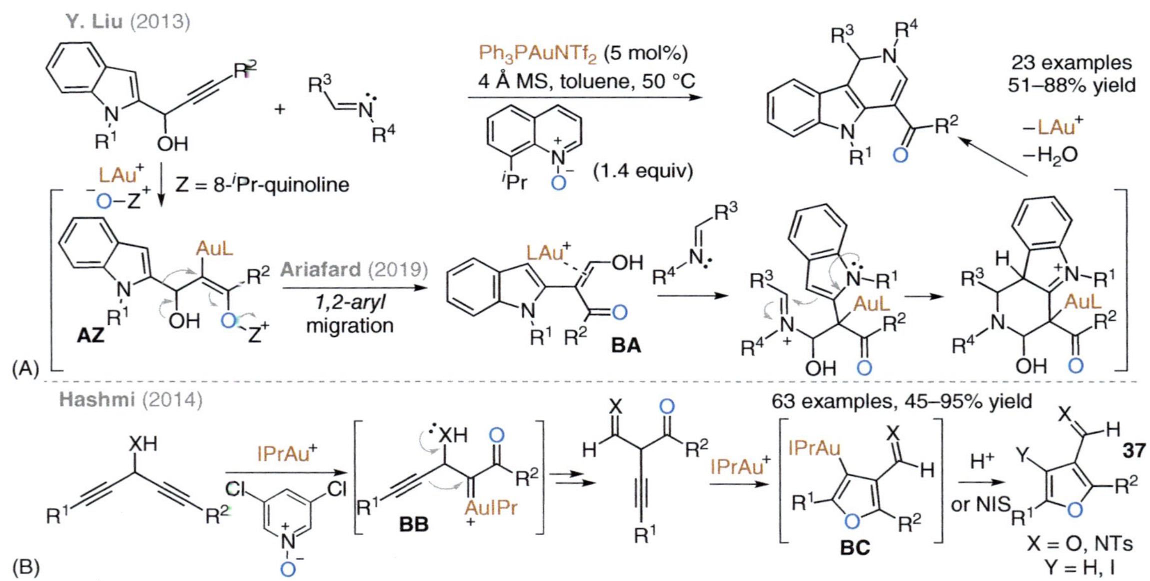

Scheme 7.30 Gold-catalyzed oxidation of secondary propargylic alcohols and amides. Source: (A) Modified from Wang et al. [91]. (B) Based on Wang et al. [77, 92, 93].

2014, Hashmi reported the gold-catalyzed oxidation of 1,1-dialkynyl carbinols and their tosylamide counterparts (Scheme 7.30B) [77, 92, 93]. The proposed mechanism entails a 1,2-alkynyl migration of the α-oxo gold carbene intermediate followed by cyclization. The method provides rapid access to 2,5-disubstituted 3-formylfurans **37**. Interestingly, no sign of the typically facile 1,2-hydride migration has been observed. DFT calculations later determined that the transition state for the 1,2-hydride shift is significantly higher in energy than that of the corresponding 1,2-alkynyl shift, with a difference of as much as 5.4 kcal mol^{-1}. With the addition of *N*-iodosuccinimide (NIS), *in situ* iodination of the furylgold(I) intermediate **BC** occurs, thereby affording the 4-iodo-3-formylfuran products and their imine counterparts.

Azepines and their derivatives are an important class of medium-sized nitrogen heterocycles featured in many bioactive natural products as a key structural motif. Y. Liu and Y. Li reported the synthesis of functionalized azepines via the ring expansion of 1,2-dihydropyridine to an oxidatively generated α-oxo gold carbene (Scheme 7.31) [94]. The reaction tolerates a wide range of substitution patterns and the yields are mostly >80%. DFT studies support a carbenoid pathway, where the attack of electron-rich dihydropyridine double bond and the expulsion of pyridine occur in a concerted manner and the bicycle **BD** is most likely formed as an intermediate during the ring expansion.

Stereoelectronic biases have been extensively exploited in the regioselective generation of α-oxo gold carbenes from internal alkyne substrates. However, there is no efficient approach to accessing the sterically/electronically disfavored α-oxo gold carbene regioisomer. In 2015, L. Zhang reported a solution by using the alkynyl sulfone **38** as the substrate [95]. As shown in Scheme 7.32, due to the strong stereoelectronic bias of the C—C triple bond of **38**, α-oxo gold carbene **BF** should be generated with excellent regioselectivity. Subsequent stepwise desulfonylative 1,2-aryl migration would give the carbene **BE**, which is the minor regioisomer when generated from arylalkynes. **BE** can be further oxidized or undergoes cyclopropanation reactions. Comparative experiments with diazo precursor to the carbene **BF** led to similar outcomes, thereby supporting its formation.

The α-oxo gold carbene or its carbenoid precursor can also undergo 1,2-insertion into a carbon–heteroatom bond [96]. Recently, X. Xu and L. Zhang reported the Wolff rearrangement of the α-oxo gold carbene **BG** (Scheme 7.33) [97]. With TBS-terminated alkynes as substrates, a series of silylketenes **38** was synthesized and subsequently trapped by thiophenol in one pot. Relative to a previously reported method featuring a rhodium-catalyzed Wolff rearrangement [98], this reaction requires fewer synthetic steps and avoids the usage of hazardous diazo substrates.

7.3.8 Remote C(sp^3)–H Functionalizations by Carbene/Carbenoid Intermediates

In 2009, L. Zhang reported intramolecular alkyne oxidation by a tethered amine *N*-oxide, which is generated in one pot, as shown in Scheme 7.34A [99]. In combination with the preparation of the substrate from a secondary amine and

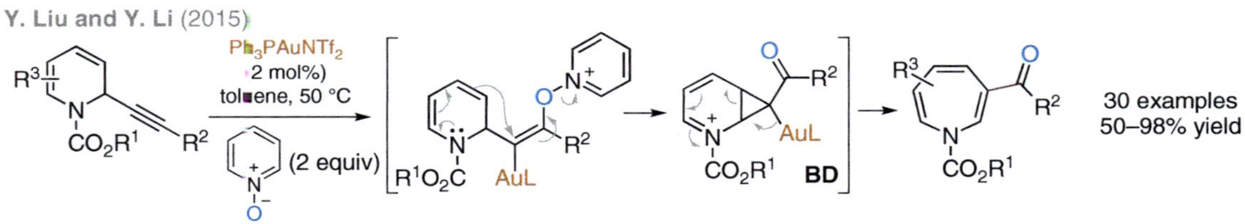

Scheme 7.31 Ring expansion of 1,2-dihydropyridines to access functionalized azepines. Source: Modified from Chen et al. [94].

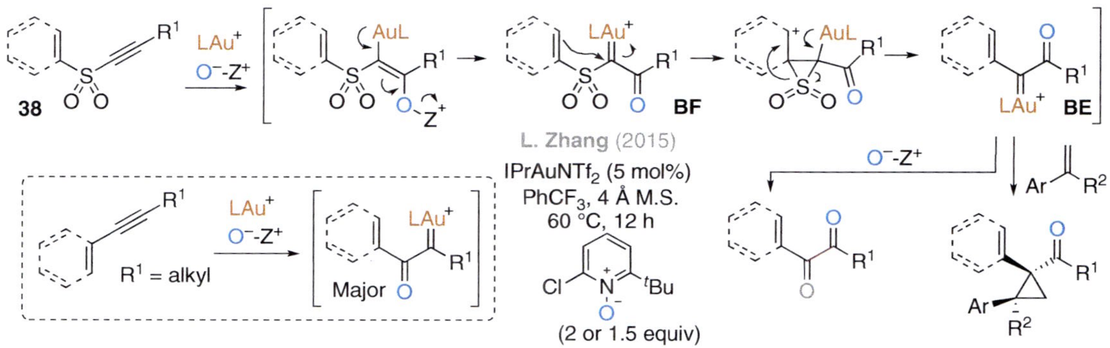

Scheme 7.32 A desulfonylative 1,2-migration to access stereoelectronically disfavired gold carbenes. Source: Modified from Chen and Zhang [95].

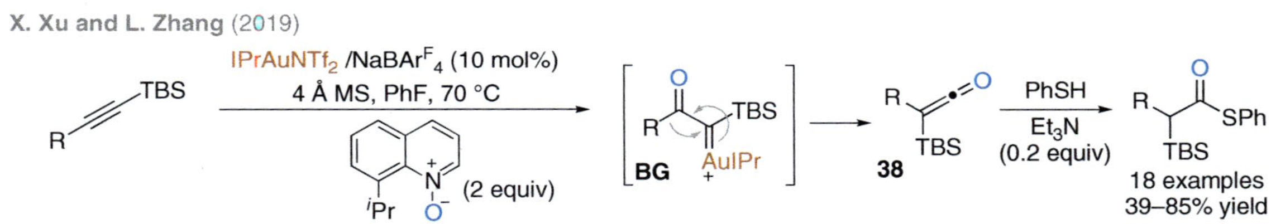

Scheme 7.33 Access to silylketenes through the Wolff rearrangement of gold carbenes. Source: Modified from Zheng et al. [97].

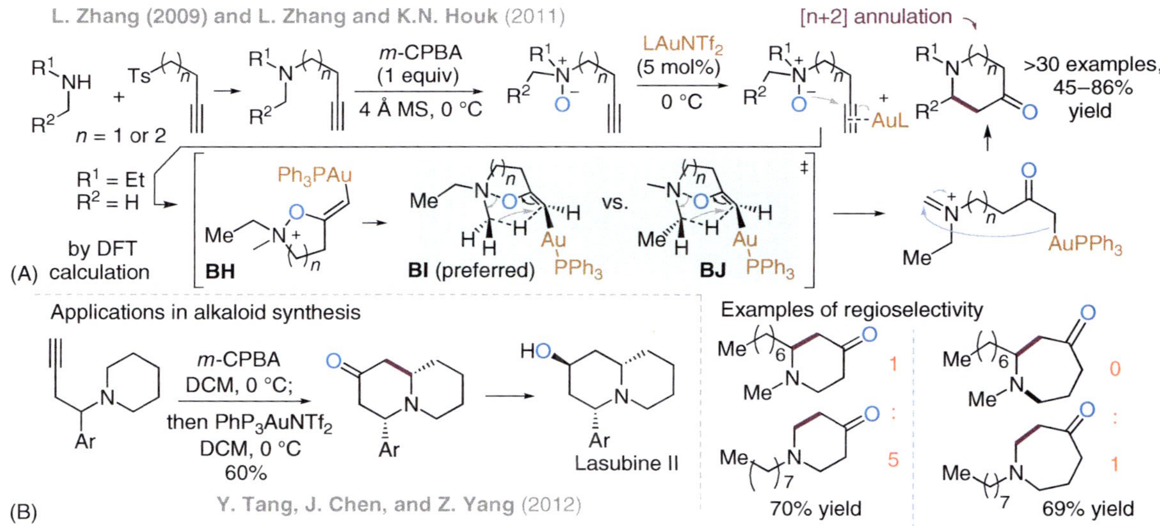

Scheme 7.34 Synthesis of N-hetereocyclic ketones via gold-catalyzed alkyne oxidation by tethered *tert*-amine *N*-oxides. Source: (A) Based on Cui et al. [99] and Houk and coworkers [100]. (B) Modified from Shan et al. [23].

but-3-yn-1-yl tosylate, the overall two-step process constitutes an [4+2] annulation. When the secondary amine has two different alkyl substituents, the apparent C–H insertion occurs selectively at the smaller group, as shown in the selected examples. This chemistry was subsequently extended to the synthesis of azepan-4-ones in a two-step, [5+2] annulation, where substantially higher regioselectivities were observed than in the [4+2] counterpart [101]. The reaction mechanism, supported by DFT calculations reported in 2011 [100], reveals that the intermediate **BH** might not undergo fragmentation to generate the proposed α-oxo gold carbene intermediate. Instead, an energetically favored hetero-retro-ene pathway, as outlined by the transition states **BI** and **BJ**, occurs. Subsequent cyclization of the iminium gold enolate would furnish the product and regenerate the cationic gold catalyst. In the hetero-retro-ene transition states, the preference of **BI**, which places a smaller group (i.e. the Me group) at the (pseudo)axial position of the *N,O*-containing ring, is consistent with the observation that smaller groups on the nitrogen are preferred in this remote C–H functionalization. The proposed mechanistic pathway is supported by the failure to perform the same reaction using the corresponding diazo ketones.

The synthetic utilities of the [4+2] annulation were demonstrated in the synthesis of (+)-lentiginosine from L-(+)-tartaric acid [102] and the highly diastereoselective synthesis of (±)-cermizine C [99] by L. Zhang. Later in 2012, this chemistry was employed as a crucial step to the total synthesis of lasubine II and (±)-decinine by Y. Yang, J. Chen, and Z. Yang (Scheme 7.34B) [23].

In 2012, R.-S. Liu reported the gold-catalyzed oxidation of *ortho*-substituted phenylacetylenes to form indan-1-one products [103]. As shown in Scheme 7.35A, the reaction undergoes an apparent 1,5-C–H insertion, and furnish cyclopentenone products. Mechanistic studies including a comparison to the experimental outcomes using the corresponding diazo ketone substrates support a carbenoid pathway, where the initial adduct **BK** undergoes a 1,5-hydride shift before cyclization.

In 2015, L. Zhang revealed that the acceptor–acceptor gold carbenes generated from ynones can insert into unactivated C(sp^3)—H bonds (Scheme 7.35B) [104]. With bulky ^{t}BuBrettPhosAuNTf$_2$ as the catalyst, the C–H insertions exhibit poor regioselectivities, delivering a cyclopentanone and two cyclobutanone products. On the other hand, the selective insertion into the γ-C—H bonds was realized by using the previously reported *P,N*-bidentate ligand **L1** [31]. Under this set of conditions, spiro-, fused-, and bridged bicyclic or monocyclic cyclopentanones were isolated in good yields. The selective insertion into the β-C(sp^3)—H bonds was later realized [105]. Bulky phosphine ligands such as BuBrettPhos are essential to achieve this desired regioselectivity. DFT calculations offer insight into the steric effect of the ligand. One common feature of both types of C–H insertion is the presence of a quaternary carbon center in the substrate backbone in order to harness the rate-enhancing Thorpe-Ingold effect. In its absence, the reaction is not efficient, with the yield of one reported cyclopentanone case being <50%. Mechanistically, an acceptor/acceptor-type gold carbene is the proposed reactive intermediate, which is supported by the similar product distribution when the α-diazo β-diketone corresponding to **BL** was treated with BuBrettPhosAuNTf$_2$.

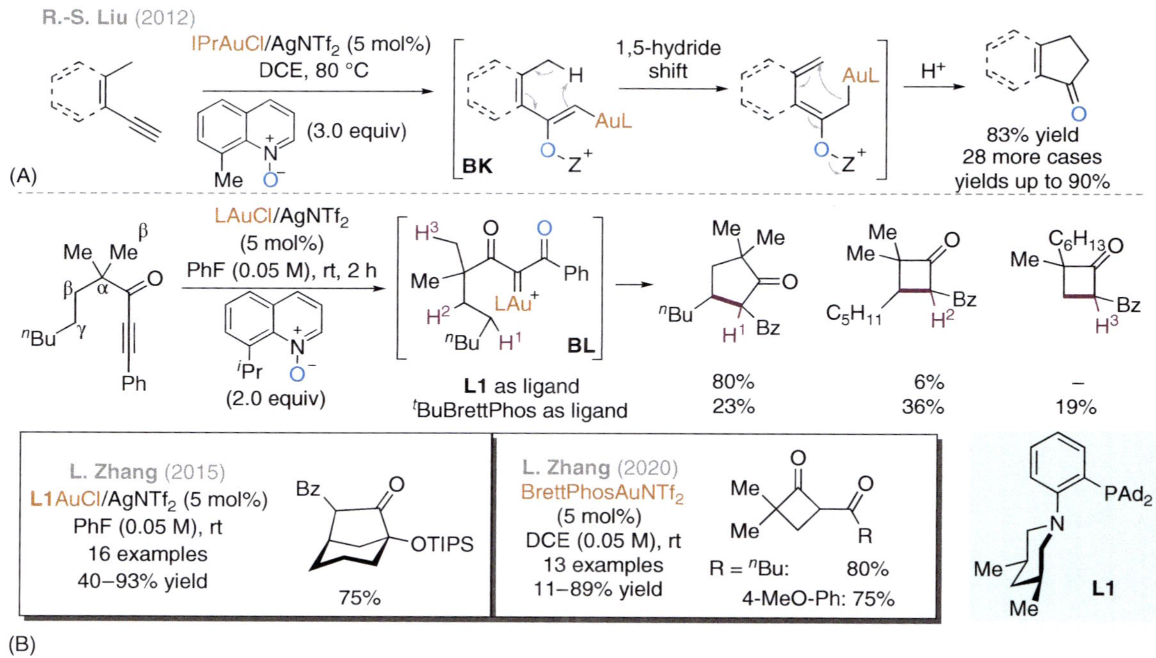

Scheme 7.35 Synthesis of cyclic ketones via remote C–H functionalization. Source: (A) Modified from Bhunia et al. [103]. (B) Based on Wang et al. [104] and Zheng et al. [105].

7.4 Conclusion

The oxidation of alkynes by *N*-oxides or sulfoxides in the presence of a gold catalyst involves the intermediacy of α-oxo gold carbenes or its carbenoid precursors. These versatile intermediates undergo a diverse range of transformations characteristic of metal carbene chemistry. They include being trapped by protic heteronucleophiles, formation of ylides, electrophilic aromatic substitutions, cyclopropanation of alkenes, and C–H insertions. A range of valuable synthetic methods have bene developed, some of which are employed as key steps in natural product total synthesis. Comparing to their counterpart based on hazardous diazo carbonyl substrates, these methods offer improved operational safety and step economy. Further advance in this subfield of gold catalysis is expected to broaden the scope of the alkyne/substrate types, unearth new reactivities, and improve the atom economy of the oxidants.

References

1 Prieto, A., Fructos, M.R., Mar Díaz-Requejo, M. et al. (2009). *Tetrahedron* 65: 1790–1793.
2 Pawar, S.K., Wang, C.-D., Bhunia, S. et al. (2013). *Angew. Chem. Int. Ed.* 52: 7559–7563.
3 Zheng, Z., Ma, X., Cheng, X. et al. (2021). Homogeneous Gold-Catalyzed Oxidation Reactions. *Chem. Rev.* 121: 8979–9038.
4 Shapiro, N.D. and Toste, F.D. (2007). *J. Am. Chem. Soc.* 129: 4160–4161.
5 Li, G. and Zhang, L. (2007). *Angew. Chem. Int. Ed.* 46: 5156–5159.
6 Fang, R. and Yang, L. (2012). *Organometallics* 31: 3043–3055.
7 Lu, B., Li, Y., Wang, Y. et al. (2013). *J. Am. Chem. Soc.* 135: 8512–8524.
8 Davies, P.W. and Albrecht, S.J.C. (2009). *Angew. Chem. Int. Ed.* 48: 8372–8375.
9 Cuenca, A.B., Montserrat, S., Hossain, K.M. et al. (2009). *Org. Lett.* 11: 4906–4909.
10 Barrett, M.J., Davies, P.W., and Grainger, R.S. (2015). *Org. Biomol. Chem.* 13: 8676–8686.
11 Li, C.-W., Pati, K., Lin, G.-Y. et al. (2010). *Angew. Chem. Int. Ed.* 49: 9891–9894.
12 Xu, C.-F., Xu, M., Jia, Y.-X., and Li, C.-Y. (2011). *Org. Lett.* 13: 1556–1559.
13 Matsuda, T., Nishida, Y., Yamanaka, K., and Sakurai, Y. (2015). *Tetrahedron* 71: 869–874.
14 Schießl, J., Stein, P.M., Stirn, J. et al. (2019). *Adv. Synth. Catal.* 361: 725–738.
15 Ye, L., Cui, L., Zhang, G., and Zhang, L. (2010). *J. Am. Chem. Soc.* 132: 3258–3259.
16 Cai, J., Wang, X., Qian, Y. et al. (2019). *Org. Lett.* 21: 369–372.
17 Wei, H., Bao, M., Dong, K. et al. (2018). *Angew. Chem. Int. Ed.* 57: 17200–17204.
18 Ye, L., He, W., and Zhang, L. (2010). *J. Am. Chem. Soc.* 132: 8550–8551.
19 Xu, M., Ren, T.-T., and Li, C.-Y. (2012). *Org. Lett.* 14: 4902–4905.

20 Shu, C., Liu, R., Liu, S. et al. (2015). *Chem. Asian J.* 10: 91–95.
21 Zhao, Q., Henrion, G., and Gagosz, F. (2019). *Bioorg. Med. Chem.* 27: 2616–2620.
22 Fu, J., Shang, H., Wang, Z. et al. (2013). Gold-catalyzed rearrangement of allylic oxonium ylides: efficient synthesis of highly functionalized dihydrofuran-3-ones. *Angew. Chem., Int. Ed.* 52: 4198–4202.
23 Shan, Z.-H., Liu, J., Xu, L.-M. et al. (2012). *Org. Lett.* 14: 3712–3715.
24 Han, M., Bae, J., Choi, J., and Tae, J. (2013). *Synlett* 24: 2077–2080.
25 Murai, M., Kitabata, S., Okamoto, K., and Ohe, K. (2012). *Chem. Commun.* 48: 7622–7624.
26 Ji, K., Nelson, J., Zhang, L., and Beilstein, J. (2013). *Org. Chem.* 9: 1925–1930.
27 Sun, N., Chen, M., and Liu, Y. (2014). *J. Org. Chem.* 79: 4055–4067.
28 Luo, Y., Ji, K., Li, Y., and Zhang, L. (2012). *J. Am. Chem. Soc.* 134: 17412–17415.
29 He, W., Xie, L., Xu, Y. et al. (2012). *Org. Biomol. Chem.* 10: 3168–3171.
30 Shi, S., Wang, T., Yang, W. et al. (2013). *Chem. Eur. J.* 19: 6576–6580.
31 Ji, K., Zhao, Y., and Zhang, L. (2013). *Angew. Chem. Int. Ed.* 52: 6508–6512.
32 Xie, L., Liang, Z., Yan, D. et al. (2013). *Synlett* 24: 1809–1812.
33 Wu, C., Liang, Z., Yan, D. et al. (2013). *Synthesis* 45: 2605–2611.
34 Wu, G., Zheng, R., Nelson, J., and Zhang, L. (2014). *Adv. Synth. Catal.* 356: 1229–1234.
35 Zhang, Y., Xue, Y., Li, G. et al. (2016). *Chem. Sci.* 7: 5530–5536.
36 He, W., Li, C., and Zhang, L. (2011). *J. Am. Chem. Soc.* 133: 8482–8485.
37 Wu, Y., Peng, S., Ouyang, Y. et al. (2012). *Acta Chim. Sinica* 70: 1367–1370.
38 Dubovtsev, A.Y., Dar'in, D.V., and Kukushkin, V.Y. (2019). *Adv. Synth. Catal.* 361: 2926–2935.
39 Wang, Q., Hoffmann, S., Schießl, J. et al. (2020). *Eur. J. Org. Chem.* 2020: 2384–2388.
40 Rassadin, V.A., Boyarskiy, V.P., and Kukushkin, V.Y. (2015). *Org. Lett.* 17: 3502–3505.
41 Karad, S.N. and Liu, R.-S. (2014). *Angew. Chem. Int. Ed.* 53: 5444–5448.
42 Ye, L., He, W., and Zhang, L. (2011). *Angew. Chem. Int. Ed.* 50: 3236–3239.
43 Shu, C., Li, L., Xiao, X.-Y. et al. (2014). *Chem. Commun.* 50: 8689–8692.
44 Shu, C., Li, L., Yu, Y.-F. et al. (2014). *Chem. Commun.* 50: 2522–2525.
45 Huple, D.B., Ghorpade, S., and Liu, R.-S. (2013). *Chem. Eur. J.* 19: 12965–12969.
46 Talbot, E.P.A., Richardson, M., Mckenna, J.M., and Toste, F.D. (2014). *Adv. Synth. Catal.* 356: 687–691.
47 Li, J., Ji, K., Zheng, R. et al. (2014). *Chem. Commun.* 50: 4130–4133.
48 Santos, M.D. and Davies, P.W. (2014). *Chem. Commun.* 50: 6001–6004.
49 Baker, T. and Davies, P.W. (2019). *Eur. J. Org. Chem.* 2019: 5201–5204.
50 Zeng, X., Liu, S., Shi, Z. et al. (2016). *Angew. Chem. Int. Ed.* 55: 10032–10036.
51 Zeng, X., Lu, Z., Liu, S. et al. (2017). *Adv. Synth. Catal.* 359: 4062–4066.
52 Yang, J.-M., Zhao, Y.-T., Li, Z.-Q. et al. (2018). *ACS Catal.* 8: 7351–7355.
53 Cui, L., Zhang, G., Peng, Y., and Zhang, L. (2009). *Org. Lett.* 11: 1225–1228.

54 Wang, Y., Ji, K., Lan, S., and Zhang, L. (2012). *Angew. Chem. Int. Ed.* 51: 1915–1918.
55 Henrion, G., Chavas, T.E.J., Le Goff, X., and Gagosz, F. (2013). *Angew. Chem. Int. Ed.* 52: 6277–6282.
56 Yang, L.-Q., Wang, K.-B., and Li, C.-Y. (2013). *Eur. J. Org. Chem.* 2013: 2775–2779.
57 Pan, F., Liu, S., Shu, C. et al. (2014). *Chem. Commun.* 50: 10726–10729.
58 Lin, M., Zhu, L., Xia, J. et al. (2018). *Adv. Synth. Catal.* 360: 2280–2284.
59 Li, L., Shu, C., Zhou, B. et al. (2014). *Chem. Sci.* 5: 4057–4064.
60 Qian, D. and Zhang, J. (2012). *Chem. Commun.* 48: 7082–7084.
61 Ji, K., Yang, F., Gao, S. et al. (2016). *Chem. Eur. J.* 22: 10225–10229.
62 Wagh, S.B., Sharma, P., Patil, M.D., and Liu, R.-S. (2019). *Org. Chem. Front.* 6: 226–230.
63 Liu, J., Zhu, L., Wan, W., and Huang, X. (2020). *Org. Lett.* 22: 3279–3285.
64 Li, J., Yang, F., Ma, Y.-T., and Ji, K. (2019). *Adv. Synth. Catal.* 361: 2148–2153.
65 JiméNez-NúÑEz, E. and Echavarren, A.M. (2008). *Chem. Rev.* 108: 3326–3350.
66 Vasu, D., Hung, H.-H., Bhunia, S. et al. (2011). *Angew. Chem. Int. Ed.* 50: 6911–6914.
67 Ji, K. and Zhang, L. (2014). *Org. Chem. Front.* 1: 34–38.
68 Wang, K.-B., Ran, R.-Q., Xiu, S.-D., and Li, C.-Y. (2013). *Org. Lett.* 15: 2374–2377.
69 Ji, K., Zheng, Z., Wang, Z., and Zhang, L. (2015). *Angew. Chem. Int. Ed.* 54: 1245–1249.
70 Ghorpade, S., Su, M.-D., and Liu, R.-S. (2013). *Angew. Chem. Int. Ed.* 52: 4229–4234.
71 Qian, D. and Zhang, J. (2011). *Chem. Commun.* 47: 11152–11154.
72 Qian, D., Hu, H., Liu, F. et al. (2014). *Angew. Chem. Int. Ed.* 53: 13751–13755.
73 Barrett, M.J., Khan, G.F., Davies, P.W., and Grainger, R.S. (2017). *Chem. Commun.* 53: 5733–5736.
74 Uetake, Y., Niwa, T., and Nakada, M. (2014). *Tetrahedron Lett.* 55: 6847–6850.
75 Ji, K. and Zhang, L. (2018). *Adv. Synth. Catal.* 360: 647–651.
76 Nösel, P., Dos Santos Comprido, L.N., Lauterbach, T. et al. (2013). *J. Am. Chem. Soc.* 135: 15662–15666.
77 Wang, T., Shi, S., Rudolph, M., and Hashmi, A.S.K. (2014). *Adv. Synth. Catal.* 356: 2337–2342.
78 Rode, N., Marinelli, F., Arcadi, A. et al. (2018). *Adv. Synth. Catal.* 360: 4790–4794.
79 Shu, C., Shi, C.-Y., Sun, Q. et al. (2019). *ACS Catal.* 9: 1019–1025.
80 Ji, K., Liu, X., Du, B. et al. (2015). *Chem. Commun.* 51: 10318–10321.
81 Zheng, Z. and Zhang, L. (2015). *Org. Chem. Front.* 2: 1556–1560.
82 Lu, B., Li, C., and Zhang, L. (2010). *J. Am. Chem. Soc.* 132: 14070–14072.
83 Kong, K., Enquist, J.A., Mccallum, M.E. et al. (2013). *J. Am. Chem. Soc.* 135: 10890–10893.
84 Bian, Z., Marvin, C.C., and Martin, S.F. (2013). *J. Am. Chem. Soc.* 135: 10886–10889.

85 Davies, P.W., Cremonesi, A., and Martin, N. (2011). *Chem. Commun.* 47: 379–381.

86 Ji, K., D'souza, B., Nelson, J., and Zhang, L. (2014). *J. Organomet. Chem.* 770: 142–145.

87 Nejrotti, S., Prina Cerai, G., Oppedisano, A. et al. (2017). *Eur. J. Org. Chem.* 2017: 6228–6238.

88 Landman, I.R., Acuña-Bolomey, E., Scopelliti, R. et al. (2019). *Org. Lett.* 21: 6408–6412.

89 Hashmi, A.S.K., Wang, T., Shi, S., and Rudolph, M. (2012). *J. Org. Chem.* 77: 7761–7767.

90 Zhao, J., Liu, J., Xie, X. et al. (2015). *Org. Lett.* 17: 5926–5929.

91 Wang, L., Xie, X., and Liu, Y. (2013). *Angew. Chem. Int. Ed.* 52: 13302–13306.

92 Wang, T., Huang, L., Shi, S. et al. (2014). *Chem. Eur. J.* 20: 14868–14871.

93 Wang, T., Shi, S., Hansmann, M.M. et al. (2014). *Angew. Chem. Int. Ed.* 53: 3715–3719.

94 Chen, M., Chen, Y., Sun, N. et al. (2014). *Angew. Chem. Int. Ed.* 54: 1200–1204.

95 Chen, H. and Zhang, L. (2015). *Angew. Chem. Int. Ed.* 54: 11775–11779.

96 Chen, M., Sun, N., Xu, W. et al. (2015). *Chem. Eur. J.* 21: 18571–18575.

97 Zheng, Y., Zhang, J., Cheng, X. et al. (2019). *Angew. Chem. Int. Ed.* 58: 5241–5245.

98 Marsden, S.P. and Pang, W.-K. (1999). *Chem. Commun.*: 1199–1200.

99 Cui, L., Peng, Y., and Zhang, L. (2009). *J. Am. Chem. Soc.* 131: 8394–8395.

100 Noey, E.L., Luo, Y., Zhang, L., and Houk, K.N. (2012). *J. Am. Chem. Soc.* 134: 1078–1084.

101 Cui, L., Ye, L., and Zhang, L. (2010). *Chem. Commun.* 46: 3351–3353.

102 Cui, L. and Zhang, L. (2010). *Sci. China Chem.* 53: 113–118.

103 Bhunia, S., Ghorpade, S., Huple, D.B., and Liu, R.-S. (2012). *Angew. Chem. Int. Ed.* 51: 2939–2942.

104 Wang, Y., Zheng, Z., and Zhang, L. (2015). *J. Am. Chem. Soc.* 137: 5316–5319.

105 Zheng, Z., Wang, Y., Ma, X. et al. (2020). *Angew. Chem. Int. Ed.* 59: 17398–17402.

8

Transition-Metal-Catalyzed Carbene Transformations for Polymer Syntheses

Eiji Ihara and Hiroaki Shimomoto

Ehime University, Department of Materials Science and Biotechnology, Graduate School of Science and Engineering, Matsuyama, 790-8577, Japan

8.1 Introduction

Needless to say, the most important transition-metal-catalyzed carbene transformation for polymer syntheses is metathesis polymerization of cyclic olefins (ring-opening metathesis polymerization, ROMP), alkynes, nonconjugated dienes (acyclic diene metathesis polymerization, ADMET), and related compounds bearing unsaturated carbon–carbon double or triple bonds, where transition-metal-carbene-catalyzed olefin metathesis plays an essential role in their initiation and propagation mechanisms [1–5]. Meanwhile, another type of polymerization has been developed where the metal–carbene intermediates playing an essential role in the initiation and propagation are derived from diazocarbonyl compounds [6–11]. Because the former transition-metal-catalyzed metathesis polymerization has been well established and described in some review articles, this chapter does not deal with the subject. Instead, we herein focus on the latter type of polymerization, consisting of two main categories: (i) C1 polymerization of diazoacetates and (ii) polycondensation of bis(diazocarbonyl) compounds.

8.2 Transition-Metal-Catalyzed C1 Polymerization of Diazoacetates

Diazo compounds have indeed a long history as a monomer for polymer syntheses. In particular, polymerization of diazoalkanes such as diazomethane was intensively investigated, and main group Lewis acid such as BF_3 was found to polymerize diazoalkanes effectively to yield poly(alkylmethylene)s (Scheme 8.1) [12]. For example, in the case of diazomethane as a monomer, CH_2 carbenes generated from the monomer were connected linearly, resulting in the formation of a C–C main chain to yield a high-molecular weight polymethylene, which is equivalent to linear polyethylene. This mode of polymerization is generally called

Transition Metal-Catalyzed Carbene Transformations, First Edition.
Edited by Jianbo Wang, Chi-Ming Che, and Michael P. Doyle.

"C1 polymerization," which contrasts with "vinyl polymerization," where the C–C main chain is constructed from C2 unit (vinyl group). Despite the efficiency as a method for C–C polymer syntheses, because of the high explosiveness of diazoalkanes, the polymerization had not been established as a general method for polymer synthesis.

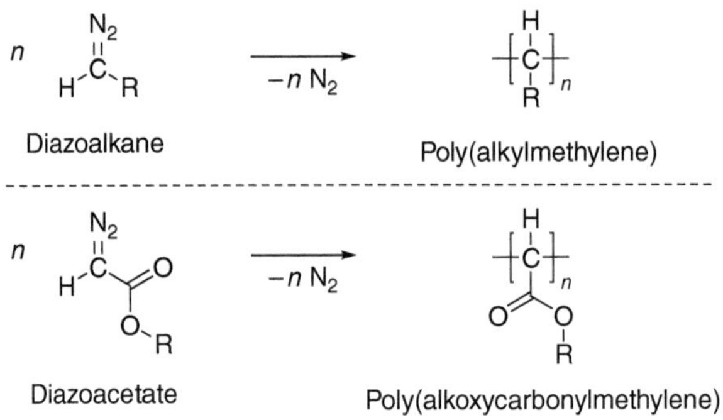

Scheme 8.1 C1 polymerization of diazo compounds. Source: Modified from Kantor and Osthoff [12].

Meanwhile, although a variety of diazocarbonyl compounds have been gaining their utility and importance in the field of organic synthesis, they had not been used as a monomer for polymer syntheses until the publication of C1 polymerization of allyl diazoacetate by Liu et al. in 2002, where the monomer was polymerized by Cu powder to afford a poly(allyloxycarbonylmethylene) with number-average-molecular weight (M_n) of a few thousands [13]. So far, Pd and Rh are the only transition metals that are able to afford high molecular weight ($M_n > 10\,000$) polymers from the C1 polymerization of diazoacetates. Among diazocarbonyl compounds, diazoacetates are the only group that can be transformed into high-molecular weight polymers by the polymerization (Scheme 8.1).

In this section, we mainly focus on the development of Pd-based initiating systems for the polymerization of diazoacetates, which has been achieved in our research group in the last two decades. The Rh-initiated diazoacetate polymerization developed by de Bruin and coworkers has been described in some review articles [6–10].

8.2.1 $PdCl_2$-Initiated Polymerization

In 2003, we published our first report on the Pd-initiated polymerization of diazoacetates, where the initiator employed was a combination of $PdCl_2$ and an amine such as triethylamine and pyridine (Scheme 8.2) [14]. Although the C1 polymerization indeed proceeded to yield poly(alkoxycarbonylmethylene)s, the M_n of the products were about a few thousands.

The $PdCl_2$-based initiating system was used for the polymerization of diazoketones, affording low M_n polymers as well, where the polymer main chains in part

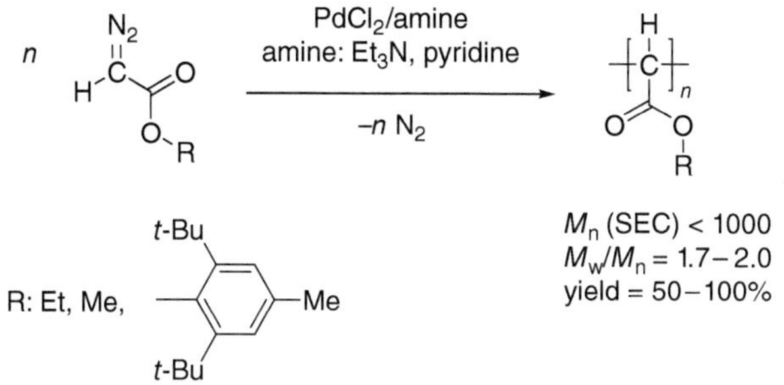

Scheme 8.2 $PdCl_2$-initiated polymerization of diazoacetates. Source: Modified from Ihara et al. [14].

contained azo groups (—N=N—) generated via a monomer insertion mode without the elimination of N_2 (Scheme 8.3) [15]. Copolymerization of diazoketones with an alkyne proceeded to yield random copolymers (Scheme 8.4) [16].

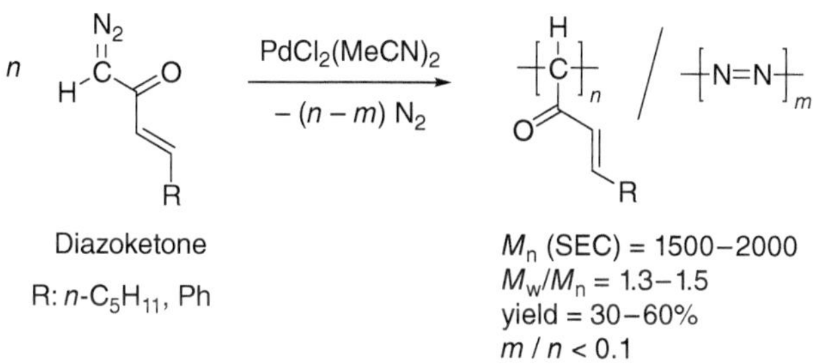

Scheme 8.3 $PdCl_2$-initiated polymerization of diazoketones. Source: Modified from Ihara et al. [15].

8.2.2 (NHC)Pd(nq)/Borate-Initiated Polymerization

In 2009, we reported the first Pd-based initiating system that can afford high M_n polymers from diazoacetates. The initiating system consists of an N-heterocyclic carbene (NHC) and naphthoquinone (nq)-ligated zero-valent Pd complex and tetraarylborate [(NHC)Pd(nq)/borate system] (Scheme 8.5) [17]. Although we have not succeeded in characterizing the chain end structures of the polymers obtained with this system, later studies on related systems suggested that the polymerization should be initiated by a Pd–Ph species generated by the transmetalation of a Ph group from the borate to the Pd(0) with oxidation of the metal center to Pd(II) [18]. The detail of the transformation, which should be called "*oxidative transmetalation,*" will be described later in this chapter (see Section 8.2.4).

In contrast to the aforementioned $PdCl_2$-based system, the (NHC)Pd(nq)/borate system can afford polymers with a relatively high M_n in good yield. In addition to standard primary alkyl groups such as ethyl and benzyl, secondary alkyl groups such as cyclohexyl and cholesteryl can be used as an ester substituent to afford high M_n

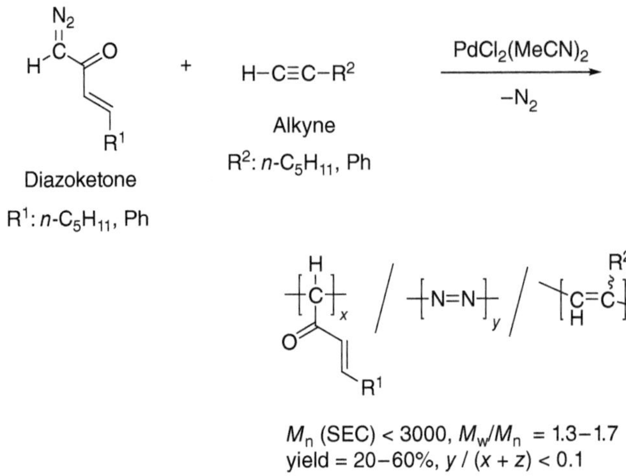

Scheme 8.4 $PdCl_2$-initiated copolymerization of diazoketone with alkyne. Source: Modified from Ihara et al. [16].

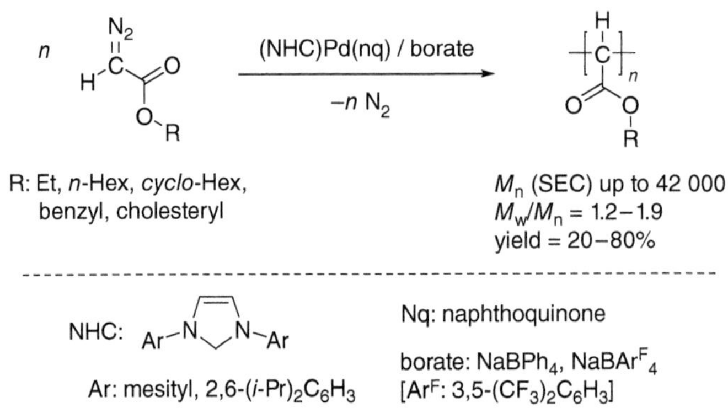

Scheme 8.5 (NHC)Pd(nq)/borate-initiated polymerization of diazoacetates. Source: Modified from Ihara et al. [17].

polymers [19]. While the tacticity of the obtained polymers was atactic, they were soluble in common organic solvents such as chloroform and THF.

8.2.3 π-AllylPdCl-Based System-Initiated Polymerization

In 2012, we reported that a combination of π-allylPdCl and $NaBPh_4$ or π-allylPdCl alone can act as an effective initiator for the diazoacetate polymerization [20]. The π-allylPdCl-based systems are active for a variety of monomers with primary and secondary ester substituents, affording polymers with M_n up to a few tens of thousands (Scheme 8.6). The number of polymer chains (N_p) produced per Pd often exceeds 1.0, suggesting that chain transfer occurs during the propagation, and thus, high M_n polymer production is difficult with these systems.

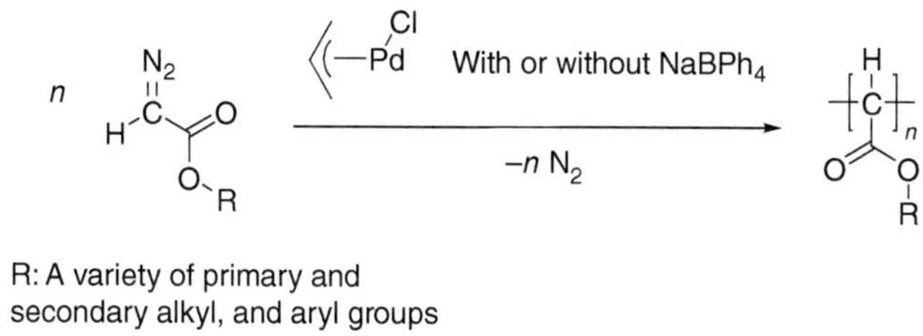

Scheme 8.6 π-AllylPdCl-based system-initiated polymerization of diazoacetates.

The initiation and termination mechanisms of the π-allylPdCl-based systems were revealed by MALDI-TOF-MS analyses of the chain end structures of the products. In the case of the π-allylPdCl/borate ($NaBPh_4$) system, the MS analyses indicated that there was only one predominant component in the product with respect to chain end structures and it contained Ph and H at its initiation and termination chain ends, respectively. The results indicated that as shown in Scheme 8.7, the polymerization was initiated with insertion of the carbene–carbon of the Pd–carbene moiety generated by the reaction of Pd with a monomer into a Pd–Ph linkage, where the Ph group was derived from transmetalation from the borate to Pd. The generation of the Pd–carbene intermediate was confirmed by DFT calculation in a related $[\pi\text{-allylPd(OAc)}]_2$ system reported by Toste and coworkers recently [21]. The polymerization should be terminated by protonolysis of the Pd—C bond at the propagating chain end with an acidic aqueous solution in a work-up procedure.

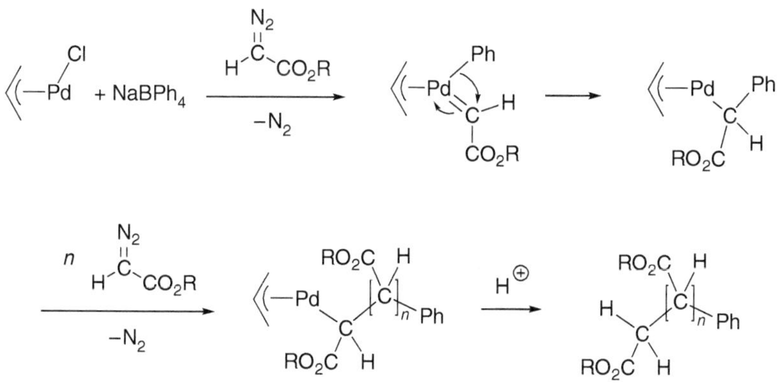

Scheme 8.7 Mechanism of π-allylPdCl/$NaBPh_4$-initiated polymerization.

When π-allylPdCl alone was used as an initiator, the polymerization was initiated with insertion of the carbene-carbon into the Pd—Cl bond of the initiator, which was confirmed by detecting the presence of Cl at the initiating chain end by MALDI-TOF-MS analyses. Two major components were observed in this case, with two different terminating chain ends; one component had a H at the chain end, while the other component had a ring-opened THF employed as a solvent with a HO group as described in Scheme 8.8. Although the Pd–OR group generated after the THF ring opening has the initiating ability for the diazoacetate polymerization, we have not observed the incorporation of ring-opened THF in the middle of the polymer chain. Accordingly, the reaction of the propagating chain end with THF should occur only after the consumption of the monomer.

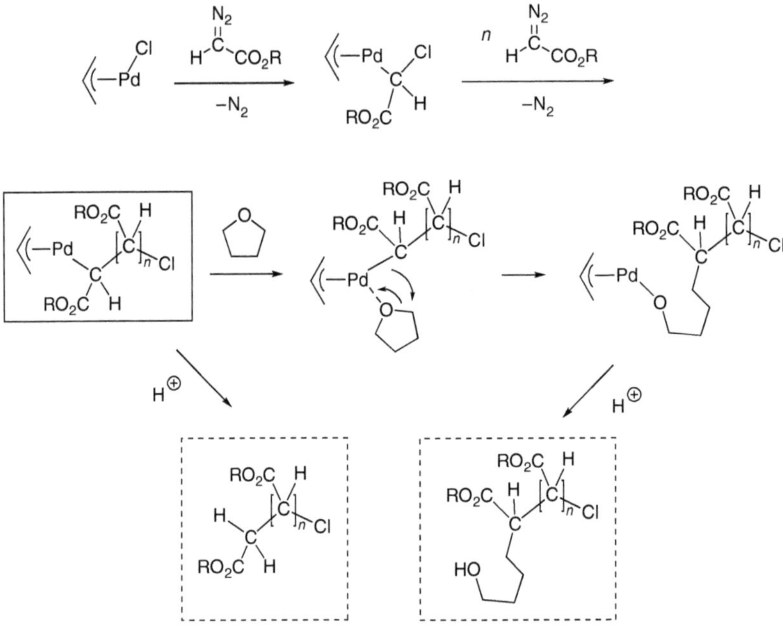

Scheme 8.8 Mechanism of π-allylPdCl-initiated polymerization.

Beside H and ring-opened THF, we have often observed a cyclic-ketone structure as another terminating chain end, which should be derived from a back-biting reaction as described in Scheme 8.9. An analogous reaction has been recognized as a major side reaction in the anionic polymerization of alkyl methacrylates [22]. This reaction could be one of the reasons for the N_p values higher than 1.0 in the π-allylPdCl-based system-initiated polymerization, because the resulting Pd–OR species can reinitiate the diazoacetate polymerization.

We have found that alcohols and H_2O can be used as a chain transfer agent for the π-allylPdCl-based system-initiated diazoacetate polymerization. When the polymerization was conducted in the presence of alcohol or H_2O, in addition to lowering of M_n of the products, we confirmed by MALDI-TOF-MS analyses that the major component of the products contained RO or HO and H at the initiating and terminating chain ends, respectively. The proposed chain transfer mechanism based on the chain end structures is illustrated in Scheme 8.10. Because the polymeric products were obtained even in the presence of a large excess of alcohol or H_2O, the chain transfer should be much slower than propagation.

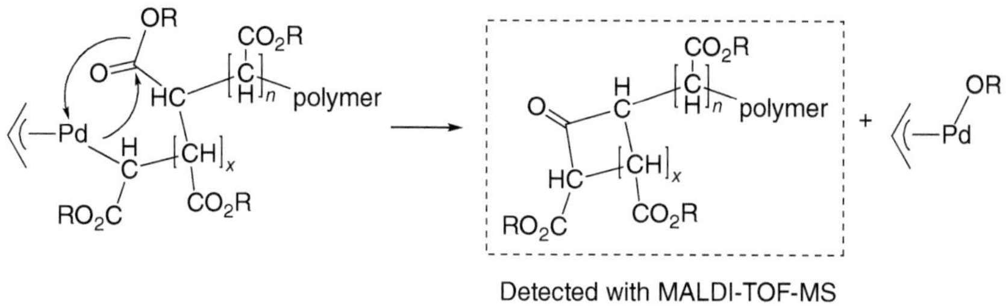

Scheme 8.9 Back-biting in diazoacetate polymerization.

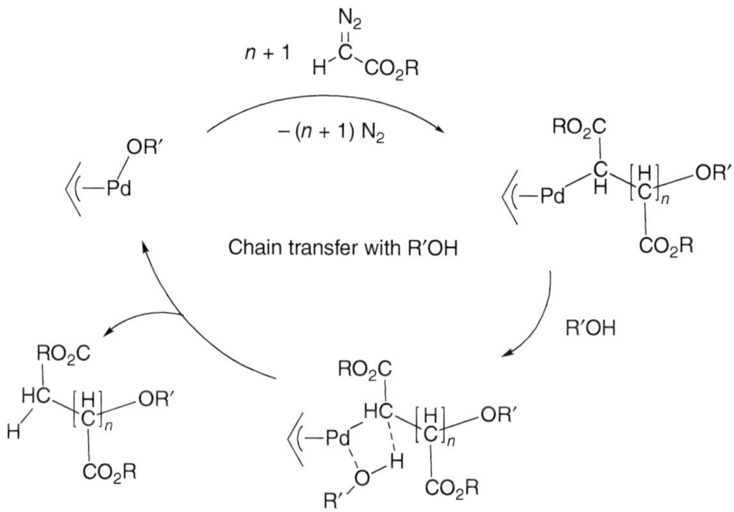

Scheme 8.10 Alcohol and H_2O-mediated chain transfer in diazocetate polymerization.

When we used cyclotriphosphazene-containing diazoacetates as a monomer for the π-allylPdCl/borate-initiated polymerization, we have found that the polymerization proceeded in a controlled manner, affording polymers with narrow molecular weight distribution (Scheme 8.11) [23]. As expected, MALDI-TOF-MS analyses revealed the presence of Ph and H at initiating and terminating chain ends, respectively, of the predominant component of the product. In accord with the controlled homopolymerization behavior, block copolymerization of two kinds of cyclotriphosphazene-containing monomers with different substituents on P atoms proceeded to yield a well-defined block copolymer; this is the first example of synthesis of a well-defined block copolymer by diazoacetate polymerization.

The controlled polymerization of cyclotriphosphazene-containing diazoacetates was applied for block copolymerization with *n*-hexyl diazoacetate, and the resulting copolymer was revealed to show lamellar-type self-assembling behavior, where the cyclotriphosphazene- and *n*-hexyl-containing blocks served as hard and soft segments, respectively (Scheme 8.12) [24].

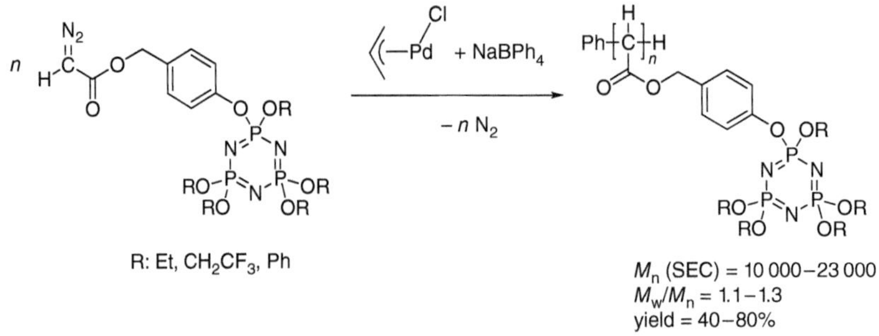

Scheme 8.11 π-AllylPdCl/NaBPh$_4$-initiated polymerization of cyclotriphosphazene-containing diazoacetates. Source: Modified from Shimomoto et al. [23].

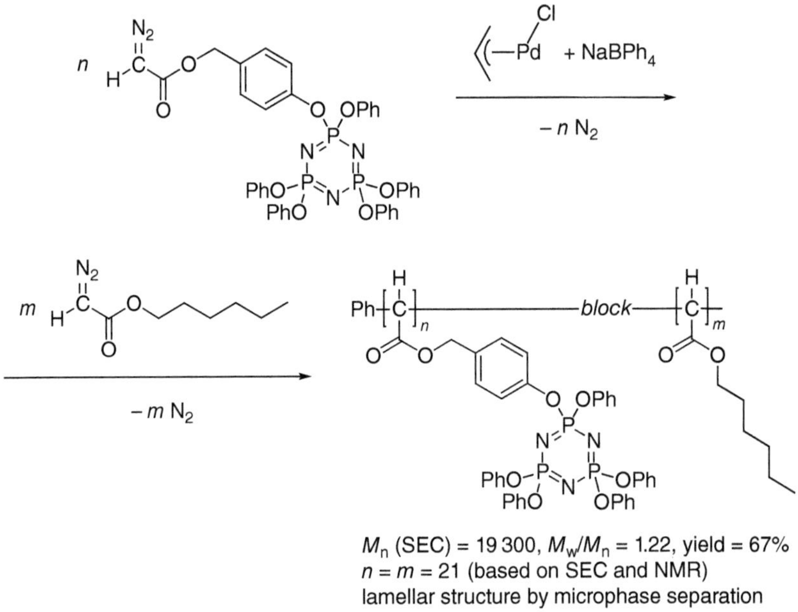

Scheme 8.12 Controlled block copolymerization of phosphazene-containing diazoacetates. Source: Modified from Kato et al. [24].

The sterically bulky cyclotriphosphazene-containing ester groups would prevent the above-described chain transfer (back-biting) and termination (insertion of ring-opened THF) to bring about the controlled polymerization. To confirm that point, we prepared benzyl-ether-type dendron-containing diazoacetates with a series of generations (zero [benzyl diazoacetate (BDA)], first, and second) and examined their polymerization behavior with the π-allylPdCl/borate system [25]. As a result, M_w/M_n values decreased with the increase in the dendron generation from 1.40 to 1.04 as shown in Scheme 8.13, clearly indicating the steric effect of the ester substituent on the controlled behavior of the polymerization.

Introduction of *t*-butoxycarbonyl groups at 3,5-positions of the periphery Ph rings of the dendron units and Lewis acid-mediated removal of the *t*-Bu groups

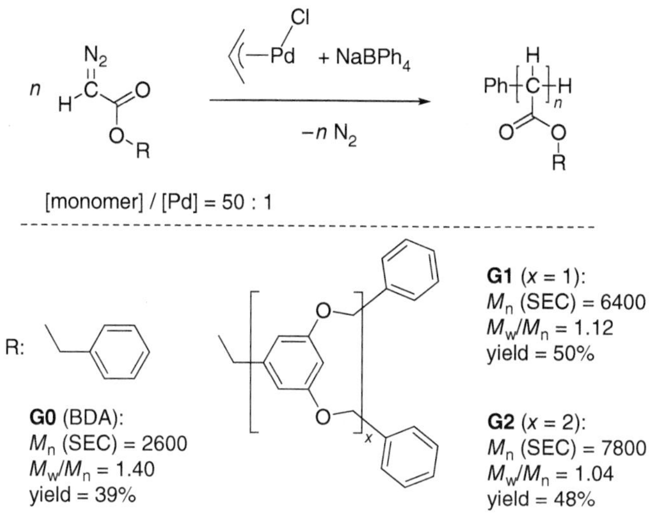

Scheme 8.13 Polymerization of dendron-containing diazoacetates.

after their polymerization led to the formation of polymers with densely packed carboxy groups around a rigid polymer main chain. Compared to their vinyl polymer counterparts, the dendron-supported-carboxy-containing poly(alkoxy-carbonylmethylene)s showed higher acidity in their pH-responsive behavior, which can be ascribed to the denser packing of the carboxy groups.

In 2018, Wu and coworkers reported that a combination of π-allylPdCl and a bidentate phosphine ligand was effective for controlling polymerization of diazoacetates with a variety of ester substituents (Scheme 8.14) [26]. In particular, when a chiral ligand was employed, the resulting polymers exhibited large values of optical rotation and a strong Cotton effect. Furthermore, depending on the chirality of the ligand, the sign of the optical rotation and the Cotton effect became opposite, suggesting that the optical activity was derived from screw-sense of the helical polymer main chains.

In 2019, Toste and coworkers reported that the use of carboxylates instead of Cl as a ligand (X) on dimeric $(\pi\text{-allylPdX})_2$ complexes resulted in a considerable improvement of the controllability for the diazoacetate polymerization (Scheme 8.15) [21]. Their theoretical calculation indicated that the polymerization proceeded with a dinuclear mechanism, which significantly lowered the activation energy for the initiation compared to a mononuclear mechanism and thus, contributed to the realization of the controlled polymerization.

8.2.4 $(nq)_2Pd$/Borate- and (cod)PdCl(Cl-nq)/Borate-Initiated Polymerization

In 2019, we reported that a zero-valent Pd complex with two nq molecules as ligands in conjunction with $NaBPh_4$ [$(nq)_2Pd$/borate system] polymerized diazoacetates effectively (Scheme 8.16) [18]. Compared with the previously reported Pd-based

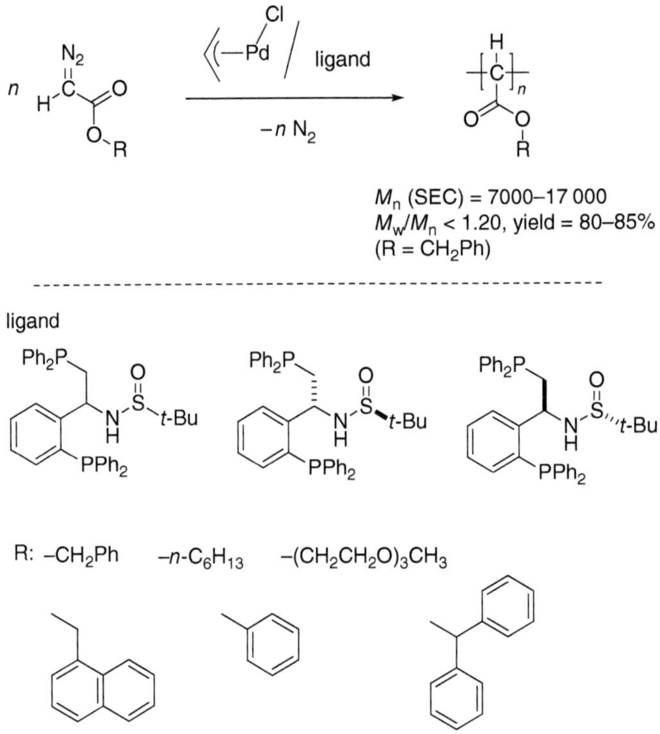

Scheme 8.14 Living and helix-sense-selective polymerization of diazoacetates (Wu and coworkers). Source: Modified from Chu et al. [26].

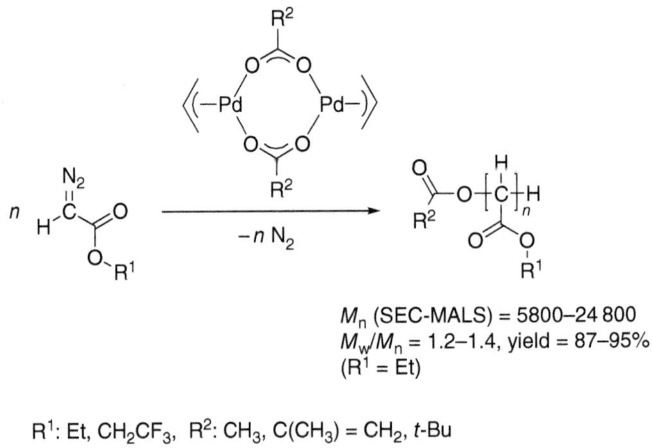

Scheme 8.15 Controlled polymerization of diazoacetates (Toste and coworkers). Source: Modified from Zhukhovitskiy et al. [21].

systems, both M_n and yield of the polymers from ethyl diazoacetate (EDA) were significantly higher, reaching up to 36 000 (SEC) and 70%, respectively. Similar results were obtained with BDA and cyclohexyl diazoacetate (*c*-HDA), demonstrating the generality of this initiating system. SEC-MALS analyses indicated that the SEC analysis based on PMMA standard calibration underestimated the molecular weight, and the absolute M_n of the polymers obtained with the $(nq)_2Pd$/borate system reached up to 50 000 corresponding to degree of polymerization (DP) of c. 600 in the case of EDA polymerization.

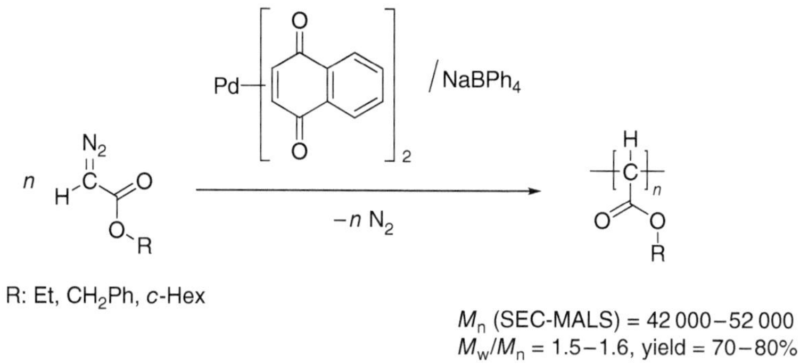

Scheme 8.16 $(nq)_2Pd$/borate-initiated polymerization of diazoacetates. Source: Modified from Shimomoto et al. [18].

MALDI-TOF-MS analyses of the product revealed that the polymerization was initiated with a Pd–Ph species, leaving a Ph group at the initiating chain end. We found in a literature that a Pd(0) complex with a quinone as a ligand was oxidatively transformed to a Pd(II)–Ph complex with transmetalation from a Ph-borane compound; the process was called "oxidative transmetalation" [27]. Likewise, as shown in Scheme 8.17, the initiating Pd–Ph species should be generated via oxidative transmetalation. As a result, the resulting Pd(II) complex possessed Ph and η^3-type anionic ligand derived from the nq ligand. The presence of the η^3-type anionic ligand in the active Pd complex of the $(nq)_2Pd$/borate system was consistent with the activity of the above-described π-allylPdCl-based initiating systems [20], and thus, demonstrate the importance of the presence of η^3-type ligand for Pd complexes to have the initiating ability. Moreover, the active species of the aforementioned (NHC)Pd(nq)/borate system [17, 19] should also be generated via oxidative transmetalation as well.

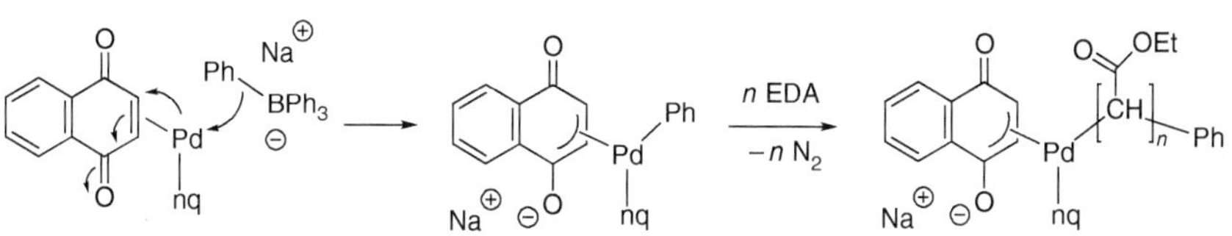

Scheme 8.17 Mechanism of $(nq)_2Pd$/borate-initiated polymerization.

While we examined structural variation in the nq-based Pd complexes to improve the initiating ability of nq-containing Pd complexes, we have found that 2,3-dichloronaphthoquinone (dichlone, Cl_2-nq) reacted with a precursor Pd(0) complex, $Pd_2(dba)_3(CHCl_3)$ (dba, dibenzalacetone), in a different manner than unsubstituted nq in the presence of cyclooctadiene (cod), affording a new Pd(II) complex via oxidative addition of Cl_2-nq as described in Scheme 8.18 [18]. The unique structure of the complex was unambiguously determined by an X-ray crystallographic analysis. More importantly, (cod)PdCl(Cl-nq) activated with $NaBPh_4$ [(cod)PdCl(Cl-nq)/borate system] polymerized EDA in a stereoselective manner affording highly syndiotactic polymers despite in low yield of 20–30%. MALDI-TOF-MS analyses of the products indicated that the polymerization was initiated with a Pd–Ph species generated by the transmetalation from borate to the Pd. These results suggested that the structure of the Pd complex with cod, Cl, Cl-nq, and Ph or propagating chain can form a suitable coordination environment for the stereoselective polymerization, although the cod ligand could be expelled from the Pd center during the propagation.

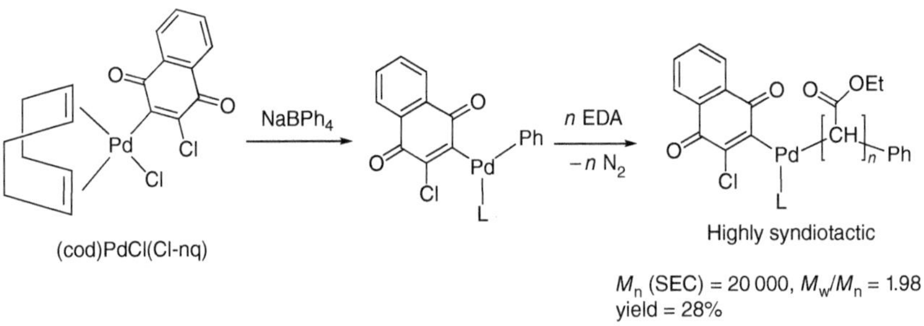

Scheme 8.18 Stereoselective polymerization of EDA by (cod)PdCl(Cl-nq)/borate system. Source: Modified from Shimomoto et al. [18].

8.2.5 Preparation of Polymers with Densely Packed Functional Groups Around Polymer Main Chain

The most striking structural feature of poly(alkoxycarbonylmethylene)s obtained from diazoacetates is that the ester substituents are densely packed around the polymer main chain. The density of the ester substituents around the C–C main chain can be simply estimated to be twofold higher in poly(alkoxycarbonylmethylene)s than in poly(alkyl acrylate)s, vinyl polymer counterparts. Thus, we can expect that the higher density of the substituents will enhance the properties derived from the ester substituents in comparison to the vinyl polymers.

8.2.5.1 Hydroxy Group-Containing Polymers

Incorporation of a hydroxy group to each ester substituent of poly(alkoxycarbonylmethylene) will afford unique water-soluble polymers with a higher density of hydroxy groups than those of hydroxy-containing poly[alkyl (meth)acrylate]s such as poly(2-hydroxyethyl methacrylate) [poly(2-HEMA)]. Because hydroxy groups would react with propagating chain ends causing chain transfer, we first prepared monomers with a silyl-protected hydroxy group and polymerized the monomers with the π-allylPdCl/borate system (Scheme 8.19) [28]. After the polymerization, the protecting group was completely removed to afford water-soluble polymers with densely packed hydroxy groups. We have found that direct polymerization of unprotected hydroxy-containing monomers can also proceed effectively, and post-polymerization silylation of the hydroxy groups of the resulting polymers afforded silyl-protected polymers with M_n in a similar range of the products of the polymerization of the protected monomers. These results indicate that the rate of the chain transfer is much slower than that of the propagation, leading to the predominant formation of linear polymers.

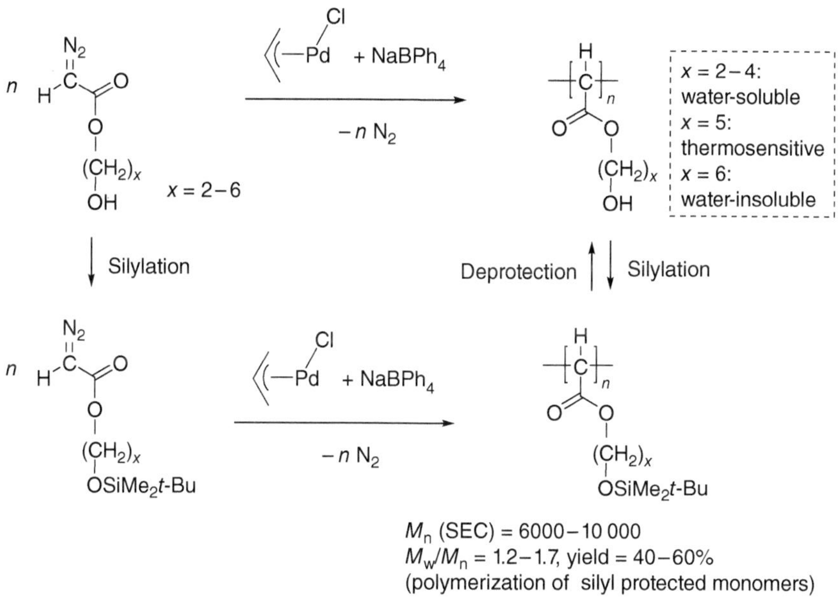

Scheme 8.19 Preparation of HO-containing C1 polymers. Source: Modified from Shimomoto et al. [28].

Among hydroxy-containing polymers with a series of spacer lengths (x = 2–6) between the ester linkage and OH, the polymer with a $(CH_2)_5$ linker exhibited thermosensitivity, being water-soluble at lower temperatures and water-insoluble at higher temperatures than the cloud point, while polymers with a longer spacer were not water-soluble at any temperatures. While a poly(alkoxycarbonylmethylene) with a $(CH_2)_4$ spacer (repeating unit composition of $C_6H_{10}O_3$) was water-soluble at any temperatures, hydroxy-containing vinyl polymers with the same repeating unit composition, poly(2-HEMA) and poly(2-hydroxypropyl acrylate) [poly(2-HPA)], have been reported to show thermosensitivity, suggesting that densely packed hydroxy groups in the poly(alkoxycarbonylmethylene) indeed enhance the hydrophilic nature of the polymers (Scheme 8.20).

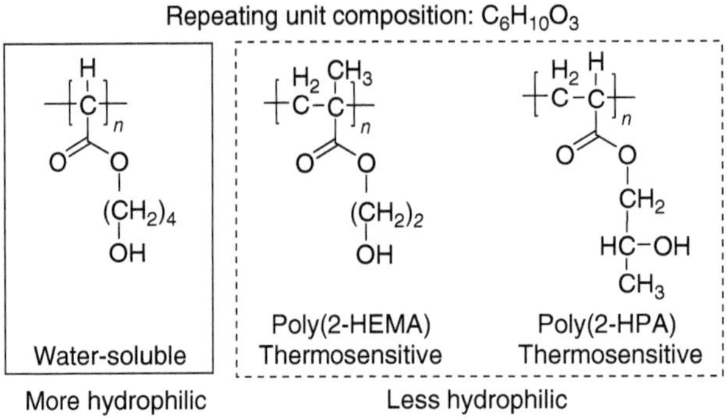

Scheme 8.20 Thermo-responsive behavior of HO-containing polymers.

Rh-initiated polymerization of the silyl-protected monomers was conducted by de Bruin and coworkers to yield syndiotactic hydroxy-containing poly(alkoxycarbonylmethylene)s after deprotection [29]. Again, a polymer with a $(CH_2)_5$ linker showed thermosensitive behavior; interestingly, the cloud point of the syndiotactic polymer was much lower than that of the atactic polymer.

8.2.5.2 Oligo(oxyethylene)-Containing Polymers

As another group of hydrophilic substituents, a series of oxyethylene chains $[(CH_2CH_2O)_n]$ were introduced to diazoacetates and the monomers were polymerized by the π-allylPdCl/borate system (Scheme 8.21) [30]. In this case, the polymers with the number of oxyethylene units of 1–3 exhibited thermosensitivity. In the comparison with the poly(alkyl methacrylate)s and poly(alkyl acrylate)s with the same oxyethylene side chains, poly(alkoxycarbonylmethylene)s showed higher cloud points, indicating that the dense-packing of the oxyethylene chains enhanced the hydrophilicity of the polymers. Although we expected that the dense-packing of the oxyethylene chains would constitute an effective path for Li ion conduction around the polymer main chain, the Li conductivity measurement of the oxyethylene-containing poly(alkoxycarbonylmethylene)s revealed only

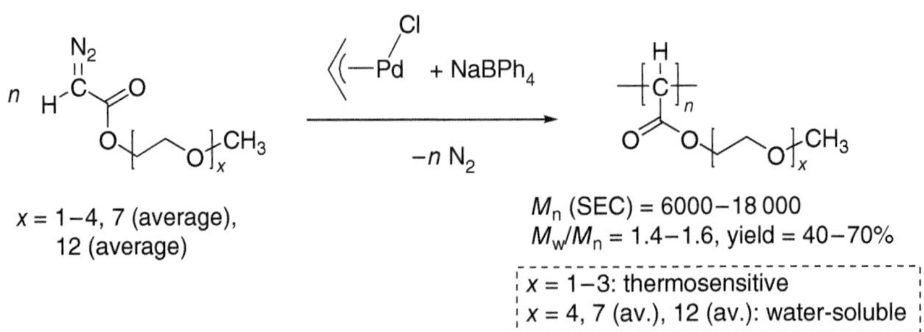

Scheme 8.21 Preparation of oligo(oxyethylene)-containing C1 polymers. Source: Modified from Shimomoto et al. [30].

a moderate increase of the ion conductivity compared to their vinyl polymer counterparts.

8.2.5.3 Pyrene-Containing Polymers

Because of the dense-packing of the ester substituents around the polymer main chain, the ester substituents should be in close contact with each other in a sterically congested environment in comparison with vinyl polymers with the same substituents. We can expect that introduction of photo-functional groups as the substituents could lead to unique properties because of the close contact. For example, introduction of pyrene would result in a much higher rate of the excimer formation, because the close contact of pyrenes in poly(alkoxycarbonylmethylene) is favorable for the excimer formation.

As shown in Scheme 8.22, we prepared a series of pyrene-containing diazoacetates with a series of spacer lengths [none, CH_2 or $(CH_2)_4$] and polymers therefrom, and investigated the effect of the structural factors on the excimer formation rate in comparison with their vinyl polymer counterparts [31]. For accurate evaluation of the excimer formation rate, we employed ultrafast time-resolved near-infrared absorption spectroscopy. As a result, it was revealed that the excimer formation rate was much faster in poly(substituted methylene)s than in their vinyl polymer counterparts. In addition, the polymers with shorter spacer lengths (none or CH_2) showed much faster rate than that with a longer spacer [$(CH_2)_4$]. The characteristics

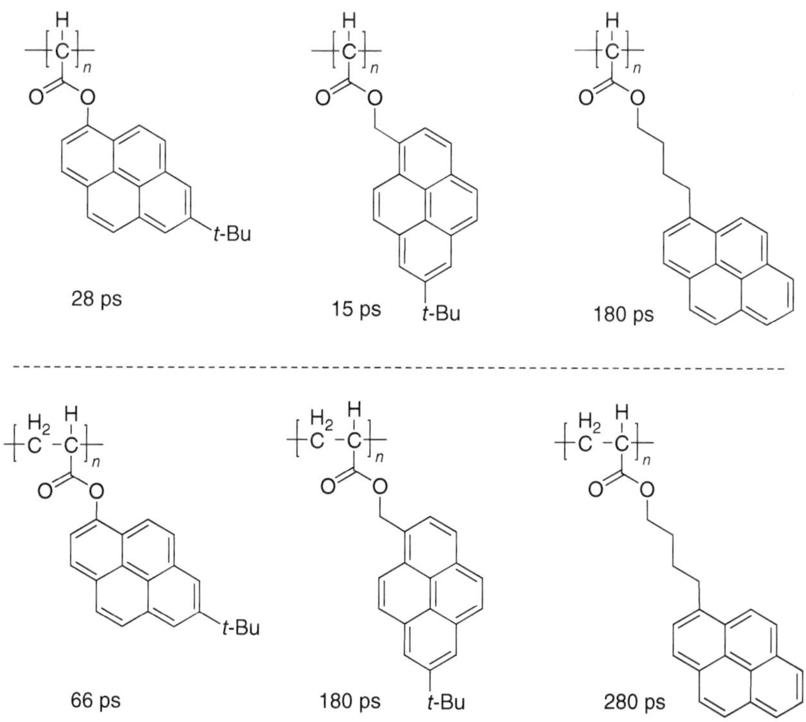

Scheme 8.22 Time-constants for excimer formation of pyrene-containing polymers.

of poly(alkoxycarbonylmethylene)s with respect to the photo-functionality could be utilized for further applications as functional polymeric materials.

8.2.5.4 Fluoroalkyl and Fluoroaryl Group-Containing Polymers

A variety of fluoroalkyl and fluoroaryl groups were employed as an ester substituent of diazoacetate, and the effect of dense-packing of the fluorine-containing groups on the polymer properties was investigated for the poly(alkoxycarbonylmethylene)s obtained by the C1 polymerization (Scheme 8.23) [32].

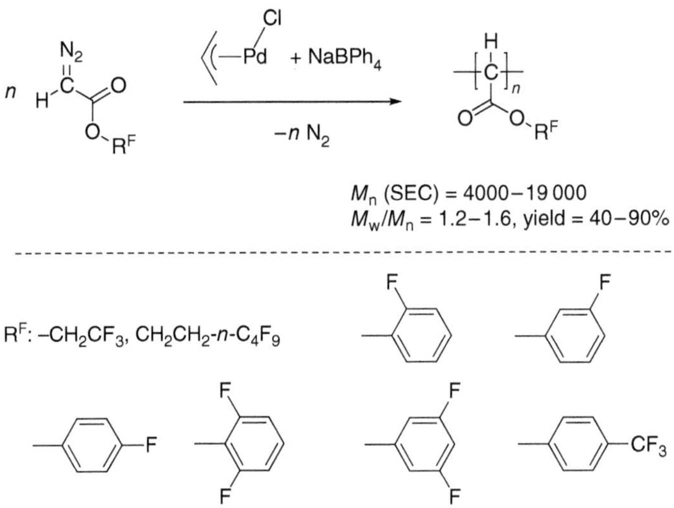

Scheme 8.23 Preparation of F-containing C1 polymers. Source: Modified from Shimomoto et al. [32].

While polymerization of diazoacetate with a CH_2CF_3 group afforded soluble polymers with $M_n > 10\,000$, a monomer with a CH_2CH_2–n–C_4F_9 group gave a product insoluble in common organic solvents. However, the latter monomer can be copolymerized with EDA to afford soluble copolymers.

Polymerization of a series of fluoroaryl diazoacetates afforded soluble polymers, except monomers bearing fluorine at ortho-positions, 2-F-phenyl, and 2,6-F_2-phenyl diazoacetates, which yielded polymers soluble only in DMSO. On the other hand, vinyl polymer counterparts of these ortho-fluorophenyl substituents were soluble in common organic solvents, suggesting that the insolubility of the former polymers should arise from dense-packing of the ortho-fluorophenyl containing groups. One possibility is that the F atoms are in close contact with other phenyl groups in nearby positions, and the interaction between the F atoms and the phenyl rings significantly reduces mobility of the polymer chain.

Fluorine-containing aryl groups could be highly active for substitution with a nucleophile. In particular, substitution with an amine yielding an *N*-alkylcarbamoyl group is of importance, because the carbamoyl groups cannot be incorporated by direct polymerization of *N*-alkyl diazoacetamide. In addition, because poly(*N*-alkylacrylamide)s such as poly(*N*-isopropylacrylamide) have been well known to have unique properties such as thermosensitivity [33], the preparation of their C1 polymer counterparts, poly(*N*-alkylcarbamoylmethylene)s, via post-polymerization modification should be worth attempting. The reaction of a F-containing poly(aryloxycarbonylmethylene) bearing a 3,5-F_2-phenyl group with a primary amine was examined. As a result, it turned out that the double substitution of the activated ester groups with the primary amine proceeded to afford five-membered cyclic imide as a predominant component as shown in Scheme 8.24 [32]. Interesting to note is that the mol% of the *N*-alkylcarbamoyl group incorporation (13.5%) was close to the value of the unreacted hydroxy unit observed in the vinylon prepared by the reaction of poly(vinyl alcohol) with formaldehyde.

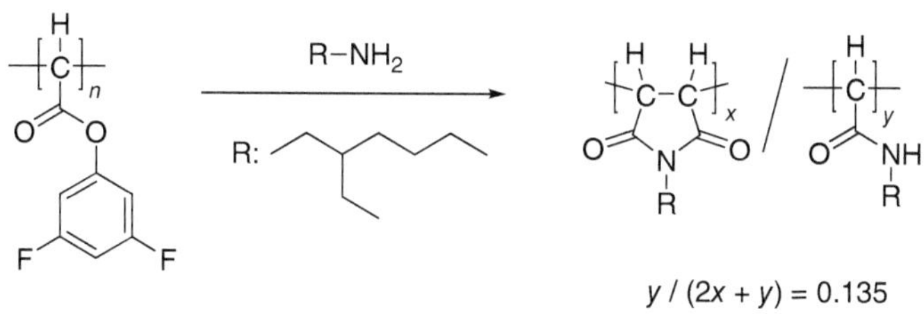

Scheme 8.24 Post-polymerization modification of F-containing C1 polymer. Source: Modified from Shimomoto et al. [32].

8.3 Polycondensation of Bis(diazocarbonyl) Compounds

Diazocarbonyl compounds have been utilized as useful reagents in a variety of organic reactions because of their unique reactivities [34]. The reactivity thus can also be utilized for polycondensation of bis(diazocarbonyl) compounds (AA-type monomer) with other bifunctional monomers (BB-type monomer) in AA + BB type polycondensation, where B is a functional group that can efficiently react with a diazocarbonyl group. Such polycondensation had not been reported until our first report on the polycondensation using a bis(diazocarbonyl) compound and diol as monomers [35]. Interestingly, the polycondensation proceeded in an unexpected manner, where THF used as a solvent was incorporated as a third component in the resulting polymer main chain. Since the first report, we have explored the utility of bis(diazocarbonyl) compounds as a monomer for three-, two-, and single-component polycondensation and indeed succeeded in preparing polymers with some unprecedented main chain structures (Scheme 8.25).

8.3.1 Three-Component Polycondensation of Bis(diazocarbonyl) Compound, Diol, and THF

As an initial attempt for realizing the new polycondensation using a bis(diazocarbonyl) compound as a monomer, we newly prepared a bis(diazoketone) and

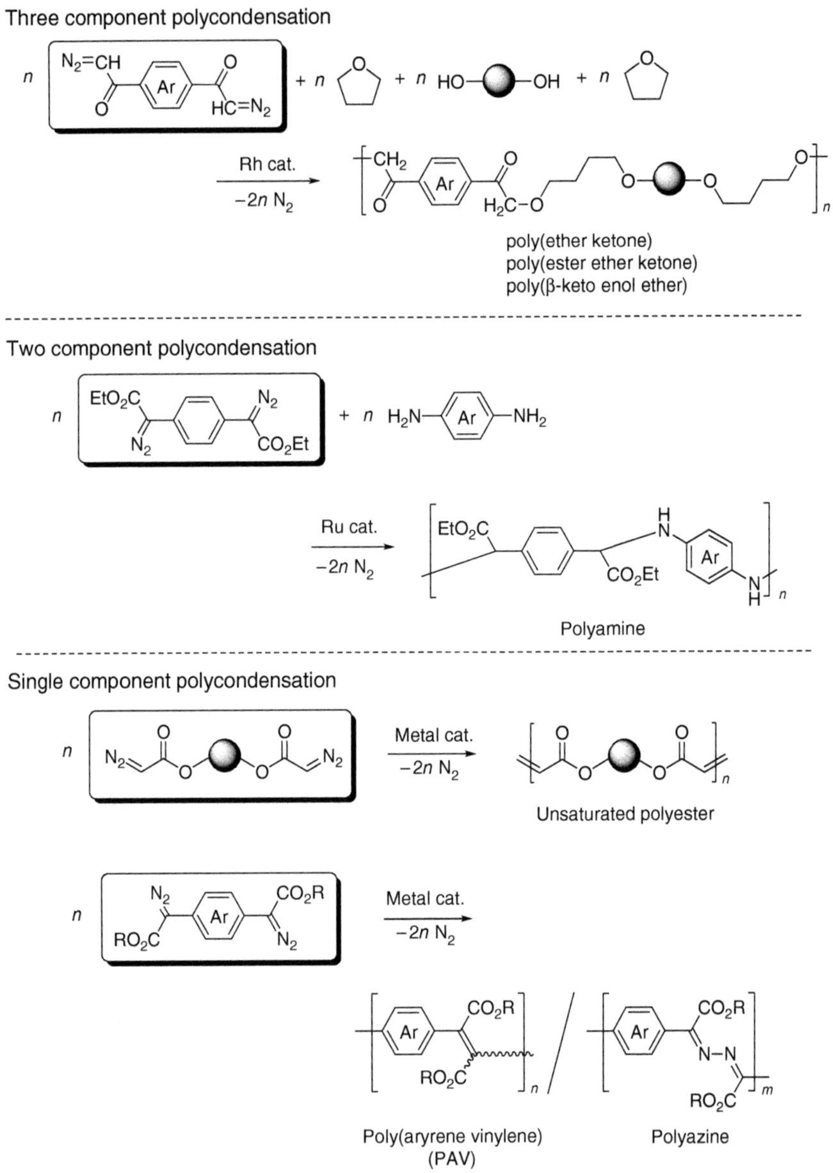

Scheme 8.25 Polycondensation of bis(diazocarbonyl) compounds to afford a variety of polymer structures.

reacted it with a variety of diols using $Rh_2(OAc)_4$ as a catalyst because insertion of a diazo-bearing carbon atom into H–O of a hydroxy group has been reported to proceed efficiently [35]. While the use of solvents such as CH_2Cl_2 and toluene afforded polymeric products with unidentifiable structures, the use of THF as a solvent afforded a well-defined product. To our surprise, the main chain structure of the product contained a ring-opened THF unit between each diazo-bearing carbon atom and alcohol-derived oxygen via insertion of the diazo-bearing carbon and ring-opened THF into H–O of the diol monomer (Scheme 8.26). Progress of the unprecedented reaction was confirmed by a model reaction of mono-functional diazoketone and alcohol in THF.

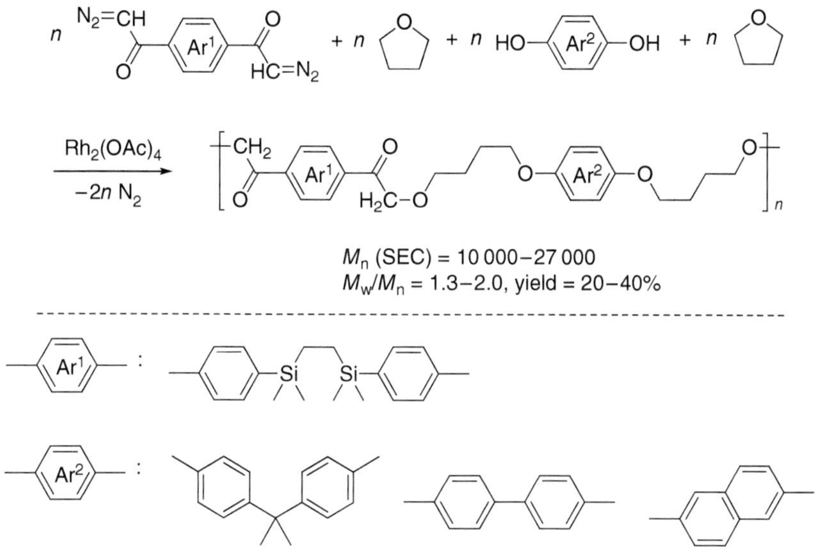

Scheme 8.26 Three-component polycondensation of bis(diazoketone), diol, and THF.

We proposed a mechanism for the reaction as described in Scheme 8.27, where a THF molecule coordinated on Rh center is attacked by HO group of a diol, resulting in its ring opening and insertion between the diazo-bearing carbon and oxygen. After our publication of the polycondensation, the unique reaction was investigated using monofunctional substrates [36, 37].

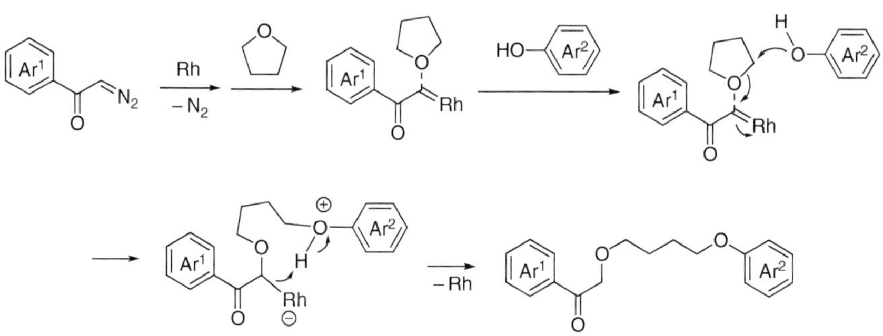

Scheme 8.27 Mechanism of THF incorporation.

The ring-opening insertion of THF was almost quantitative because THF had high tendency for the ring opening. We found that the use of six-membered cyclic ether, THP and 1,4-dioxane, with much less ring-opening tendency resulted in the formation of polymer structures containing repeating units without ring-opened cyclic ether (70–80%) as a major component and those with ring-opened THP (30%) and dioxane (20%) as minor components.

8.3.2 Three-Component Polycondensation of Bis(diazocarbonyl) Compound, Dicarboxylic Acid, and THF

The above-described three-component polycondensation can be extended to that using dicarboxylic acid in place of diol. Ring-opened THF was almost quantitatively incorporated into the polymer main chain as well (Scheme 8.28) [38]. Whereas aliphatic diol did not give well-defined products in the above-described polycondensation using diol, aliphatic diacids can be used to afford well-defined products. These results suggest that relatively high acidity is required for the polymerization to proceed and yield well-defined products.

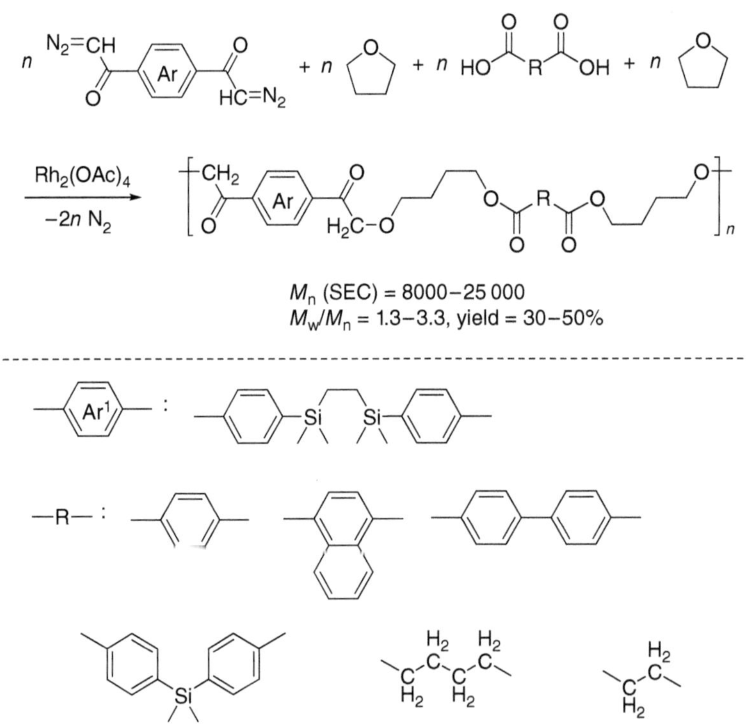

Scheme 8.28 Three-component polycondensation of bis(diazoketone), dicarboxylic acid, and THF. Source: Modified from Ihara et al. [38].

In 2020, Wang et al. reported that $Rh_2(OAc)_4$-catalyzed two-component polycondensation of bis(α-diazo-1,3-dicarbonyl) compound with a variety of dicarboxylic acid proceeded efficiently in CH_2Cl_2 [39].

8.3.3 Three-Component Polycondensation of Bis(diazocarbonyl) Compound, Enol-form of 1,3-Diketone, and THF

Transition-metal-catalyzed insertion of a diazocarbonyl compound and THF into H–O of an enol-form of 1,3-diketone was reported in 2017 [36], and we attempted to utilize the reaction for the three-component polycondensation of bis(diazoketone), enol form of 1,3-diketone, and THF (Scheme 8.29) [40].

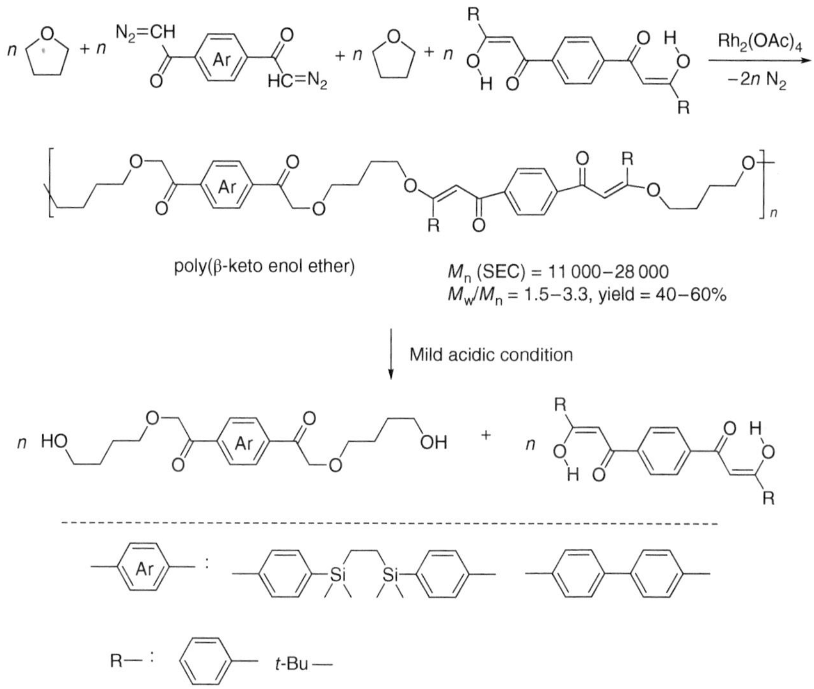

Scheme 8.29 Preparation and mild acid degradation of poly(β-keto enol ether)s. Source: Modified from Shimomoto et al. [40].

The Rh-catalyzed polymerization proceeded in an expected manner, affording polymers bearing β-keto enol ether framework in the polymer main chain. Because the β-keto enol ether framework can be readily cleaved with a catalytic action of acid to give an alcohol and the original enol unit, the polymer could be an environmentally friendly material with high acid degradability. The acid-catalyzed degradation indeed proceeded under a very mild acidic condition to afford a combination of two well-defined low-molecular weight products in high yield. The polymer was so sensitive toward acidic conditions that even standing in $CDCl_3$ solution in an NMR sample tube at room temperature led to almost complete degradation in a few days without adding external acid, while the polymer was intact in toluene and DMSO solutions.

8.3.4 Two-Component Polycondensation of Bis(diazocarbonyl) Compound with Aromatic Diamine

Polycondensation of bis(diazoacetate) and aromatic diamine proceeded via N–H insertion of a diazo-bearing carbon atom to afford a new type of polyamines (Scheme 8.30) [41]. The Ru-catalyzed polymerization proceeded efficiently, yielding well-defined polyamines with a variety of aromatic frameworks between the two amino groups in the polymer main chain. Glass transition temperature (T_g) of the product gradually changed in a temperature range of 88–173 °C, depending on the structure of a linker (X) between the two phenyl rings in the aromatic diamine monomers.

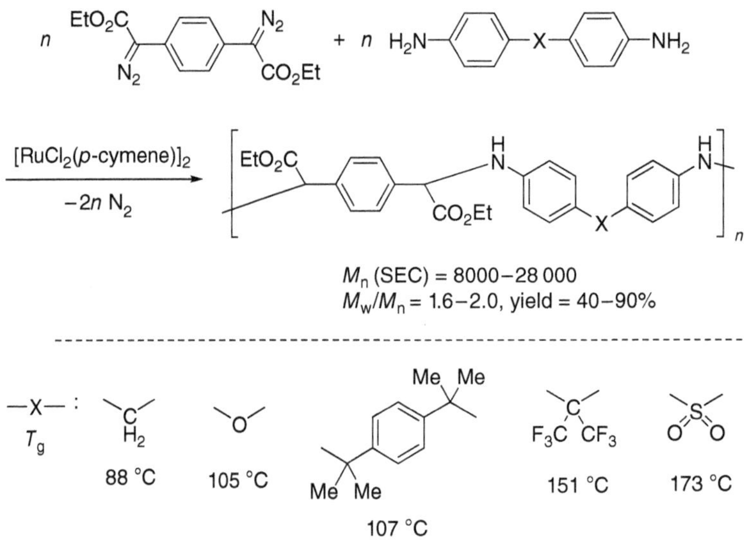

Scheme 8.30 Polycondensation of bis(diazocarbonyl) compound with aromatic diamine. Source: Modified from Shimomoto et al. [41].

8.3.5 Single-Component Polycondensation of Bis(diazocarbonyl) Compound to Afford Unsaturated Polyesters

Transition-metal-catalyzed C=C bond formation between two diazo-bearing carbon atoms of diazocarbonyl compounds is a well-known reaction [34]. We expected that the single-component polycondensation of bis(diazoacetate)s utilizing this C=C bond formation as propagation would yield unsaturated polyesters as shown in Scheme 8.31 [42]. Ruthenium catalysts (Grubbs first and second), $Rh_2(OAc)_4$, and $Cu(acac)_2$ behaved as effective catalysts for the polycondensation. The cis/trans-C=C selectivity in the products depended on the kind of the catalyst employed, while Ru-catalysts showed high cis-selectivity (75–94%), and $Rh_2(OAc)_4$ and $Cu(acac)_2$ showed low trans-selectivity (56–68%). It was revealed that the initially predominant cis-C=Cs in the product obtained by the Ru catalysts was almost quantitatively transformed to thermodynamically more stable trans-C=Cs by treatment with diethylamine. Independently from us, Liu and coworkers reported analogous single-component polycondensation of bis(diazocarbonyl) compounds [43].

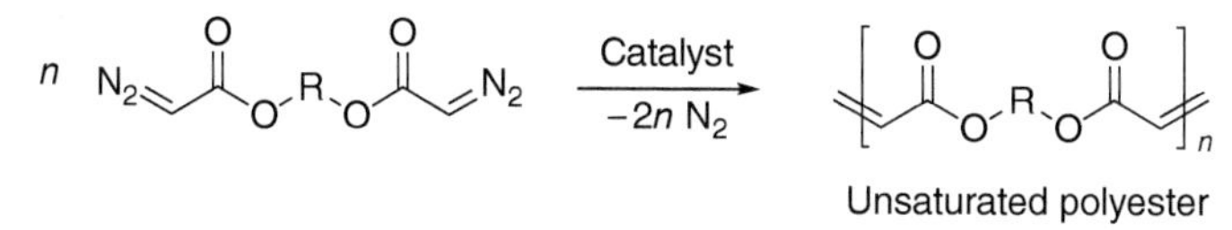

Catalyst : Ru (Grubbs 1st, 2nd), $Rh_2(OAc)_4$, $Cu(acac)_2$

M_n (SEC) = 2000–22 000
M_w/M_n = 1.5–12, yield = 20–60%
cis : trans = 94 : 6–32 : 68

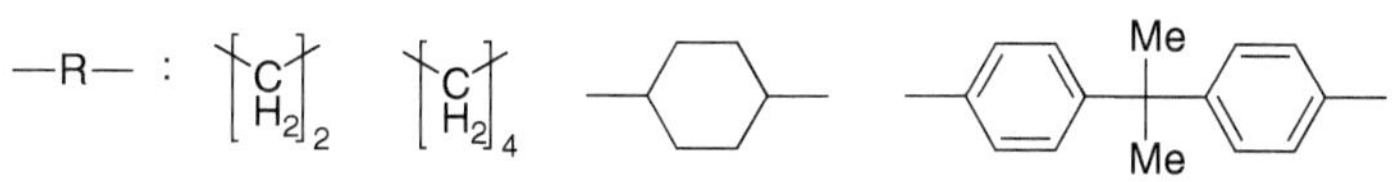

Scheme 8.31 Polycondensation of bis(diazocarbonyl) compound to afford unsaturated polyester. Source: Modified from Shimomoto et al. [42].

8.3.6 Single-Component Polycondensation of Bis(diazocarbonyl) Compound to Afford Poly(arylene vinylene)s (PAV)

The above-described C=C bond forming single-component polycondensation of bis(diazocarbonyl) compounds was applied to poly(arylene vinylene) (PAV) synthesis via modification of monomer structures as shown in Scheme 8.32 [44].

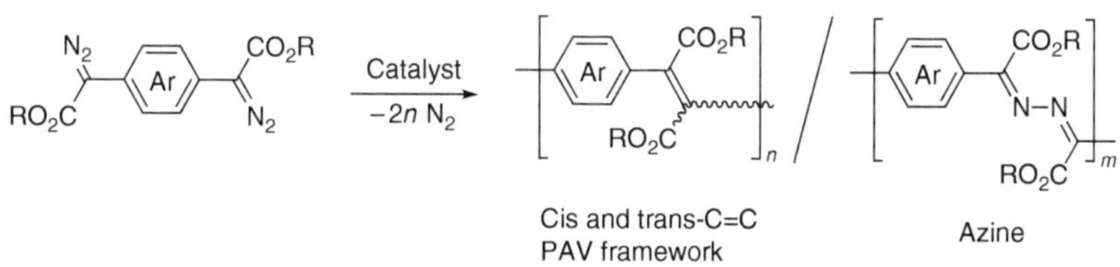

Catalyst : RuCl(cod)Cp* (trans selective), $Cu(acac)_2$ (non-selective),
$Rh_2(OAc)_4$ (azine selective)

M_n (SEC) = 9000–90 000
M_w/M_n = 1.3–3.1, yield = 20–80%

—R : n-C_6H_{13}, n-$C_{12}H_{25}$

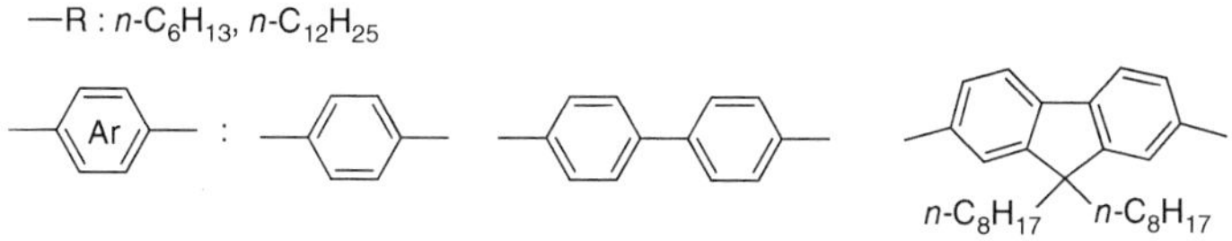

Scheme 8.32 Polycondensation of bis(diazocarbonyl) compound to afford poly(arylene vinylene). Source: Modified from Shimomoto et al. [44].

By conducting some model reactions using a monofunctional diazoacetate, we found that RuCl(cod)Cp* and $Cu(acac)_2$ were effective catalysts for the C=C bond forming reaction with highly trans-selective (c. 90%) and nonselective manners, respectively. In contrast, when $Rh_2(OAc)_4$ was employed as a catalyst, azine framework (=N—N=) with incomplete N_2 elimination was predominantly obtained with a high selectivity (c. 90%) instead of C=C bonds.

When the polymerization of bis(diazoacetate)s with a series of aromatic linker structures was conducted with the Ru and Cu catalysts, PAVs were obtained with cis-C=C:trans-C=C:azine ratios reflecting those observed in the model reactions. A PAV containing a fluorene linker with a 90% trans-C=C composition exhibited a hole mobility of $4 \times 10^{-6}\ cm^2\ V^{-1}\ s^{-1}$.

8.4 Concluding Remarks

This chapter has described polymerization of a variety of diazocarbonyl compounds, particularly focusing on the results achieved in our research group in the last two decades. The readers would appreciate that diazocarbonyl compounds have high utility as monomers for polymerization, yielding polymers with unique structures and properties because of their distinctive reactivities. This research field will further flourish with the progress of the chemistry of diazocarbonyl compounds.

References

1 Grubbs, R.H. and Khosravi, E. (eds.) (2015). *Handbook of Metathesis*, Polymer Synthesis, 2e, vol. Vol. 3. Weinheim: Wiley-VCH.
2 Bielawski, C.W. and Grubbs, R.H. (2007). *Prog. Polym. Sci.* 32: 1.
3 Schrock, R.R. (2014). *Acc. Chem. Res.* 47: 2457.
4 Baughman, T.W. and Wagener, K.B. (2005). *Adv. Polym. Sci.* 176: 1.
5 Buchmeiser, M.R. (2000). *Chem. Rev.* 100: 1565.
6 Ihara, E. (2010). *Adv. Polym. Sci.* 231: 191.
7 Jellema, E., Jongerius, A.L., Reek, J.N.H., and de Bruin, B. (2010). *Chem. Soc. Rev.* 39: 1706.
8 Franssen, N.M.G., Walters, A.J.C., Reek, J.N.H., and de Bruin, B. (2011). *Catal. Sci. Technol.* 1: 153.
9 Cahoon, C.R. and Bielawski, C.W. (2018). *Coord. Chem. Rev.* 374: 261.
10 Ihara, E. and Shimomoto, H. (2019). *Polymer* 174: 234.
11 Shimomoto, H. (2020). *Polym. J.* 52: 269.
12 Kantor, S.W. and Osthoff, R.C. (1953). *J. Am. Chem. Soc.* 75: 931.
13 Liu, L., Song, Y., and Li, H. (2002). *Polym. Int.* 51: 1047.
14 Ihara, E., Haida, N., Iio, M., and Inoue, K. (2003). *Macromolecules* 36: 36.
15 Ihara, E., Fujioka, M., Haida, N. et al. (2005). *Macromolecules* 38: 2101.
16 Ihara, E., Nakada, A., Itoh, T., and Inoue, K. (2006). *Macromolecules* 39: 6440.
17 Ihara, E., Ishiguro, Y., Yoshida, N. et al. (2009). *Macromolecules* 42: 8608.

18 Shimomoto, H., Ichihara, S., Hayashi, H. et al. (2019). *Macromolecules* 52: 6976.
19 Ihara, E., Takahashi, H., Akazawa, M. et al. (2011). *Macromolecules* 44: 3287.
20 Ihara, E., Akazawa, M., Itoh, T. et al. (2012). *Macromolecules* 45: 6869.
21 Zhukhovitskiy, A.V., Kobylianskii, I.J., Thomas, A.A. et al. (2019). *J. Am. Chem. Soc.* 141: 6473.
22 Odien, G. (2004). *Principles of Polymerization*, 416–420. New York: John Wiley & Sons, Inc.
23 Shimomoto, H., Asano, H., Itoh, T., and Ihara, E. (2015). *Polym. Chem.* 6: 4709.
24 Kato, F., Chandra, A., Tokita, M. et al. (2018). *ACS Macro Lett.* 7: 37.
25 Shimomoto, H., Hohsaki, R., Hiramatsu, D. et al. (2020). *Macromolecules* 53: 6369.
26 Chu, J.-H., Xu, X.-H., Kang, S.-M. et al. (2018). *J. Am. Chem. Soc.* 140: 17773.
27 Yamamoto, Y. (2010). *Adv. Synth. Catal.* 352: 478.
28 Shimomoto, H., Itoh, E., Itoh, T. et al. (2014). *Macromolecules* 47: 4169.
29 Tromp, D.S., Lankelma, M., de Valk, H. et al. (2018). *Macromolecules* 51: 7248.
30 Shimomoto, H., Shimizu, K., Takeda, C. et al. (2015). *Polym. Chem.* 6: 8124.
31 Takaya, T., Oda, T., Shibazaki, Y. et al. (2018). *Macromolecules* 51: 5430.
32 Shimomoto, H., Kudo, T., Tsunematsu, S. et al. (2018). *Macromolecules* 51: 328.
33 Schild, H.G. (1992). *Prog. Polym. Sci.* 17: 163.
34 Doyle, M.P., Mckervey, M.A., and Ye, T. (1998). *Modern Catalytic Methods for Organic Synthesis with Diazo Compounds*. New York: John Wiley & Sons, Inc.
35 Ihara, E., Saiki, K., Goto, Y. et al. (2010). *Macromolecules* 43: 4589.
36 Chen, W., Wang, H., Zhang, Y. et al. (2017). *Tetrahedron Lett.* 58: 3262.
37 Lu, L., Chen, C., Jiang, H., and Yin, B. (2018). *J. Org. Chem.* 83: 14385.
38 Ihara, E., Hara, Y., Itoh, T., and Inoue, K. (2011). *Macromolecules* 44: 5955.
39 Wang, X., Ding, Y., Tao, Y. et al. (2020). *Polym. Chem.* 11: 1708.
40 Shimomoto, H., Mori, T., Itoh, T., and Ihara, E. (2019). *Macromolecules* 52: 5761.
41 Shimomoto, H., Mukai, H., Bekku, H. et al. (2017). *Macromolecules* 50: 9233.
42 Shimomoto, H., Hara, Y., Itoh, T., and Ihara, E. (2013). *Macromolecules* 46: 5483.
43 Xiao, L., Li, Y., Liao, L., and Liu, L. (2013). *New J. Chem.* 37: 1874.
44 Shimomoto, H., Moriya, T., Mori, T. et al. (2020). *ACS Omega* 5: 4787.

9

Metal-Catalyzed Quinoid Carbene (QC) Transfer Reactions

Hai-Xu Wang[1], *Vanessa K.-Y. Lo*[1], *and Chi-Ming Che*[1,2,3]

[1] *The University of Hong Kong, State Key Laboratory of Synthetic Chemistry and Department of Chemistry, Pokfulam Road, Hong Kong, P. R. China*
[2] *HKU Shenzhen Institute of Research & Innovation, Yuexing 2nd Rd, Shenzhen, P. R. China*
[3] *Laboratory for Synthetic Chemistry and Chemical Biology Limited, Units 1503-1511, 15/F, Building 17W, Hong Kong Science and Technology Parks, New Territories, Hong Kong, P. R. China*

9.1 Introduction

Metal–quinoid carbene (QC) species, also known as metallaquinones, are a class of carbocyclic metal–carbene complexes that are key intermediates in catalytic QC transfer/insertion reactions in which a transferred QC moiety can subsequently undergo aromatization to form a functionalized phenol group (Scheme 9.1) [1, 2]. Due to its electrophilic nature, the metal–QC species is regarded as "an umpoled phenol" by Baran and Dao [3]; it is also a rare example of metal–carbene species that acts as an aryl synthon [4]. Although quinone diazides (also known as diazoquinones), the diazo precursors of QC, were identified more than a century ago [5], transition metal-catalyzed QC transfer reaction was not reported until 1988 [6]. With the electron-withdrawing carbonyl group being conjugated to the carbene carbon, QC can be classified as an acceptor-type carbene exhibiting electrophilic carbene reactivity. Very recently, metal–QC species were reported to display redox reactivities (Scheme 9.1), which provides a new direction in C–H functionalization and small molecule activation reactions [2]. In this chapter, examples of isolated metal–QC complexes are first reviewed in the context of their syntheses, structures, and reactivities. A summary of transition metal-catalyzed QC transfer/insertion reactions classified based on their reaction types, i.e. cyclopropanation, C–H insertion, nucleophilic addition, etc., is also presented.

9.2 Metal–Quinoid Carbene (QC) Complexes and Stoichiometric Reactivity

Up to now, there have only been three reports on isolated metal–QC complexes, including the pincer-type $PC_{carbene}P$ Ru– (**1**) and Fe–QC (**2**) complexes (also termed

Hai-Xu Wang and Vanessa K.-Y. Lo contributed equally to this work.

Transition Metal-Catalyzed Carbene Transformations, First Edition.
Edited by Jianbo Wang, Chi-Ming Che, and Michael P. Doyle.

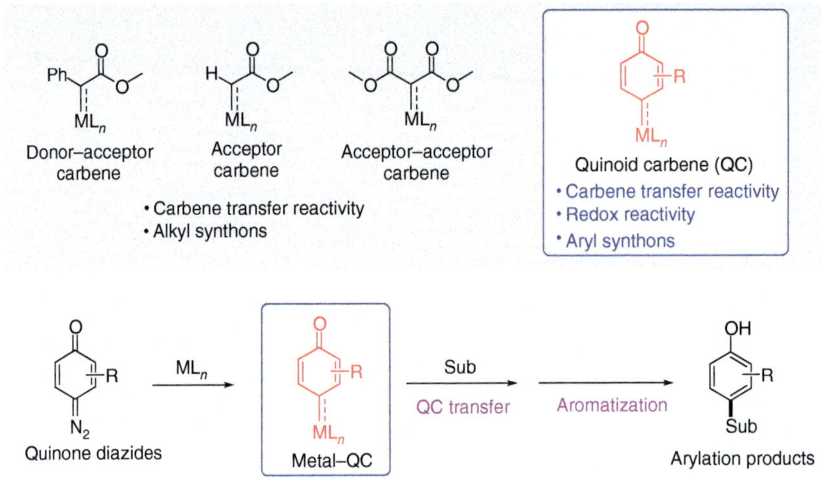

Scheme 9.1 Comparison among quinoid carbene (QC) and acceptor-type carbenes (top) and the role of metal–QC intermediates as aryl synthons in organic synthesis (bottom). Source: Based on Othman et al. [1] and Wang et al. [2].

ruthenaquinone and *ferraquinone*, respectively) by Milstein and coworkers [7, 8] and a series of Ru(II) porphyrin QC complexes (**3**) by Che and coworkers [9]. The chelation-stabilized complexes **1–2** were prepared by deprotonation and/or dehydrogenation of their phenolic metal–PCP precursors (Scheme 9.2A). Ru(II) porphyrins have previously been reported to stabilize a wide range of reactive carbene ligands [10–15], and the corresponding QC complexes **3** were prepared by using quinone diazides as the carbene precursors (Scheme 9.2B). All these metal–QC complexes are low-spin diamagnetic species and show characteristic ^{13}C NMR (nuclear magnetic resonance) signals of $\delta(M{=}C_{carbene})$ and $\delta(C{=}O)$ in the downfield region (>170 ppm, Scheme 9.2D); stretching frequencies of the QC moieties can also be observed in their infrared (IR) spectra (1560–1670 cm^{-1}). The only known X-ray crystal structures of metal–QC complexes were reported for **3** which reveal short Ru–$C_{carbene}$ distances (1.83–1.93 Å) as well as a *quinoid* form of the QC planes (Scheme 9.2C) [9]. In contrast, an *aromatic* form was obtained upon crystallization of **1**, which is attributable to solvent-induced tautomerization [7]. The electrochemical properties of **3** were investigated [9]. These complexes display two reversible reduction waves with $E_{1/2}$ at −1.26 to −0.67 V versus $Fc^{+/0}$ in dichloromethane (DCM) (Scheme 9.2E), with the first 1e$^-$ reduction potentials being close to that of *p*-benzoquinone in a similar aprotic medium. Based on the first 1e$^-$ oxidation potentials of these Ru porphyrin complexes, it can be inferred that the coordinated QC ligands act as strong π-acceptors and, depending on the substituents on QCs, their π-acidity can be comparable to or higher than that of a carbonyl ligand [9, 12, 15].

These isolated metal–QC complexes show interesting stoichiometric/catalytic reactivities, which can be categorized into carbene transfer reactivity and redox

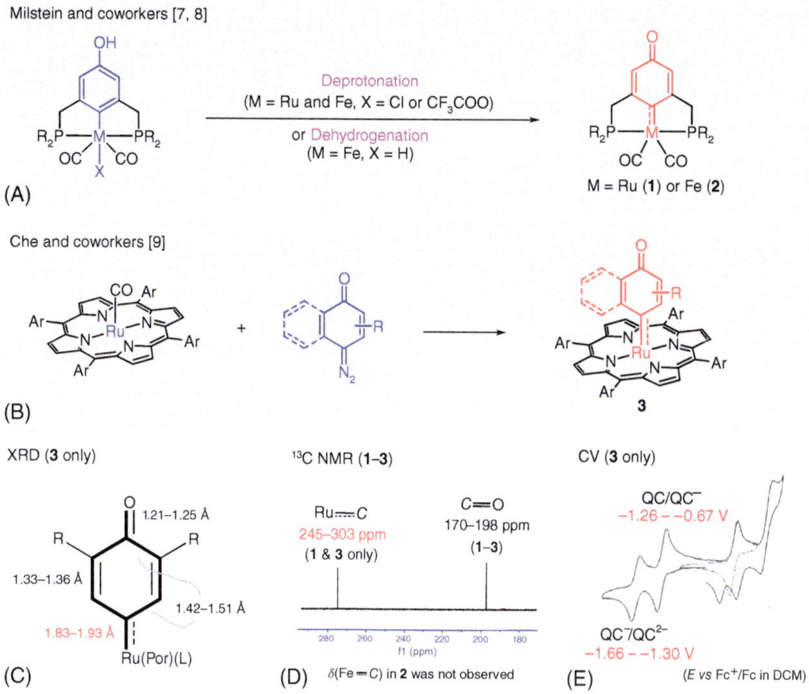

Scheme 9.2 Synthesis of pincer-type Ru- and Fe-QC complexes (A) and Ru(II) porphyrin QC complexes (B), as well as their characterization by X-ray crystallography (C), ^{13}C NMR (D), and electrochemistry (E). Source: (A) Modified from Ashkenazi et al. [7] and Dauth et al. [8]. (B–E) Modified from Wang et al. [9].

reactivity. The carbene transfer reactivity is mechanistically relevant to catalytic QC transfer reactions. Che and coworkers reported a stoichiometric intermolecular QC transfer reaction between **3** and nitrosoarenes (ArNO) to form nitrone products in good to high yields (Scheme 9.3A) [9]. By combining this step with the formation of **3** from quinone diazides, the authors developed a catalytic nitrone formation reaction mediated by Ru(II) porphyrins (see Section 9.3.4). Detailed kinetic studies of both stoichiometric and catalytic reactions unexpectedly showed a reverse carbene substituent effect – electron-donating groups on QC enhance its reactivity with the nucleophilic ArNO substrate (Scheme 9.3A). Density functional theory (DFT) calculations reveal that this might be due to an early transition state structure which features a tilted QC group partially breaking the Ru–carbene π bond. For QC that bears electron-donating substituents, a weaker Ru–carbene π bond is formed, and hence a lower energy is needed to reach the transition state [9]. This is also the first structure–reactivity relationship study on the carbene substituent effect in stoichiometric carbene transfer reactions using isolated metal–carbene complexes.

Metal–QC species exhibit a unique redox reactivity. While the QC ligand is structurally similar to other acceptor-type carbene ligands, such as those bearing carbonyl group(s), it features rather different electronic properties due to its unique

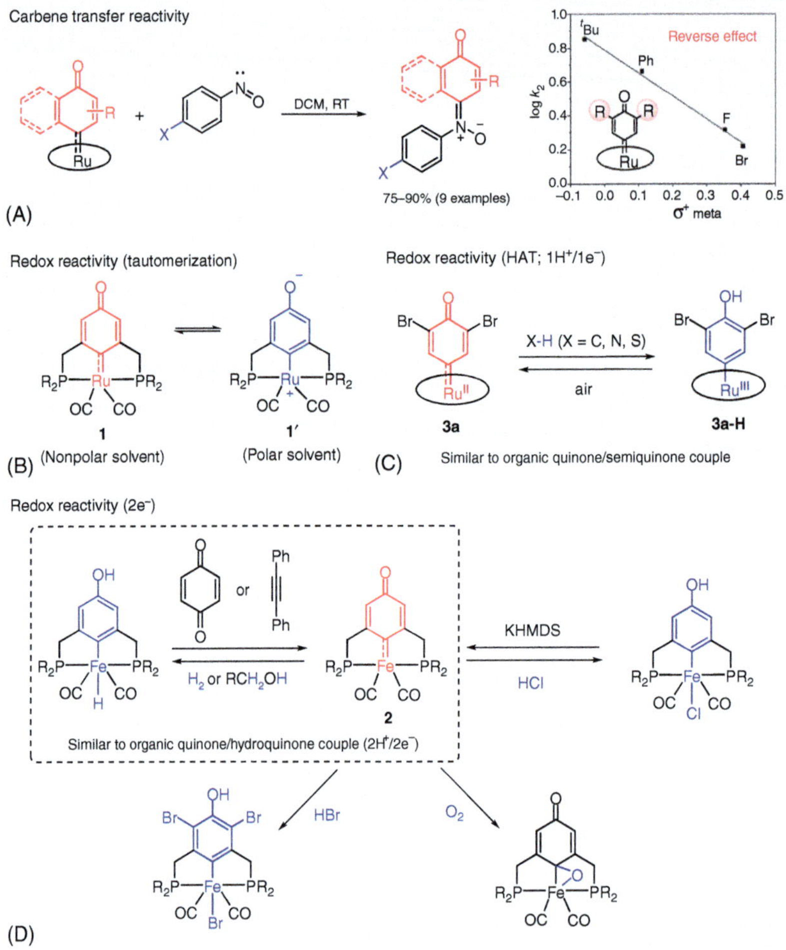

Scheme 9.3 Carbene transfer (A) and redox reactivities (B–D) of isolated metal–QC complexes. Source: (A, C) Modified from Wang et al. [9]. (B) Modified from Ashkenazi et al. [7]. (D) Modified from Dauth et al. [8].

molecular geometry. For donor–acceptor carbenes, the carbonyl substituents are often orthogonal to the carbene plane in the solid-state structures, rendering the metal–carbene π^* orbitals mainly localized at the carbene carbons [16–19]. In metal–QC complexes, however, the metal center is coplanar with the QC plane; consequently, the carbonyl group is in direct conjugation with the metal d_π orbital and acts as a π-acceptor [9]. Redox communication between metal center and QC ligand was first demonstrated by the quinoid-aromatic tautomerism of **1** reported by Milstein and coworkers (Scheme 9.3B) [7]. Later, the same group reported a $2H^+/2e^-$ ferraquinone/ferrahydroquinone redox couple of **2**, and this couple is reminiscent of the organic quinone/hydroquinone couple (Scheme 9.3D) [8]. Related formal 1,6-addition reactions of **2** with alcohols or HCl were also reported.

In comparison, a $1H^+/1e^-$ Ru(II)–QC/Ru(III)–aryl (**3a**/**3a–H**) couple was reported for **3a**, which is similar to that of the organic quinone/semiquinone [9]. The conversion of **3a** to **3a–H** proceeds through an *O*-centered hydrogen atom transfer (HAT) reaction with weak C—H or X—H bonds, while the reverse transformation can occur via aerobic oxidation (Scheme 9.3C). Such a C–H activation pathway is often found in metal-oxo chemistry and relatively less common for reactive metal–carbene intermediates. Inspired by the **3a**/**3a–H** redox couple, a catalytic aerobic oxidation of 1,4-cyclohexadiene to benzene, with up to 48 turnover numbers using **3a** as the catalyst, was developed [9]. The HAT reactivity of metal–QC species was later applied in a catalytic QC C(sp^3)–H insertion reaction using an Ir(III) catalyst (see Section 9.3.3) [20].

9.3 Metal-Catalyzed QC Transfer Reactions

9.3.1 Cyclopropanation Reactions

Reports on metal-catalyzed QC transfer reactions (Scheme 9.4) first appeared in 1988. In the search for a synthetic route to the A-ring structure of the antitumor antibiotic CC-1065, Sundberg et al. reported the use of

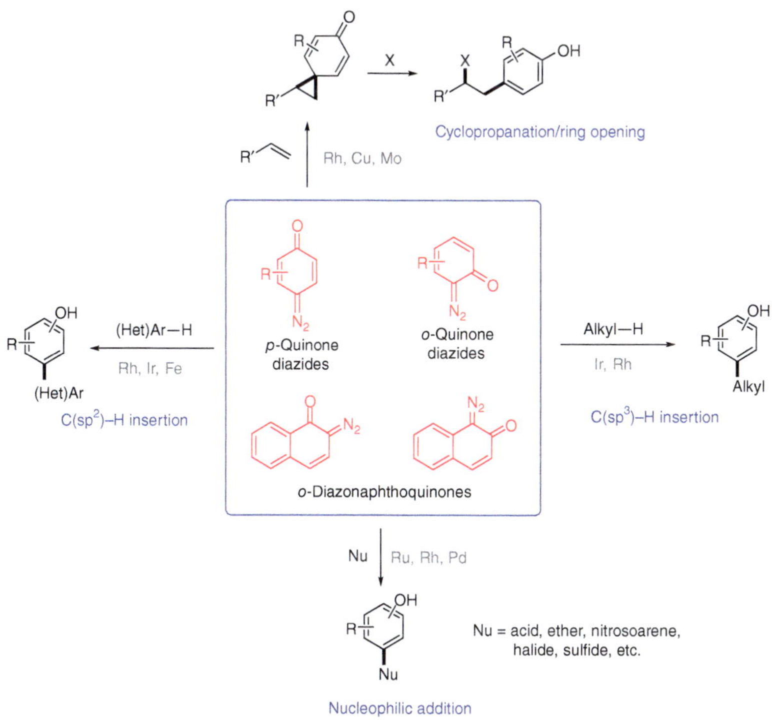

Scheme 9.4 Summary of metal-catalyzed QC transfer reactions.

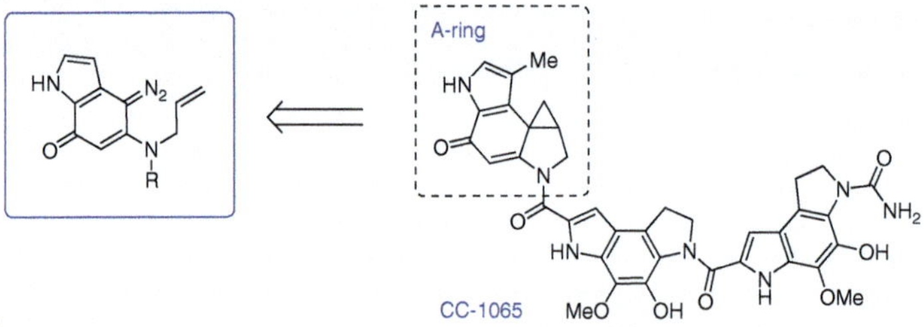

(A)

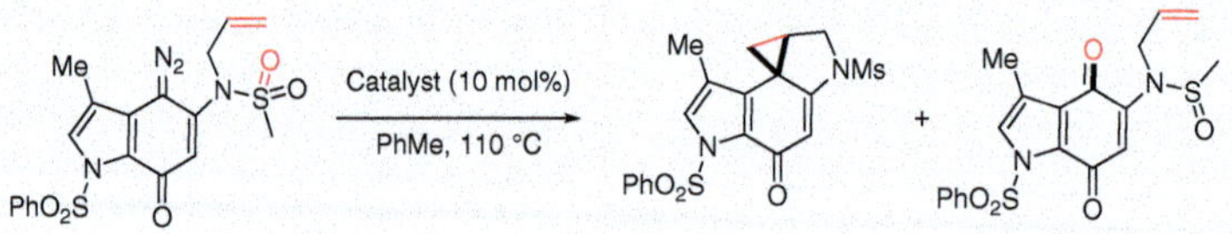

Catalyst	Reaction time (h)	Yield of cyclopropane (%)	Yield of quinone (%)
$[Cu(sal)_2]$	0.5	52	10
$[Cu(acac)_2]$	0.5	56	25
$[Mo(CO)_6]$	2.0	41	15

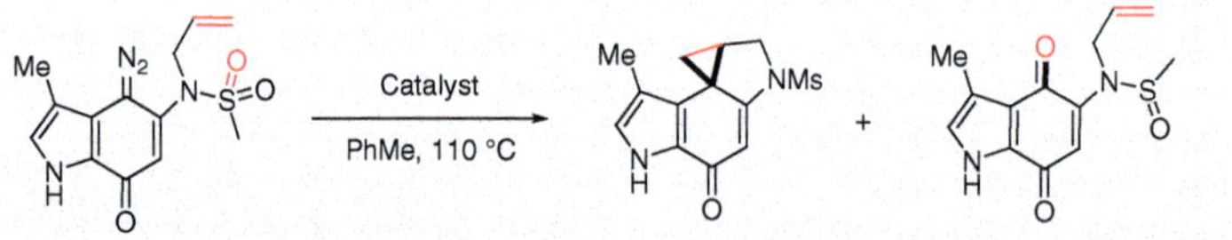

Catalyst (loading)	Reaction time (h)	Yield of cyclopropane (%)	Yield of quinone (%)
$[Cu(sal)_2]$ (10 mol%)	0.5	43	29
$[Cu(sal)_2]$ (30 mol%)	0.5	47	29
$[Cu(sal)_2]$ (100 mol%)	0.5	44	10
$[Cu(acac)_2]$ (20 mol%)	0.5	56	38
$[Mo(CO)_6]$ (35 mol%)	2.0	25	22
$[Cu(CO)(tfpd)(^{n}BuNH_2)]$ (20 mol%)	4.0	63 (DCM, 40 °C)	0

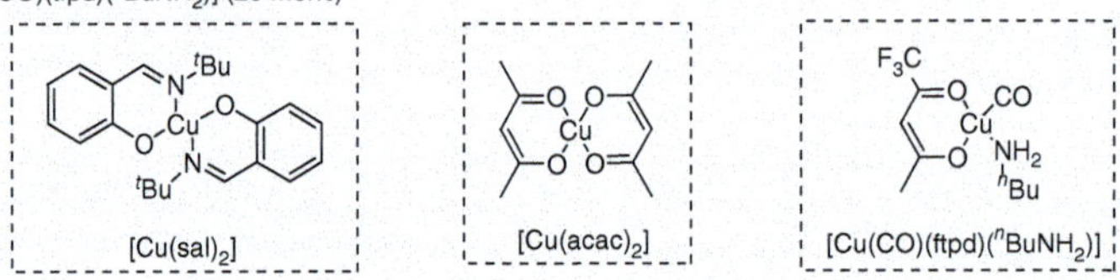

(B)

Scheme 9.5 Retrosynthesis of the A ring of CC-1065 (A) and catalyst screening for the intramolecular QC cyclopropanation reactions (B–C). Source: (A–B) Modified from Sundberg et al. [6]. (C) Modified from Sundberg and Pitts [21].

4-diazo-4,7-dihydroindol-7-ones as carbene precursors for metal-catalyzed intramolecular cyclopropanation reaction (Scheme 9.5A) [6]. Metal catalysts, such as $[Cu(sal)_2]$ (sal = *N-tert*-butyl(salicylaldiminato)), $[Cu(acac)_2]$ (acac = acetoacetate), and $[Mo(CO)_6]$, were found to be catalytically active for this reaction in refluxing toluene (Scheme 9.5B). Quinone was observed as a byproduct formed by oxidation of the metal–QC intermediate by the sulfonamide S=O group. In order to lower the reaction temperature, Sundberg and Pitts extended their study to Cu(I) catalysts. The best result was obtained by using 20 mol% of $[Cu(CO)(tfpd)(^{n}BuNH_2)]$

(tfpd = trifluoropentanedionato) as the catalyst and in refluxing DCM, leading to the desired cyclopropane in 63% isolated yield without the formation of quinone (Scheme 9.5C) [21]. Attempts to achieve asymmetric intramolecular cyclopropanation by using chiral amine additives were not successful, suggesting that this reaction proceeds via a stepwise mechanism.

The synthesis of the difluorocyclopropane analog of the A-ring of CC-1065 was reported by Boger and Jenkins using the same synthetic strategy as that reported by Sundberg [22]. The key step of this synthesis is the [$Rh_2(OAc)_4$]-catalyzed intramolecular QC transfer to a 1,1-difluoroalkene in refluxing toluene, leading to the desired difluorosubstituted cyclopropane in 74% isolated yield (Scheme 9.6).

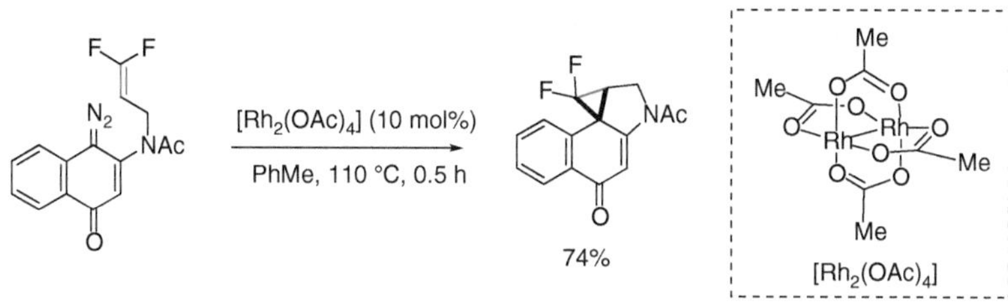

Scheme 9.6 Synthesis of the difluorocyclopropane analog of the A-ring of CC-1065 by a Rh-catalyzed intramolecular QC cyclopropanation reaction. Source: Modified from Boger and Jenkins [22].

Baran and Dao reported a [$Rh_2(esp)_2$]-catalyzed intermolecular cyclopropanation of alkenes using *p*-quinone diazides as carbene sources [3]. As a model study, intermolecular cyclopropanation of cyclooctene with *p*-quinone diazides, catalyzed by 5 mol% of [$Rh_2(esp)_2$] (esp = α,α,α′,α′-tetramethyl-1,3-benzenedipropionate), proceeded smoothly in the presence of 4 Å molecular sieves (MS) at room temperature and led to the desired cyclopropanes in up to 66% yields (Scheme 9.7A). These cyclopropanes are prone to C–C cleavage-induced aromatization upon addition of nucleophiles, which was applied to an intramolecular cyclopropanation reaction for the construction of steroidal structures (Scheme 9.7B).

9.3.2 C(sp^2)–H Insertion Reactions

C(sp^2)–H insertion reactions are among the most studied reactions with QC, since the biaryl products generated from these reactions are of great synthetic interest. There are two major advantages of this methodology that make it a potential alternative/complement to conventional cross-coupling reactions: (i) direct QC insertion leads to functionalized phenol groups which are challenging to be installed by conventional coupling methods, since either special ligands (with limited examples) or additional protection/deprotection steps are required; (ii) halide groups in the substrates, which are unreactive toward QC, together with the newly formed phenol groups in the products, can provide reactive sites for orthogonal functionalization using conventional cross-coupling reactions. QC C(sp^2)–H insertion reactions generally proceed via three pathways: (i) nucleophilic

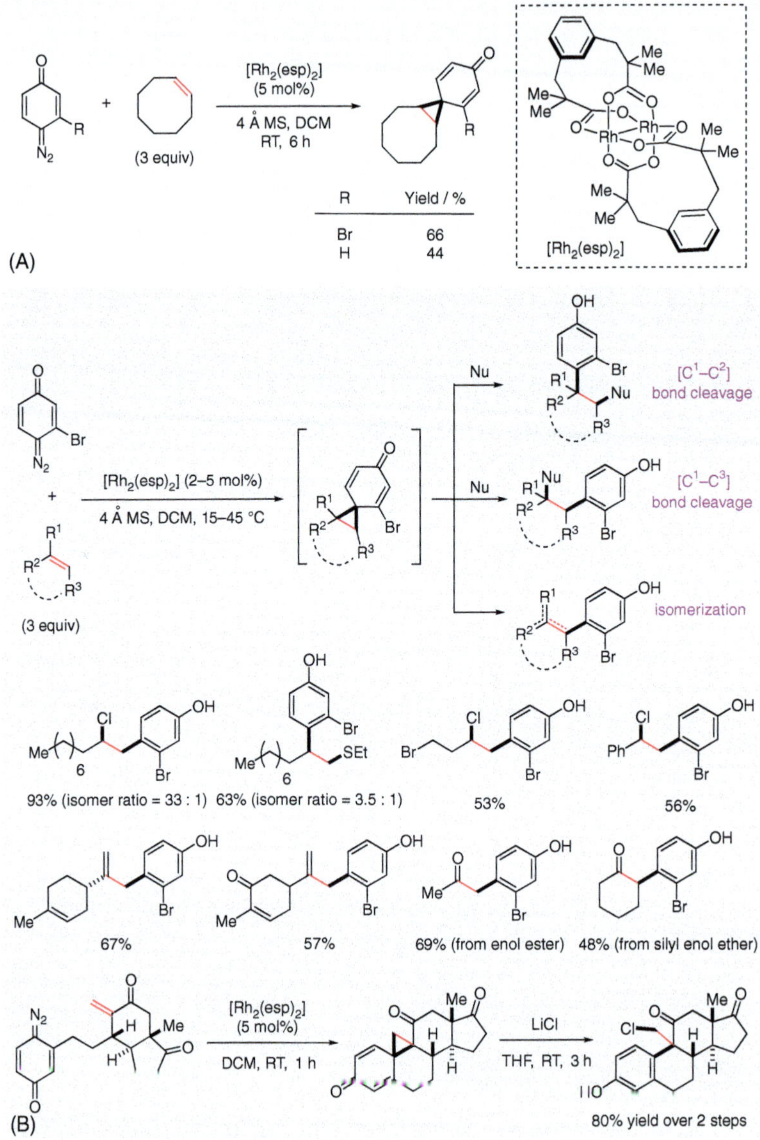

Scheme 9.7 Rh-catalyzed intermolecular QC cyclopropanation reaction (A) and cyclopropanation/ring-opening cascade reaction (B). Source: Modified from Dao and Baran [3].

attack of arene substrates on metal–QC intermediates followed by aromatization; (ii) cyclometallation of directing-group (DG)-containing substrates followed by coupling with QC; and (iii) arene cyclopropanation followed by ring opening.

Reactions via direct nucleophilic attack of arene substrates were first investigated by Kitamura and coworkers [1, 23] and Lee and coworkers [24, 25] using dirhodium catalysts. Kitamura et al. reported an intramolecular QC C–H insertion cyclization method that could be used to construct the $C(sp^2)$—$C(sp^2)$ bond of

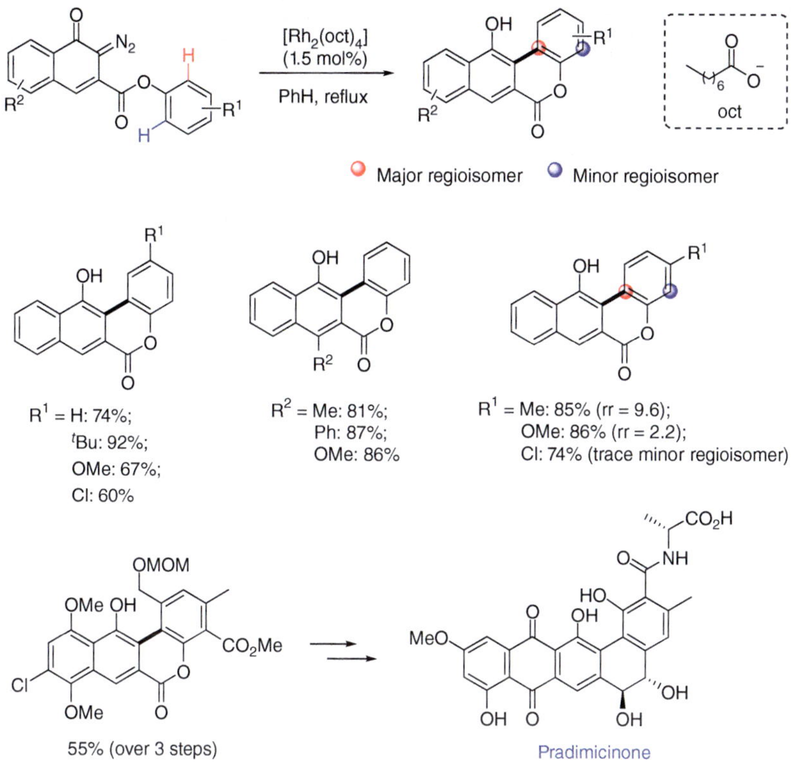

Scheme 9.8 Rh-catalyzed intramolecular QC C(sp^2)–H insertion reaction. Source: Modified from Kitamura et al. [23].

six-membered β-phenylnaphthalene lactone products (Scheme 9.8) [23]. [$Rh_2(oct)_4$] (oct = octanoate) was the optimized catalyst for this transformation in benzene at 90 °C giving the cyclization products in moderate to high yields. In cases where two different C—H bonds are available for the QC insertion, moderate to excellent regioselectivity was obtained, which is mainly dictated by steric factor. This methodology was further applied as a key step in the formal synthesis of the natural product, pradimicinone. Moreover, the authors observed the formation of an intermolecular C(sp^2)-H insertion product with the benzene solvent upon changing the catalyst to [$Rh_2(OCOCF_3)_4$], although only one example was reported. Lee and coworkers reported the intermolecular version of this methodology and discovered that [$Rh_2(OPiv)_4$] (OPiv = pivalate) could catalyze QC C–H insertion at the C3 position of indoles and benzofurans, as well as the C2 position of pyrroles and furans, with reaction temperature down to room temperature for most substrates (Scheme 9.9A) [24]. The reactions proceeded with excellent regioselectivity and high product yields, and the observed regioselectivity at the most nucleophilic reactive sites led the authors to propose a direct nucleophilic attack mechanism. Later, Lee and coworkers further discovered that $Fe(OTf)_3$ could also act as a catalyst

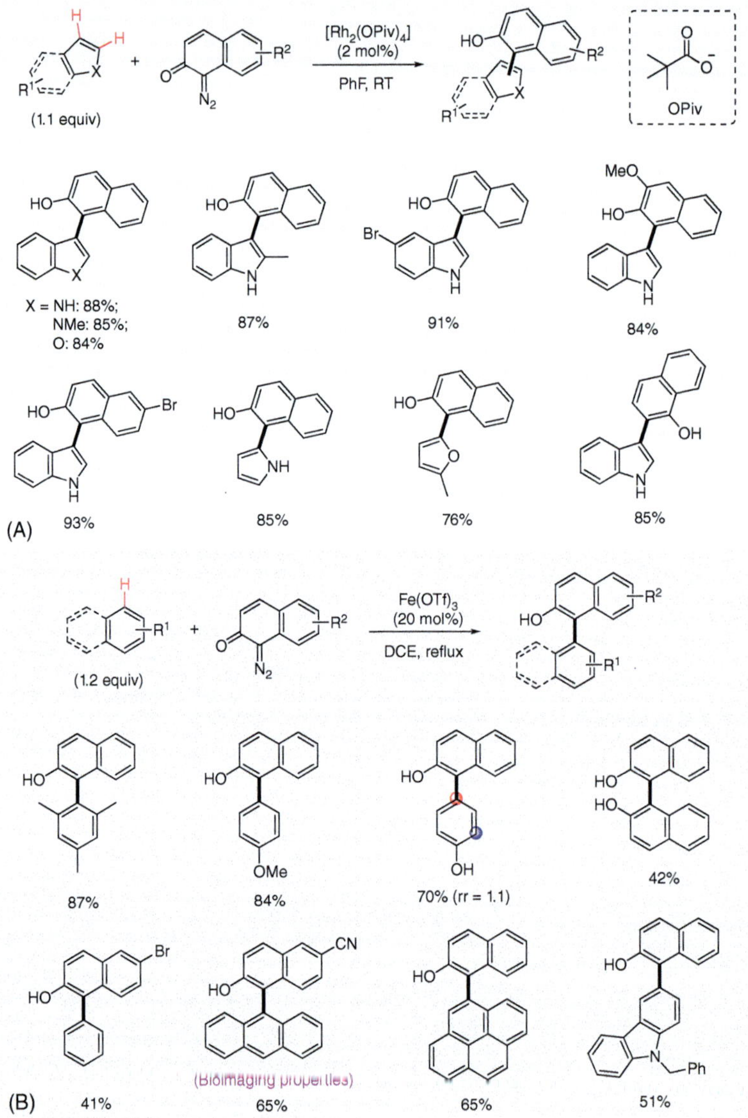

Scheme 9.9 Rh- (A) and Fe-catalyzed (B) intermolecular QC C(sp^2)–H insertion reactions. Source: (A) Modified from Baral et al. [24]. (B) Modified from Magar et al. [25].

for intermolecular QC C(sp^2)-H insertion into electron-rich arene molecules, including anthracene and substituted benzenes (Scheme 9.9B); electron-deficient arenes such as chlorobenzene were inactive in this reaction [25]. Several biaryl products derived from anthracene were demonstrated to display luminescent and bioimaging properties.

Despite the advances described above, the scope of quinone diazides in QC C(sp^2)-H insertion reactions was limited to *o*-diazonaphthoquinones. For intermolecular reactions which are of higher synthetic interest, only electron-rich heterocycles were reported, while simple arenes, especially their regioselective functionalization, remained little explored. Recently, Che and coworkers developed a protocol using $[Rh_2(esp)_2]$ as the catalyst for intermolecular QC insertion reactions with a broad substrate scope including electron-deficient arenes (Scheme 9.10) [26]. Both *p*- and *o*-quinone diazides, as well as various arenes and

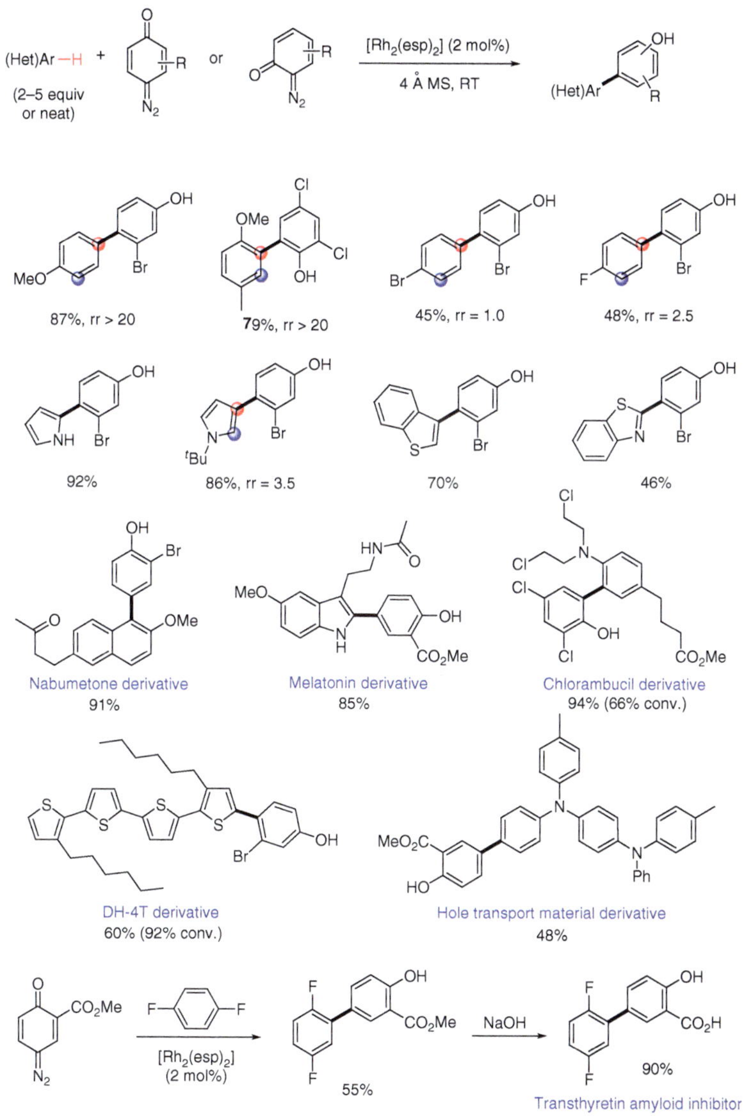

Scheme 9.10 Rh-catalyzed intermolecular QC C(sp^2)–H insertion reaction and its application in late-stage functionalization. Source: Modified from Wu et al. [26].

heteroaromatics, are compatible with this method, providing biaryl products in up to 92% yields. Moreover, the Rh catalyst, together with 4 Å MS, was found to confer excellent regioselectivity on electron-rich arenes (regioisomeric ratio (rr) = 16 : 1 to >20 : 1), though lower rr values were observed with electron-deficient arenes. This strategy was applied to the late-stage functionalization of bioactive and functional molecules, as well as in a simple two-step synthesis of a transthyretin amyloid inhibitor with good yield. A direct nucleophilic attack pathway was proposed based on product distribution. This putative mechanism was corroborated by a secondary KIE (kinetic isotope effect) value using anisole/anisole-d_5 as substrates, as well as by DFT calculations.

To promote regioselective C–H functionalization, DGs are commonly employed as synthetic handles which can override the intrinsic stereoelectronic effect of the substrate. Wang and coworkers first introduced this strategy to Cp*Rh(III)- and Cp*Ir(III)-catalyzed (Cp* = pentamethylcyclopentadienyl) QC C(sp^2)–H insertion reactions to construct biaryl moieties (Scheme 9.11) [27]. By changing the metal catalyst and loading of quinone diazides, mono- or bis-arylation products were obtained with moderate to high chemoselectivity and good to high yields. The authors discovered that several DGs, such as pyridines, pyrimidine and pyrazole,

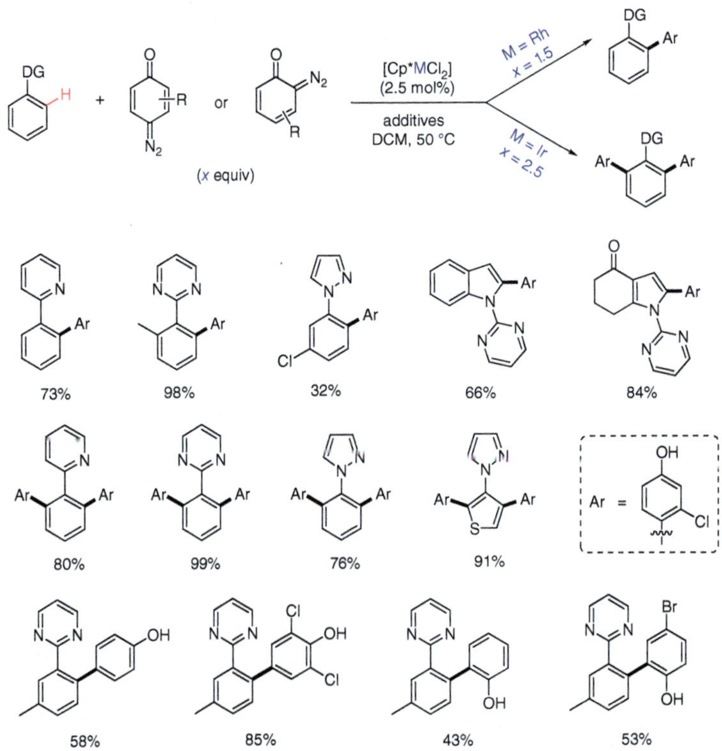

Scheme 9.11 Rh- and Ir-catalyzed intermolecular QC C(sp^2)–H insertion reaction of DG-containing substrates. Source: Modified from Zhang et al. [27].

are suitable for this transformation. In addition to simple arenes, arylation of heteroarenes, such as thiophene, indoles, and pyrroles, was also achieved; both *p*- and *o*-quinone diazides are well tolerated. Following this work, a number of studies were reported to extend the scope of this methodology using similar catalytic systems (Rh(III) or Ir(III) catalysts $[Cp^{*}MCl_2]_2$), with DGs ranging from pyridine, amides, phosphine oxides, *N*-oxides, carboxylic acid, and oxime (Scheme 9.12) [28–37]. Notably, when benzamides were used as substrates, Cui and Chen discovered a cascade C–H activation/cyclization reaction giving rise to a series of lactone products by harnessing the reactivity of the newly installed *ortho*-phenol group [28]. Later, Li and coworkers reported a similar cascade cyclization reaction using benzoic acids and oxime as substrates, with the cyclization step taking place at the carbene carbon to give spirocyclic products [37]. In addition to arene $C(sp^2)$—H bonds, Samanta and Ghosh also reported the arylation of aldehyde $C(sp^2)$—H bonds using quinoline as DG, providing facile access to various diarylketones [34].

Notably, with the recent advances in chiral cyclopentadienyl (Cp) ligands, several reports on enantioselective, ligand-directed QC $C(sp^2)$–H insertion reactions have emerged (Scheme 9.13) [30, 33, 36]. Waldmann, Antonchick, and coworkers developed a series of new chiral Cp ligands, *Jas*Cp, for the preparation of Rh(I) precatalysts. In the presence of $(BzO)_2$ as oxidant, active Rh(III) catalysts were formed *in situ*, which mediated the enantioselective QC $C(sp^2)$–H insertion reactions using amides as DGs (Scheme 9.13A) [30, 36]. Benzamides were converted to axially chiral biaryls in up to 93% yields and with up to 90% enantiomeric excess (ee) [30]. It should be noted that only C–H arylation products were reported by Waldmann's group. In contrast, a cascade C–H activation/cyclization reaction was reported by Cui and Chen using similar coupling partners (benzamides + *o*-diazonaphthoquinones) and a similar Rh(III) catalyst (Scheme 9.12) [28]; the difference in reaction products is presumably due to the higher reaction temperature employed in Cui's protocol. Very recently, asymmetric arylation of more challenging substrates, including (benzo)thiophenes, (benzo)furans and protected indoles, to afford five-membered ring atropisomers in up to 96% yields and with up to 94% ee were reported by Waldmann, Antonchick, and coworkers [36]. However, for the above reactions reported by Waldmann's group, the diazo partners are limited to *o*-diazonaphthoquinones bearing an ester group. Alternatively, Cramer and coworkers achieved an enantioselective QC C–H insertion reaction using phosphine oxides as DGs and Ir(III) catalysts supported by binaphthyl-substituted Cp ligands coupled with a chiral carboxylic acid co-catalyst (Scheme 9.13B) [33]. When $P(O)Ar_2R$ ($Ar \neq R$) substrates were employed, C–H arylation on one of the Ar groups led to desymmetrization of the phosphorus atom to afford *P*-chiral biaryl phosphine oxides in up to 98% yields and with up to 97% ee. Various *o*-quinone diazides were compatible with this method and the enantioselectivity remained uniformly high for most substrates (95–97% ee). This is consistent with a plausible catalytic cycle whereby the enantiodetermining step is the formation of an iridacycle which occurs prior to QC insertion. When symmetrical $P(O)Ar_3$ and the bulkier *o*-diazonaphthoquinones were used as coupling partners, biaryl products with a stable atropchiral axis were obtained in up to 95% yields and with up to 92% ee.

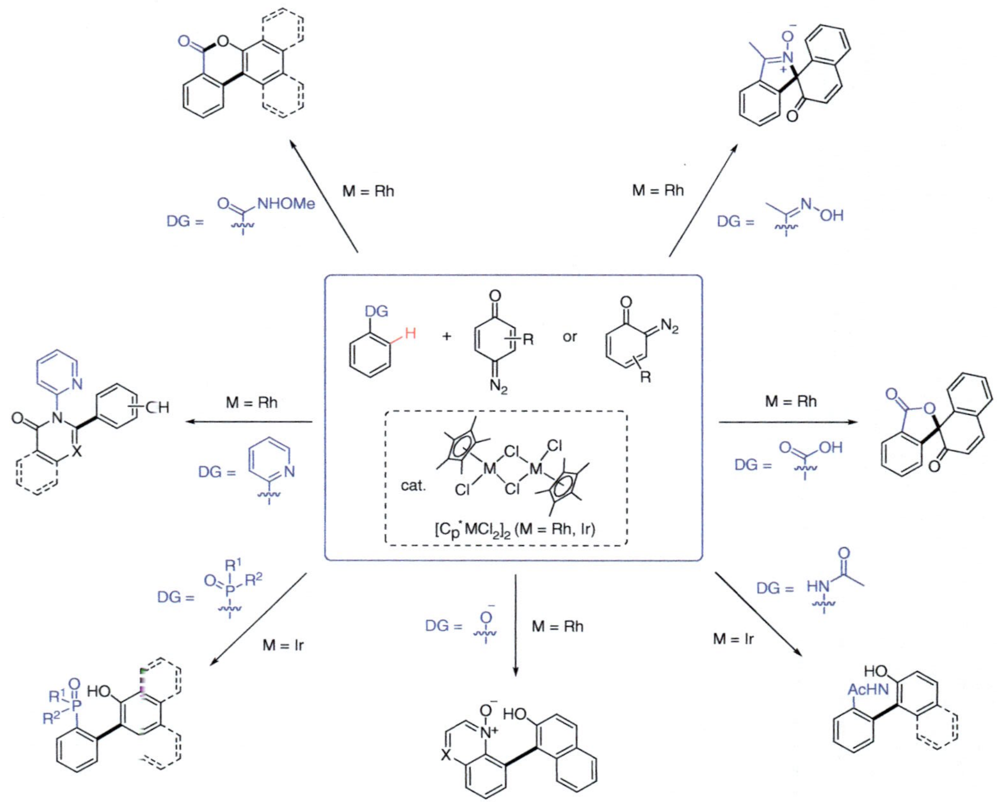

Scheme 9.12 Summary of Rh- and Ir-catalyzed intermolecular QC C(sp^2)–H insertion reactions of DG-containing substrates.

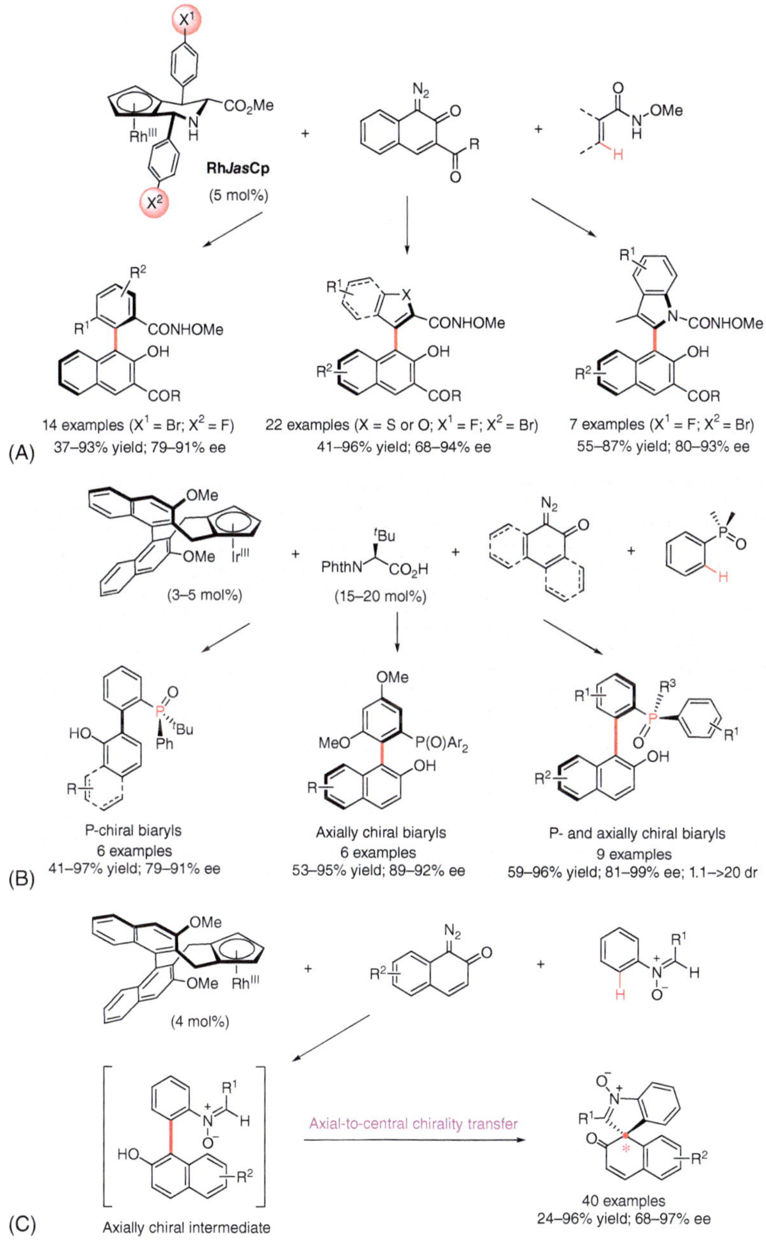

Scheme 9.13 Asymmetric intermolecular QC $C(sp^2)$–H insertion reactions of DG-containing substrates by Rh (A and C) and Ir (B) catalysts bearing chiral Cp ligands. Source: (A) Modified from Jia et al. [30]. (B) Modified from Jang et al. [33]. (C) Modified from Kong et al. [38].

Furthermore, the authors explored the combination of $P(O)Ar_2R$ substrates and *o*-diazonaphthoquinones to obtain both *P*- and axially chiral products in up to 96% yields and 99% ee, with diastereoselectivity ranging from 1.1 : 1 to >20 : 1 diastereomeric ratio (dr) depending on the coupling partners. These chiral phosphine oxides could be readily reduced to the corresponding phosphorus(III) compounds bearing *P*- and/or axial chirality, which are highly desirable ligand backbones for asymmetric catalysis. Recently, by using Cramer's chiral Cp ligand, Li and coworkers reported a Rh(III)-catalyzed enantioselective C–H activation/cyclization reaction of *o*-diazonaphthoquinones to construct chiral spirocycles (Scheme 9.13C) [38]. After screening a series of cyclizable DGs, the authors found that nitrone-containing substrates could undergo a cascade C–H activation/cyclization reactions with a high degree of axial-to-center chirality transfer (up to 97% ee). Mechanistic studies and DFT calculations supported an axially chiral biaryl species as the intermediate, which could then undergo oxidative cyclization in the presence of Ag(I) salt.

A third pathway for formal QC $C(sp^2)$–H insertion reactions proceeds through cyclopropanation followed by ring opening/rearrangement, with only two reported examples using substrates bearing electron-rich C=C moieties (Scheme 9.14) [3, 39]. In the course of the total synthesis of diazonamide A, Wood and coworkers discovered that a C3-substituted indole could undergo formal C–H QC insertion at the C2 position with an *o*-quinone diazide using $[Rh_2(OAc)_4]$ as the catalyst, and the reaction was conceived to involve a cyclopropane intermediate (Scheme 9.14A) [39]. In the QC cyclopropanation reaction described by Baran and Dao (see Section 9.3.1), it was found that 3,4-dihydro-2*H*-pyran can be arylated at the C5 position with a *p*-quinone diazide, which is initiated by a $[Rh_2(esp)_2]$-catalyzed cyclopropanation reaction (Scheme 9.14B) [3].

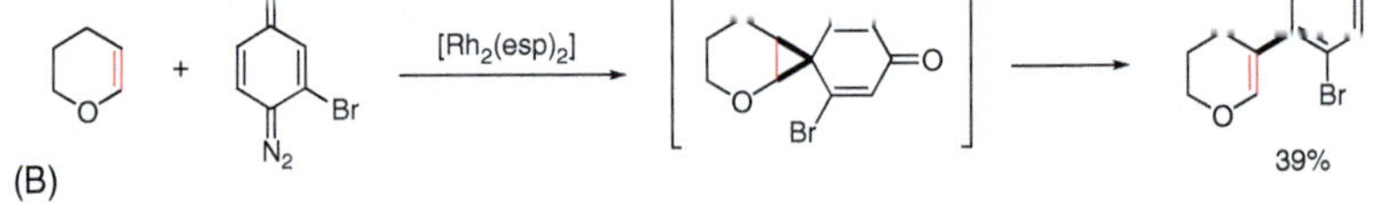

Scheme 9.14 Rh-catalyzed formal intermolecular QC $C(sp^2)$–H insertion reactions initiated by cyclopropanation. Source: (A) Modified from Sawada et al. [39]. (B) Modified from Dao and Baran [3].

9.3.3 $C(sp^3)$–H Insertion Reactions

Compared to metal-catalyzed QC insertion into $C(sp^2)$—H bonds, reports on the analogous QC insertion into $C(sp^3)$—H bonds are sparse. Kitamura et al.

reported a [$Rh_2(OAc)_4$]-catalyzed intramolecular QC C(sp^3)–H insertion using *o*-diazonaphthoquinone as the QC source, leading to a tricyclic product in 40% yield (Scheme 9.15A) [40]. 1,2-Naphthalenediol was also formed in 42% yield via intermolecular O–H insertion of residual water in solvent (see Section 9.3.4), which could be effectively suppressed by the addition of 4 Å MS. This reaction was adopted as a key step in the total synthesis of eleuthoside A, a natural cytotoxic product extracted from an Indonesian medicinal plant [41]. *o*-Diazonaphthoquinone was also reported to undergo a [$Rh_2(oct)_4$]-catalyzed intramolecular QC insertion to the secondary C—H bond of an ethyl group, leading to eleutherol in 63% yield (Scheme 9.15B).

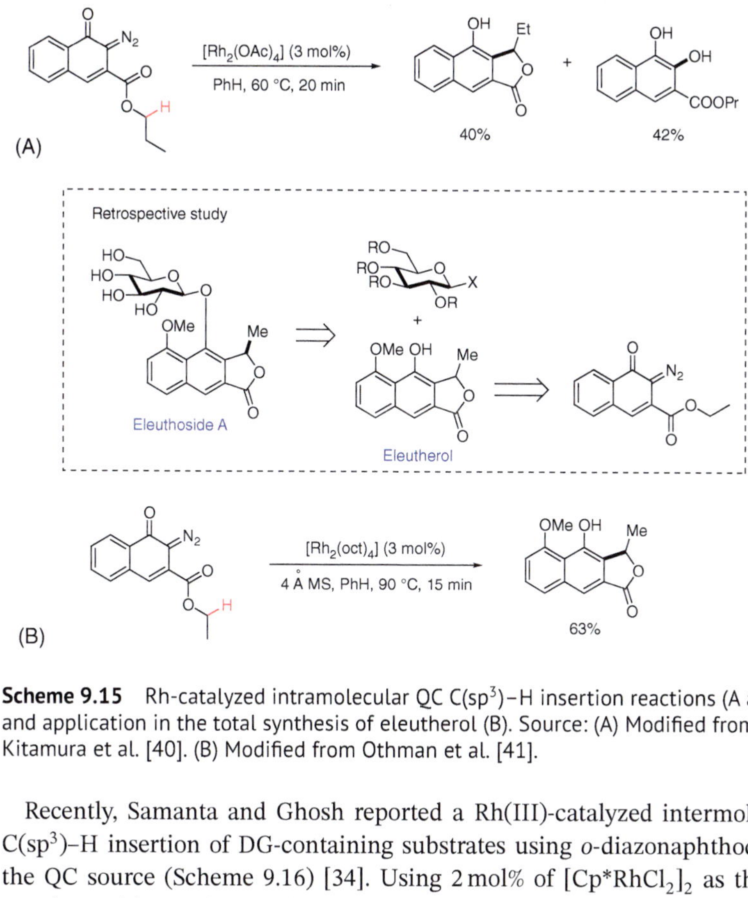

Scheme 9.15 Rh-catalyzed intramolecular QC C(sp^3)–H insertion reactions (A and B) and application in the total synthesis of eleutherol (B). Source: (A) Modified from Kitamura et al. [40]. (B) Modified from Othman et al. [41].

Recently, Samanta and Ghosh reported a Rh(III)-catalyzed intermolecular QC C(sp^3)–H insertion of DG-containing substrates using *o*-diazonaphthoquinone as the QC source (Scheme 9.16) [34]. Using 2 mol% of [Cp^*RhCl_2]$_2$ as the catalyst, together with $AgSbF_6$ and pivalic acid as co-catalysts, *o*-diazonaphthoquones underwent QC insertion to the methyl C(sp^3)—H bond of 8-methylquinolines in good to excellent yields. This QC insertion reaction showed excellent tolerance to alkene groups and cyclopropanation was not observed. A KIE of ~6.8 suggests that the C(sp^3)—H bond cleavage (i.e. cyclometallation) is the turnover-limiting step in the catalytic cycle.

Despite the success of the above two examples, the development and application studies of QC C(sp^3)–H insertion reaction still remain a great challenge. For direct

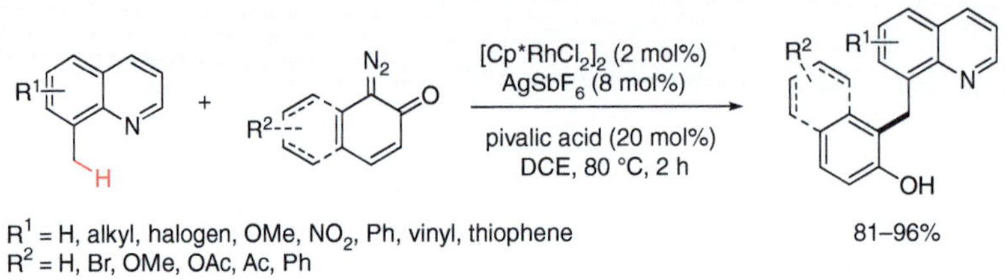

Scheme 9.16 Rh-catalyzed intermolecular QC C(sp^3)–H insertion reaction of DG-containing substrates. Source: Modified from Ghosh and Samanta [34].

electrophilic carbene C(sp^3)–H insertion pathway, QC generally exhibits decreased reactivity compared to other acceptor-type carbenes, while for DG-containing substrates, cyclometallation is intrinsically less favored for C(sp^3)—H bonds. As a result, only specific classes of substrates are applicable to these reactions. To achieve an efficient QC C(sp^3)–H insertion reaction, especially the more desirable intermolecular transformation, it is necessary to develop new catalytic systems. Inspired by the HAT reactivity of Ru–QC complexes [9], Che and coworkers recently reported an iridium(III) porphyrin-catalyzed intermolecular QC C(sp^3)–H insertion reaction that proceeds via a radical mechanism (Scheme 9.17) [20]. Using *p*-quinone diazides as the QC source, QC insertion to activated C—H bonds (such as allylic and benzylic C—H bonds, and α-C—H bonds of O and N atoms) could be achieved at reaction temperature down to room temperature and with as short as 40-minutes reaction time. The regioselectivity of this QC C(sp^3)–H insertion is mainly governed by steric effect. For example, when 1-methylcyclohexene or 3-methylcyclohexene was used as the substrate, primary C–H insertion was found to be favored over secondary/tertiary C–H insertion (Scheme 9.17A). Mechanistic studies on the reaction intermediates and elementary steps in the plausible catalytic cycle supported a mechanism that involves an initial H atom abstraction of the metal–QC intermediate with aliphatic substrate, followed by radical rebound to afford the C–H arylation product (Scheme 9.17B). While previous examples of radical mechanism in carbene C–H insertion reactions are limited to intramolecular reactions via the "metal carbene radical" intermediates (or "metalloalkyl radicals"), this reaction reported by Che and coworkers constitutes a rare example of a radical mechanism in metal-catalyzed *intermolecular* carbene C–H insertion reactions via a metal–carbene intermediate.

9.3.4 Nucleophilic Addition and Miscellaneous Reactions

There are several reports on the reaction of O-containing species (e.g. carboxylic acid, acid anhydride, water, alcohol, ether) with metal–QC intermediates generated from *o*-diazonaphthoquinones by Kitamura and coworkers [40, 42–45]. The general mechanism involves nucleophilic attack of the metal–QC intermediate by the O

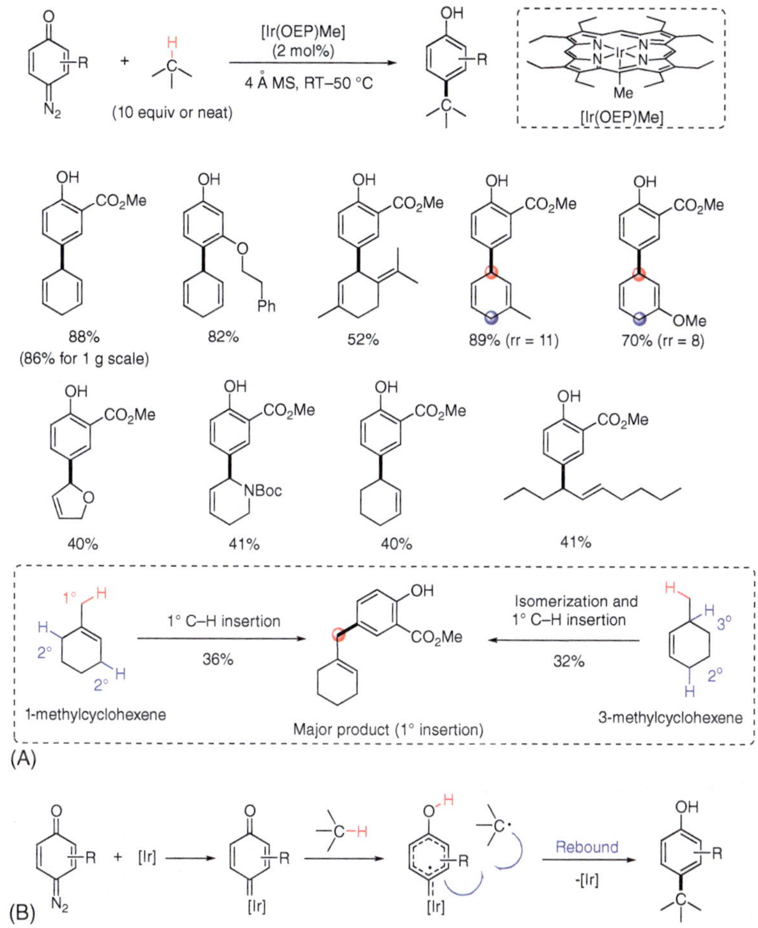

Scheme 9.17 Ir-catalyzed intermolecular QC C(sp^3)–H insertion reaction (A) which proceeds through a unique radical mechanism (B). Source: Modified from Wang et al. [20].

atom to form an oxonium ylide species, followed by aromatization and regeneration of the metal catalyst (Scheme 9.18).

Kitamura et al. reported a Pd(OAc)$_2$-catalyzed intermolecular QC O–H insertion of acetic acid using *o*-diazonaphthoquinones as the QC source, affording 1,2-naphthalenediol monoacetates in moderate to good yields (Scheme 9.19A) [42]. Extension of this reaction to a QC O–H insertion reaction with water and alcohols using [Rh$_2$(OAc)$_4$] as catalyst was reported to afford the corresponding 1,2-naphthalenediol derivatives in up to quantitative yields (Scheme 9.19B) [40]. For the QC O–H insertion reaction with propargyl alcohol, the phenoxide ion formed after aromatization of the quinone moiety was found to attack the alkyne group, giving the tricyclic products in 72–76% yields (Scheme 9.19B).

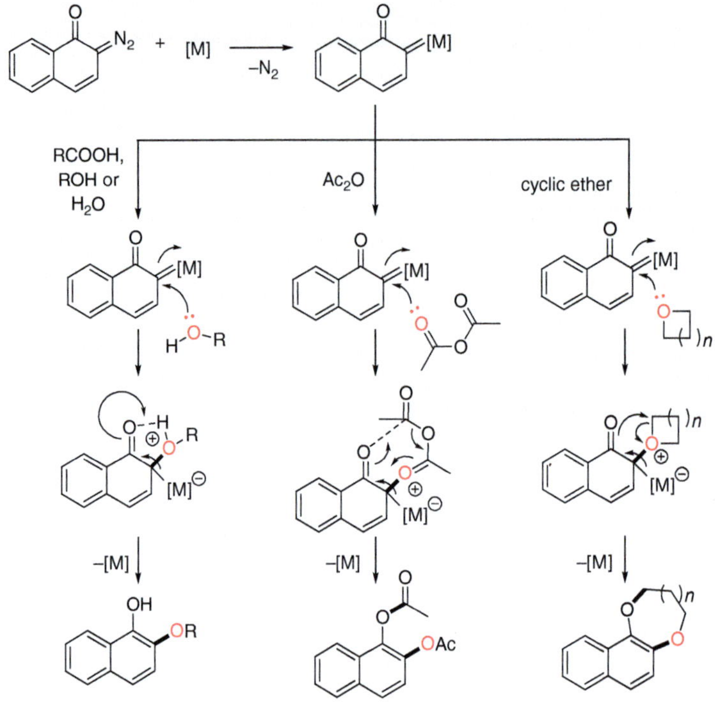

Scheme 9.18 General mechanisms for the intermolecular nucleophilic attack of metal–QC intermediates by O-containing nucleophiles.

o-Diazonaphthoquonines were reported to react with acetic anhydride in the presence of 1 mol% of $[Rh_2(OAc)_4]$ to provide 1,2-naphthalenediol diacetates in moderate to good yields (Scheme 9.20) [43]. A bis(diazo) compound, 2,4-bis(diazo)-1,2,3,4-tetrahydronaphthalene-1,3-dione, could also undergo this $[Rh_2(OAc)_4]$-catalyzed QC acetoxylation reaction with acetic anhydride, giving 1,2,3,4-tetra(acetoxy)naphthalene in 41% yield (Scheme 9.20).

The reaction of *o*-diazonaphthoquinones with small cyclic ethers gave macrocyclic ethers in the presence of 2.5 mol% of $Pd(OAc)_2$ [44]. When excess oxetane or tetrahydropyrane was used as the substrate, one *o*-diazonaphthoquinone molecule reacted with multiple cyclic ether molecules, leading to the formation of 15-membered ring structures (Scheme 9.21).

In addition to the above substrates, Kitamura et al. reported a Pd(II)-catalyzed QC reaction with carbonyl compounds using diazophenanthrenequinone as the QC source, leading to 9,10-phenanthrenedioxyacetals in 50–90% yields (Scheme 9.22) [45]. The proposed mechanism of this reaction involves nucleophilic attack of the carbonyl group to the metal–QC intermediate to give an oxonium ylide. The phenoxide ion generated from aromatization of the quinone moiety then attacks the carbonyl moiety to form the desired acetal via 5-*exo-tet* cyclization (Scheme 9.22).

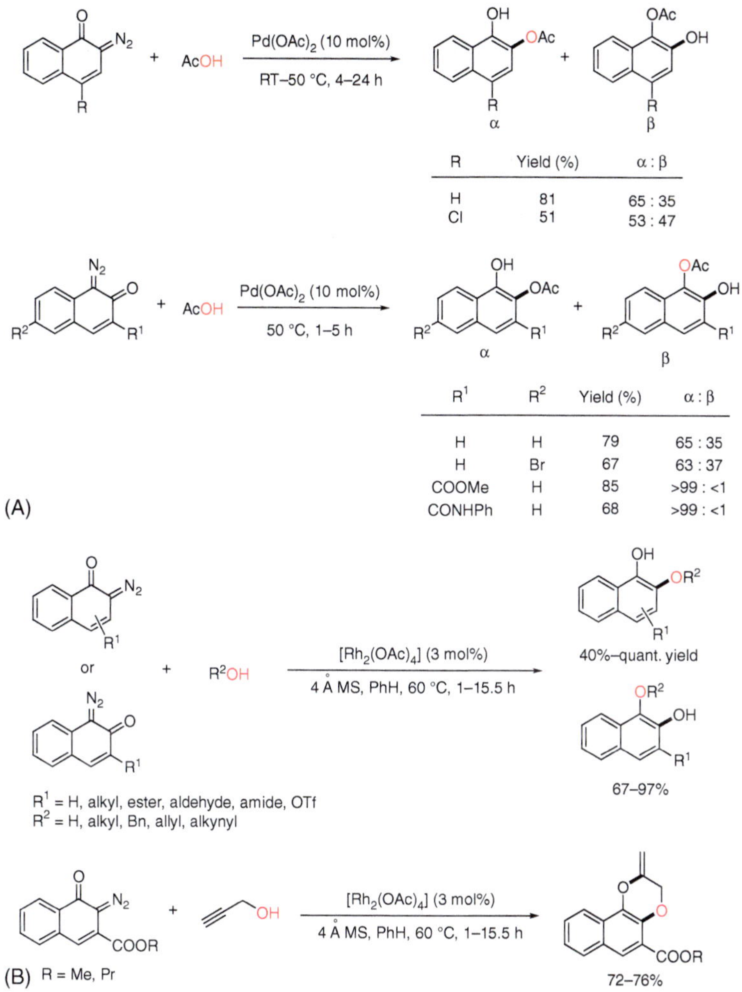

R	Yield (%)	α : β
H	81	65 : 35
Cl	51	53 : 47

R^1	R^2	Yield (%)	α : β
H	H	79	65 : 35
H	Br	67	63 : 37
COOMe	H	85	>99 : <1
CONHPh	H	68	>99 : <1

Scheme 9.19 Pd-catalyzed intermolecular reaction between *o*-diazonaphthoquinones and acetic acid (A) and Rh-catalyzed intermolecular reaction between *o*-diazonaphthoquinones and alcohols (B). Source: (A) Modified from Kitamura et al. [42]. (B) Modified from Kitamura et al. [40].

Kitamura et al. reported the Rh-catalyzed cycloaddition of *o*-diazonaphthoquinones with enol ethers, giving dihydronaphthofurans in up to 82% yields [46]. Subsequent addition of tetrabutylammonium fluoride (TBAF) led to ring opening of the dihydronaphthofurans to give α-naphthyl esters in moderate to excellent yields (Scheme 9.23A). A mechanism involving nucleophilic attack of the alkene moiety of enol ether to the metal–QC intermediate followed by aromatization was proposed. This reaction was applied to the synthesis of the BCD ring fragment of kosinostatin aglycone, a class of natural products with antibiotic and cytotoxic activities (Scheme 9.23B) [47].

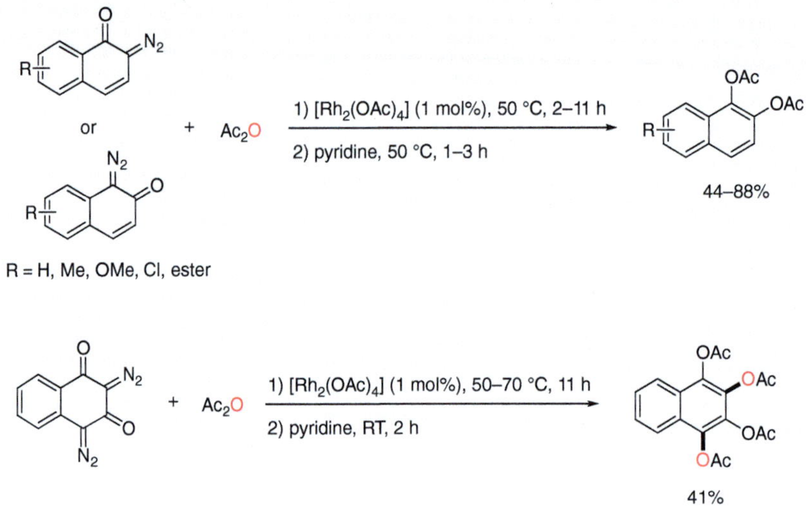

Scheme 9.20 Rh-catalyzed intermolecular QC acetoxylation reactions with acetic anhydride. Source: Modified from Kitamura et al. [43].

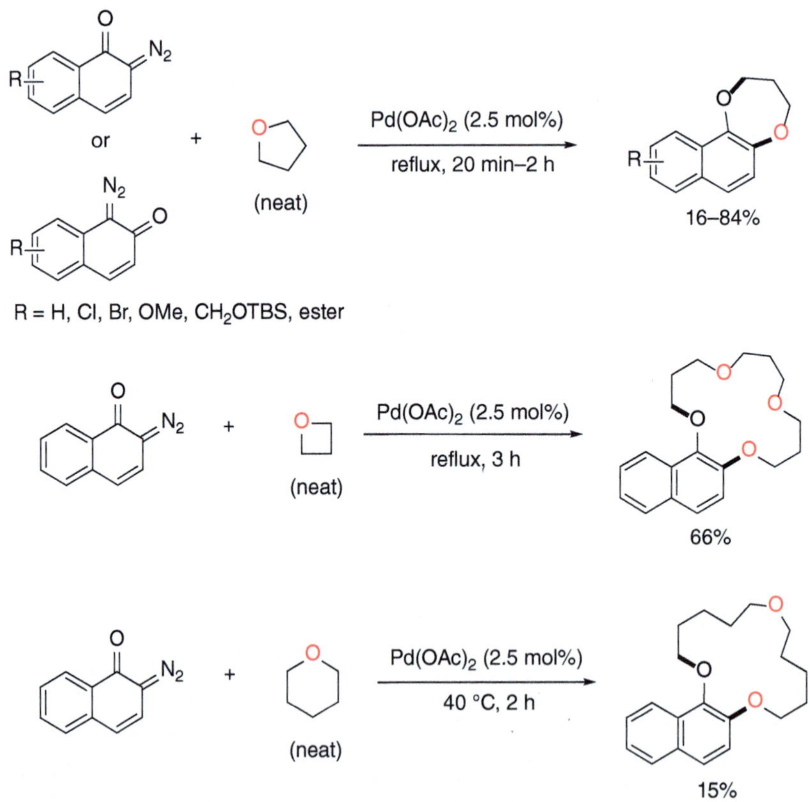

Scheme 9.21 Pd-catalyzed intermolecular reaction between *o*-diazonaphthoquinones and cyclic ethers followed by macrocyclization. Source: Kitamura et al. [44].

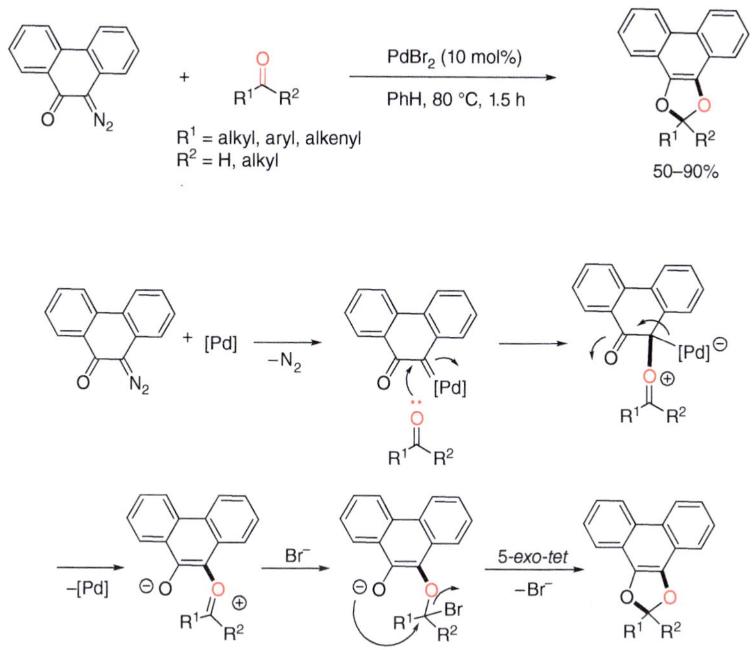

Scheme 9.22 Pd-catalyzed intermolecular reaction of ketones with metal–QC intermediates followed by cyclization. Source: Modified from Kitamura et al. [45].

In addition to the reactions of the metal–QC intermediate with O- and C-containing nucleophiles, reaction with halides has also been explored. Lee and coworkers reported the Rh-catalyzed reaction of *o*-diazonaphthoquinones with benzyl halides or acid halides, forming functionalized halonaphthalenyl ethers and esters in moderate to good yields (Scheme 9.24) [48].

During their study on Ru–QC complexes, Che and coworkers discovered that Ru(II) porphyrins can catalyze the reaction between various *p*-quinone diazides and nitrosoarenes to give nitrone products in moderate yields (Scheme 9.25) [9].

Kitamura et al. reported that in the presence of 10 mol% of $Pd(OAc)_2$, *o*-diazonaphthoquinones underwent cross-coupling with arylboronic acids to afford 2-aryl-1-naphthols in moderate to good yields (Scheme 9.26) [49].

Zhou and coworkers reported the intermolecular reaction of *p*-quinone diazides and allyl sulfides catalyzed by $[Rh_2(esp)_2]$ (Scheme 9.27) [50]. Depending on the substituents on the allyl sulfides, the sulfide ylides generated from the nucleophilic attack of allyl sulfides to the metal–QC intermediates could then undergo (i) chemoselective tautomerization/[2,3]-sigmatropic rearrangement reaction (for electron-withdrawing groups); (ii) Doyle–Kirmse rearrangement/Cope rearrangement reaction (for aryl substituents); or (iii) Doyle–Kirmse rearrangement/elimination reaction (for alkyl substituents), leading to alkenyl sulfides or multi-substituted phenols in moderate to high yields. Under similar reaction

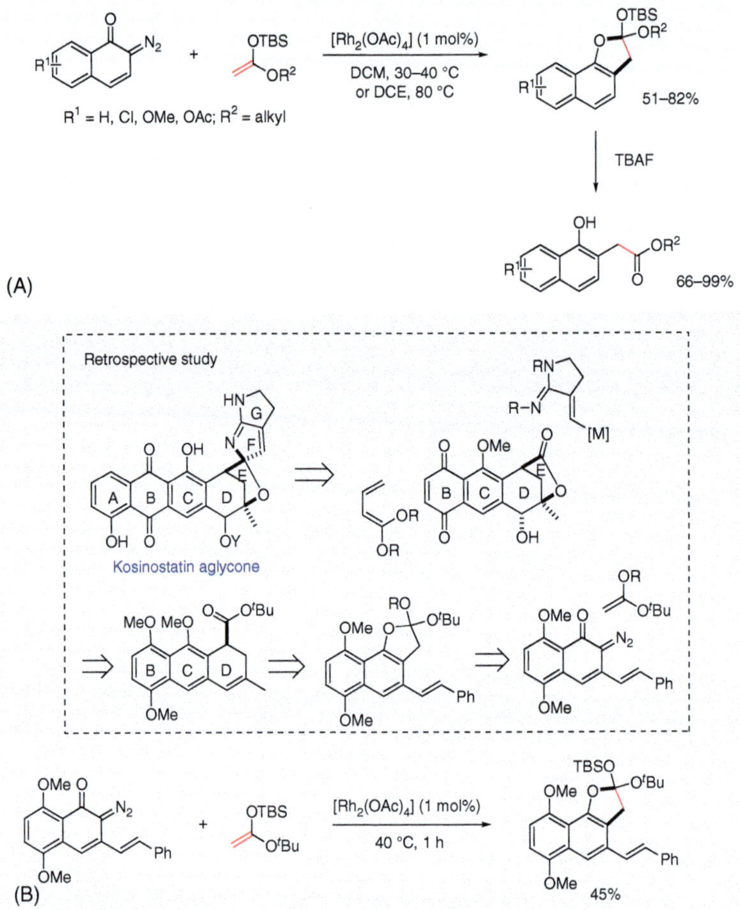

Scheme 9.23 Rh-catalyzed intermolecular reaction between *o*-diazonaphthoquinones and silyl enol ethers (A) and its application in the total synthesis of kosinostatin aglycone (B). Source: (A) Modified from Kitamura et al. [46]. (B) Modified from Kitamura et al. [47].

conditions, propargyl sulfides reacted with *p*-quinone diazides affording the corresponding allenyl sulfides in moderate to high yields (Scheme 9.27).

Recently, Ye, Lu, and coworkers reported an example of QC transfer reaction without the use of quinone diazides [51]. In their work, Au(I)–QC intermediates were generated via oxidative cyclization of terminal diynes with one molecule of quinoline *N*-oxide, while another quinoline *N*-oxide molecule oxidized the Au–QC species to form *p*-quinone products (Scheme 9.28). Although the reaction between a terminal alkyne and oxidant is a common route to α-carbonyl carbene intermediates with soft metal ions, such as Au(I), the authors proposed that using 1,2-diyne as the substrate could lead to the formation of Au–QC intermediates via a carbene/alkyne metathesis pathway. Using a Au(I) catalyst containing a bulky BrettPhos ligand, various *p*-naphthoquinones, carbazole-*p*-quinones, and indole-*p*-quinones were produced in moderate to high yields under mild reaction

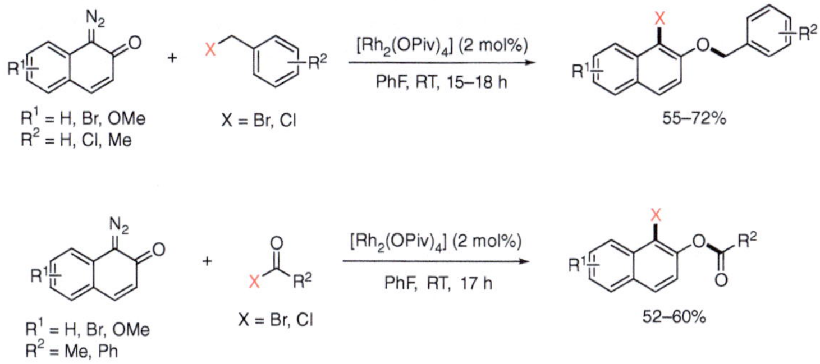

Scheme 9.24 Rh-catalyzed intermolecular reactions between *o*-diazonaphthoquinones and alkyl/acyl halides. Source: Modified from Baral et al. [48].

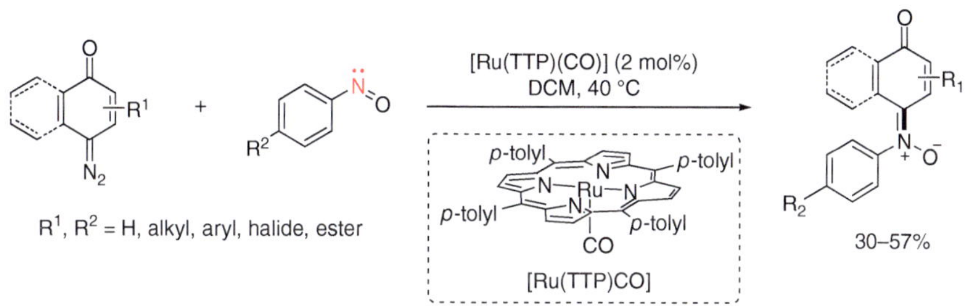

Scheme 9.25 Ru-catalyzed intermolecular reaction between QC and nitrosobenzenes. Source: Modified from Wang et al. [9].

Scheme 9.26 Pd-catalyzed intermolecular QC arylation reactions with arylboronic acids. Source: Modified from Kitamura et al. [49].

conditions. The presence of the plausible Au–QC intermediate was corroborated by control experiments using halide sources as well as by DFT calculations.

9.4 Conclusion

The versatility of metal-catalyzed carbene transfer reactions has motivated chemists to explore new carbene functionalities and increase the breadth of metal carbene chemistry. In these endeavors, the past decade has witnessed the rapid development

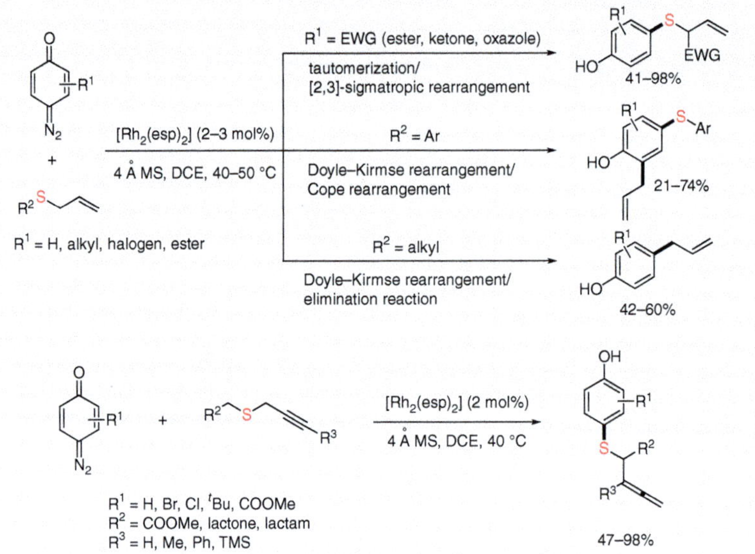

Scheme 9.27 Rh-catalyzed intermolecular reactions between QC and allylic/propargylic sulfides and subsequent rearrangement reactions. Source: Modified from Yan et al. [50].

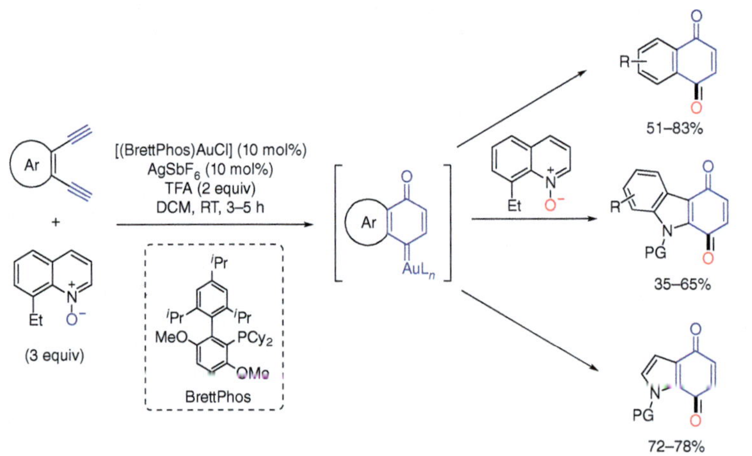

Scheme 9.28 Au-catalyzed oxidation of QC generated by oxidative cyclization of terminal diynes. Source: Modified from Shu et al. [51].

of many new carbene types, including enol carbene [52], donor-type carbene [53], and QC. While initially deemed a novel arylating reagent and a subclass of acceptor-type carbene, metal–QC species generally display diminished electrophilic carbene reactivities in cyclopropanation and C–H insertion reactions when compared to the conventional acceptor-type carbenes. Nonetheless, the unique redox reactivity of metal–QCs, as observed in both stoichiometric reactions of isolated metal–QC complexes [7–9] (Section 9.2) and the Ir-catalyzed C(sp^3)–H insertion reaction (Section 9.3.3) [20], is unprecedented compared to other metal–carbene species. The new properties/reactivities of QC are fundamentally rooted in the structural and electronic features of metal–QC complexes, the key intermediates in catalytic QC transfer reactions. Similar knowledge can be learned from the recently developed metal–dialkylcarbene chemistry: while dialkylcarbene groups are generally perceived to be highly reactive in carbene transfer and other reactions [54–58], recent studies showed that a dialkylcarbene ligand bearing a diamondoid skeleton can confer remarkable stability to a series of metal–carbene complexes and even act as a robust ancillary ligand in promoting challenging catalytic reactions [15], and these unexpected features can be rationalized by the electron-donating property and strong *trans* effect of the dialkylcarbene ligand. Therefore, when studying new types of carbenes bearing novel carbene substituents, it is of great importance to consider the carbene substituent effect since it may have a dramatic impact on the properties/reactivities of the metal–carbene species.

Acknowledgment

This work is supported by grants from the Innovation and Technology Commission (HKSAR, China) to the State Key Laboratory of Synthetic Chemistry, The University of Hong Kong, Hong Kong Research Grants Council (17304019 and 17302020) and Shenzhen Science and Technology Innovation Commission (JCYJ20170412140251576). We also acknowledge the funding support from "Laboratory for Synthetic Chemistry and Chemical Biology" under the Health@InnoHK Program launched by Innovation and Technology Commission, The Government of Hong Kong Special Administrative Region of the People's Republic of China.

References

1 Othman, D.I. and Kitamura, M. (2016). *Heterocycles* 92: 1761.
2 Wang, H.-X., Wu, K., and Che, C.-M. (2021). *Synlett* 32: 249.
3 Dao, H.T. and Baran, P.S. (2014). *Angew. Chem. Int. Ed.* 53: 14382.
4 Ford, A., Miel, H., Ring, A. et al. (2015). *Chem. Rev.* 115: 9981.
5 Griesbeck, A.G. and Zimmermann, E. (2006). Product class 11: quinone diazides. In: *Category 4, Compounds with Two Carbon Heteroatom Bonds*, 1e, vol. 28 (ed. A.G. Griesbeck). Stuttgart: Georg Thieme Verlag.
6 Sundberg, R.J., Baxter, E.W., Pitts, W.J. et al. (1988). *Org. Chem.* 53: 5097.
7 Ashkenazi, N., Vigalok, A., Parthiban, S. et al. (2000). *J. Am. Chem. Soc.* 122: 8797.
8 Dauth, A., Gellrich, U., Diskin-Posner, Y. et al. (2017). *J. Am. Chem. Soc.* 139: 2799.
9 Wang, H.-X., Wan, Q., Wu, K. et al. (2019). *J. Am. Chem. Soc.* 141: 9027.

10 Galardon, E., Maux, P.L., Toupet, L., and Simonneaux, G. (1998). *Organometallics* 17: 565.
11 Che, C.-M., Huang, J.-S., Lee, F.-W. et al. (2001). *J. Am. Chem. Soc.* 123: 4119.
12 Li, Y., Huang, J.-S., Xu, G.-B. et al. (2004). *Chem. Eur. J.* 10: 3486.
13 Le Maux, P., Roisnel, T., Nicolas, I., and Simonneaux, G. (2008). *Organometallics* 27: 3037.
14 Deng, Q.-H., Chen, J., Huang, J.-S. et al. (2009). *Chem. Eur. J.* 15: 10707.
15 Wang, H.-X., Wan, Q., Low, K.-H. et al. (2020). *Chem. Sci.* 11: 2243.
16 Werlé, C., Goddard, R., Philipps, P. et al. (2016). *J. Am. Chem. Soc.* 138: 3797.
17 Werlé, C., Goddard, R., Philipps, P. et al. (2016). *Angew. Chem. Int. Ed.* 55: 10760.
18 Tindall, D.J., Werlé, C., Goddard, R. et al. (2018). *J. Am. Chem. Soc.* 140: 1884.
19 Shishkov, I.V., Rominger, F., and Hofmann, P. (2009). *Organometallics* 28: 1049.
20 Wang, H.-X., Richard, Y., Wan, Q. et al. (2020). *Angew. Chem. Int. Ed.* 59: 1845.
21 Sundberg, R.J. and Pitts, W.J.J. (1991). *Org. Chem.* 56: 3048.
22 Boger, D.L. and Jenkins, T.J. (1996). *J. Am. Chem. Soc.* 118: 8860.
23 Kitamura, M., Takahashi, S., and Okauchi, T. (2015). *J. Org. Chem.* 80: 8406.
24 Baral, E.R., Lee, Y.R., and Kim, S.H. (2015). *Adv. Synth. Catal.* 357: 2883.
25 Magar, K.B.S., Edison, T., and Lee, Y.R. (2017). *Eur. J. Org. Chem.* 2017: 7046.
26 Wu, K., Cao, B., Zhou, C.-Y., and Che, C.-M. (2018). *Chem. Eur. J.* 24: 4815.
27 Zhang, S.-S., Jiang, C.-Y., Wu, J.-Q. et al. (2015). *Chem. Commun.* 51: 10240.
28 Chen, R. and Cui, S. (2017). *Org. Lett.* 19: 4002.
29 Das, D., Poddar, P., Maity, S., and Samanta, R. (2017). *J. Org. Chem.* 82: 3612.
30 Jia, Z.-J., Merten, C., Gontla, R. et al. (2017). *Angew. Chem. Int. Ed.* 56: 2429.
31 Liu, Z., Wu, J.-Q., and Yang, S.-D. (2017). *Org. Lett.* 19: 5434.
32 Ghosh, B., Biswas, A., Chakraborty, S., and Samanta, R. (2018). *Chem. Asian J.* 13: 2388.
33 Jang, Y.-S., Woźniak, Ł., Pedroni, J., and Cramer, N. (2018). *Angew. Chem. Int. Ed.* 57: 12901.
34 Ghosh, B. and Samanta, R. (2019). *Chem. Commun.* 55: 6886.
35 Li, X., Wang, J., Xie, X. et al. (2020). *Chem. Commun.* 56: 3441.
36 Shaaban, S., Li, H., Otte, F. et al. (2020). *Org. Lett.* 22: 9199.
37 Han, X., Kong, L., Feng, J., and Li, X. (2020). *Chem. Commun.* 56: 5528.
38 Kong, L., Han, X., Liu, S. et al. (2020). *Angew. Chem. Int. Ed.* 59: 7188.
39 Sawada, T., Fuerst, D.E., and Wood, J.L. (2003). *Tetrahedron Lett.* 44: 4919.
40 Kitamura, M., Otsuka, K., Takahashi, S., and Okauchi, T. (2017). *Tetrahedron Lett.* 58: 3508.
41 Othman, D.I.A., Otsuka, K., Takahashi, S. et al. (2018). *Synlett* 29: 457.
42 Kitamura, M., Kisanuki, M., Sakata, R., and Okauchi, T. (2011). *Chem. Lett.* 40: 1129.
43 Kitamura, M., Kisanuki, M., and Okauchi, T. (2012). *Eur. J. Org. Chem.* 2012: 905.
44 Kitamura, M., Kisanuki, M., Kanemura, K., and Okauchi, T. (2014). *Org. Lett.* 16: 1554.

45 Kitamura, M., Fujimura, R., Nishimura, T. et al. (2020). *Eur. J. Org. Chem.* 2020: 5319.

46 Kitamura, M., Araki, K., Matsuzaki, H., and Okauchi, T. (2013). *Eur. J. Org. Chem.* 2013: 5045.

47 Kitamura, M., Kubo, K., Yoshinaga, S. et al. (2014). *Tetrahedron Lett.* 55: 1653.

48 Baral, E.R., Lee, Y.R., Kim, S.H., and Wee, Y.-J. (2016). *Synthesis* 48: 579.

49 Kitamura, M., Sakata, R., and Okauchi, T. (2011). *Tetrahedron Lett.* 52: 1931.

50 Yan, S., Rao, J., and Zhou, C.-Y. (2020). *Org. Lett.* 22: 9091.

51 Shu, C., Shi, C.-Y., Sun, Q. et al. (2019). *ACS Catal.* 9: 1019.

52 Xu, X. and Doyle, M.P. (2014). *Acc. Chem. Res.* 47: 1396.

53 Zhu, D., Chen, L., Fan, H. et al. (2020). *Chem. Soc. Rev.* 49: 908.

54 Reddy, A.R., Zhou, C.-Y., Guo, Z. et al. (2014). *Angew. Chem. Int. Ed.* 53: 14175.

55 Le, P.Q. and May, J.A. (2015). *J. Am. Chem. Soc.* 137: 12219.

56 Reddy, A.R., Hao, F., Wu, K. et al. (2016). *Angew. Chem. Int. Ed.* 55: 1810.

57 Wang, H., Zhou, C.-Y., and Che, C.-M. (2017). *Adv. Synth. Catal.* 359: 2253.

58 Hao, F., Reddy, A.R., Zhou, C.-Y., and Che, C.-M. (2018). *Adv. Synth. Catal.* 360: 1433.

10

Asymmetric Rearrangement and Insertion Reactions with Metal–Carbenoids Promoted by Chiral *N,N′*-Dioxide or Guanidine-Based Catalysts

Xiaobin Lin, Xiaohua Liu, and Xiaoming Feng

Sichuan University, College of Chemistry, Chengdu, 610064 China

10.1 Introduction

Carbene is a class of highly reactive molecules containing divalent carbon atoms – that is, carbon atoms that utilize only two of the four bonds that are capable of forming with other atoms. Carbene species have attracted substantial attention due to their high reactivity and diversity in organic synthesis [1–5]. The free carbenes, generated by photolysis or pyrolysis, exhibit limited synthetic applications due to the lack of selectivity in most chemical transformations. Among the various synthetic approaches to carbene species, the decomposition of carbene precursors under the assistance of transition metals represents one of the most attractive methods because of its powerful ability to rapidly form various metal–carbenoids. The metal–carbenoids could undergo plenty of conversions with high reactivity and selectivity, such as X–H (X = C, N, O, S, etc) insertion, and ylide formation reactions with a range of substrates [1–5].

During the past two decades, the chiral *N,N′*-dioxide/metal complexes and guanidine-based catalysts developed by our group have shown excellent performance on asymmetric synthesis and catalysis [6–13]. In this chapter, we summarized in detail our efforts in asymmetric rearrangement and insertion reactions of metal-carbenoids mediated by chiral *N,N′*-dioxide/metal complexes and bifunctional guanidine-based catalytic system. Moreover, representative examples reported by other groups will also be discussed. To keep this survey manageable, this chapter is organized by the catalyst and reaction type.

10.2 The Introduction of Chiral *N,N′*-Dioxide/Metal Complexes and Guanidine Catalysts

Chiral ligands occupy a preeminent position in asymmetric catalysis [14]. In the past two decades, we have developed several types of amino acid-derived chiral

Transition Metal-Catalyzed Carbene Transformations, First Edition.
Edited by Jianbo Wang, Chi-Ming Che, and Michael P. Doyle.

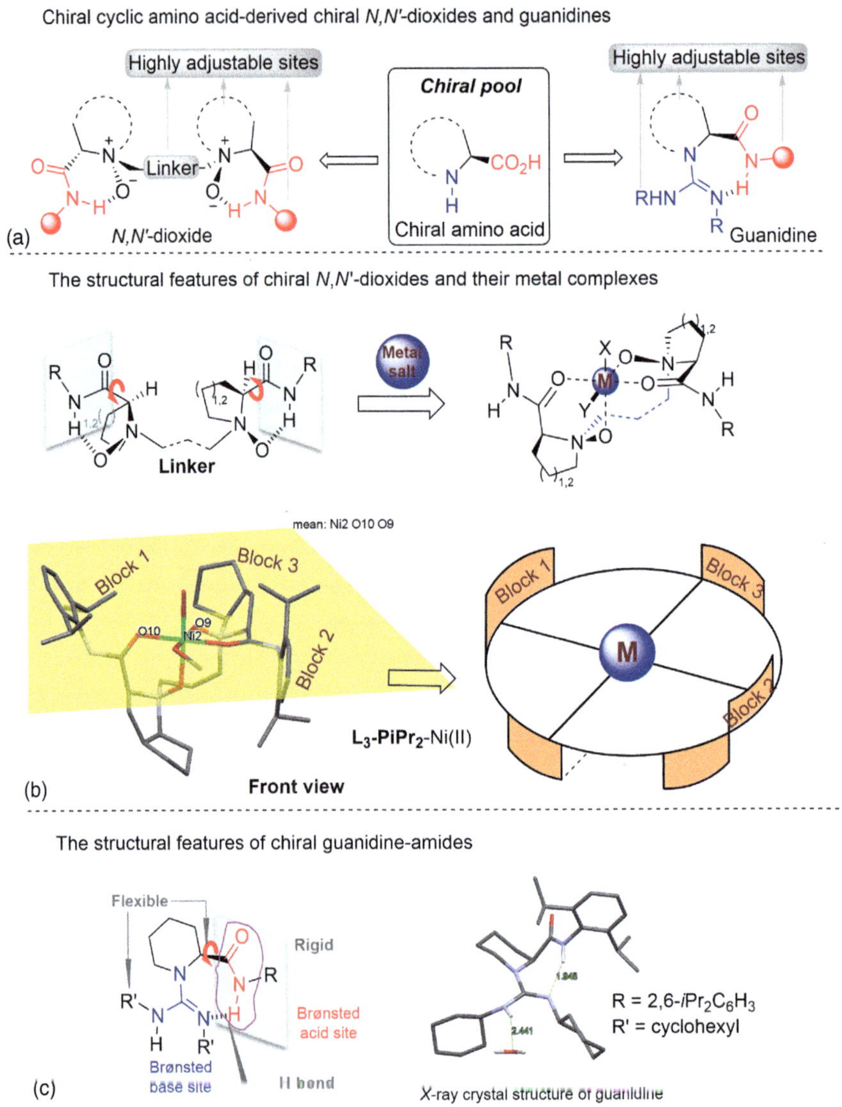

Figure 10.1 The structural features of chiral *N,N'*-dioxide/metal complexes and guanidines. (a) Chiral cyclic amino acid-derived chiral *N,N'*-dioxides and guanidines. (b) The structural features of chiral *N,N'*-dioxides and their metal complexes. (c) The structural features of chiral guanidine-amides.

ligands and organocatalysts that are particularly efficient in various asymmetric transformations, including chiral C_2-symmetric *N,N'*-dioxide amides, guanidine-amides, and others [12]. They can be straightforwardly synthesized from readily available cyclic amino acids and amines (Figure 10.1a). Modulation of chiral amino acid skeletons, amines, or other units enabled a library of chiral *N,N'*-dioxides and guanidines (for structures of chiral *N,N'*-dioxides and guanidines used in this chapter, see Figure 10.2). As illustrated in Figure 10.1b, unlike other classic "privileged" chiral ligands with rigid frameworks, chiral *N,N'*-dioxide amides are

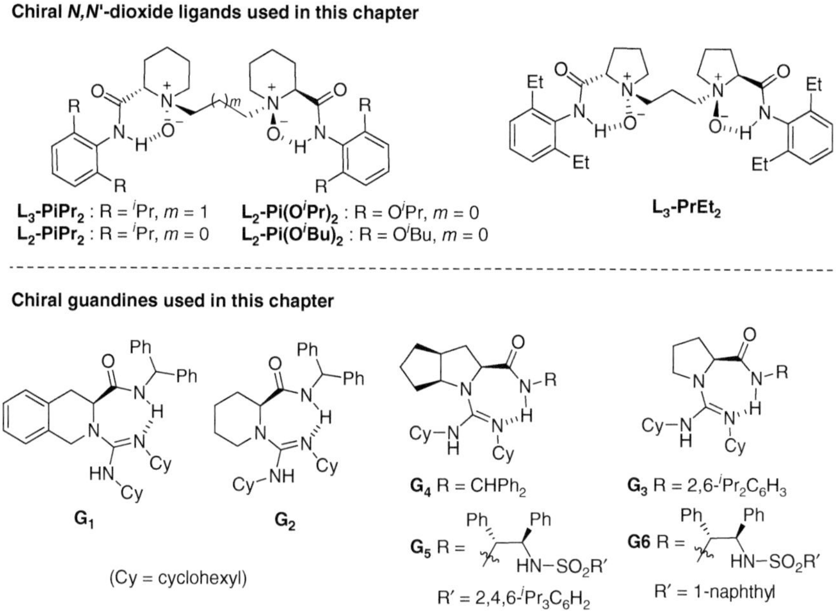

Figure 10.2 Chiral *N,N′*-dioxides and guanidines used in this chapter.

full of conformationally flexible subunits. Nevertheless, this type of compounds are found to serve as powerful neutral tetradentate oxygen ligands to coordinate more than 20 types of metal ions, ranging from main group metals to transition metals, and rare earth metals. Taking the chiral *N,N′*-dioxide/Ni(II) complex for example, X-ray crystal structure analysis indicated a slightly distorted octahedral structure was formed, wherein the metal Ni^{2+} located in the core and the four kinds of quadrants of the chiral structure are clearly discriminated spatially (Figure 10.1b). The chiral *N,N′*-dioxide/metal complexes were proved to be versatile and exhibited excellent catalytic efficiency in more than 50 asymmetric transformations, including enantioselective Michael additions, cycloadditions, rearrangement reactions, and so on. The more detailed discussions on their synthesis, coordination chemistry, and some applications have been illustrated in several reviews [6–12].

Guanidine is a unique moiety in nature, and structurally rigid mono- and bicyclic chiral guanidines have been widely used as superbases in asymmetric catalysis. We developed a type of chiral guanidine-amide compounds that are characterized by an open-chain CN_3 structure with a combination of readily modified amide units. X-ray diffraction analysis of guanidine-amide showed that hydrogen-bonds exist between the nitrogen atoms of the guanidine unit and the amide unit or exogenous water (Figure 10.1c). It confirms that these molecules meet the requirement of bifunctional catalysts, and the conformation of the acyclic guanidine is under control of the near hydrogen-bond donor. The poly-nitrogen-containing structure of guanidine-amide makes it a strong hydrogen-bonding donor and acceptor as well as an ingenious ligand to various metals. The asymmetric synthesis and catalysis with the guanidine-amide catalysts have also been summarized by our group [12, 13].

10.3 Chiral *N*,*N*′-Dioxide/Metal Complexes-Catalyzed Rearrangement Reactions

Rearrangement reactions have undoubtedly experienced a great boost in organic chemistry during the past decades, due to their pronounced ability in C—C bond and C—X (X = N, O, S, I, etc) bond reorganizations. Among them, the catalytic asymmetric [2,3]- and [3,3]-sigmatropic rearrangements of sulfonium ylides are effective synthetic methods for the sophisticated and optically active sulfur-containing molecules. In recent decades, with the development of chiral technology, several catalytic asymmetric [2,3]-sigmatropic rearrangement reactions of sulfonium ylides have been achieved by multifarious chiral catalyst systems, among which the chiral bisoxazoline–copper(I) complex and the chiral dirhodium(II) complex are two most frequently employed metal catalysts. The great achievements in [2,3]-sigmatropic rearrangement reactions have been well documented by Wang and coworkers [15, 16], Smith and coworkers [17], Maulide and coworkers [18], and Zhu and coworkers [19] over the past 10 years.

Mechanistically, for example, the catalytic asymmetric Doyle–Kirmse reaction as a type of [2,3]-sigmatropic rearrangement is generally considered as below (Scheme 10.1). Initially, metal-carbene is generated by decomposition of diazo compound with a chiral transition metal catalyst, and then it reacts with allylic/propargylic sulfide via discrimination of the heterotopic lone pairs on sulfur to form sulfonium ylide selectively. Then, the following rearrangement of chiral sulfonium ylide proceeds spontaneously accompanied by S-to-C chirality transmission to achieve enantioselective [2,3]-σ rearrangement reaction. However, great challenges remain in this field. First, it is difficult to well discriminate the heterotopic lone pairs of sulfur with chiral metal complex to afford the enantioenriched sulfonium ylide. Second, chiral metal-bonded ylide is in equilibrium with free ylide, which is more likely to undergo the following rearrangement. In addition, if the inversion at sulfur of free non-racemic ylide is much faster than sigmatropic rearrangement,

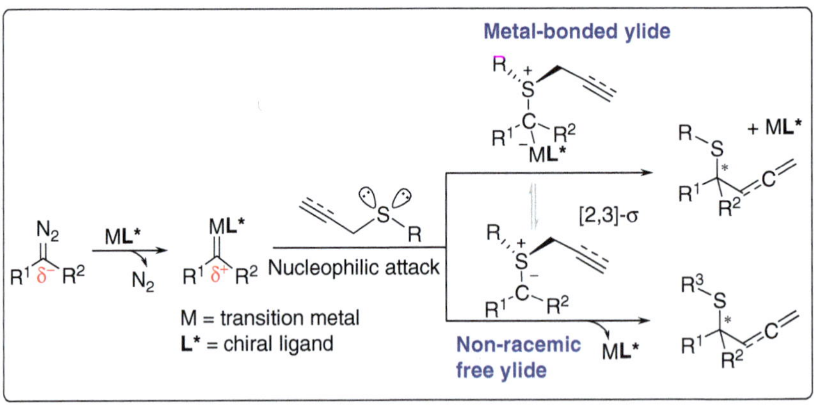

Scheme 10.1 General mechanism of asymmetric Doyle–Kirmse reaction.

the S-to-C chirality transmission is difficult to maintain, leading to deteriorated enantioselectivity.

Since the pioneering work by Uemura and coworkers in 1995 [20], studies on chiral metal complexes-mediated asymmetric Doyle–Kirmse reaction of α-diazo compounds have been widely reported, which are summarized and shown as Scheme 10.2. Highly reactive and enantioselective Doyle–Kirmse reaction is a challenge because enantio-version in each step over the entire process will contribute to the final stereoselectivity. To confront these difficulties, Wang's groups and ours have developed kinds of new substrates as well as novel catalytic models for this purpose. With these strategies, a wide range of sulfides have been successfully synthesized in high efficiency with excellent diastereo- and enantioselectivities.

In 2005, the Wang group achieved a highly diastereoselective Doyle–Kirmse reaction by employing a double asymmetric induction strategy of the combination of Oppolzer's camphor sultam auxiliary and Cu(I) catalyst with chiral or achiral Salen ligands $\mathbf{L_1}$ and $\mathbf{L_2}$ (Scheme 10.3) [27]. The chiral auxiliary of the rearranged products could be readily removed by reduction with $LiAlH_4$ to deliver the final products **4** and **7** with good overall yields and enantioselectivities (50–96% ee). Performing the reaction with (*S*,*S*)- or (*R*,*R*)-$\mathbf{L_2}$ provided the major product with the same absolute configuration, which suggested that the stereoselectivity of the reaction was determined by the chiral auxiliary rather than the chiral copper complex. It is worthy of noting that both allyl and propargyl sulfides were compatible in the catalytic system, affording plenty of enantioenriched tertiary chiral sulfides **4** and **7**.

In 2017, Wang and coworkers developed the highly enantioselective asymmetric Doyle–Kirmse reaction of trifluoromethyl allyl and propargyl sulfides **8** and **9** with diazoacetates **10**, respectively, in the presence of catalytic amount of chiral Rh(II)- or Cu(I)-based catalysts. This protocol provided an attractive and efficient method for the enantioselective construction of chiral $C(sp^3)—SCF_3$ bonds (Scheme 10.4) [21]. The lack of enantioselectivity (<6% ee) with symmetric diallyl sulfide **13** and the consistent low dr values for the reaction of cinnamyl(trifluoromethyl)sulfane catalyzed by different Rh(II) catalysts indicated that the [2,3]-sigmatropic rearrangement of sulfonium ylide proceeded via a free ylide rather than a metal-bonded ylide. The presence of trifluoromethylthio (SCF_3) functional group in the sulfides was proposed to be the key for the success of this catalytic system, since its corresponding free sulfonium ylide exhibits sufficient configurational stability, which guaranteed the complete transfer of the good facial selectivity obtained in the initial sulfonium formation step to the final product via a predicable envelope-like transition state.

α-Diazo compounds constitute the prototypical metal carbene precursors, and the structure of α-diazo compounds exhibited a profound influence on the reactivity and selectivity of carbene species [28]. Therefore, their innovation spurs the development of carbene species and chiral transition metal carbene catalysts [3, 4]. In stark contrast to previous works [20, 21, 27], in 2018, our group developed a conceptual alternative for achieving a highly enantioselective Doyle–Kirmse reaction wherein new donor–acceptor α-diazo pyrazoleamides and a novel dual-tasking chiral Ni(II) complex synergistically emerged [22]. In view of the high affinity of pyrazoleamide group to transition metal ions [29–32], the pyrazole unit was embedded into the

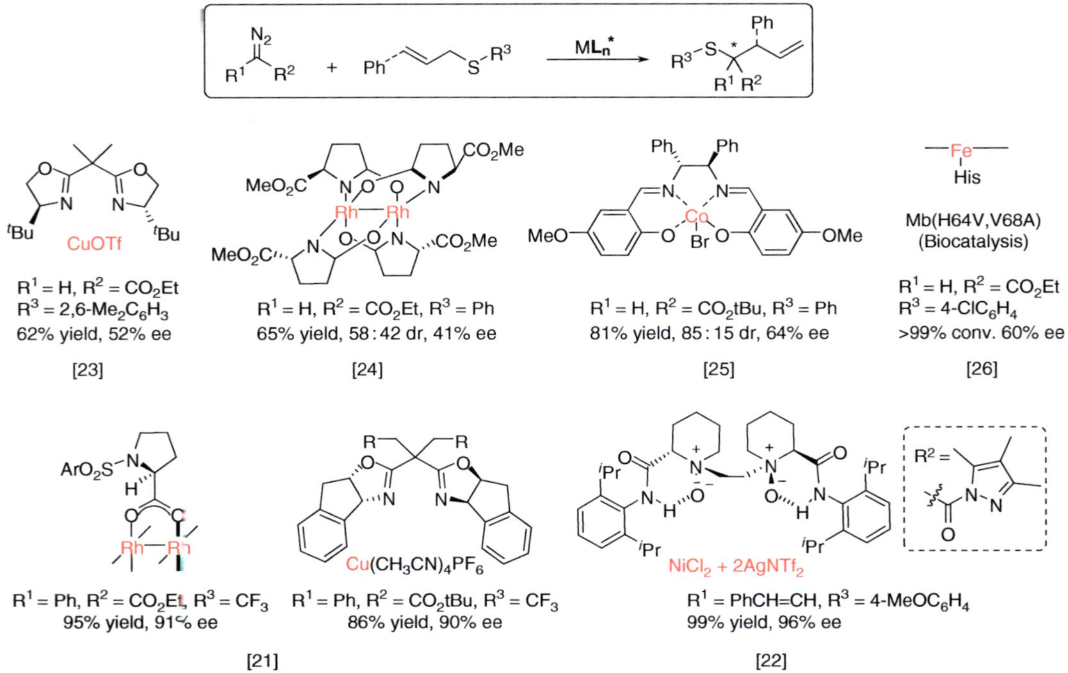

Scheme 10.2 Enantioselective Doyle–Kirmse reaction in various catalytic systems.

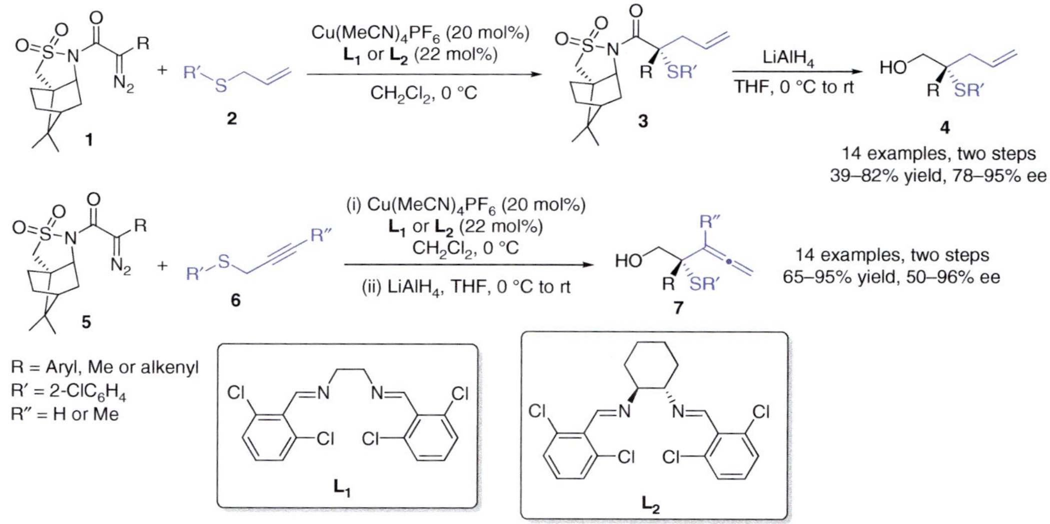

Scheme 10.3 Stereoselective Doyle–Kirmse reaction. Source: Modified from Ma et al. [27].

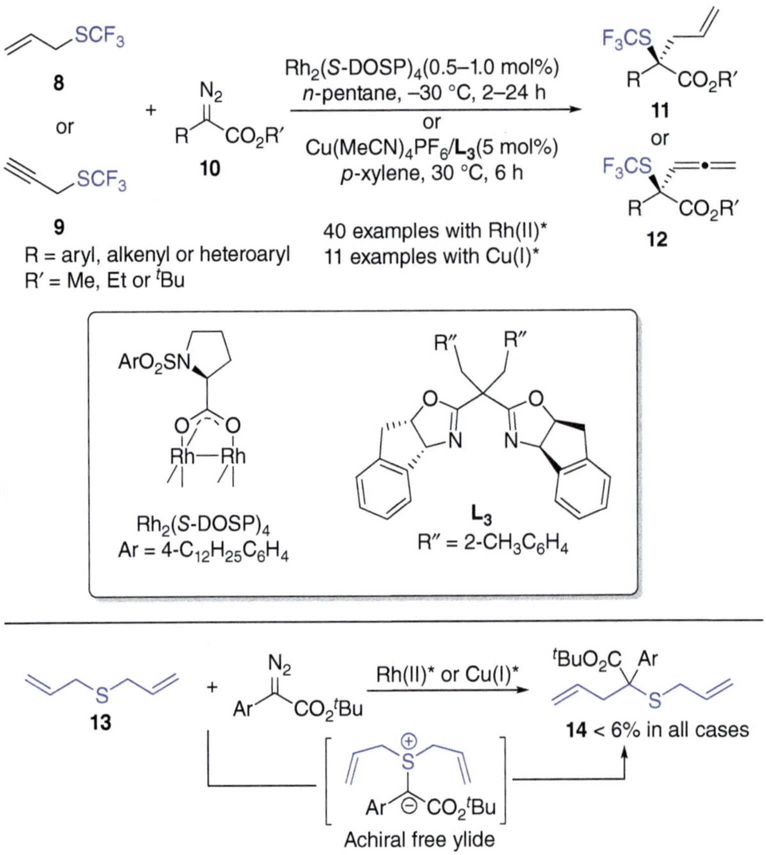

Scheme 10.4 Chiral Rh(II)- and Cu(I)-catalyzed Doyle–Kirmse reaction. Source: Modified from Zhang et al. [21].

α-diazo carbonyl compound instead of classical ester group with an aim to enhance the chiral discrimination ability in the sulfonium ylide formation step and participate in the final [2,3]-sigmatropic arrangement by coordination with chiral metal complex (Scheme 10.5).

The ^{13}C NMR (nuclear magnetic resonance) spectra of classical α-diazo compounds showed that the chemical shifts of N_2-attached carbon varied broadly, and high field shift was observed when pyrazoleamide unit was introduced instead of ester group (around 66 ppm versus 77 ppm) [33, 34]. Note: the chemical shift of diphenyldiazomethane is 62.6 ppm [35]. It indicates that the diazo pyrazoleamides might be more nucleophilic than the corresponding α-diazo esters, and its reactivity is in the middle of α-diazo ester and aryldiazoalkanes, but the stability of the related carbenoid is enhanced due to the flanked acceptor group (pyrazoleamide) and donor group (vinyl or aryl group).

As illustrated in Scheme 10.6, the chiral $\mathbf{L_2}$-$\mathbf{PiPr_2}$ ligand of Ni(II) complex proved to be an efficient catalyst after careful evaluation of different transition metals and

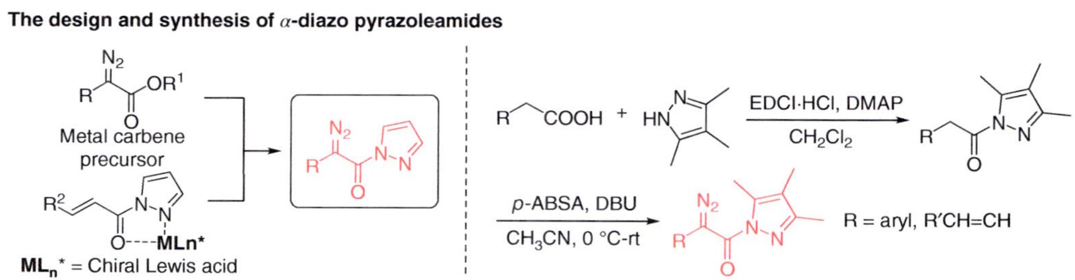

Scheme 10.5 The design and synthesis of α-diazo pyrazoleamides. Source: Modified from Lin et al. [22].

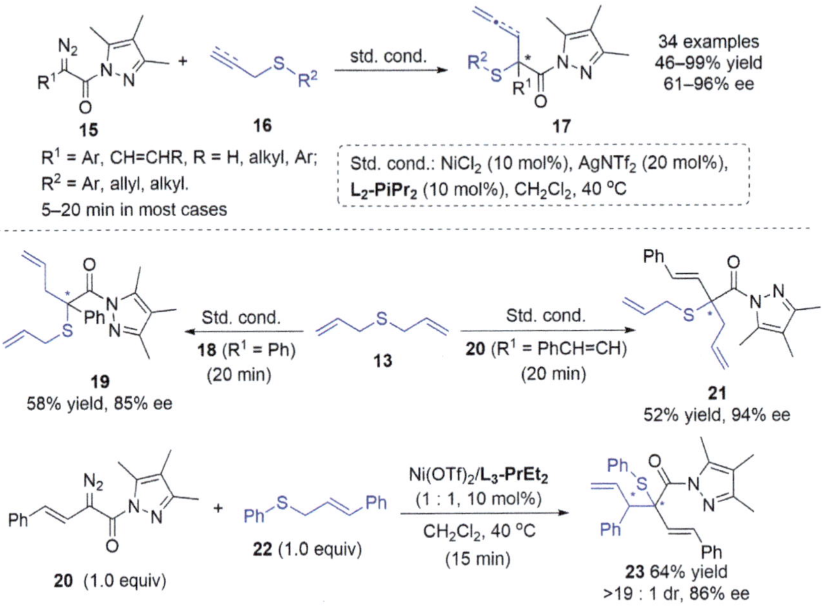

Scheme 10.6 Chiral Ni(II) complex-catalyzed asymmetric Doyle–Kirmse reaction.

chiral *N,N′*-dioxide ligands [22]. A wide range of aryl- or vinyl-substituted α-diazo pyrazoleamides **15** and sulfides **16** were well tolerated under ambient conditions, yielding a plethora of optically active sulfides **17** in high reactivity and enantioselectivity. The pyrazoleamide group of the product could be easily transformed to the corresponding ester, alcohol, and amide in high efficiency. Notably, different from previous reports [21, 27], high ee values of the corresponding products **19** and **21** of symmetric diallylic sulfide **13**, as well as excellent diastereoselectivity of the corresponding product **23** of cinnamyl(phenyl)sulfane **22**, were observed, implying that the [2,3]-sigmatropic rearrangement of sulfonium ylide might be a Lewis-acid-bonded ylide pathway in this catalytic system, rather than free ylide species. Conversely, low ee values (<6%) were observed when α-diazo pyrrolylamide and α-diazo ester were employed under the optimized reaction conditions. The results highlighted the necessity of pyrazoleamide unit and supported its coordination with chiral nickel(II) complex in the stereocontrol step.

According to the X-ray diffraction analysis of crystal structure of chiral Ni(II)/$\mathbf{L_2}$-**PiPr**$_2$ complex, the central metal nickel(II) was found to coordinate to six oxygen atoms of the ligand $\mathbf{L_2}$-**PiPr**$_2$ and solvents (tetrahydrofuran and water) as an octahedron. Asymmetric induction model was proposed based on the control experiments (Figure 10.3). Initially, chiral nickel(II) carbenoid is generated from Ni(II)/$\mathbf{L_2}$-**PiPr**$_2$ complex and α-diazo pyrazoleamide **20**. Then it reacts with allylic sulfide and the chiral catalyst is released and re-coordinated to pyrazoleamide group. The spontaneous [2,3]-sigmatropic rearrangement of the

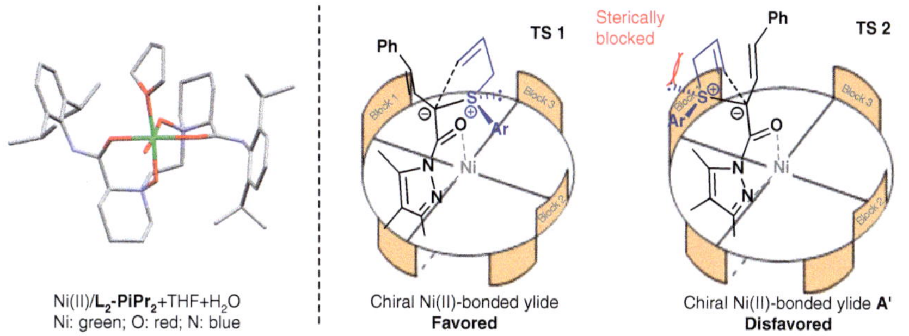

Figure 10.3 Asymmetric induction model for dual-tasking chiral Ni(II) complex-catalyzed asymmetric Doyle–Kirmse reaction. Source: Based on Lin et al. [22].

sulfonium ylide proceeds in a favored envelop-like transition state via a chiral nickel-bonded ylide intermediate, which is in analogy to a chiral auxiliary strategy (**TS 1**, Figure 10.3) [27]; thus, the dominant enantiomer was afforded with high enantioselectivity. According to this mode, high enantioselectivity of the reaction with symmetric allylic sulfide **13** and good diastereoselectivity of the reaction with cinnamyl(phenyl)sulfane **22** could be obtained as well. In this case, the chiral Ni(II)/$\mathbf{L_2}$-$\mathbf{PiPr_2}$ complex serves as a dual-tasking catalyst, which enables the generation of nickel(II) carbene species as well as controls the stereochemistry of the [2,3]-sigmatropic rearrangement as a Lewis acid.

The similar strategy was extended to the enantioselective [2,3]-sigmatropic rearrangement of allylic selenides **24** with α-diazo pyrazoleamides **15** (Scheme 10.7) [33]. In contrast to the sulfur, the element selenium possesses larger atomic radii and stronger nucleophilicity and the selenides are more air-sensitive compounds. In consideration of bond energies, the carbon–selenium bond is weaker than the carbon–sulfur bond, but the configuration stability of selenium ylide is stronger than sulfonium ylide. These raised new issues for asymmetric [2,3]-sigmatropic rearrangement of allylic selenides. Interestingly, with regard to allylic selenides **24**, the chiral Co(II) complexes of *N,N′*-dioxide $\mathbf{L_2}$-$\mathbf{Pi(O^iPr)_2}$ or $\mathbf{L_3}$-$\mathbf{PiPr_2}$ showed better performance than the corresponding chiral *N,N′*-dioxide/Ni(II) complexes in this transformation. The reaction offers an efficient methodology for the synthesis of optically active selenides **25** with high reactivity (up to 99% yield) and enantioselectivity (up to 97% ee). Experimental studies indicated the allylic selenides were more reactive than allylic sulfides regardless of the use of the chiral *N,N′*-dioxide/Co(II) or Ni(II) catalysts. When the symmetric allylic selenide **26** was adopted to react with α-diazo pyrazoleamide **18**, the corresponding product **27** was generated in 75% yield as a racemate. Thus, a relatively different catalytic model was proposed in consideration of the distinct result from the above reaction [22]. The enantioselective nucleophilic attack of allylic selenide to the chiral cobalt(II) carbenoid was considered as the enantio-determining step, giving the selenium ylide with more stable configuration, followed by a [2,3]-sigmatropic rearrangement with the chirality transmission from Se* to carbon.

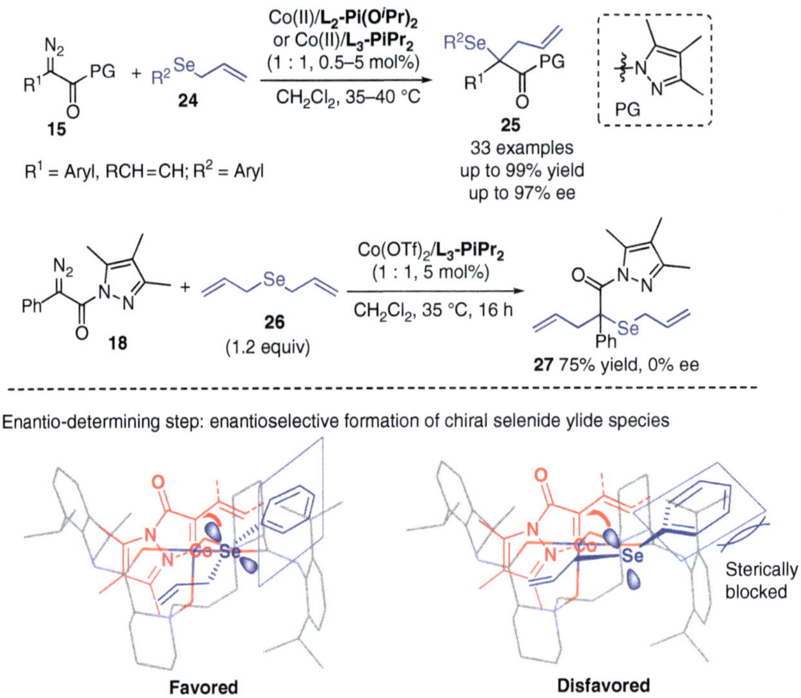

Scheme 10.7 Chiral Co(II) complexes catalyzed [2,3]-sigmatropic rearrangements of selenium ylides.

Highly asymmetric catalytic [2,3]-sigmatropic rearrangements of sulfonium ylides, including [2,3] Stevens rearrangement and Sommelet–Hauser reaction, could also be performed well by the use of α-diazo pyrazoleamides **28** and **31** in the presence of chiral *N,N′*-dioxide–metal complexes (Scheme 10.8) [36]. In comparison with the asymmetric Doyle–Kirmse reaction, the [2,3] Stevens rearrangement and Sommelet–Hauser reaction are more complicated, in which an elusive 1,3-proton transfer process is involved before the rearrangement step (Figure 10.5). Remarkably, the catalytic asymmetric Sommelet–Hauser reaction meets with difficult steps as the dearomatization and rearomatization. When Sommelet–Hauser reaction of α-(aryl)diazo esters was conducted with chiral Rh_2(*S*-DOSP)$_4$ catalyst, no enantioselectivity was observed [37]. Eventually, using chiral Ni(II)/**L$_2$-PiPr$_2$** complex as the catalyst in the presence of 20 mol% of 3-phenylpropanoic acid as an additive, the catalytic asymmetric [2,3] Stevens rearrangement of vinyl-substituted α-diazo pyrazoleamides **28** with sulfides **29** was established, affording various chiral 1,6-dicarbonyls **30** with C–S stereocenters in decent yields with high enantio- and diastereoselectivities. With the same catalyst, the enantioselective Sommelet–Hauser reaction of α-(aryl)diazo pyrazoleamides **30** with sulfides **31** was also achieved, delivering the di- or tri-substituted arenes **32** bearing a C—S bond stereocenter in moderate efficiency.

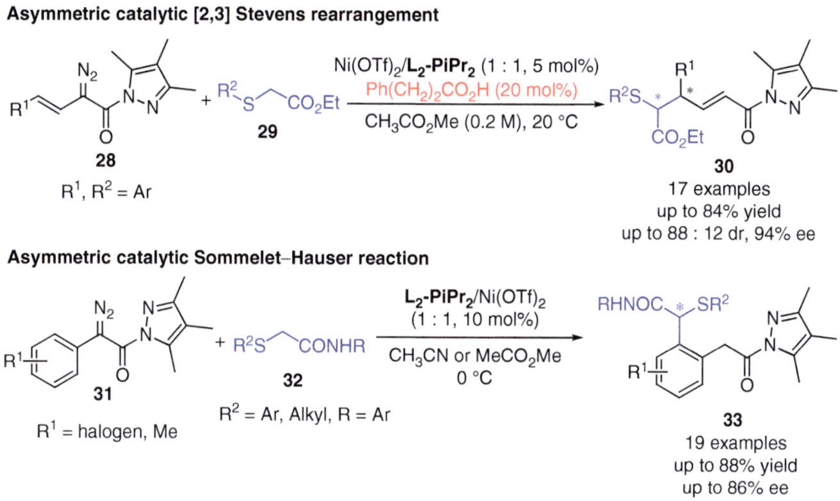

Scheme 10.8 Chiral Ni(II) complex-catalyzed [2,3] Stevens rearrangement and Sommelet–Hauser reaction. Source: Modified from Lin et al. [36].

It is noteworthy that performing the reactions under strict water-free conditions resulted in a sharp decrease in the reactivity. Moreover, the catalyst showed no enantiocontrol in the reactions of symmetric thiodiacetates. The deuterium-labeling experiments of [2,3] Stevens rearrangement and Sommelet–Hauser reaction indicated that it was an intermolecular 1,3-proton transfer rather than an intramolecular pathway, in which the carbonyl acid or water could act as a proton shuttle. In comparison with water, the carbonyl acid was considered as a milder proton shuttle, which enables the slower tautomerization process and higher enantioselectivity. The catalytic models for the formation of chiral sulfonium ylide, 1,3-proton transfer, and [2,3]-sigmatropic rearrangement process were proposed to elucidate the origin of the stereoselective control in [2,3] Stevens rearrangement and Sommelet–Hauser reaction (Figures 10.4 and 10.5). First, the trap of chiral nickel carbenoid with the sulfide afforded the chiral sulfonium ylide through a fine discrimination of the heterotopic lone pairs on sulfur to generate species **A** with a (*R*)-configured sulfur center as a priority (Figure 10.4). Then the chiral nickel(II)-bonded favorable sulfonium ylides **A1** and **A2** underwent an intermolecular 1,3-proton transfer process aided by carbonyl acid or water to generate the chiral sulfonium ylides **B1** or **B2**. Next, the chiral nickel(II)-directed [2,3]-sigmatropic rearrangement of intermediates **B1** and **B2** took place via the dominating transition states **TS1** and **TS2,** respectively, assisted by chiral ligand $\mathbf{L_2}$**-PiPr**$_\mathbf{2}$ and the *S* charity of the chiral sulfonium ylides, affording the desired sulfides with good enantiocontrol.

The Wang group realized chiral Cu(I)/bisoxazoline-catalyzed enantioselective Sommelet–Hauser rearrangement of α-diazo esters **34** with sulfides **35** (Scheme 10.9) [38]. This catalytic system can tolerate a wide range of diazo ester substrates **34** and sulfides **35**, providing an efficient strategy to access chiral sulfur-containing

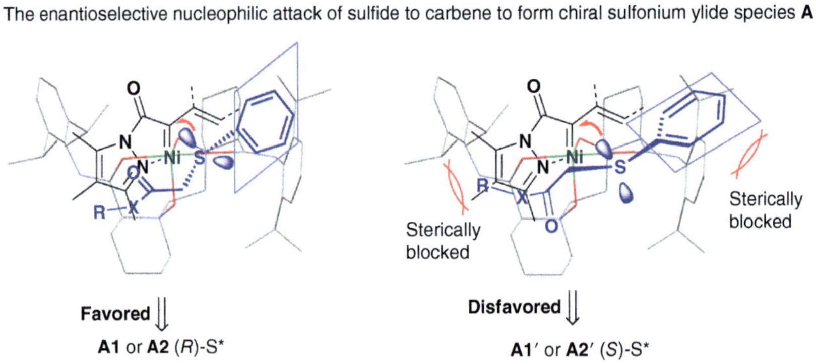

Figure 10.4 Asymmetric induction models for [2,3]-sigmatropic rearrangements.

compounds **36** bearing benzylic stereocenters. The reaction of symmetric sulfide **13** with various Cu(I)/bisoxazoline complexes provided uniformly low enantioselectivities (≤8% ee in all cases), which are indicative of a free ylide process. The results of the deuterium-labeling experiments in this catalytic system were similar with the earlier findings in the Sommelet–Hauser reaction [36], and a reversible water-assisted 1,3-proton transfer process was also put forward (Scheme 10.9). In the stereochemical model, the *R*-configuration sulfonium ylide **A1** was generated by the reaction of sulfide **35** and chiral copper(I) carbene in a favored transition state, followed by the diastereoselective proton transfer through a water-involved six-membered chair-like transition state **TS1**. The chiral-free ylide **B** then undergoes a [2,3]-sigmatropic rearrangement via an envelope-like configuration shown as **TS2** and **TS2′**. As a result, the high stereoselectivity was transferred from S* center to C center in good efficiency.

Since the seminal work by Kwart and Hackett in 1962 [39], the thio-Claisen rearrangement, as a variant of Claisen rearrangement, has been widely applied in the synthesis of numerous thiocarbonyl compounds and sulfur-containing heterocycles. However, catalytic asymmetric version of thio-Claisen rearrangement remained elusive. Recently, our group has disclosed a chiral nickel(II)-catalyzed highly enantioselective thio-Claisen rearrangement of sulfonium ylide generated from α-diazo pyrazoleamides **37** and thioindoles **38** (Scheme 10.10) [40]. The corresponding C-3-substituted indole derivatives were unstable, which were further oxidized by *m*-chloroperoxybenzoic acid (*m*-CPBA) to afford the final products **39** in high yields (two steps, up to 95%) with excellent enantioselectivities (up to 96% ee).

A feasible catalytic mechanism was proposed to rationalize the high efficiency in this transformation based on the X-ray single-crystal structure of the complex $Ni(BF_4)_2$/$\mathbf{L_2}$-$\mathbf{Pi(O^iBu)_2}$ (Figure 10.6). Analogous to the previous works [22, 33, 36], the chiral sulfonium ylide **B** was generated after the enantioselective nucleophilic attack of thioindole **38a** to the nickel carbene intermediate **A**, followed by a water-assisted diastereoselective proton transfer process controlled by the chiral Ni(II)/$\mathbf{L_2}$-$\mathbf{Pi(O^iBu)_2}$ complex as well as the asymmetric induction of chiral sulfur

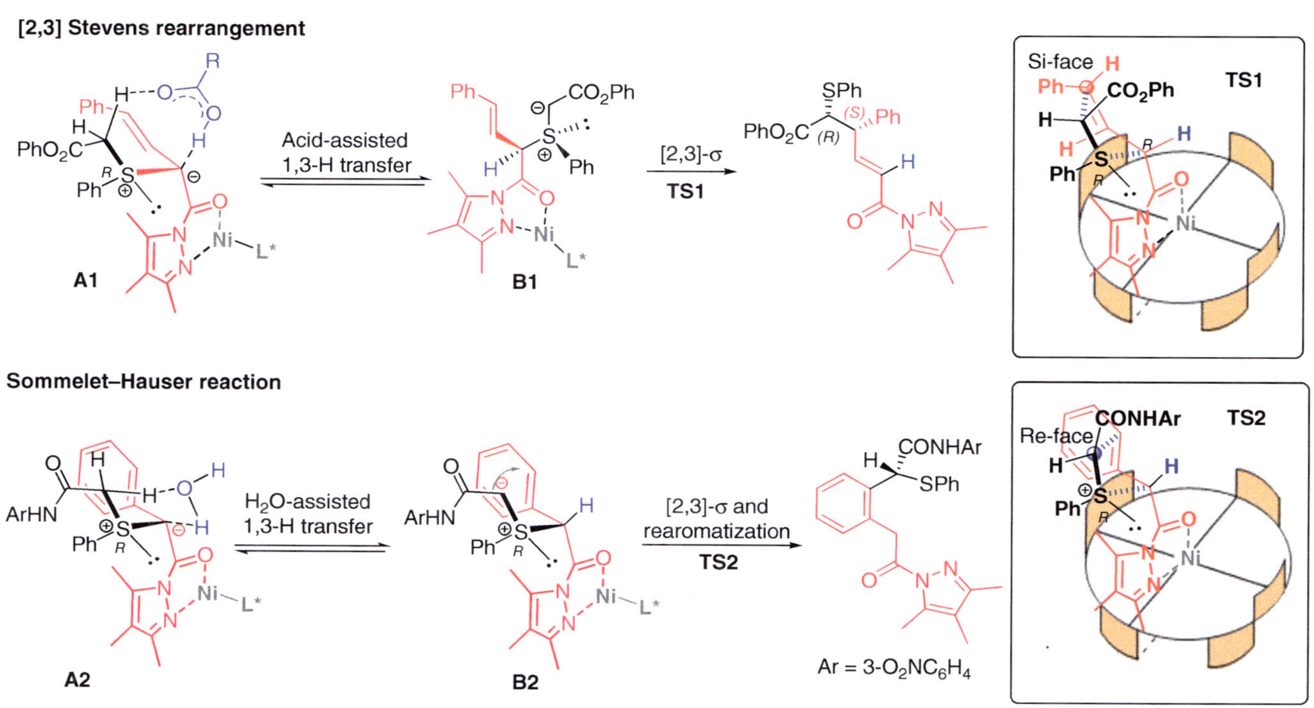

Figure 10.5 Catalytic models for [2,3] Stevens rearrangement and Sommelet–Hauser reaction.

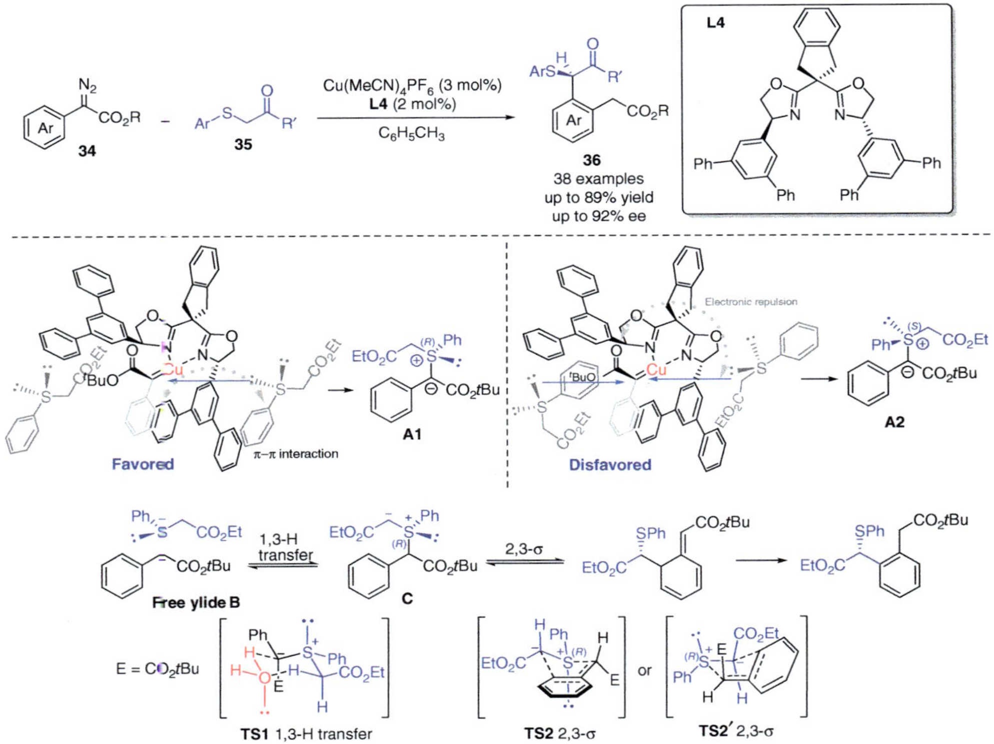

Scheme 10.9 Cu(I)/chiral bisoxazoline-catalyzed Sommelet–Hauser reaction. Source: Modified from Li and Wang [38].

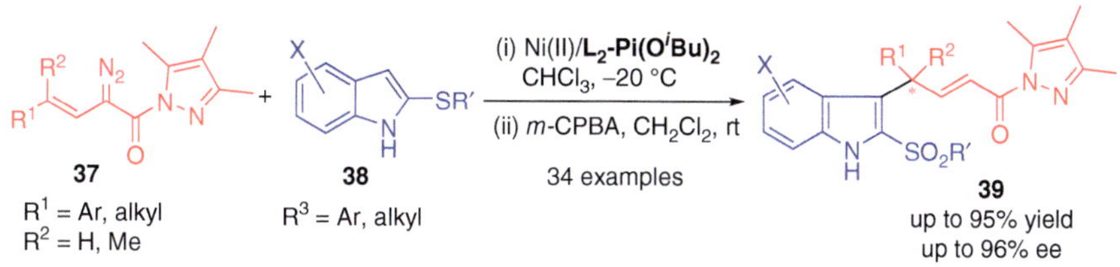

Scheme 10.10 Chiral Ni(II) complex-catalyzed thio-Claisen rearrangement. Source: Modified from Yang et al. [40].

center. Subsequently, the [3,3]-sigmatropic rearrangement of the nickel-bonded intermediate **C** took place in a favored chair conformation to produce intermediate **D**, thus leading to the (*R*,*E*)-configured product **39a** after isomerization in excellent stereoselectivity.

10.4 Chiral Guanidine-Based Catalyst-Mediated Asymmetric Carbene Insertion Reactions

The asymmetric insertion reactions of carbene species have drawn a great deal of attention due to their wide-ranging applications in organic synthesis [28, 41–45]. The related achievements have been well elaborated in many reviews [1–5, 28, 43–45]. Furthermore, catalytic X–H insertion reactions into diazo substrates (where X is any heteroatom) are a powerful yet underutilized class of transformations. The chiral transition metal complexes-mediated asymmetric carbene insertion reactions have also been generally summarized [5]. The general acceptable mechanisms of carbene insertion into diazo compounds was depicted as Scheme 10.11. The insertion of heteroatom–hydrogen (X—H) bond involves an ylide formation from the reaction of metal–carbenoids with heteroatoms of the other substrate, which is followed by a proton transfer process. Recently, the reaction partners including water, alcohols, amines, carbonyl acids, boranes, and thiols have been extensively explored. In this part, we mainly discuss the asymmetric carbene insertion reactions of α-diazo carbonyl compounds promoted by chiral guanidine-based catalysts developed by our research group.

Early in 2014, our group reported the example of enantioselective N–H insertion reactions of primary and secondary anilines with palladium(0) and chiral guanidine as the catalyst (Scheme 10.12) [46]. A series of anilines **40** and alkyl α-diazo esters **41** were suitable substrates in this transformation, and the corresponding products **42** were obtained in modest to good reactivity (up to 99% yield) and high enantiopurity (up to 94% ee). The electrospray ionization high resolution mass spectrometer (ESI HRMS) analysis as well as fluorescence and UV/vis absorption spectroscopy experiments confirmed the coordination of the guanidine with Pd(0), which was further demonstrated by the density functional theory (DFT) calculations from Su and coworkers [47]. According to the mechanistic investigations, the N–H insertion process could be accelerated by the chiral guanidine/Pd(0) complex catalyst with lower

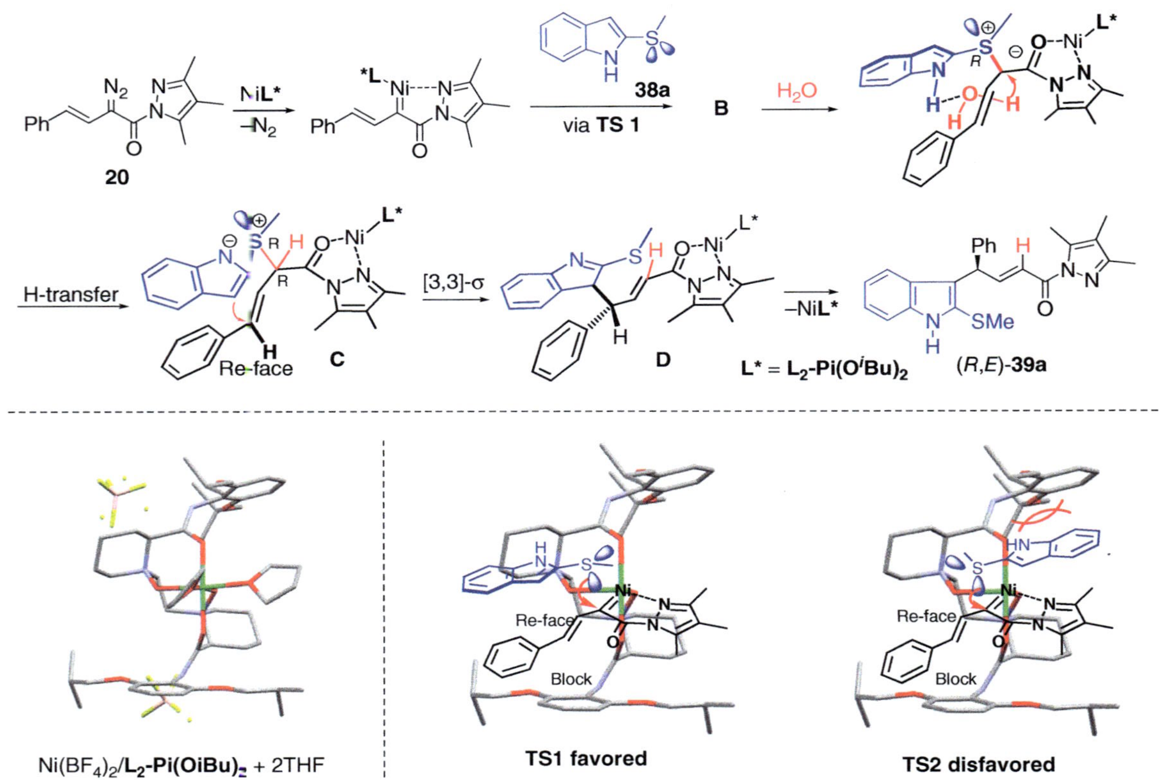

Figure 10.6 Proposed mechanism for asymmetric thio-Claisen rearrangement.

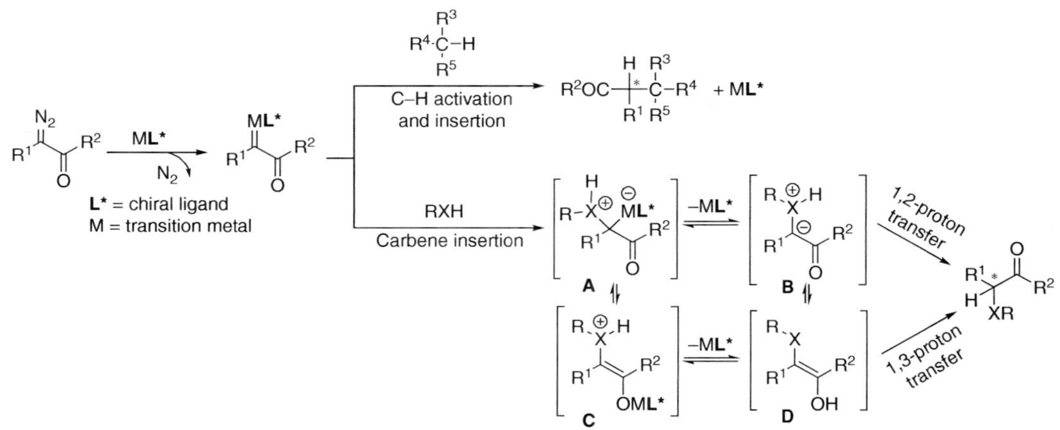

Scheme 10.11 General mechanism of carbene insertion reactions.

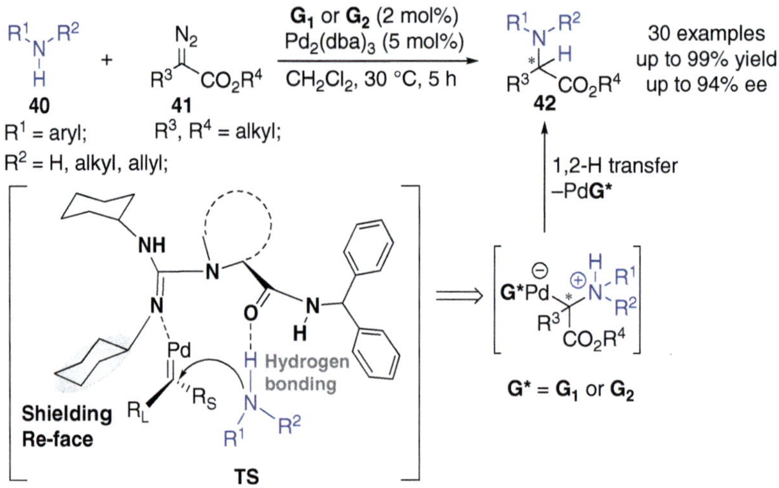

Scheme 10.12 Chiral guanidines and palladium(0) catalyzed enantioselective N–H insertion reactions. Source: Modified from Zhu et al. [46].

activation energy via forming reactive Pd–carbene intermediate. The large steric hindrance of cyclohexyl group in guanidine **G*** (**G_1** or **G_2**) provides sufficient steric shielding around Pd–carbene intermediate from the *Re*-face attack by amines **40**. The combination of a hydrogen bonding from amide unit of the guanidine **G*** as well as the steric repulsion between the ester group of α-diazo esters **41** and bulky $CH(Ph)_2$ group in guanidine ligand **G*** plays an important role to direct the nucleophilic attack of amine to the Pd–carbene from the favorable *Si*-face (**TS**), then followed by a 1,2-proton transfer process, yielding the predominant *S*-configuration product.

Chiral bifunctional guanidines, with strong basic property, have been found to be effective proton shuttles in hydrogen transmission process. In 2016, we disclosed an example of catalytic enantioselective O–H insertion of α-diazo carbonyls **43** to carboxylic acids **44** (Scheme 10.13) [48]. In this reaction, the chiral bifunctional guanidine **G_3** and dirhodium(II) carboxylate were proved to be efficient catalyst combination, and the corresponding α-acyloxy carbonyl compounds **45** were delivered with good enantioselectivities (up to 95% ee) and yields (up to 99%). In the mechanistic studies, there are no interactions between chiral guanidine **G_3** and $Rh_2(OAc)_4$ based on the circular dichroism and NMR spectrum. By contrast, strong interaction between guanidine **G_3** and carboxylic acid was detected. The control experiments in Scheme 10.13 show that the chiral amide catalyst **L_5** was demonstrated ineffective to the stereoselectivity in this reaction, and this result highlighted the crucible role played by guanidine moiety in the catalyst. The earlier findings along with previous report [49] led us to assume that *in situ* generated guanidium carboxylate served as a chiral proton shuttle for the stereo-determinant protonation of an enol intermediate or oxonium ylide. This conjecture was later confirmed by DFT calculations from Su and coworkers [50]. The detailed reaction mechanism was illustrated in Scheme 10.13, bottom. The reaction of $Rh_2(OAc)_4$ with α-diazo

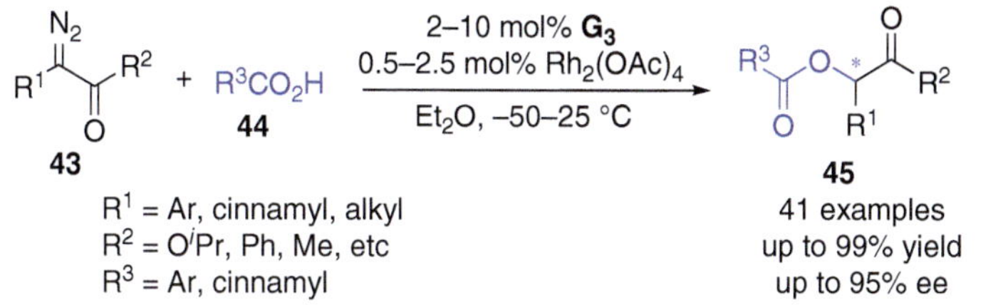

Conditions	Yield (%)	ee (%)
$Rh_2(OAc)_4$ (2.5 mol%), 1 h	99	0
$Rh_2(OAc)_4$ (2.5 mol%), **L_5** (10 mol%), 15.5 h	91	–6
$Rh_2(OAc)_4$ (2.5 mol%), **G_3** (2 mol%), 3 h	99	90
$Rh_2(OAc)_4$ (2.5 mol%), **G_3** (10 mol%), 3 h	84	92
$Rh_2(OAc)_4$ (2.5 mol%), **G_3** (20 mol%), 3 h	72	92

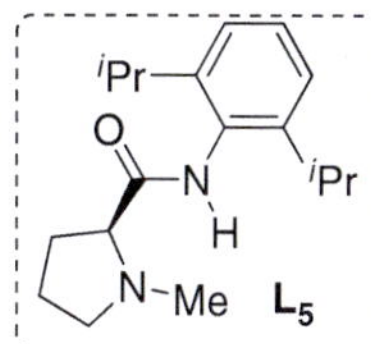

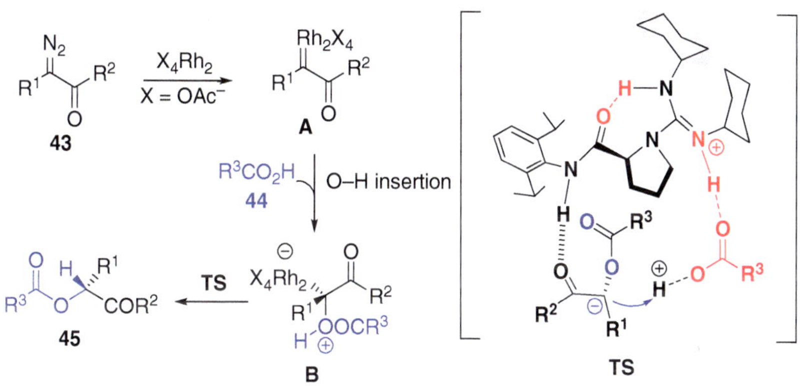

Scheme 10.13 Catalytic enantioselective O–H insertion of α-diazo carbonyls to carboxylic acids. Source: Modified from Tan et al. [48].

carbonyl **43** gave rise to Rh-carbene species with the extrusion of nitrogen. Then the O–H insertion of carbonyl acid **44** to the Rh-carbene intermediate **A** occurred to provide the free oxonium ylide **B**. Subsequently, an intermolecular 1,2-proton transfer process proceeded in an enantioselective manner (**TS**) with the assistance of chiral guanidinium carboxylate, affording the final products **45** in high enantioselectivity. According to the DFT calculations [50], the repulsions between the phenyl group of the enol intermediate and the cyclohexyl as well as the *ortho*-substituted isopropyl group of chiral guanidine **G_3** played important roles in controlling stereoselectivity.

Later, the enantioselective carbene insertion into the $N(sp^2)$—H bond of benzophenone imine **46** was realized by the cooperative catalysis of chiral guanidine **G_6** and dirhodium(II) carbonate (Scheme 10.14) [51]. In the presence of chiral guanidine **G_4** and $Rh_2(esp)_2$, alkyl- and aryl-substituted amino acetates **47** were obtained in good outcomes (up to 99% yield and up to 91% ee) under mild reaction conditions. The products **56** could also be converted to corresponding amino acetate hydrochlorides **48** in high efficiency, accompanied with the diarylketone recovered.

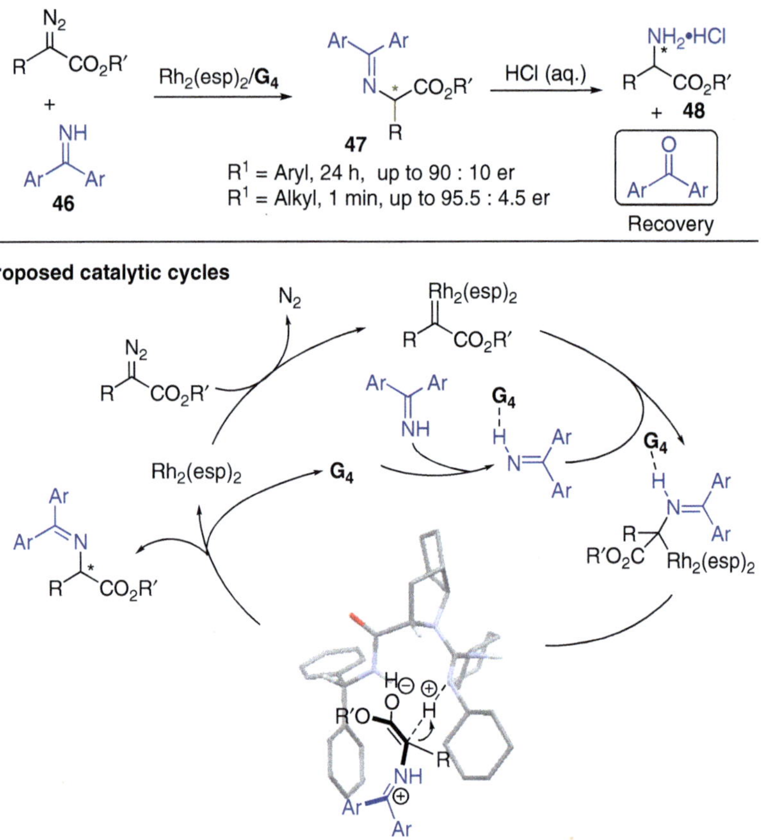

Scheme 10.14 Asymmetric carbene insertion into N–H bond of benzophenone imine co-catalyzed by chiral guanidine and Rh(II). Source: Modified from Yang et al. [51].

According to proposed catalytic cycles, the reaction involved an enantioselective 1,2-H transfer process under the guidance of chiral bifunctional guanidine $\mathbf{G_4}$ as an effective proton shuttle, which is akin to our previous report [48].

In 2016, we reported an asymmetric C(sp)–H insertion of carbene into terminal alkynes using chiral cationic guanidinium salts ($\mathbf{G_1}$·XH, X = Cl, Br) and copper(I) complex (Scheme 10.15) [52]. The scope of the reaction was studied by taking a variety of α-diazo esters **50** as carbene precursors and 1-alkynes **49** as the substrates. It was observed that a great deal of disubstituted allenoates **51** were formed in excellent yields (94–99%) with moderate to high enantioselectivities. The presence of chiral guanidinium salt (5 mol%) could significantly accelerate the reaction, showing an obvious ligand-acceleration effect. Excess amount of metal salts was crucial for the high conversion. Experimental studies indicated that chiral guanidine-promoted isomerization of the corresponding alkynoate to form the chiral allenoate product was unlikely in the current reaction system.

Asymmetric multicomponent reactions (AMCRs) are efficient and economical methodologies for generating enantioenriched compounds with complex structures

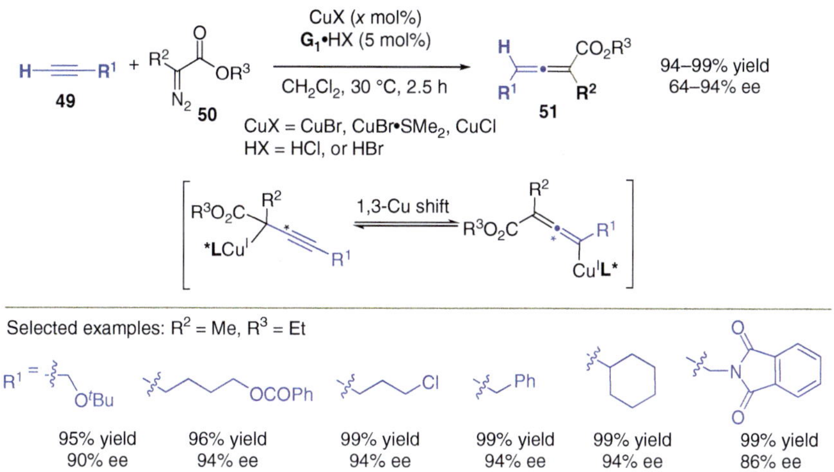

Scheme 10.15 Asymmetric C(sp)–H insertion of carbene into terminal alkynes. Source: Modified from Tang et al. [52].

and multiple stereocenters. On foundation of the outcome in asymmetric C(sp)–H insertion of carbene into 1-alkynes, the catalytic AMCR of α-diazo esters **52**, terminal alkynes **49**, and isatins **53** was realized with the chiral guanidinium salt/CuBr/YBr_3 catalytic system. This protocol provides a direct and facile route to the chiral tetrasubstituted allenoates **54** in good yields (87–98%) with high diastereo- and enantioselectivities (Scheme 10.16) [53]. The coordination between guanidinium salt and the copper salt was further confirmed by NMR studies. The addition of catalytic amount of Lewis acid YBr_3 was turned out to be necessary for maintaining higher reactivity instead of stoichiometric or substoichiometric copper(I) salt. This improvement could be rationalized by combined-acid systems. Similarly, the reactivity of asymmetric catalytic C(sp)–H insertions of α-diazo esters **52** to terminal alkynes **49** could be further improved by using ytterbium

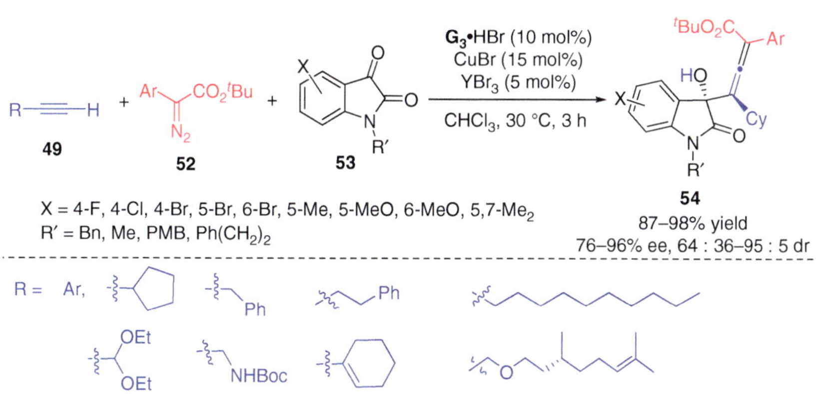

Scheme 10.16 Asymmetric three-component reaction for the synthesis of tetrasubstituted allenoates. Source: Modified from Tang et al. [53].

trihalide or hydrohalic acid or both as the additives [52]. To get more insight into the reaction pathway, a series of control experiments were conducted. First, the alkynoate was subjected into the reaction conditions; however, the corresponding allenoate was not detected at all. Therefore, an allenoate-Cu(I) species [47] was probably the key reactive intermediates, whose stereochemistry was supposed to be heavily reliant on the enantioselective 1,3-Cu transfer process controlled by the chiral guanidinium ligand. Finally, the chiral allenoate-Cu(I) intermediate was captured by hydrogen-activated isatin **53** via the asymmetric induction of chiral guanidine-amide to afford the enantioenriched tetrasubstituted allenoate **54**.

We have delineated a chiral guanidinium/Cu(I)-catalyzed AMCR of α-aryl diazoacetates **55** with trimethylsilyl cyanide, *tert*-butanol, and *N*-phenylmaleimides **56** (Scheme 10.17) [54]. The transformation involves hydrocyanation/Michael reaction steps, furnishing an array of cyanide-containing pyrrolidine-2,5-diones **58** in excellent yields (up to 99% yield) with decent enantio- and diastereoselectivities (up to 96% ee and 19 : 1 dr). The formation of chiral copper(I) species **A** was the first step of the catalytic cycle, followed by a C–H insertion of hydrocyanic acid (HCN), which was generated from tertiary butanol and trimethylsilyl cyanide, affording the intermediate **B** with a keto-enol tautomerism equilibrium with **B′**. The copper-bonded species **B**, **B′**, or **C** was supposed to be the active nucleophile to perform the following addition reaction. The sulfonamide unit of the guanidinium salt G_4·HBr might activate the maleimide **57** via hydrogen-bonding interaction. Then a diastereo- and enantioselective conjugate addition occurs to give intermediate **D**, and the final product was generated after protonation.

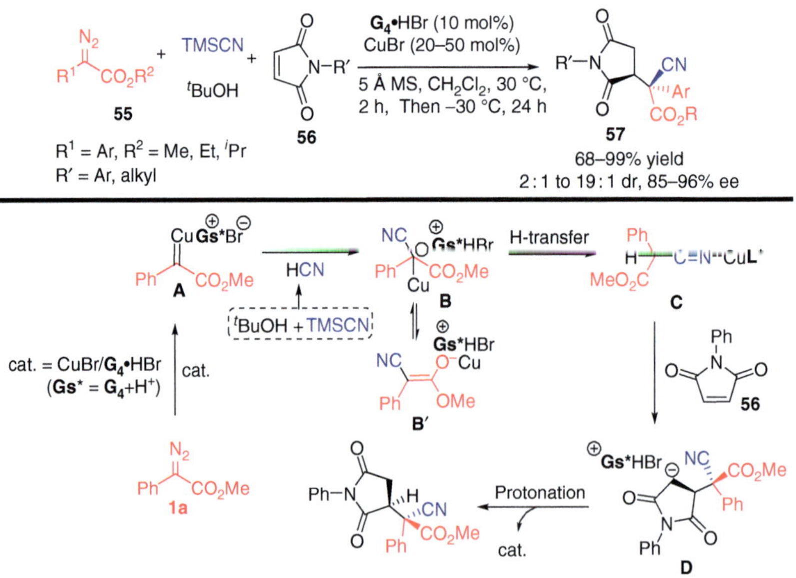

Scheme 10.17 Asymmetric hydrocyanation/Michael reaction of α-diazoacetates. Source: Modified from Ruan et al. [54].

10.5 Conclusion and Outlook

The present chapter mainly describes our recent contributions on the field of asymmetric carbene transformations. As illustrated earlier, through the development of new donor–acceptor α-diazo pyrazoleamides and employment of novel dual-tasking chiral *N,N'*-dioxide/Ni(II) or Co(II) complexes as the catalysts, several highly enantioselective [2,3]- and [3,3]-sigmatropic rearrangements of sulfonium ylides have been well established. Mechanistic studies indicated that the key for the success lied in the excellent performance on fine discrimination of heterotopic lone pairs on sulfur or selenium, as well as high stereocontrol in proton transfer and rearrangement processes. In addition, the bifunctional guanidine-amides, as co-catalysts or chiral ligands, show excellent enantiocontrol in C(sp)–H, N–H, and O–H insertions of carbene species. The diversity of reaction types, high reactivity, substrate tolerance, and mild conditions of these reactions render the chiral *N,N'*-dioxide/metal complexes and guanidine-based catalysts very splendid and attractive promoters in asymmetric carbene transformations.

While the achievements made to date are notable, further developing novel and efficient transformations of carbene species using chiral *N,N'*-dioxide/metal complexes and guanidine-based catalysts remain interesting and urgent. On the other hand, their further application in the synthesis of drugs and natural products awaits.

References

1 Doyle, M.P., Duffy, R., Ratnikov, M., and Zhou, L. (2010). *Chem. Rev.* 110: 704.
2 Zhu, S.-F. and Zhou, Q.-L. (2012). *Acc. Chem. Res.* 45: 1365.
3 Gillingham, D. and Fei, N. (2013). *Chem. Soc. Rev.* 42: 4918.
4 Ford, A., Miel, H., Ring, A. et al. (2015). *Chem. Rev.* 115: 9981.
5 Gurmessa, G.T. and Singh, G.S. (2017). *Res. Chem. Intermed.* 43: 6447.
6 Liu, X.H., Lin, L.L., and Feng, X.M. (2011). *Acc. Chem. Res.* 44: 574.
7 Liu, X.H., Lin, L.L., and Feng, X.M. (2014). *Org. Chem. Front.* 1: 298.
8 Liu, X.H., Zheng, H.F., Xia, Y. et al. (2017). *Acc. Chem. Res.* 50: 2621.
9 Zheng, K., Liu, X.H., and Feng, X.M. (2018). *Chem. Rev.* 118: 7586.
10 Wang, Z., Liu, X.H., and Feng, X.M. (2020). *Aldrichimica Acta* 53: 3.
11 Wang, M.Y. and Li, W. (2020). *Chin. J. Chem.* 39, https://doi.org/10.1002/cjoc.202000508.
12 Liu, X.H., Dong, S.X., Lin, L.L., and Feng, X.M. (2018). *Chin. J. Chem.* 36: 791.
13 Dong, S.X., Feng, X.M., and Liu, X.H. (2018). *Chem. Soc. Rev.* 47: 8525.
14 Zhou, Q.-L. (ed.) (2011). *Privileged Chiral Ligands and Catalysts*. Weinheim, Germany: Wiley-VCH.
15 Zhang, Y. and Wang, J. (2010). *Coord. Chem. Rev.* 254: 941.
16 Sheng, Z., Zhang, Z., Chu, C. et al. (2017). *Tetrahedron* 73: 4011.
17 West, T.H., Spoehrle, S.S.M., Kasten, K. et al. (2015). *ACS Catal.* 5: 7446.
18 Neuhaus, J.D., Oost, R., Merad, J., and Maulide, N. (2018). *Top. Curr. Chem.* 376: 15.

19 Wu, H., Wang, Q., and Zhu, J.-P. (2019). *Eur. J. Org. Chem.* 2019: 1964.

20 Nishibayashi, Y., Ohe, K., and Uemura, S. (1995). *J. Chem. Soc. Chem. Commun.*: 1245.

21 Zhang, Z., Sheng, Z., Yu, W. et al. (2017). *Nat. Chem.* 9: 970.

22 Lin, X.B., Tang, Y., Yang, W. et al. (2018). *J. Am. Chem. Soc.* 140: 3602.

23 McMillen, D.W., Varga, N., Reed, B.A., and King, C. (2000). *J. Org. Chem.* 65: 2532.

24 Kitagaki, S., Yanamoto, Y., Okubo, H. et al. (2001). *Heterocycles* 54: 623.

25 Fukuda, T., Irie, R., and Katsuki, T. (1999). *Tetrahedron* 55: 649.

26 Tyagi, V., Sreenilayam, G., Bajaj, P. et al. (2016). *Angew. Chem. Int. Ed.* 55: 13562.

27 Ma, M., Peng, L., Li, C. et al. (2005). *J. Am. Chem. Soc.* 127: 15016.

28 Davies, H.M.L. and Morton, D. (2011). *Chem. Soc. Rev.* 40: 1857.

29 Pavlyuk, O., Teller, H., and McMills, M.C. (2015). *Chem. Rev.* 115: 9922.

30 Yao, Q., Wang, Z., Zhang, Y.H. et al. (2015). *J. Org. Chem.* 80: 5704.

31 Zhang, Y., Liao, Y.T., Liu, X.H. et al. (2016). *Chem. Eur. J.* 22: 15119.

32 Wang, G.J., Tang, Y., Zhang, Y. et al. (2017). *Chem. Eur. J.* 23: 554.

33 Lin, X.B., Tan, Z., Yang, W.K. et al. (2020). *CCS Chem.* 2: 1423.

34 Oksana, P., Henrik, T., and Mark, C.M. (2009). *Tetrahedron Lett.* 50: 2716.

35 Javed, M.I. and Brewer, M. (2008). *Org. Synth.* 85: 189.

36 Lin, X.B., Yang, W., Yang, W.K. et al. (2019). *Angew. Chem. Int. Ed.* 58: 13492.

37 Liao, M.Y., Peng, L.L., and Wang, J. (2008). *Org. Lett.* 10: 693.

38 Li, S.-S. and Wang, J. (2020). *J. Org. Chem.* 85: 12343.

39 Kwart, H. and Hackett, C.M. (1962). *J. Am. Chem. Soc.* 84: 1754.

40 Yang, W., Lin, X.B., Zhang, Y.Y. et al. (2020). *Chem. Commun.* 56: 10002.

41 Bonge, H.T. and Hansen, T. (2010). *Eur. J. Org. Chem.* 4355.

42 Zhao, X., Zhang, Y., and Wang, J. (2012). *Chem. Commun.* 48: 10162.

43 Giri, R., Shi, B.-F., Engle, K.M. et al. (2009). *Chem. Soc. Rev.* 38: 3242.

44 Davies, H.M.L. and Manning, J.R. (2008). *Nature* 451: 417.

45 Davies, H.M.L. and Loi, Ø. (2004). *Synthesis* 2004: 2595.

46 Zhu, Y., Liu, X.H., Dong, S.X. et al. (2014). *Angew. Chem. Int. Ed.* 53: 1636.

47 Li, J., Su, Z.S., Wang, J.M., and Hu, C.W. (2018). *J. Catal.* 364: 426.

48 Tan, F., Liu, X.H., Hao, X.Y. et al. (2016). *ACS Catal.* 6: 6930.

49 Liang, Y., Zhou, H.L., and Yu, Z.X. (2009). *J. Am. Chem. Soc.* 131: 17783.

50 Li, J., Meng, X.X., Hu, C.W., and Su, Z.S. (2019). *J. Org. Chem.* 84: 15020.

51 Yang, J., Ruan, P.R., Yang, W. et al. (2019). *Chem. Sci.* 10: 10305.

52 Tang, Y., Chen, Q.G., Liu, X.H. et al. (2015). *Angew. Chem. Int. Ed.* 54: 9512.

53 Tang, Y., Xu, J., Yang, J. et al. (2018). *Chem* 4: 1658.

54 Ruan, S., Zhong, X., Chen, Q.G. et al. (2020). *Chem. Commun.* 56: 2155.

11

Multi-Component Reaction via *gem*-Difunctionalization of Metal Carbene

Mengchu Zhang, Xinfang Xu, and Wenhao Hu

Sun Yat-sen University, School of Pharmaceutical Sciences, Guangdong Provincial Key Laboratory of Chiral Molecule and Drug Discovery, Guangzhou 510006, China

11.1 Introduction

Multicomponent reactions (MCRs) are generally defined as reactions in which three or more starting materials are involved to form a product [1]. In MCRs, a molecule could be assembled via a one convergent chemical step in one pot by simply mixing the corresponding starting materials, which features advantages in step economy, time, and cost-saving over traditional multistep approaches for the assembly of target compounds. Thus, massive chemical entities with structural diversity have been described in the literature through MCRs. Moreover, considerably complex molecules could be synthesized by MCRs. The structure–activity relationships (SARs) can be rapidly generated using MCRs, since all property determining moieties are introduced in one step [2]. Although the first MCR has been disclosed as early as 1851, it took more than 100 years until Ugi discovered his fantastic four-component reaction [3], and the potential of MCRs in synthetic chemistry has been recognized with the development of a variety of novel MCRs [4]. Beyond these advances, the discovery of new MCRs is mainly based on the serendipitous observations so far. Therefore, the exploration of effective strategies for the rational design of MCRs is highly desirable, which would provide general principle for the MCR discovery, and more importantly, enable synthetic applications for the diversity-oriented synthesis (DOS).

Metal carbenes, catalytically derived from diazo compounds or other diverse carbene precursors, are among the most important reactive synthons in modern organic synthesis [5]. They can undergo a variety of synthetically important transformations such as X–H insertion, cyclopropanation, cycloadditions, and others [6]. Recently, metal carbenes have been successfully applied to MCRs for assembly of complex molecules [7–9]. The MCRs in this area could be divided mainly in three types: (i) formation of carbonyl or azomethine ylides with carbonyl compounds or imines, followed by 1,3-dipolar cycloadditions with different types of dipolarophiles to afford corresponding heterocycles [7]; (ii) MCRs involving

Transition Metal-Catalyzed Carbene Transformations, First Edition.
Edited by Jianbo Wang, Chi-Ming Che, and Michael P. Doyle.

metal carbene migratory insertion processes, followed by various terminating approaches [8]; (iii) the metal carbene initially attacked by nucleophiles to generate ylides or zwitterionic intermediates, followed by electrophilic interceptions with different electrophiles (Scheme 11.1) [9]. In this chapter, we mainly focused on the third version of MCRs, which features a *gem*-difunctionalization of metal carbene species, leading to the polyfunctionalized molecules with different types of carbene precursors (Scheme 11.2), especially with different diazo compounds [10]. In this chapter, a comprehensive review covering various reactive intermediates trapped by different electrophiles via Mannich-, Aldol-, Michael-type, and other additions has been summarized (Scheme 11.3), and focusing on the representative examples, including the pioneering works, asymmetric catalytic versions, and synthetic applications from our group and others.

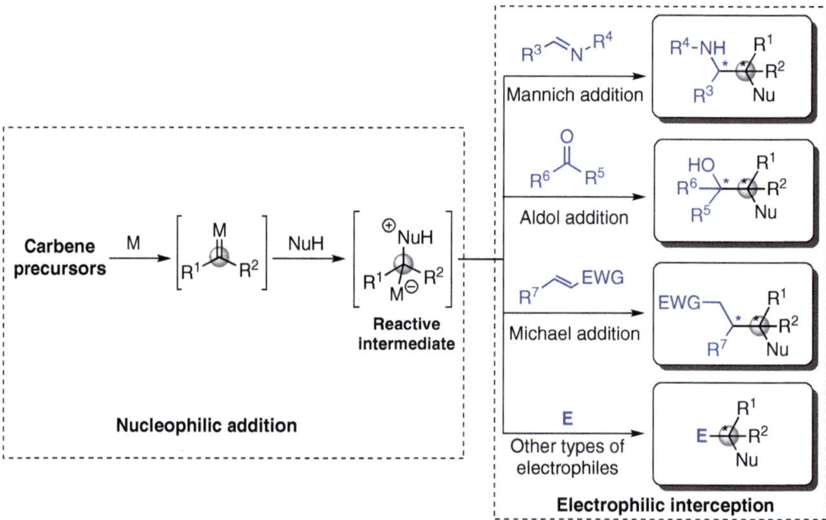

Scheme 11.1 General reaction pathways of multicomponent reactions through interception of active intermediates. Sources: Based on Guo and Hu [9a]; Zhang and Hu [9b].

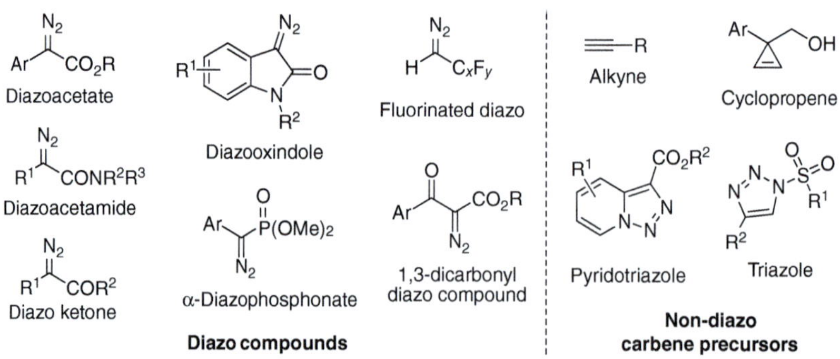

Scheme 11.2 Metal carbene precursors used in multicomponent reactions (MCRs). Source: Based on Doyle et al. [10a]; Ford et al. [10b]; Marinozzi et al. [10c]; Ceias et al. [10d].

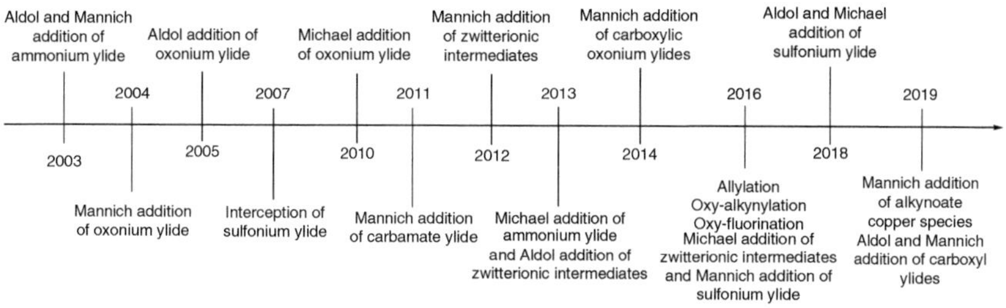

Scheme 11.3 Summary of multicomponent reactions via different types of interception processes.

11.2 Mannich-Type Interception

11.2.1 Interception of Ammonium Ylide

The pioneering work of the MCRs via interception of *in situ* formed active ammonium ylide intermediates that generated from metal carbene and arylamines was reported by Hu and Doyle in 2003 [11]. By treating methyl phenyldiazoacetate **1a** with arylamines **2** and imines **3** in the presence of dirhodium acetates, a novel three-component reaction has been achieved, providing a practical method for the synthesis of 1,2-diaryl-1,2-diaminopropanoates **4** in good to high yields with up to 97 : 3 dr (Scheme 11.4). The two-component products **5** and **6** that generated through metal carbene N–H insertion and aziridination could be obtained under the same catalytic conditions in minor amount. However, these compounds could not lead to the three-component **4** under the same conditions, which is consistent with the proposed reaction mechanism involving trapping of the reactive ammonium ylide via Mannich-type addition.

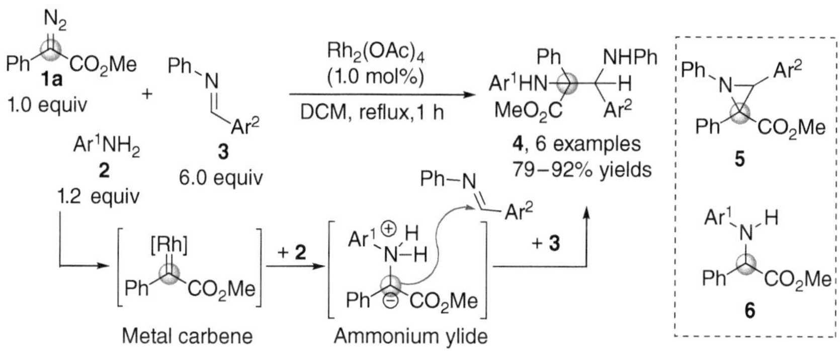

Scheme 11.4 Interception of ammonium ylide via Mannich-type addition.

In 2011, the asymmetric catalytic version of this three-component reaction with carbamates **7** as the nucleophiles has been realized by a $Rh_2(OAc)_4$ and chiral Brønsted acid **8** cooperative catalysis method [12]. In this work, the highly reactive protic carbamate ammonium ylides were successfully trapped by imine electrophiles in a diastereoselectively switchable enantioselective manner controlled by the chiral Brønsted acids, providing a rapid and efficient access to both *syn*- and *anti*-α-substituted α,β-diamino acid derivatives **9** with a high-level control of chemo-, diastereo-, and enantioselectivity (Scheme 11.5). Recently, the detailed mechanistic studies have been conducted by Sunoj through density functional theory (DFT) computations [13]. The calculated results indicate that both $Rh_2(OAc)_4$ and the chiral phosphoric acid (PA) cooperatively participated in the stereocontrolling step of the reaction through the addition of dirhodium-bound enol to the activated imine.

With the development catalytic methods, this three-component reaction via Mannich-type interception of ammonium ylide intermediates has become a practical method for the synthesis of α,β-diamino derivatives (Scheme 11.6). A variety of reagents, including diazoacetates (**10**) [14] and fluorinated diazo compounds (**11**) [15], phosphoramidates (**12**) [16] and sulfonamides (**13**) [17], *N*-sulfonyl aldimines (**14**) [18] and azodicarboxylates (**16**) [19], are all well tolerated under corresponding catalytic conditions, providing the functionalized amines with structural diversity.

In 2014, a highly diastereo- and enantioselective four-component reaction of a diazo ketone **17** with ethyl glyoxylate **18** and two molecules of anilines was reported by Hu [20]. Under $Rh_2(OAc)_4$ and chiral phosphoric acid co-catalyzed conditions, this transformation proceeds through a Mannich-type trapping of the *in situ* generated ammonium ylide with iminoester that derived from aniline and ethyl glyoxylate (Scheme 11.7). With this method, useful chiral α,β-diamino acid derivatives were efficiently formed from readily available starting materials in good yields and with good to excellent diastereo- and enantioselectivities.

In 2019, a tandem intermolecular ammonium ylide formation/intramolecular Mannich-type cyclization reaction of aryl diazoacetates **1** with *ortho*-aminophenyl imine was developed by Hu and coworkers [21]. The reaction provides an efficient protocol for the direct synthesis of trisubstituted tetrahydroquinoxalines **20** bearing a quaternary carbon center in moderate to good yields with excellent diastereoselectivities (Scheme 11.8).

11.2.2 Interception of Oxonium Ylide

In 2005, Hu and coworkers reported the first example of Mannich-type interception of oxonium ylide in the presence of $Rh_2(OAc)_4$ [22]. In this work, a three-component reaction of aryl diazoacetates with alcohols and aldehydes was disclosed, which not only provided the experimental evidence of alcoholic oxonium ylide formation for O–H insertion, but also offered a convenient access to a series of important compounds in organic and medicinal chemistry, α-alkoxy-β-amino acid derivatives **22**, in good to high yields with good diastereoselectivities (Scheme 11.9).

In the above work, only the imine that derived from electron-deficient benzaldehydes, which are stronger electrophiles, could successfully trap the active oxonium ylide intermediates. In 2007, Hu and coworkers found that imines derived from 2-aminophenol could smoothly intercept the oxonium ylide intermediates [23]. Moreover, the chiral *N*-(*tert*-butylsulfinyl)imines **23** were also found

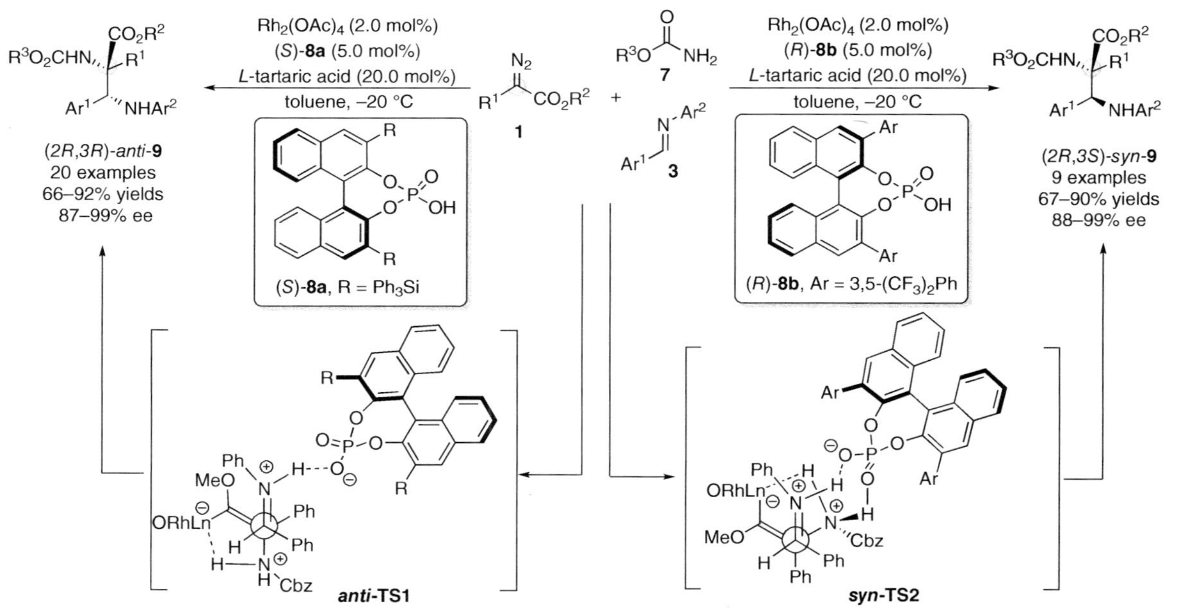

Scheme 11.5 Enantioselective trapping of carbamate ammonium ylides with imines via $Rh_2(OAc)_4$ and chiral Brønsted acid cooperative catalysis.

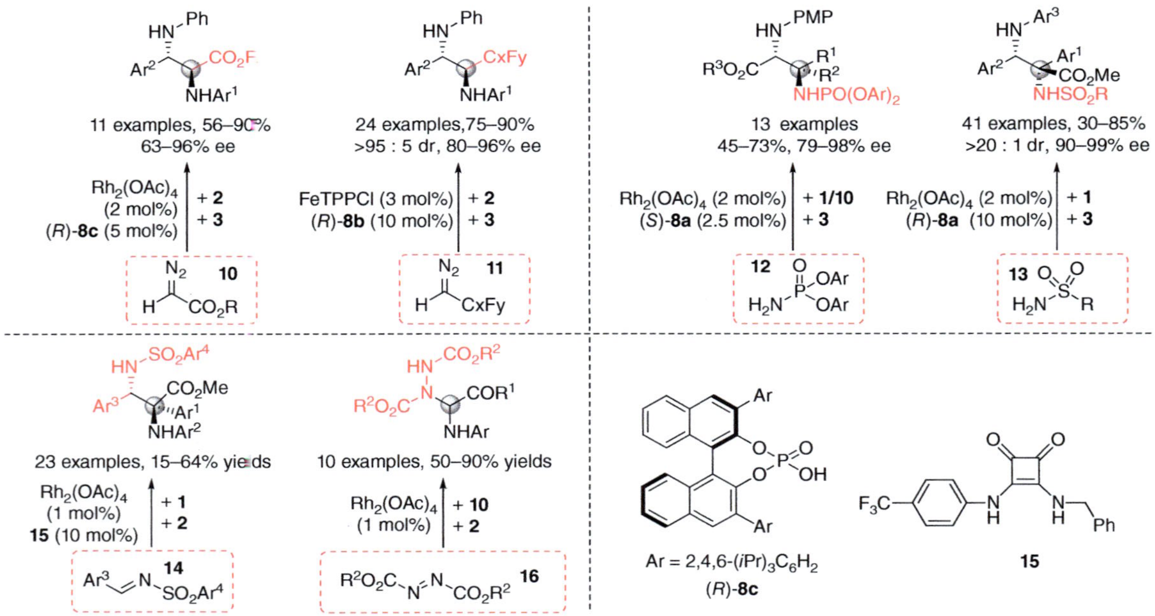

Scheme 11.6 Synthesis of α,β-diamino cerivatives via Mannich-type interception of ammonium ylide intermediates.

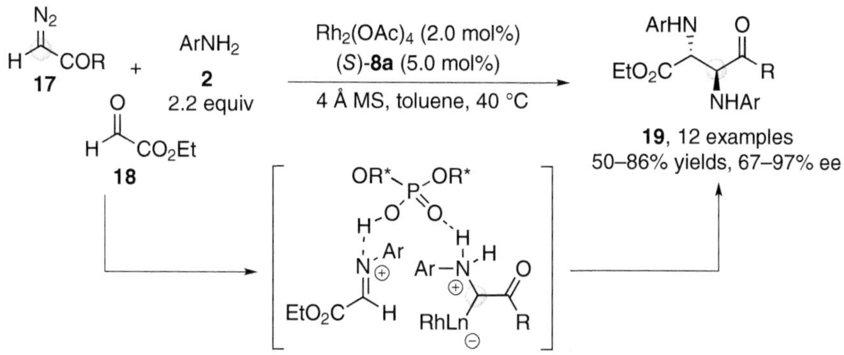

Scheme 11.7 Mannich-type interception of ammonium ylide with *in situ* generated imines.

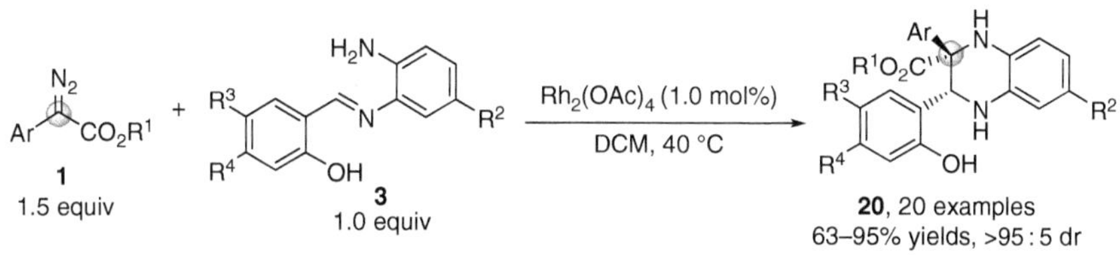

Scheme 11.8 Intramolecular Mannich-type interception of ammonium ylide.

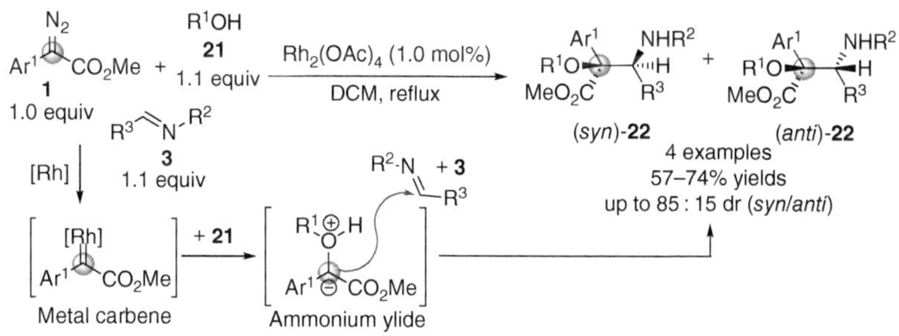

Scheme 11.9 Interception of oxonium ylide via Mannich-type addition.

compatible in this reaction, giving β-amino-α-hydroxyesters with high optical purity (Scheme 11.10).

One year later, Hu et al. realized the first enantioselective three-component reaction of diazo compounds with alcohols and imines in the presence of $Rh_2(OAc)_4$ and a chiral Brønsted acid [24]. The use of bulky alcohols is essential in this reaction, which dramatically increased the dr and ee values. This reaction provides an effective access to *syn*-β-amino-α-alkoxyl acid derivatives bearing a quaternary carbon stereogenic center in high yields with excellent enantioselectivities (Scheme 11.11).

In 2009, Hu and coworkers reported that water and ethyl diazoacetate are matched components for generating highly active nucleophilic oxonium ylide in the presence of a dirhodium acetate catalyst. Simultaneously trapping of this active intermediate with aryl imines delivered β-aryl isoserine derivatives with

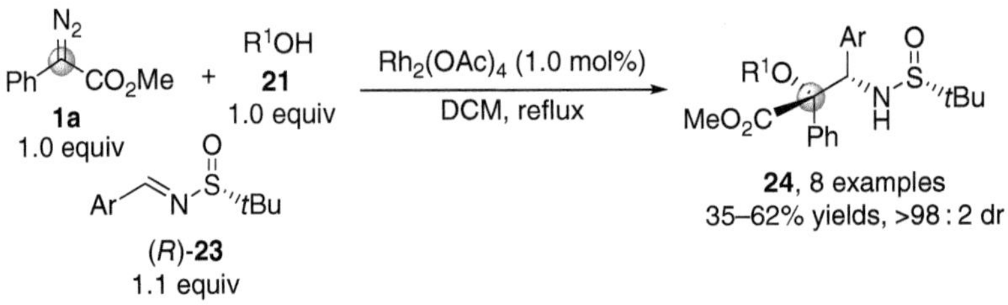

Scheme 11.10 Interception of oxonium ylide with *N*-(*tert*-butylsulfinyl)imines.

high diastereoselectivities [25]. Later, they applied the metal/organo cooperative catalysis strategy to this water involved reaction [26], enabling an enantioselective three-component reaction of diazo compounds with water and imines to produce chiral β-amino-α-hydroxy acid derivatives **26** with good diastereoselectivities and high enantioselectivities (Scheme 11.12).

This Mannich-type interception process could be expanded to various of reactive electrophiles in the presence of $Rh_2(OAc)_4$ (Scheme 11.13). In 2006, Hu and coworkers reported a reaction of aryl diazoacetate with methanol and diethyl azodicarboxylate (DEAD) in the presence of dirhodium acetate, providing the three-component adduct **27** in 66% yield (Scheme 11.13A) [27]. Later, two heterocyclic imines, isatin- and pyrazolinone-derived ketimines **28** and **30**, were applied to this reaction by Reddy and coworkers [28] and Hu and coworkers [29] independently, delivering the oxindole-derived α-alkoxy-β-amino acids **29** and pyrazolone derivatives **31** (Scheme 11.13B), respectively.

Beyond the readily available imines, the *in situ* generated ones are also compatible in this three-component reaction (Schemes 11.14 and 11.15). In 2016, Hu and coworkers disclosed a highly chemo- and enantio-selective one-pot reaction that involved an oxidative functionalization of $C(sp^3)$—H bonds adjacent to nitrogen atoms of *N*-aryl glycine esters [30]. Mechanistically, dichlorodicyanobenzoquinone (DDQ) would react with the *N*-aryl glycine esters to provide imine **33**. Meanwhile, the rhodium catalyst promoted the formation of protic oxonium ylide from diazo compound and water. Then, under the catalysis of chiral phosphoric acid, the Mannich-type addition between these two active intermediates led to a variety of highly functionalized β-hydroxyl-α-amino acid derivatives **34** in good to high yields (Scheme 11.14A). Two years later, the same group reported an enantioselective formal S_N1 reaction [31]. The active *N*-acyl ketiminium ions **36** were formed via the dehydration of **35** in the presence of chiral phosphoric acid, followed by a Mannich-type addition with *in situ* generated oxonium ylides to enable the formal S_N1 process. This reaction provides an efficient assess for the synthesis of isoindolinone derivatives with two contiguous quaternary stereogenic centers in high yields with excellent enantioselectivities (Scheme 11.14B).

In 2019, Hu and coworkers reported an asymmetric counter-anion-directed aminomethylation reaction with α-aminomethyl ether **38** as the imine precursor for the interception of oxonium ylide [32a]. This method offers an efficient and convenient access to furnish optically active α-hydroxyl-β-amino acids in high yields

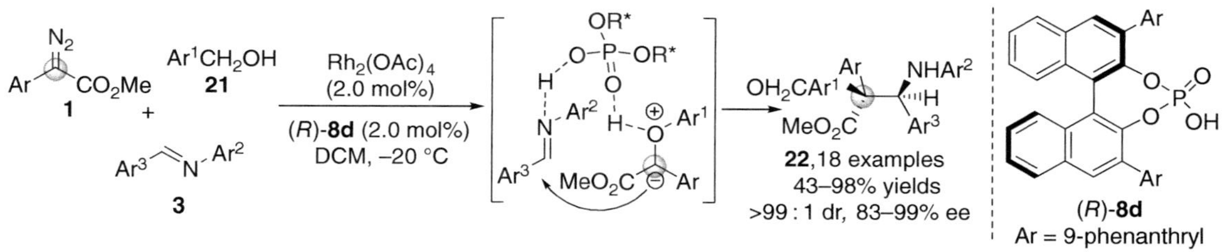

Scheme 11.11 Asymmetric catalytic interception of oxonium ylide via Mannich-type addition.

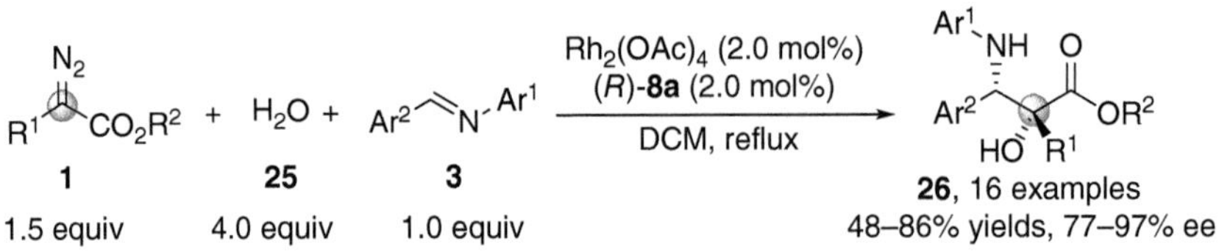

Scheme 11.12 Asymmetric three-component reaction of diazo compounds with water and imines.

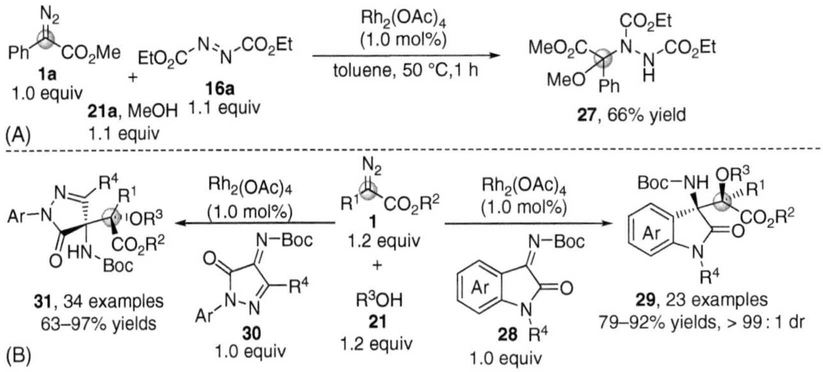

Scheme 11.13 Interception of oxonium ylide with different Mannich addition acceptors. Source: Modified from Guo et al. [27].

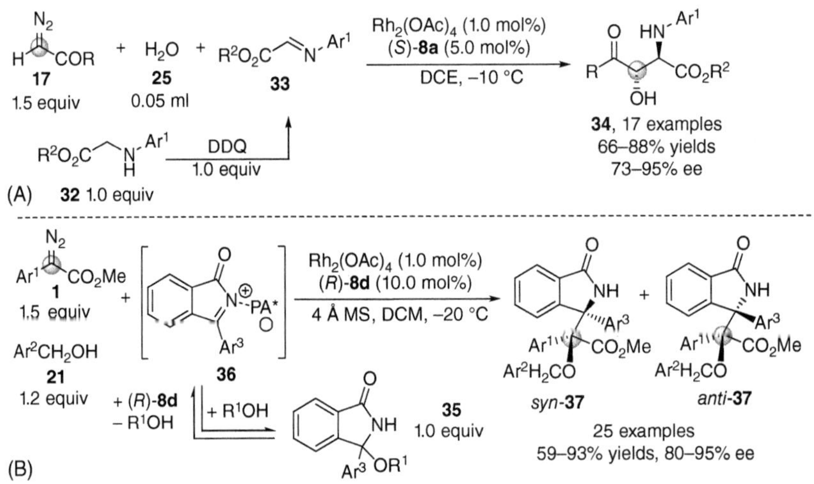

Scheme 11.14 Interception of oxonium ylide with *in situ* generated imines.

with high to excellent enantioselectivities (Scheme 11.15). Density functional thoery (DFT) calculations indicated that the transformation proceeds through trapping the *in situ* generated enol species with methylene iminium ion intermediate, and the asymmetric induction was realized by chiral pentacarboxycyclopentadiene anion via H-bonding and electrostatic interaction. In the same year, the copper-catalyzed

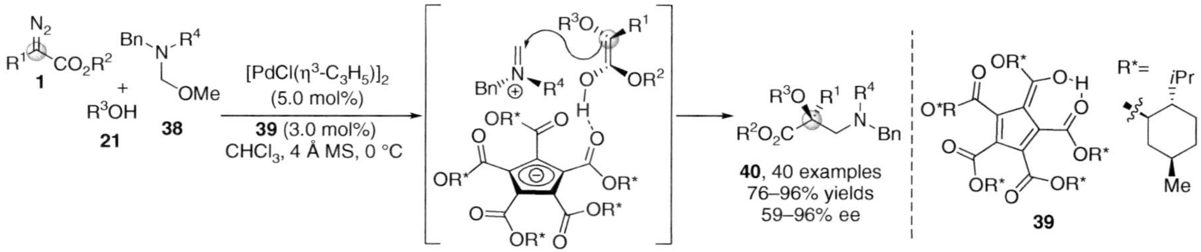

Scheme 11.15 Asymmetric aminomethylation reaction of diazo compound with alcohol and α-aminomethyl ether.

racemic version was disclosed by Sun and coworkers [32c]. In this reaction, the readily available *N*,*O*-acetals have been introduced as efficient bifunctionalization reagents for the concurrent incorporation of an alkoxy and an iminium species onto metal carbene intermediates.

Recently, Hu and coworkers introduced 1,3,5-triazines **42** as the precursor of formaldimines for the interception of oxonium ylide [33]. By applying the effective rhodium/chiral phosphoric acid cooperative catalysis, an enantioselective three-component reaction of α-diazo ketones **41** with alcohols and 1,3,5-triazines has been realized (Scheme 11.16). A dual hydrogen bonding between the chiral phosphoric acid catalyst and two distinct active intermediates was proposed to be crucial for the efficient electrophile-based enantiocontrol. This reaction features an enantioselective α-aminomethylation of carbonyl compounds, which constitutes a powerful protocol for the introduction of aminomethyl species to organic molecules, generating a series of chiral β-amino-α-hydroxy ketones with high efficiency and enantioselectivity.

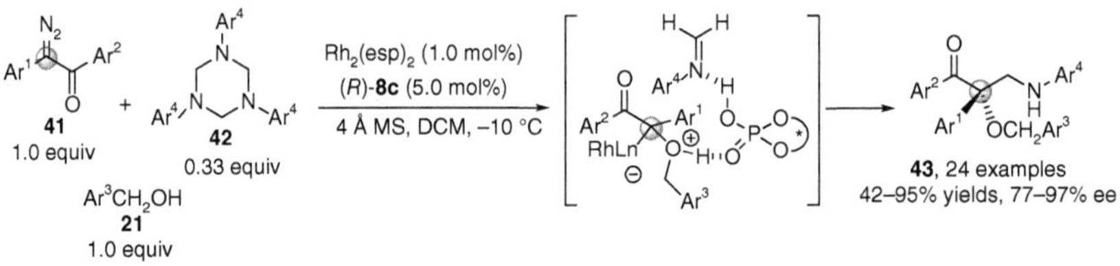

Scheme 11.16 Asymmetric three-component reaction of α-diazo ketones with alcohols and 1,3,5-triazines.

The oxonium ylide generated from different functionalized diazo compounds or other type of carbene precursors have been found well tolerated under the powerful $Rh_2(OAc)_4$ and chiral Brønsted acid cooperative catalysis system [34]. For examples, diazoketones [34a], cyclopropyl diazoacetates [34b], alkynyl diazoacetates [34c], 3-diazooxindoles [34d], and pyridotriazoles [34e] have been successfully introduced into the three-component reaction, delivering the corresponding polyfunctionalized products with adjacent quaternary/tertiary stereocenters in good to high yields with excellent enantioselectivities (Scheme 11.17). Notably, further functionalization or cyclization could be realized with these assembled molecules, especially with alkynyl species, catalytic intramolecular cyclization, or intermolecular cycloaddition could lead various heterocycles with different ring sizes and structural diversity [34e].

In 2018, Hu and coworkers successfully expanded this ylide interception process to the non-diazo carbene precursor, alkynes [35a]. An atom-economic and enantioselective Mannich-type reaction of 3-butynol **49** with nitrones **50** has been realized via gold(I)/chiral phosphoric acid cooperative catalysis, which affords dihydrofuran-3-ones **51** in good yields and with excellent enantioselectivities (Scheme 11.18). The reaction is initiated by gold-catalyzed alkyne oxidation, followed by modification of the resulting gold carbene species with the tethered hydroxy group to form gold enolate species and terminated with an enantioselective

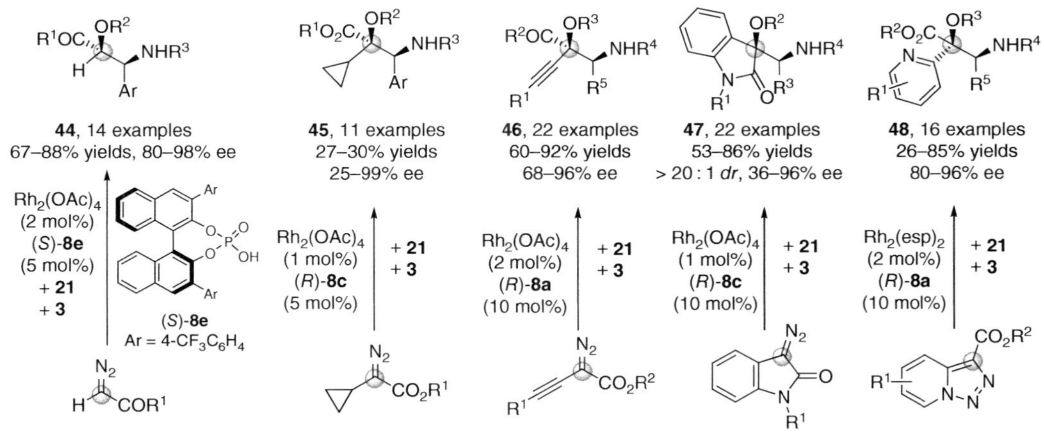

Scheme 11.17 Asymmetric three-component reaction with various functionalized diazo compounds.

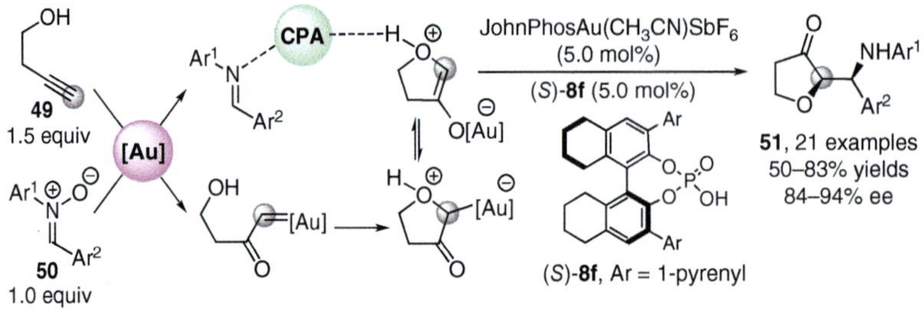

Scheme 11.18 Asymmetric Mannich-type reaction of 3-butynol with nitrones.

Mannich-type addition with the assistance of chiral phosphoric acid. This novel pattern of alkyne transformation involves a chemical bond cleavage, fragment modification, and reassembly process. Notably, the analogous gold-catalyzed racemic version has been disclosed by Liu in the same year with 2-ethynylphenols, acyclic 4-yn-1-ols, and phenoxyethynes as the carbene precursors [35b].

Instead of using prepared imines, direct using of corresponding aldehydes and amines for the *in situ* formation of imines could realize a four-component reaction [36]. Although competition between the formation of oxonium ylide and ammonium ylide exists, Hu and coworkers have found that >95% conversion of benzaldehyde and aniline to the imine was achieved within five minutes in the presence of chiral phosphoric acid. Thus, an asymmetric four-component reaction of trapping oxonium ylide with *in situ* generated imines has been achieved by rhodium/chiral phosphoric acid cooperative catalysis [36a]. This method provides an efficient way to extend the scope of electrophilic imines including the unstable imines, producing β-amino-α-alkoxyl acid derivatives **22** in a single step with excellent control of chemo-, diastereo-, and enantioselectivities (Scheme 11.19). In 2013, an enantioselective four-component reaction of diazoketone, water, aniline, and ethyl glyoxylate was disclosed by the same group, leading to the α-amino-β-hydroxy acid derivatives in good yields with high diastereoselectivities and enantioselectivities [36b].

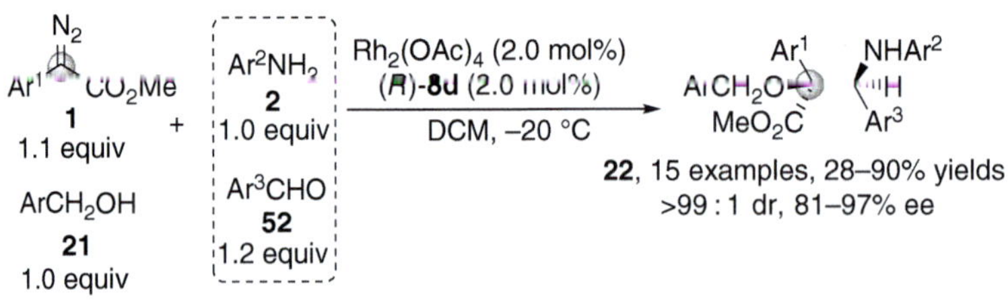

Scheme 11.19 Asymmetric four-component reaction through trapping of oxonium ylide with *in situ* generated imines.

In 2019, Wang and coworkers reported a copper-catalyzed intramolecular cyclization reaction of oxonium ylide [37]. In this reaction, the copper(I)/bisoxazoline catalyst serves a dual function: (i) reacts with the diazo compound to generate the metal carbene speices and (ii) upon formation of an oxonium ylide, it acts as a Lewis acid

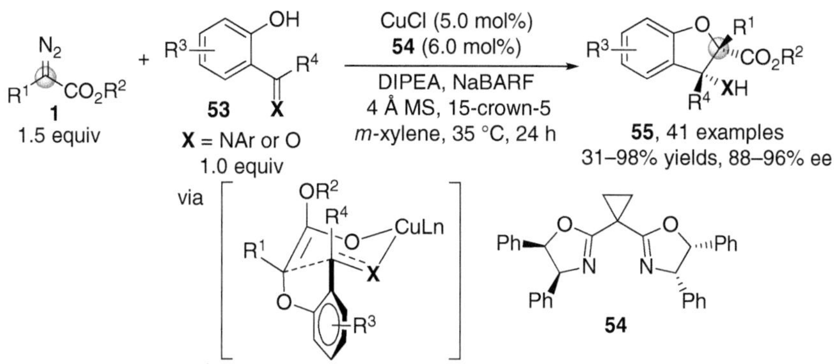

Scheme 11.20 Asymmetric annulation of diazo compounds with 2-iminyl 2-acyl-substituted phenols.

to activate the imine or ketone for diastereo- and enantioselective cyclization. This method provides a practical access for the highly diastereo- and enantioselective construction of 2,3-dihydrobenzofurans **55** bearing tetrasubstituted carbon stereocenters (Scheme 11.20).

11.2.3 Interception of Zwitterionic Intermediate

In 2012, Hu and coworkers reported that the zwitterionic intermediates, which have been proposed previously to undergo a formal C–H insertion reaction, could be smoothly trapped with electrophilic imines when activated by chiral Brøsted acid [38]. This work not only provided a direct experimental evidence for the stepwise mechanism of metal carbene formal C–H insertion reaction but also enabled a novel method for the synthesis of polyfunctionalized indole derivatives (Scheme 11.21).

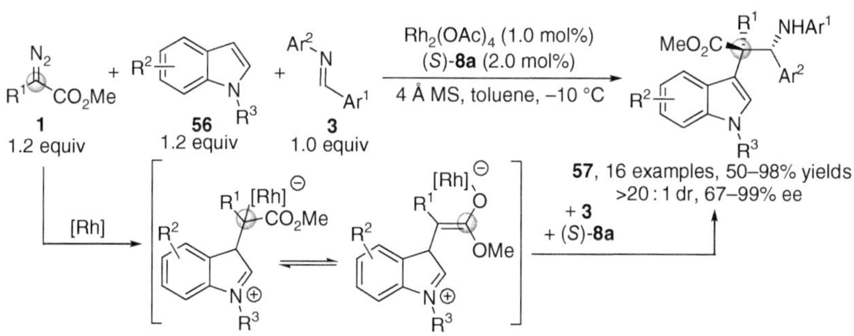

Scheme 11.21 Asymmetric interception of zwitterionic intermediate with imine.

Soon after the above report, the same group found that both unprotected pyrroles **58** [39] and N,N-disubstituted anilines **60** [40] could be introduced as the nucleophiles for the catalytic formation of corresponding zwitterionic intermediates, followed by Mannich-type addition with imines. This method provides a robust approach for the efficient construction of β-amino acid derivatives with an α,α-diaryl

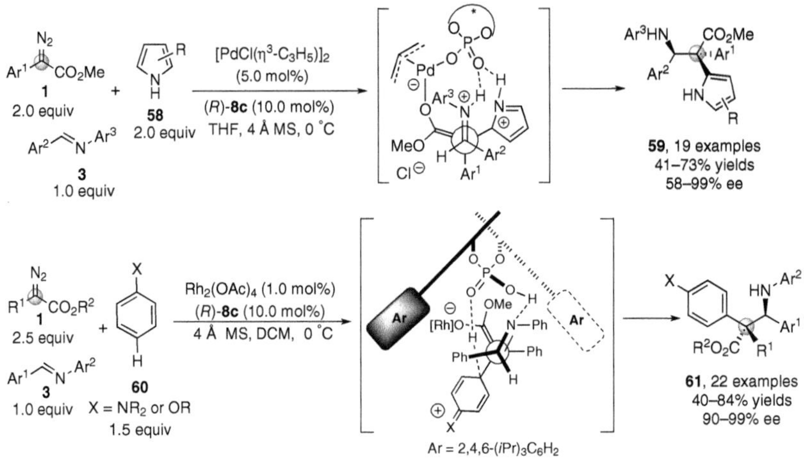

Scheme 11.22 Asymmetric catalytic interception of different zwitterionic intermediates.

benzylic quaternary stereocenter in good yields with high diastereoselectivities and excellent enantioselectivities (Scheme 11.22).

Moreover, different types of electrophilic imines, including isatin-derived ketimines [41a], *in situ* generated iminoesters [41b], and methylene iminium ions [41c], have been found to be compatible for the interception of the zwitterionic intermediates by Shi and Hu. In all these cases, the 3-diazooxindoles were used as the carbene precursor with indoles as the nucleophiles, thus, offering an extremely efficient strategy for the construction of 3,3-disubstituted 3-indol-3′-yloxindoles with an all-carbon quaternary center in good yields.

11.3 Aldol-Type Interception

11.3.1 Interception of Ammonium Ylide

In 2003, the first example of three-component reaction via Aldol-type interception of *in situ* formed reactive ammonium ylide intermediates was reported by Hu and coworkers [11]. In this work, the dirhodium-catalyzed reaction of phenyldiazoacetate **1a** with *o*-anisidine **2a** and 2-nitrobenzaldehyde **52a** gave 1,2-amino alcohol **62** in 90% yield with low diastereoselectivity (Scheme 11.23). Soon after this report, Hu and coworkers investigated this dirhodium-catalyzed Aldol-type addition with various carbonyl compounds, including different aromatic aldehydes [42], formaldehyde [43], and isatins and α-keto esters [44], affording highly substituted α-amino-β-hydroxyl carbonyl compounds in good reactivity with moderate to high stereoselectivity depending on the used electrophilic Aldol-type addition acceptors.

Further exploration of amino species indicated that carbamates [45], aliphatic amines [46], and even ammonia [47] could be applied to the formation ammonium ylide with metal carbene intermediates. In the latter two cases, the iron porphyrin catalyst Fe(TPP)Cl, rather than $Rh_2(OAc)_4$, was employed as the effective catalyst

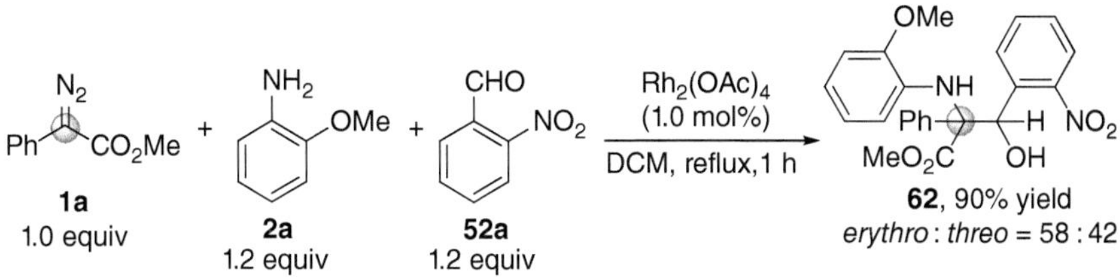

Scheme 11.23 Dirhodium-catalyzed three-component reaction of phenyldiazoacetate with *o*-anisidine and 2-nitrobenzaldehyde.

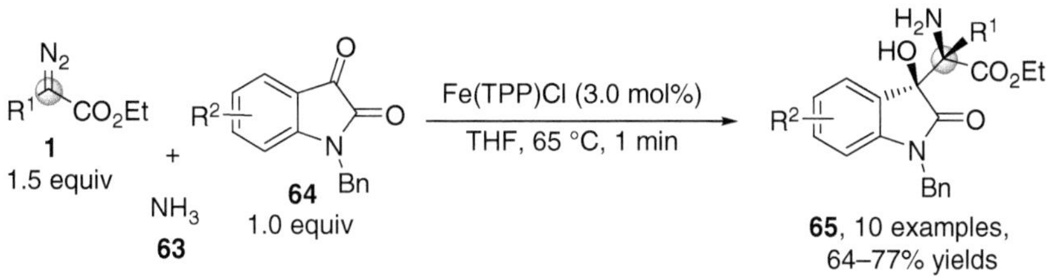

Scheme 11.24 Iron-catalyzed three-component reaction of *α*-alkyl diazoacetates with ammonia and isatins.

that allows aliphatic amines or ammonia to be used as the substrate without encountering catalyst poisoning issue (Scheme 11.24). As the synthetic applications for the construction of N-heterocycles, the dirhodium-catalyzed intramolecular versions have been independently disclosed by Hu and coworkers [48], Moody and coworkers [49], and Rastogi and coworkers [50] for the synthesis of indolines, substituted prolines, and indolones, respectively.

Further study revealed that α-alkyl diazoacetates could also smoothly delivered the corresponding three-component products in the presence of Rh (I)-complex [51]. The intramolecular α–H shift of the α-alkyl carbene intermediate was successfully suppressed by the association of this Rh(I) complex to form the active ammonium ylide with arylamines. Moreover, in this work, a Rh(I)-chiral diene complex was further identified as an effective catalyst to enable an asymmetric version of this Aldol-type interception reaction with good enantioselectivities.

In 2012, Che and coworkers reported the asymmetric catalytic three-component reaction of α-diazophosphonates **66** with anilines and electron-deficient aldehydes to give chiral α-amino-β-hydroxyphosphonates **67** [52]. By using the chiral dirhodium catalysts, Rh_2(*S*-PTAD)$_4$ or Rh_2(*S*-PTTL)$_4$, a series of β-hydroxy-α-amino phosphonates were obtained in good to high yields and with good to high enantioselectivities (Scheme 11.25). These results also provide evidence for the intermediacy of a metal-bound ammonium ylide in the nucleophilic addition step. One year later, a rhodium(II)/chiral Brøsted acid cooperative asymmetric catalytic three-component reaction of 3-diazooxindoles and anilines with glyoxylates has been disclosed by Gong and coworkers [53]. In this reaction, the *in situ* generated ammonium ylide, catalyzed by dirhodium acetate, acts as an active nucleophile to participate in an asymmetric Aldol-type reaction with glyoxylates in the presence of chiral

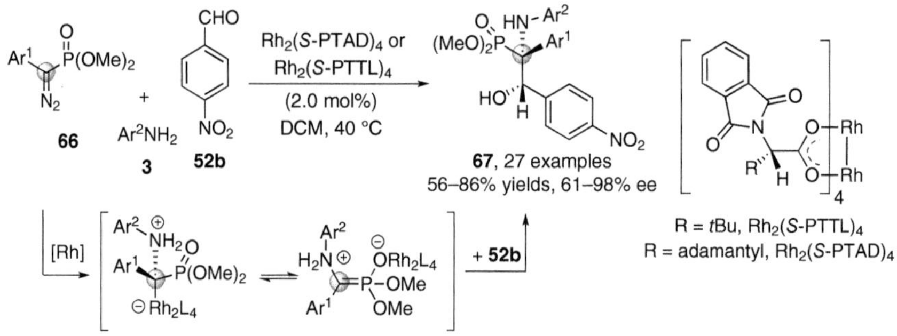

Scheme 11.25 Asymmetric three-component reaction of α-diazophosphonates with anilines and electron deficient aldehydes.

phosphoric acid. This method provides an effective route to access functionalized 3-amino oxindoles in high optical purity.

11.3.2 Interception of Oxonium Ylide

In 2005, Hu and coworkers reported the first example of an Aldol-type interception of oxonium ylide with benzaldehydes in the presence of $Rh_2(OAc)_4$ [22]. In this reaction, only the electron-deficient aromatic aldehydes gave the three-component products **68** in good to high yields with moderate diastereoselectivities, whereas, the O–H insertion product **69** was formed as major product when *p*-anisaldehyde used as the Aldol-type addition acceptor (Scheme 11.26). Later, the same group found that a variety of carbonyl compounds, including isatins [54], α-ketoesters [55], indane-1,2,3-triones [56], and alkynals [57], could be introduced to this three-component reaction, providing a convenient route to α,β-dihydroxy acid derivatives with structural diversity under corresponding catalytic conditions. Notably, Alcaide has successfully applied this method to the three-component reaction with azetidine-2,3-diones, ethyl diazoacetate, and alcohols [58]. This reaction generated the substituted β-lactam derivatives with two new adjacent stereogenic centers in a single step with good to moderate stereoselectivity. On the other hand, a Lewis acid, namely $InBr_3$, could also promote the analogous three-component reaction via Aldol-type addition, but the ylide intermediate was catalytically generated via a non-carbene process [59].

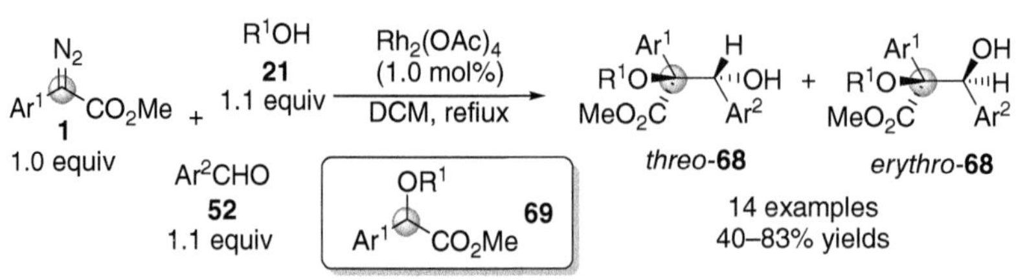

Scheme 11.26 $Rh_2(OAc)_4$-catalyzed three-component reaction of diazoacetates with alcohols and benzaldehydes.

Beyond the alcohols, Hu and coworkers introduced titanium(IV) alkoxides [60] and water [43b, 61] to this three-component reaction to form the key oxonium ylide intermediates under the rhodium or copper catalysis. Moreover, the intramolecular versions of this transformation have been successfully applied to the synthesis of 1,4-dioxan-2-ones [62] and tetrahydrofurans [63] by Hu and Moody independently. High stereoselectivities were observed in these intramolecular Aldol-type transformations; however, only low to moderate enantioselectivities were observed in the formation of tetrahydrofurans with chiral dirhodium catalyst or copper complex [63], which implied that much robust asymmetric catalytic system is needed for this type of Aldol addition.

In 2008, Hu and coworkers has developed a dirhodium/chiral Zr complex co-catalyzed asymmetric three-component reaction of aryl diazoacetates with benzyl alcohols and aldehydes [64]. The reaction occurs through the interception of *in situ* generated reactive alcoholic oxonium ylide with a chiral Zr-complex-activated aldehyde (Scheme 11.27). This method successfully expanded the substrate scope of aldehyde to the electron-neutral and electron-rich ones, providing a practical and highly enantioselective route to the construction of α,β-dihydroxy acid derivatives **68** with good substrate generality.

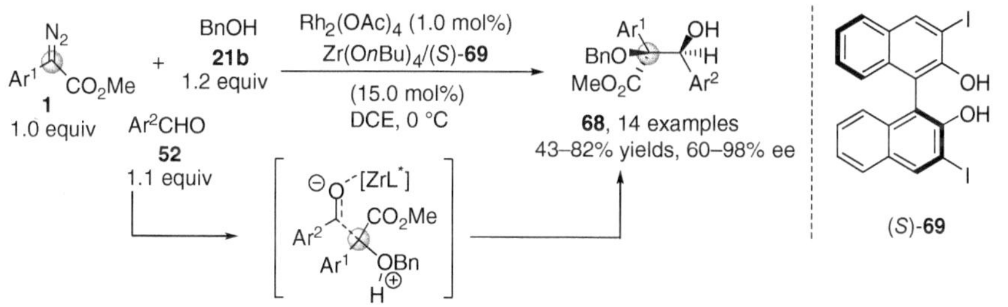

Scheme 11.27 Rh/Chiral Zr–complex co-catalyzed asymmetric three-component reaction of aryl diazoacetates with benzyl alcohol and aldehydes.

On the other hand, by using non-diazo carbene precursors, including homopropargyl alcohols [65] and cyclopropenes [66], Xu and Hu have realized the corresponding Aldol-type addition with isatins as the acceptor in the presence of gold and rhodium catalysts, respectively. In the latter case, $Rh_2(esp)_2$ promoted the ring opening of cyclopropene **70** to generate γ-hydroxyl carbene intermediate, which converted to the cyclic oxonium ylide with the tethered hydroxyl group. Finally, nucleophilic addition of the metal-free ylide intermediate to isatins **64** led to 3-heterocyclic motif substituted 3-hydroxyoxindoles **71** in good to high yields (Scheme 11.28).

11.3.3 Interception of Zwitterionic Intermediate

In 2012, Hu and coworkers reported a dirhodium-catalyzed diastereoselective reaction of diazoacetamides **72** with isatins [67]. By trapping of *in situ* generated

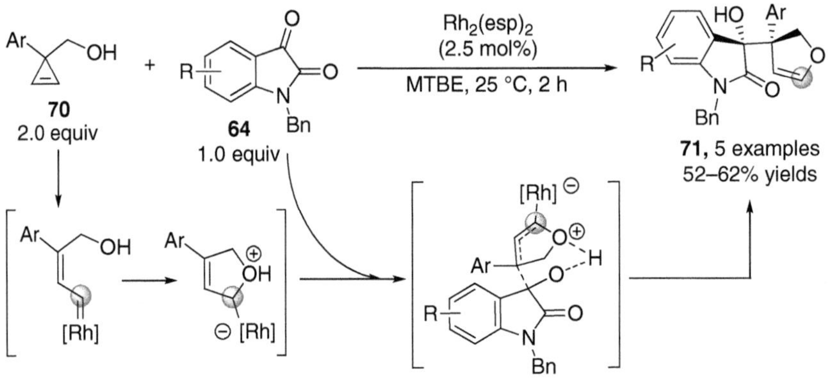

Scheme 11.28 Rh-catalyzed Aldol-type addition of 3-hydroxymethyl-cyclopropenes with isatins.

zwitterionic intermediates with isatins, this reaction produces 3,3′-bioxindole derivatives **74** in good to high yields with high diastereoselectivities (Scheme 11.29). Later, this dirhodium-catalyzed Aldol-type addition has been expanded to the three-component version with broad substrate generality. For example, formaldehyde and glyoxylate derivatives could be used as the Aldol-type addition acceptors in the presence of $Rh_2(OAc)_4$ in ethyl acetate and dichloromethane (DCM), respectively [68]. Moreover, the *in situ* generated benzopyrylium ions under acidic conditions were also found to be compatible for the interception of zwitterionic intermediates in the presence of palladium complex, furnishing C-2 functionalized 2*H*-chromene derivatives in moderate to high yields (43–75%) with good to excellent diastereoselectivities (up to >95 : 5 dr) [69]. At the same time, the intramolecular version with α-phenoxy ketones and 3-diazooxindoles led to a series of functionalized spiro-chroman-4,3′-oxindole derivatives in a highly diastereoselective manner with very good yields [70]. Despite these advances, the asymmetric version of this Aldol-type interception of zwitterionic intermediates is still under exploration.

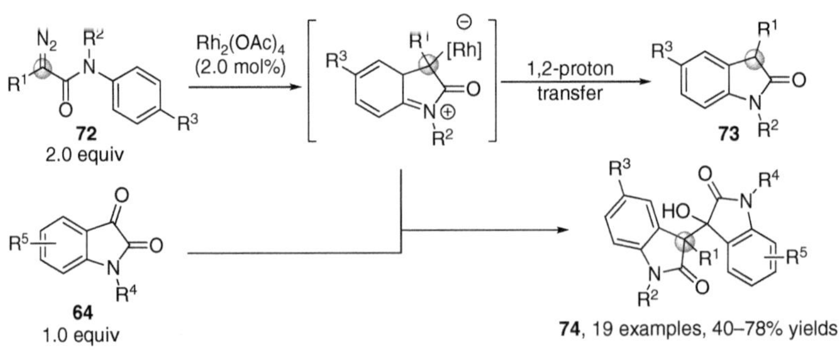

Scheme 11.29 Dirhodium-catalyzed reaction of diazoacetamides with isatins.

11.4 Michael-Type Interception

11.4.1 Interception of Ammonium Ylide

In 2013, Hu and coworkers developed a three-component reaction of diazo compounds with anilines and 4-*oxo*-enoates **75** [48a]. This reaction proceeds through dirhodium-catalyzed ammonium ylide generation/Michael addition process to form linear α-amino ester derivatives **76** in good to high yields (Scheme 11.30). The generated α-amino-ester-derived product could be readily cyclized under acidic conditions, providing the 2,3-dihydropyrrole as a single *trans*-isomer in nearly quantitative yield. Moreover, by controlling the addition sequence of the substrates, this three-component reaction could go through an *aza*-Michael addition/ylide generation/intramolecular Aldol-type addition cascade to generate pyrrolidines with the assistance of alkaline Al_2O_3. The intramolecular version of this transformation with diazomalonates and α,β-unsaturated *δ*-(*N*-aryl)amino esters, which features a [4+1]-annulation reaction, has been disclosed by Nikolaev and coworkers in 2015 [71].

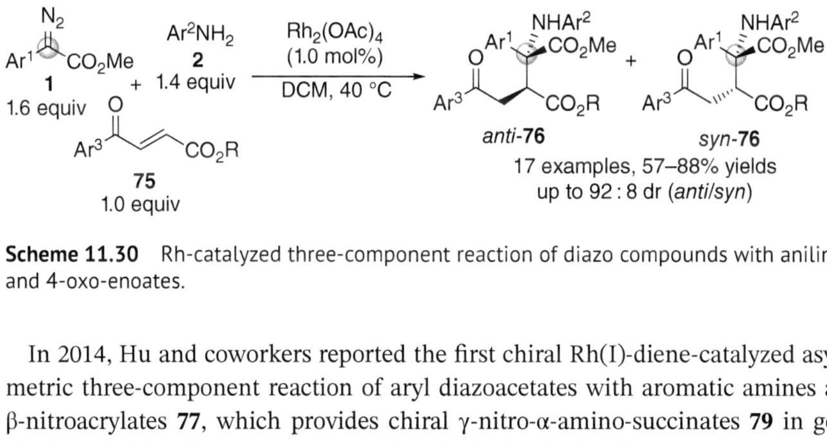

Scheme 11.30 Rh-catalyzed three-component reaction of diazo compounds with anilines and 4-oxo-enoates.

In 2014, Hu and coworkers reported the first chiral Rh(I)-diene-catalyzed asymmetric three-component reaction of aryl diazoacetates with aromatic amines and β-nitroacrylates **77**, which provides chiral γ-nitro-α-amino-succinates **79** in good yields with high diastereo- and enantioselectivities [72]. This asymmetric reaction is proposed to proceed through the enantioselective trapping of Rh(I)-associated ammonium ylides by nitroacrylates (Scheme 11.31). Later, the same group has developed a metal/chiral guanidine cooperative catalysis method for the analogous three-component reaction, resulting valuable functionalized 1,4-diketone derivatives in good yields with a moderate enantioselectivity (up to 71% ee) [73].

In 2015, Hu and coworkers reported a palladium(II)/chiral phosphoric acid catalyzed three-component reaction of aryldiazoacetates with enamines **80** and imines that affords α-amino-δ-oxo pentanoic acid derivatives **81** in good yields with excellent diastereoselectivities and high enantioselectivities [74]. In this work a novel bond-forming strategy that involves bond cleavage, fragment modification, and reconstruction of the modified fragment sequence has been proposed. This

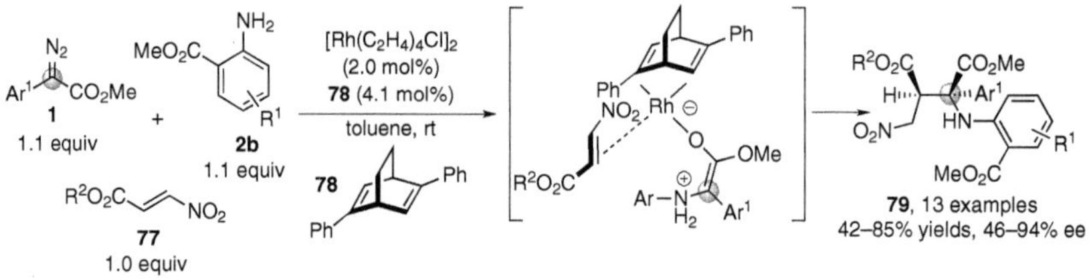

Scheme 11.31 Asymmetric trapping of chiral rhodium(I)-associated ammonium ylides with β-nitroacrylates.

innovative transformation represents a highly efficient way to build structurally diversified polyfunctional molecules in an atom and step economic fashion. The keto-iminium has been proposed as a key intermediate and a chiral palladium/phosphoric acid (PA*) complex is the real catalyst in this MCR (Scheme 11.32).

11.4.2 Interception of Oxonium Ylide

In 2009, a highly stereoselective one-pot synthesis of fully substituted tetrahydrofurans was reported by Hu and coworkers [75]. This stepwise reaction involves a rhodium-catalyzed O–H insertion of aryl diazoacetate with a secondary ally alcohol **82**, followed by a highly stereoselective intramolecular Michael type addition in the presence of piperidine (Scheme 11.33). After this work, a variety of three-component reactions that involve a Michael-type trapping of *in situ* generated oxonium ylide intermediate have been reported by the same group with corresponding Michael addition acceptors [76]. Beyond these readily available electrophilic acceptors, the *in situ* generated electrophilic species, such as 2-indolylmethanol-derived delocalized cations, and indole- or ketal-derived iminiums, have been introduced to this transformation by Shi and coworkers [77] and Hu and coworkers [78], respectively.

The first asymmetric Michael-type addition of reactive oxonium ylide was disclosed by Hu and coworkers in 2010 [79]. With a combination of $Rh_2(OAc)_4$, (*S*)-*t*Bu-box-Zn(OTf)$_2$, and TsOH, they have developed an enantioselective three-component reaction of diazo compounds with water and α,β-unsaturated 2-acyl imidazoles **84**, leading to δ hydroxyketone derivatives **86** containing a stereogenic quaternary carbon center in good yields with excellent enantioselectivities (Scheme 11.34). Recently, the same group has reported a chiral secondary amine and iridium(I) co-catalyzed version with enals as electrophiles, followed by a reduction process, generating 1,2,5-triol derivatives with vicinal chiral centers in good yields and excellent enantioselectivities [80].

Another type of asymmetric Michael addition of active oxonium ylide was developed by Schneider with a transition metal and chrial Brøsted acid synergistic catalysis method [81]. A notable feature of this reaction is the using of water, which was a leaving fragment in the formation of *ortho*-quinone methide (*o*-QM) in the presence of phosphoric acid, to trap the rhodium–carbene intermediate, thus furnishing the reactive oxonium ylide. The excellent diastereo- and enantioselectivity have been achieved with the assistance of chiral phosphoric acid

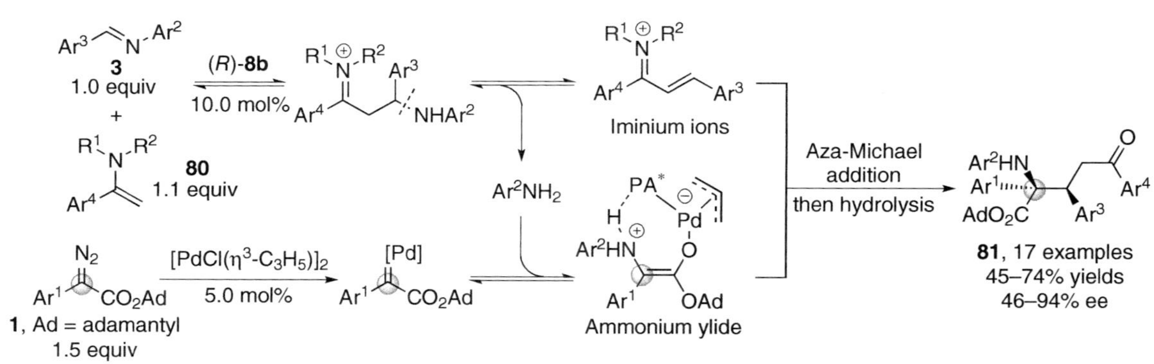

Scheme 11.32 Asymmetric three-component reaction involving a bond cleavage, fragment modification, and reassembly cascade.

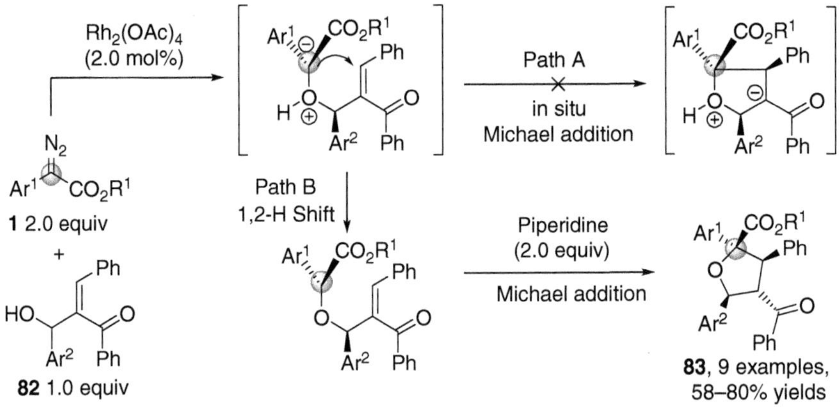

Scheme 11.33 Rhodium/piperidine relay catalysis for the one-pot synthesis of fully substituted tetrahydrofurans.

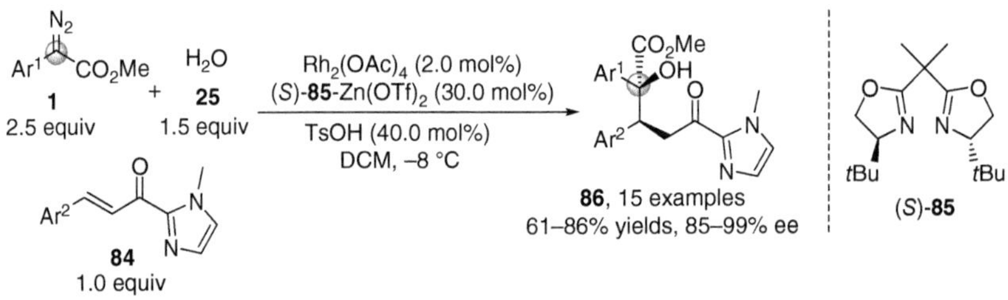

Scheme 11.34 Asymmetric Michael-type addition of reactive oxonium ylide with α,β-unsaturated 2-acyl imidazoles.

during the Michael-type addition of transient oxonium ylide with *o*-QM via H-bonding (Scheme 11.35). Recently, the intramolecular Michael-type addition of active oxonium ylide has been reported by Hu and coworkers [82]. With the dirhodium and chiral phosphoric acid synergistic catalysis, the reaction produced spirochroman-3,3-oxindoles with vicinal quaternary and tertiary stereocenters in good to excellent yields and enantioselectivities [82b].

11.4.3 Interception of Zwitterionic Intermediate

In 2015, the first Michael-type addition of active zwitterionic intermediate was developed by Gong via a cooperative catalysis system of $Rh_2(esp)_2$ with amino acid-derived chiral phosphine [83]. The reaction goes through a sequential C–H functionalization/asymmetric allylation or Michael addition process, affording a diverse range of enantioenriched 3,3′-indolyloxindole derivatives **93**. Notably, according to the observed positive effect of LiCl, the author proposed an interaction model that the lithium enolate might be able to form a chiral thiourea–chloride complex through Li–Cl coordination to assist the stereochemical control (Scheme 11.36). Soon after this work, Zhou and coworkers [84a] and Hu and coworkers [84b] independently reported the asymmetric Michael-type addition via analogous catalytic

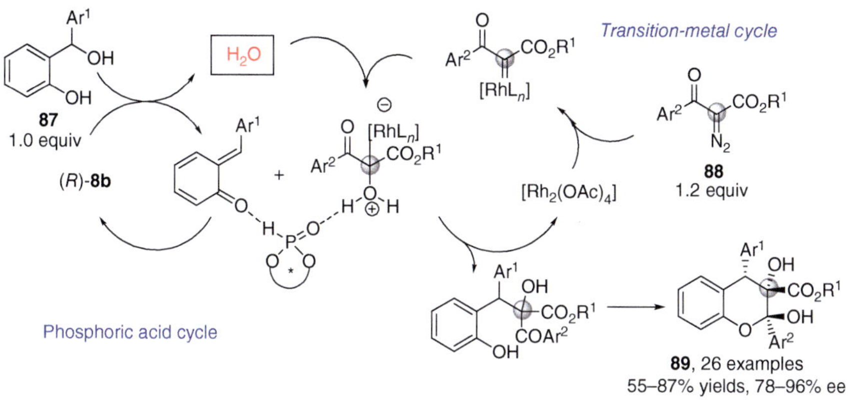

Scheme 11.35 Asymmetric Michael-type addition of reactive oxonium ylide with *ortho*-quinone methide.

strategy, using nitroenynes and Morita–Baylis–Hillman (MBH) carbonates as the Michael addition acceptors, respectively.

One year later, Hu and coworkers reported the first example of iridium(I) and chiral secondary amine co-catalyzed enantioselective three-component reaction of aryldiazoacetates with indoles and enals [85]. This reaction is proposed to go through the enantioselective trapping of iridium(I)-associated zwitterionic intermediates via a selective 1,4-addtition with the chiral amine activated enals (Scheme 11.37). This method generates 3-substituted indole derivatives **94** in good yields and with moderate to excellent stereoselectivities. Recently, the analogous two-component version has been disclosed by Hu and coworkers [86a] and Xu and coworkers [86b], delivering 4-(2-oxoindolin-3-yl)-1,4-dihydropyridines and dihydrocyclopenta[*b*]indoles in the presence of palladium and copper catalysts, respectively.

11.5 Miscellaneous Transformations

11.5.1 Interception Other Types of Active Intermediates

Sulfonium ylides, carboxylic oxonium ylides, and alkynoate copper intermediates are three other types of active intermediates that have been applied to the three-component reaction via the electrophilic interception process. In 2007, Hu and Huang reported the first example of trapping catalytically generated sulfonium ylides with azodicarboxylates [87]. This reaction proceeds via an unusual 1,2-*aza* shift from a sulfonium ylide intermediate, giving *N*,*S*-ketals **96** in moderate to good yields (Scheme 11.38). Recently, Hu and coworkers have subjected the sulfonium ylide intermediates to Mannich-, Aldol-, and Michael-type additions under different catalytic conditions with corresponding acceptors [88], providing polyfunctionalized linear and cyclic sulfur-containing compounds with good efficiency and excellent stereoselectivity.

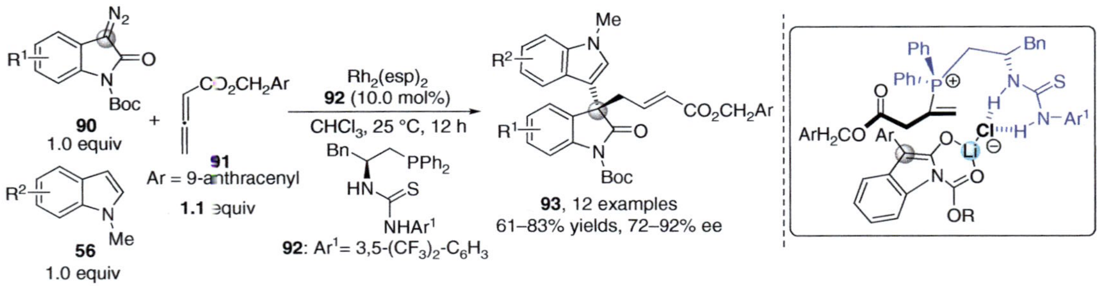

Scheme 11.36 Asymmetric Michael-type addition of zwitterionic intermediate with allenoates.

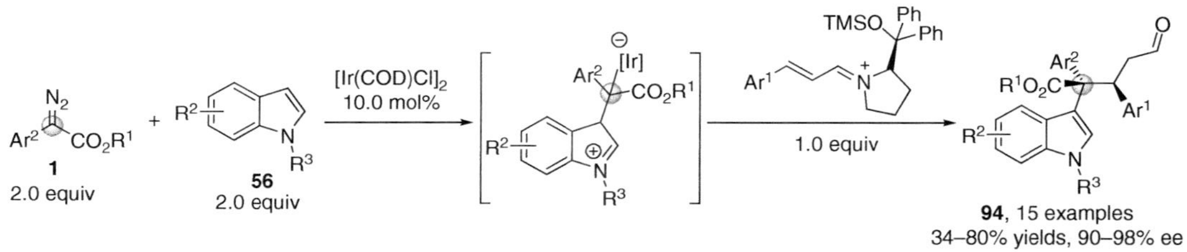

Scheme 11.37 Asymmetric Michael-type addition of zwitterionic intermediate with enals.

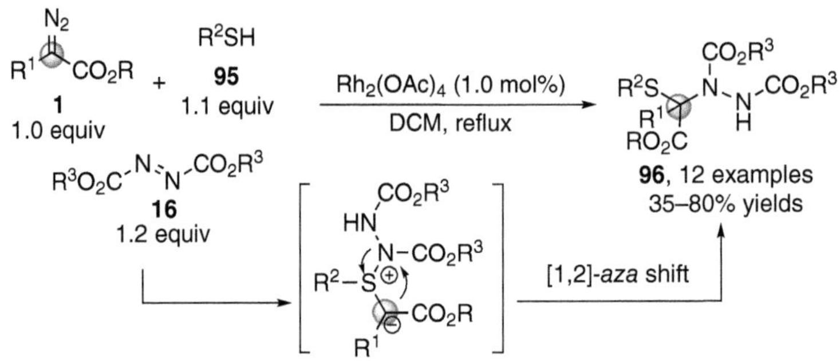

Scheme 11.38 Interception of sulfonium ylide with azodicarboxylates.

The first $Rh_2(OAc)_4$-catalyzed Mannich-type addition of carboxylic oxonium ylides with *N*-Boc imines was reported by Hu and coworkers in 2014, which offers a direct access to β-amino alcohol derivatives in good yields with moderate to good stereoselectivities [89a]. Two years later, Li and coworkers reported a copper-catalyzed version with oxidatively formed iminium cations as the Mannich addition acceptor, providing a practical access to α-(acyloxy)-β-amino ketones and α,β-diamino ketones in moderate to good yields [89b]. Recently, Hu and coworkers accomplished the rhodium-catalyzed Aldol- and Mannich-type interception reactions of carboxyl oxonium ylides with isatins and imines, respectively (Scheme 11.39) [89c]. The key carboxyl oxonium ylides have been proposed to generate from **97** through the reaction between donor–acceptor rhodium carbene intermediates and the tethered secondary amide species. Moreover, using of chiral phosphoric acid as the co-catalyst resulted in high enantioselectivity in the Mannich-type addition.

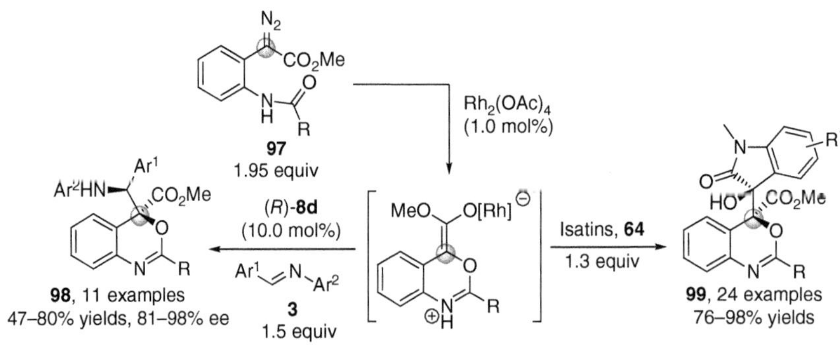

Scheme 11.39 Catalytic interception of carboxylic oxonium ylides. Source: Modified from Jia et al. [89c].

In 2019, a three-component reaction between terminal alkynes **100**, α-diazo amides **72**, and isatin ketimines **28** was realized by Hu and coworkers under copper catalysis [90a]. This reaction is proposed to proceed through a Mannich-type trapping process of an alkynoate copper species derived from terminal alkyne and

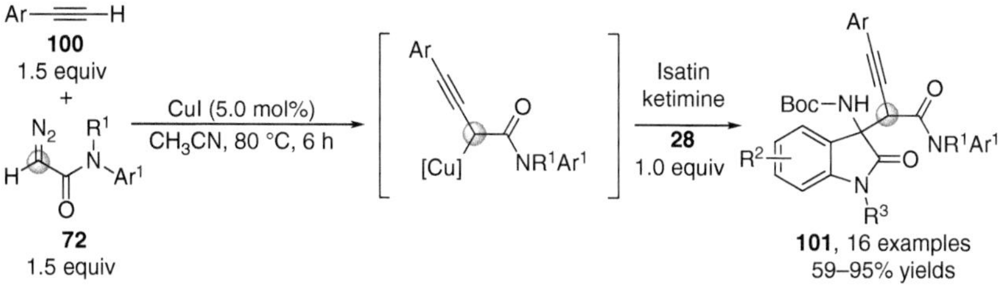

Scheme 11.40 Mannich-type addition of alkynoate copper intermediate with imines.

copper carbene intermediate. A variety of alkynyl-containing 3,3-disubstituted oxindole derivatives **101** were synthesized with high efficiency and good diastereoselectivity under mild reaction conditions (Scheme 11.40). Meanwhile, the analogous alkynoate copper species derived from terminal alkyne and trifluoromethyl diazo compound was applied to the [3+2]-cycloaddition with nitrosoarenes by the same group, and a series of fluorinated dihydroisoxazoles were effectively synthesized in high yields [90b].

11.5.2 Interception of Active Intermediates with Other Electrophiles

Beyond the carbonyl compounds, hypervalent iodine reagents have been successfully introduced as effective electrophiles for the interception of active ylide intermediates by Waser and coworkers [91] and Szabó and coworkers [92], independently. In 2016, Waser and Hari developed an atom economical oxy-alkynylation of diazo compounds with ethynylbenziodoxol(on)e (EBX) **103** [91a]. In this reaction, the intramolecular alkyne transfer of the organocopper species, which generated from metal carbene intermediate via addition with the carboxylate group of the EBX reagent, is the key step, leading to the *gem*-difunctionalization products **105** in good to high yields (Scheme 11.41). In the case with vinyl diazo compounds **102**, they obtained enynes **106** as single olefin isomers in high yields. The enantioselective catalytic version has also been developed by Waser and Hari with a chiral copper complex, giving highly functionalized α-benzoyloxy propargylic esters with excellent yields and enantioselectivities [91b].

In the same year, Szabó and coworkers disclosed a rhodium-catalyzed novel *gem*-difunctionalization of diazocarbonyl compounds with fluoro-benziodoxole **107** as the electrophilic reagent [92]. The reaction has been proposed to go through an electrophilic trapping process of the rhodium-containing oxonium ylide-type intermediates with F or CF_3 electrophiles (Scheme 11.42).

At the same time, Lautens and coworkers reported a Ru/Pd-co-catalyzed asymmetric allylic alkylation of α-diazoamides with allyl *tert*-butyl carbonate **111** [93]. The oxindole enolate species has been proposed as the key intermediate in this dual-metal-catalyzed reaction, which combines ruthenium-catalyzed C–H functionalization with palladium-catalyzed asymmetric allylic alkylation, affording chiral 3-allyl-3-aryl oxindoles **113** in up to 99% yield with up to 85% ee (Scheme 11.43).

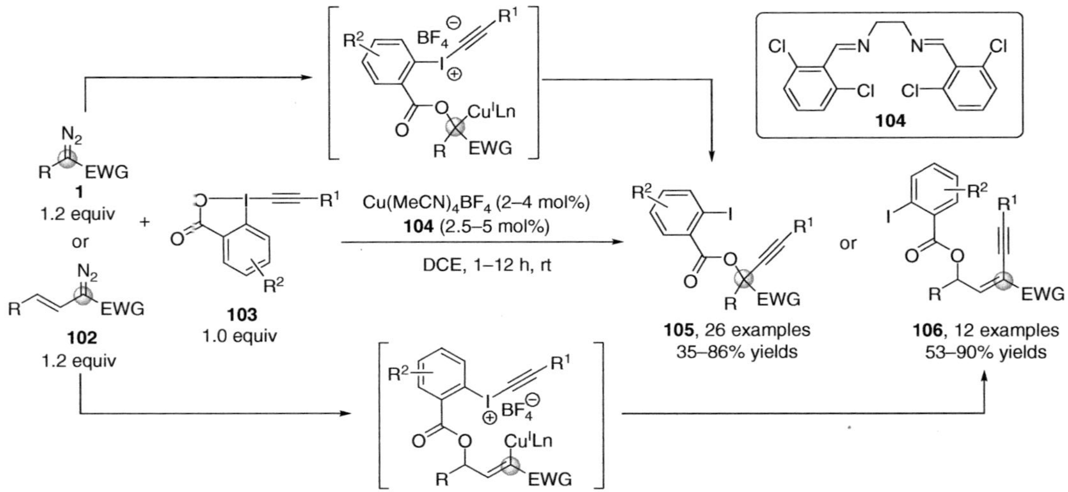

Scheme 11.41 Copper-catalyzed oxy-a.kynylation of diazo compounds with ethynylbenziodoxol(on)es.

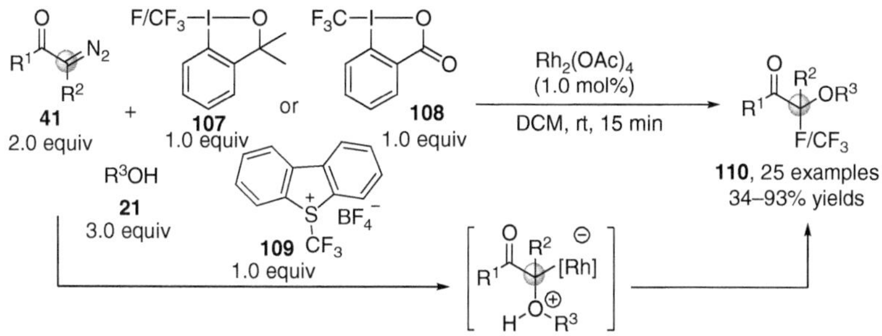

Scheme 11.42 Rhodium-catalyzed *gem*-difunctionalization of diazocarbonyl compounds with fluoro-benziodoxole reagent.

Soon after Lautens' report, Anbarasan and coworkers accomplished an unprecedented reaction that involving intramolecular interception of ammonium ylides with nonactivated double bond [94]. A reaction mechanism has also been proposed based on the experimental and DFT studies, which is imitated by the formation of ammonium ylide followed by cyclization through metallo-ene-type reaction via a six-membered cyclic transition state. This method allows the synthesis of various tri- and tetrasubstituted indolines with two contiguous quaternary carbon centers in good yields with high diastereoselectivities. On the other hand, the catalytic cyclization of ylide intermediates with tethered electrophilic alkyne species has been reported by Hu and coworkers [95], Xu and coworkers [96], and Sun and coworkers [97] independently, leading to dihydrofurans, furans, dihydropyrroles, and pyrrolidines with high efficiency and structural diversity.

11.5.3 Applications in Cascade Reactions

In 2012, Hu and coworkers disclosed a $Rh_2(OAc)_4$-catalyzed three-component tandem 1,4-conjugate addition–cyclization reaction of diazoacetophenones with anilines and unsaturated ketoesters **114** [98a]. This reaction features a formal [3+1+1] cycloaddition process, generating multi-substituted pyrrolidines in moderate to high yields (up to 84%) with high diastereoselectivities. Five years later, the same group developed a Ru(II)/chiral amine co-catalyzed asymmetric version with enals **115** [98b]. The reaction mechanism has been postulated as a sequential trapping of Ru(II)-associated ammonium ylides via a selective 1,4-addition to chiral amine-activated enals, followed by an *aza*-aldol process (Scheme 11.44). Moreover, the control experiments and DFT calculation revealed that the reversible NaOAc-facilitated *aza*-aldol process resulted in the diastereomeric conversion, forming multi-substituted pyrrolidines **117** in one step with excellent diastereoselectivities and enantioselectivities.

On the other hand, the Mannich-type interception products of ylide intermediates with imines, which possess one or two nucleophilic amino species, could also be subjected to the cascade transformations for the synthesis of different N-heterocycles [99]. For example, Katukojvala and coworkers have realized a synergistic rhodium and phosphoric acid-catalyzed three-component reaction

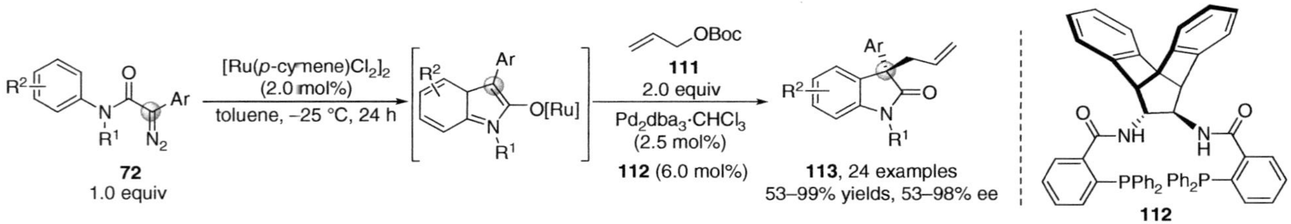

Scheme 11.43 Ru/Pd co-catalyzed asymmetric allylic alkylation of α-diazoamides with allyl *tert*-butyl carbonate.

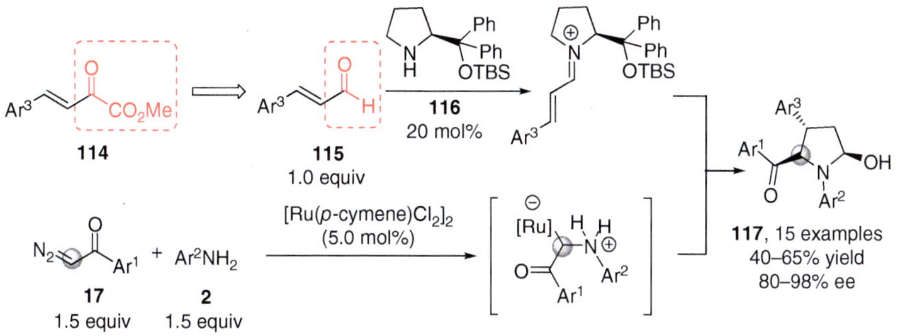

Scheme 11.44 Asymmetric Michael-type addition of ammonium ylide followed by *aza*-Aldol-type addition.

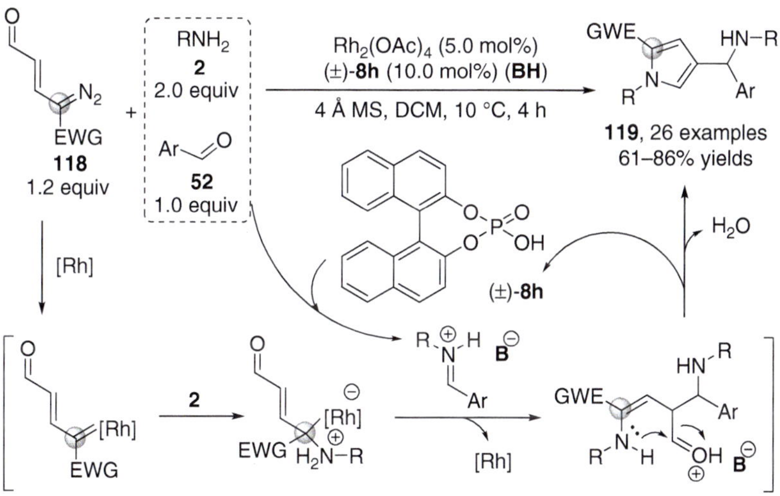

Scheme 11.45 Three-component cascade reaction of enaldiazo compounds with arylamines and aryl aldehydes.

of enaldiazo compounds **118** with arylamines and aryl aldehydes, leading to the direct synthesis of α-pyrrolylbenzylamines **119** in good to high yields [99a]. The reaction is proposed to involve a transient ammonium ylide that generated from rhodium enalcarbene, followed by a regioselective Mannich-type addition and cyclocondensation cascade (Scheme 11.45). Meanwhile, Hu and coworkers have developed various cascade reactions based on the one-pot transformations of the Mannich-type addition products, including *aza*-Michael- and *aza*-Mannich-type addition [99], leading to the corresponding N-heterocycles with high efficiency and selectivity.

In addition to the above-mentioned advances on the cascade transformations, these generated polyfunctionalized products via the catalytic *gem*-difunctionalization of metal carbene species could be applied to a variety of other types of manipulations, such as coupling reaction, cycloaddition, rearrangement, and others [100]. All these results not only show excellent functionality compatibility that allow a wide range of functionalities to be pre-installed in each reagent in these MCRs but also provide a rapid and DOS of an array of chiral polyfunctional heterocycles through a variety of subsequent cyclization reactions with these implanted functional groups. On the other hand, *gem*-difunctionalization of diazo compounds via the non-metal–carbene pathway has not been included in this part [101].

11.6 Synthetic Applications

11.6.1 Synthesis and Modification of Natural Products

The $Rh_2(OAc)_4$/chiral Brønsted acid co-catalyzed three-component reactions of diazo acetates with alcohols and imines is an effective approach for the direct synthesis of optically active β-amino-α-hydroxyl acid derivatives. In 2010, Hu successfully applied this method for the synthesis of the taxol side chain and (−)-*epi*-cytoxazone [102] (Scheme 11.46). Deprotection of **120a** with ammonium formate in the presence of Pd/C (10%) gave the free alcohol **121** with >99% ee after recrystallization, followed by reprotection with TBSCl, oxidative cleavage of the *p*-methoxyphenyl (PMP) group, benzoylation, and deprotection with tetrabutylammonium fluoride (TBAF), the α-hydroxyl ester **122** was synthesized in a yield of 46% (four steps) without any loss of ee. Finally, the optically pure taxol side chain **123** was synthesized with >99% ee after hydrolysis (Scheme 11.46b). The (−)-*epi*-cytoxazone was prepared in four steps from **120b** (Scheme 11.46c). The deprotected product was treated with triphosgene and diisopropylethylamine (DIPEA) in dichloromethane, followed by oxidation with ceric ammonium nitrate in acetonitrile and water, leading to the compound **125** in 54% yield. The subsequent reduction with superhydride gave the (4*R*,5*S*)-*epi*-cytoxazone **126** in 72% yield.

In 2013, Hu and coworkers reported a rhodium-catalyzed three-component reaction of 3-diazooxindoles with indoles and ethyl glyoxylates [103]. This reaction allows the straightforward synthesis of mixed 3,3′-bisindoles in high yields with excellent diastereoselectivities. These products could be easily converted to the key skeleton that could be applied to the total synthesis of natural alkaloid (±)-gliocladin C. Soon after this report, Gong and coworkers [104] and Han and coworkers [105] independently applied this MCR for the synthesis of indole alkaloids with 3-diazooxindoles as the carbene precursor. For example, Gong has demonstrated that a combination of either ruthenium(II) or rhodium(II) complexes with chiral quinine-derived squaramide enables a highly efficient asymmetric three-component reaction of 3-diazooxindoles with indoles and nitroalkenes [104]. This method generates optically active 3,3′-bisindoles **128**, which turned out to be applicable to the total synthesis of (−)-Folicanthine **130** in seven steps with 14.5% overall yield (Scheme 11.47).

Recently, Duan and coworkers have reported a Rh-catalyzed three-component reaction of rapamycin **131** with diazoesters and isatins [106]. This method has enabled a site- and diastereoselective functionalization of 40-OH in rapamycin by 3-hydroxyoxindole moieties through the active oxonium ylide trapping reaction with isatins, resulting in new analogs of rapamycin **132** (Scheme 11.48). Notably,

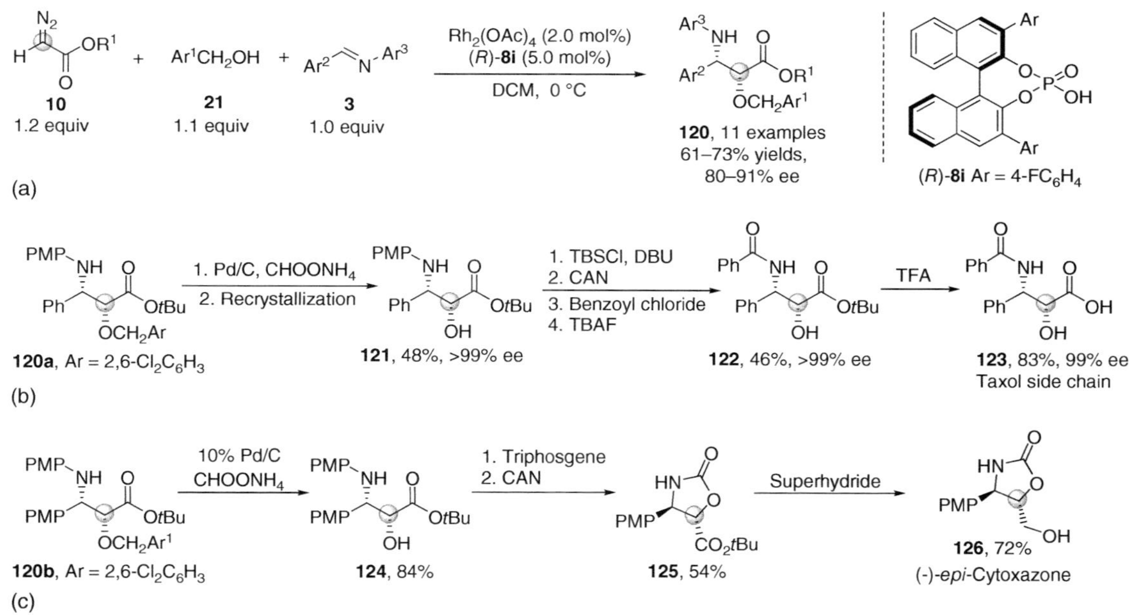

Scheme 11.46 Synthesis of taxol side chain and (−)-*epi*-cytoxazone. Source: Modified from Qian et al. [102].

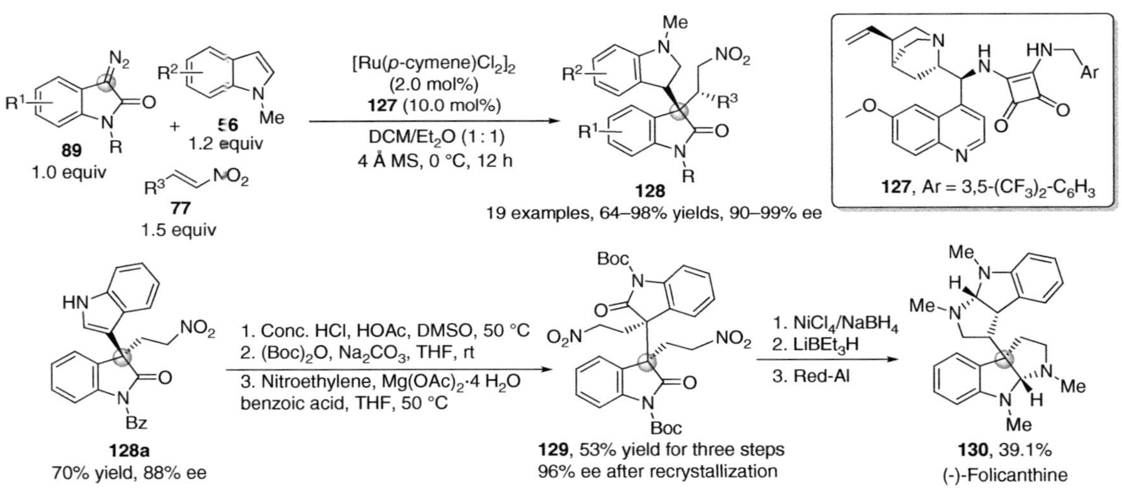

Scheme 11.47 Synthesis of (−)-Folicanthine.

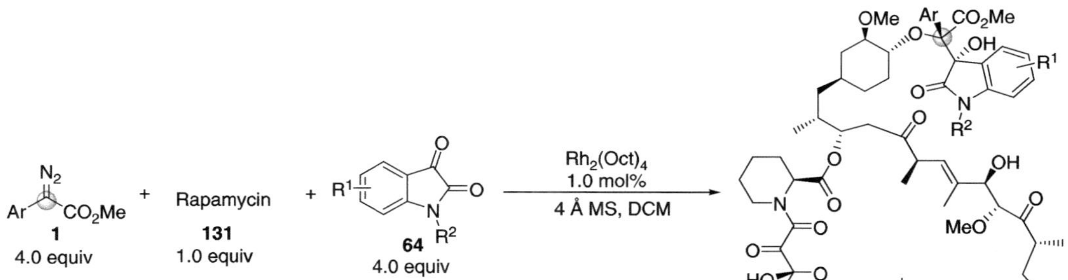

The cytotoxicity (IC_{50} in nM)	A549	Hela	SKBR3
132a, Ar = Ph, R^1 = 6-F, R^2 = H	8.3 ± 3.1	62.9 ± 4.7	9.9 ± 4.7
132b, Ar = Ph, R^1 = 6-CF_3, R^2 = H	4.7 ± 1.3	8.8 ± 1.8	5.2 ± 2.3
132c, Ar = Ph, R^1 = 7-CF_3, R^2 = H	4.8 ± 2.0	52.3 ± 12	5.5 ± 1.9
132d, Ar = Ph, R^1 = 5,6-diF, R^2 = H	7.4 ± 2.1	13.8 ± 4.1	39.7 ± 7.8
132e, Ar = Ph, R^1 = H, R^2 = Me	8.2 ± 3.7	10.5 ± 3.3	5.5 ± 2.7
132f, Ar = Ph, R^1 = H, R^2 = Ac	3.8 ± 1.1	20.3 ± 4.8	18.4 ± 8.1

Scheme 11.48 Rhodium-catalyzed three-component reaction of rapamycin with diazoesters and isatins, and cytotoxicity study.

several of these compounds have IC_{50} in lower nM range against the three cancer cell lines. In addition, these analogs showed reduced cytotoxicity against the tested cancer cell lines comparing to rapamycin. This study demonstrated the potential utility of MCRs in late-stage functionalization of complex natural products.

11.6.2 Synthesis of Bioactive Molecules

In 2013, Perumal and coworkers reported a Rh-catalyzed one-pot three-component reaction of 3-diazooxindole with benzyl alcohol and different carbonyl compounds [107]. They suggested that the reaction occurs through a metal-associated alcoholic oxonium ylide intermediate and followed by Aldol-type addition with electrophiles, such as isatin, ninhydrin, and acenaphthenequinone. Antibacterial activities of these synthesized oxindoles **133** against five Gram +ve and Gram −ve organisms, including *Staphylococcus aureus* NCIM 5021, *Escherichia coli* NCIM 2931, *Pseudomonas aeruginosa* NCIM 5029, *Salmonella typhimurium* NCIM 2501, and *Proteus vulgaris* NCIM 2813, were evaluated using microdilution method. Group-based quantitative structure–activity relationships (GQSARs) for all antibacterial activities were developed. The statistical measures with all the tested compounds were found to be satisfactory. Three compounds were found to be the most active against all the five organisms (Table 11.1).

At the same time, Hu and coworkers reported that the α-hydroxyl-δ-amino ester derivatives **135**, which obtained from 1,4-conjugate addition of β,γ-unsaturated *N*-sulfonylimines **134** via the oxonium ylides trapping process, exhibit inhibitory activity on PTP1B and SIRT1 enzymes in vitro [108a]. The initially biological investigation revealed that few products exhibited significant inhibitory activity

Table 11.1 Synthesis of 3-bisoxindoles and their antibacterial activity study.

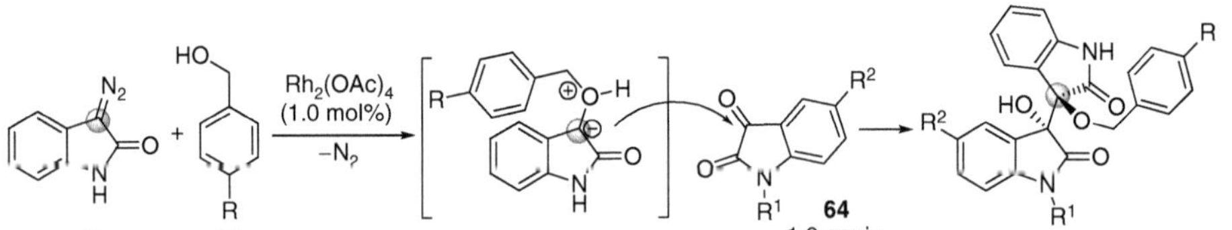

89 1.2 equiv; **21** 1.2 equiv; **133**, 82–90% yields

	133			Antibacterial activity (minimum inhibitory concentration (MIC) in μM)				
Entry	R	R^1	R^2	*S. aureus*	*P. aeruginosa*	*E. coli*	*P. vulgaris*	*S. typhi*
1	OMe	H	H	0.075	0.038	0.038	0.075	0.038
2	Cl	H	H	0.149	0.074	0.074	0.149	0.074
3	Br	H	H	0.067	0.067	0.034	0.067	0.034

Table 11.2 Synthesis of γ-dehydro-α-hydroxy-δ-amino esters and their inhibitory activity against PTP1B and SIRT1.

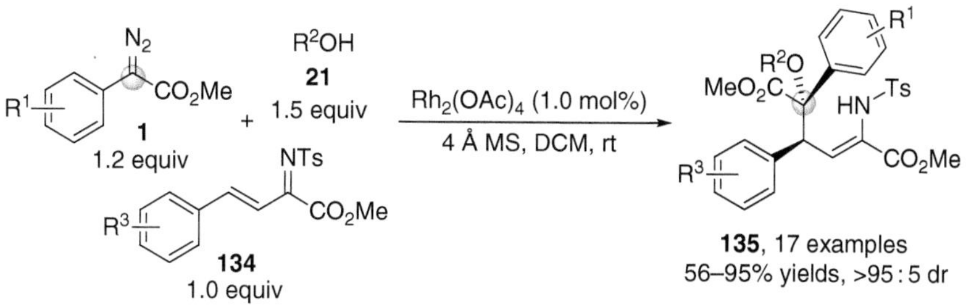

Entry	135			PTP1B IC_{50} (μM)	SIRT1 IC_{50} (μM)
	R^1	R^2	R^3		
1	4-Me	Bn	4-Cl	5.4–0.8	14.3–4.0
2	4-Br	Bn	4-Cl	17.8–2.6	—
3	H	Et	4-Cl	1.6–0.1	—
4	H	Me	4-Cl	—	15.6–1.7
5	H	Bn	4-Me	2.2–0.2	—

against both PTP1B and SIRT1 with IC_{50} values in the low micromolar range, which could be a potential novel class of PTP1B and SIRT1 inhibitors (Table 11.2). Two years later, Zhu and coworkers found another molecule, 3,3′-bisindoles that generated via the $Rh_2(OAc)_4$ catalyzed three-component reaction of 3-diazooxindoles with indoles and different electrophiles, as a novel PTP1B inhibitors with low micromole ranged inhibitory activity [108b]. The best active compound inhibited PTP1B activity with an IC_{50} of 2.79 μM. Meanwhile, it had low cytotoxicity and enhanced glucose uptake in vitro.

Recently, a $Rh_2(OAc)_4$ catalyzed highly diastereoselective Mannich-type reaction of aryl diazoacetates with propiolamides **136** and imines has been reported by Hu and coworkers [109a]. Most of these generated products **137** exhibited good anticancer activity in HCT116, BEL7402, and SMMC7721 cells (Table 11.3). Moreover, the established reaction provides a direct method to introduce a terminal alkynyl species into complex molecules for hit hunting. All these results validate the potential application of the developed method in drug discovery. In the same year, Hu and coworkers reported a three-component reaction for the synthesis of substituted 3-amino-3-hydroxymethyloxindoles, which showed anticancer potency via in vitro cytotoxicity screening [109b]. The most potent compound showed cytotoxic effect toward SJSA-1 cells (IC_{50} = 3.14 μM), as well as inhibiting the growth of other type of cancer cell lines (HCT-116, Jurkat, KB and Bel7402).

Table 11.3 Synthesis of alkynylamide-substituted α,β-diamino acid derivatives and their anticancer activity study.

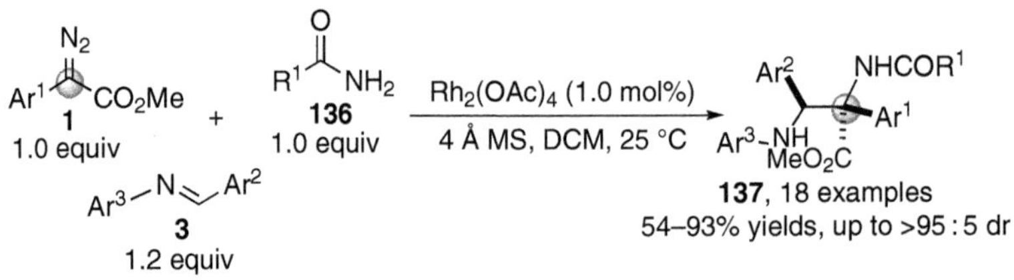

Entry	137			IC$_{50}$ (μM)		
	Ar^1	R^1	Ar^2/Ar^3	HCT116	BEL7402	SMMC7721
1	Ph	Acetylene	4-BrPh/2-OHPh	1.748	1.874	6.460
2	Ph	Acetylene	4-FPh/2-OHPh	2.973	2.005	2.902
3	Ph	Acetylene	4-MePh/2-OHPh	2.065	2.577	3.901
4	Ph	Acetylene	Ph/2-OHPh	1.580	1.540	3.191
5	Ph	Acetylene	2-furan/2-OHPh	3.762	3.141	3.305
6	3-ClPh	Acetylene	4-ClPh/2-OHPh	2.046	2.453	2.591

11.7 Conclusion

As summarized in this chapter, MCR involving electrophilic interceptions of *in situ* generated ylide/zwitterionic intermediates with different electrophiles, which features as a *gem*-difunctionalization process of metal carbene species, has been widely applied to a series of transformations and demonstrated the great potential of metal carbene species in synthesizing of polyfunctionalized molecules with structural diversity. Although distinct advances, including reaction selectivity, molecular complexity, and synthetic efficiency, have been achieved world widely in the past 17 years, continuous efforts in this area will be focused on: (i) introducing of non-diazo carbene precursors, including alkynes, cyclopropenes, enynones, and others, to realize novel MCRs with atom and step economy under mild conditions; (ii) discovery of effective catalysts and catalytic systems for the chemo- and stereoselective controls in these MCRs, especially the cooperative asymmetric catalysis strategies for the effective activation of electrophiles without interaction site; and (iii) further exploration of other types of reactive intermediate, rather than ylide/zwitterionic intermediates, to enable new transformations via different interception process with broad substrate generality. Novel methodology and synthetic applications are foreseeable in due course with this innovative strategy for the rational design and development of new MCRs.

References

1 (a) Dömling, A. (2005). *Chem. Rev.* 106: 17–89. (b) Zhu, J., Wang, Q., and Wang, M. (2015). *Multicomponent Reactions in Organic Synthesis.* Weinheim: Wiley–VCH.

2 Hulme, C. and Gore, V. (2003). *Curr. Med. Chem.* 10: 51–80.

3 Ugi, I. and Steinbruckner, C. (1960). *Angew. Chem.* 72: 267–268.

4 (a) Pellissier, H. (2020). *Adv. Synth. Catal.* 362: 2289–2325. (b) Nunes, P.S.G., Vidal, H.A., and Corrêa, A.G. (2020). *Org. Biomol. Chem.* 18: 7751–7773.

5 (a) Doyle, M.P., Ratnikov, M., and Liu, Y. (2011). *Org. Biomol. Chem.* 9: 4007–4016. (b) Nawrat, C.C. and Moody, C.J. (2011). *Nat. Prod. Rep.* 28: 1426–1444. (c) Caballero, A. and Pérez, P.J. (2017). *Chem. Eur. J.* 23: 14389–14393. (d) Pei, C., Zhang, C., Qian, Y., and Xu, X. (2018). *Org. Biomol. Chem.* 16: 8677–8685. (e) Cavitt, M.A., Phun, L.H., and France, S. (2014). *Chem. Soc. Rev.* 43: 804–818.

6 (a) Davies, H.M.L. and Morton, D. (2011). *Chem. Soc. Rev.* 40: 1857–1869. (b) Zhu, D., Chen, L., Fan, H. et al. (2020). *Chem. Soc. Rev.* 49: 908–950. (c) Zhu, S. and Zhou, Q. (2012). *Acc. Chem. Res.* 45: 1365–1377. (d) Xu, X. and Doyle, M.P. (2014). *Acc. Chem. Res.* 47: 1396–1405. (e) Qian, D. and Zhang, J. (2015). *Chem. Soc. Rev.* 44: 677–698.

7 Padwa, A. (2009). *Chem. Soc. Rev.* 38: 3072–3081.

8 (a) Xia, Y., Zhang, Y., and Wang, J. (2013). *ACS Catal.* 3: 2586–2598. (b) Xia, Y., Qiu, D., and Wang, J. (2017). *Chem. Rev.* 117: 13810–13889.

9 (a) Guo, X. and Hu, W. (2013). *Acc. Chem. Res.* 46: 2427–2440. (b) Zhang, D. and Hu, W. (2017). *Chem. Rec.* 17: 739–753.

10 (a) Doyle, M.P., Duffy, R., Ratnikov, M., and Zhou, L. (2010). *Chem. Rev.* 110: 704–724. (b) Ford, A., Miel, H., Ring, A. et al. (2015). *Chem. Rev.* 115: 9981–10080. (c) Marinozzi, M., Pertusati, F., and Serpi, M. (2016). *Chem. Rev.* 116: 13991–14055. (d) Ceias, N.R., Paterna, R., and Gois, P.M.P. (2016). *Chem. Rev.* 116: 2937–2981.

11 Wang, Y., Zhu, Y., Chen, Z. et al. (2003). *Org. Lett.* 5: 3923–3926.

12 Jiang, J., Xu, H., Xi, J. et al. (2011). *J. Am. Chem. Soc.* 133: 8428–8431.

13 Kisan, H.K. and Sunoj, R.B. (2016). *Org. Lett.* 18: 3746–3749.

14 Hu, W., Jiang, L., Zhang, D., and Wang, Z. (2013). *Synthesis* 45: 452–458.

15 Li, J., Zhang, D., Chen, J. et al. (2020). *ACS Catal.* 10: 4559–4565.

16 Jiang, J., Ma, X., Liu, S. et al. (2013). *Chem. Commun.* 49: 4238–4240.

17 Yu, S., Fu, X., Liu, G. et al. (2018). *Acta Chim. Sinica* 76: 895–900.

18 Qiu, L., Wang, D., Lv, F. et al. (2014). *Tetrahedron* 70: 1471–1477.

19 Huang, H., Wang, Y., Chen, Z., and Hu, W. (2005). *Adv. Synth. Catal.* 347: 531–534.

20 Jing, C., Xing, D., Qian, Y., and Hu, W. (2014). *Synthesis* 46: 1348–1354.

21 Guan, X., Tang, M., Liu, Z., and Hu, W. (2019). *Chem. Commun.* 55: 9809–9812.

22 Lu, C., Liu, H., Chen, Z. et al. (2005). *Org. Lett.* 7: 83–86.
23 (a) Huang, H., Guo, X., and Hu, W. (2007). *Angew. Chem. Int. Ed.* 46: 1337–1339. (b) Zhang, X., Zhang, N., Guo, X. et al. (2009). *Tetrahedron* 65: 8277–8282.
24 (a) Hu, W., Xu, X., Zhou, J. et al. (2008). *J. Am. Chem. Soc.* 130: 7782–7783. (b) Zhou, J., Xu, X., and Hu, W. (2011). *Org. Synth.* 88: 418–426.
25 Guo, Z., Shi, T., Jiang, J. et al. (2009). *Org. Biomol. Chem.* 7: 5028–5033.
26 Qian, Y., Jing, C., Shi, T. et al. (2011). *ChemCatChem* 3: 653–656.
27 Guo, Z., Huang, H., Fu, Q., and Hu, W. (2006). *Synlett* 15: 2486–2488.
28 Rajasekaran, T., Karthik, G., Sridhar, B. et al. (2014). *Eur. J. Org. Chem.*: 2221–2224.
29 Ao, C., Huang, J., Xu, X. et al. (2020). *Org. Biomol. Chem.* 18: 3466–3470.
30 (a) Guo, Z., Cai, M., Jiang, J. et al. (2010). *Org. Lett.* 12: 652–655. (b) Qiu, L., Guo, X., Qian, Y. et al. (2016). *Chem. Commun.* 52: 11831–11833.
31 Kang, Z., Zhang, D., Shou, J., and Hu, W. (2018). *Org. Lett.* 20: 983–986.
32 (a) Kang, Z., Wang, Y., Zhang, D. et al. (2019). *J. Am. Chem. Soc.* 141: 1473–1478. (b) Meng, X., Yang, B., Zhang, L. et al. (2019). *Org. Lett.* 21: 1292–1296. (c) Yu, J., Chen, L., and Sun, J. (2019). *Org. Lett.* 21: 1664–1667.
33 Che, J., Niu, L., Jia, S. et al. (2020). *Nat. Commun.* 11: 1511–1519.
34 (a) Xu, X., Qian, Y., Yang, L., and Hu, W. (2011). *Chem. Commun.* 47: 797–799. (b) Li, M., Zheng, Q., Jin, W. et al. (2016). *Tetrahedron* 72: 2929–2934. (c) Yu, S., Hua, R., Fu, X. et al. (2019). *Org. Lett.* 21: 5737–5741. (d) Jia, S., Lei, Y., Song, L. et al. (2017). *Chin. Chem. Lett.* 28: 213–217. (e) Lv, X., Liu, S., Zhou, S. et al. (2020). *CCS Chem.* 2: 1903–1912.
35 (a) Wei, H., Bao, M., Dong, K. et al. (2018). *Angew. Chem. Int. Ed.* 57: 17200–17204. (b) Sahani, R.L., Patil, M.D., Wagh, S.B., and Liu, R. (2018). *Angew. Chem. Int. Ed.* 57: 14878–14882.
36 (a) Xu, X., Zhou, J., Yang, L., and Hu, W. (2008). *Chem. Commun.* 48: 6564–6566. (b) Qian, Y., Jing, C., Liu, S., and Hu, W. (2013). *Chem. Commun.* 49: 2700–2702.
37 Liang, X., Li, R., and Wang, X. (2019). *Angew. Chem. Int. Ed.* 58: 13885–13889.
38 Qiu, H., Li, M., Jiang, L. et al. (2012). *Nat. Chem.* 4: 733–738.
39 Zhang, D., Qiu, H., Jiang, L. et al. (2013). *Angew. Chem. Int. Ed.* 52: 13356–13360.
40 Jia, S., Xing, D., Zhang, D., and Hu, W. (2014). *Angew. Chem. Int. Ed.* 53: 13098–13101.
41 (a) Mei, L., Tang, X., and Shi, M. (2014). *Org. Biomol. Chem.* 12: 1149–1156. (b) Jing, C., Xing, D., and Hu, W. (2015). *Org. Lett.* 17: 4336–4339. (c) Kang, Z., Zhang, D., Xu, X., and Hu, W. (2019). *Org. Lett.* 21: 9878–9883.
42 Wang, Y., Chen, Z., Mi, A., and Hu, W. (2004). *Chem. Commun.* 40: 2486–2487.
43 (a) Wang, C., Liu, S., Xing, D. et al. (2013). *Tetrahedron* 69: 11203–11208. (b) Wang, C., Xing, D., Wang, D. et al. (2014). *J. Org. Chem.* 79: 3908–3916.
44 Guo, X., Yue, Y., Hu, G. et al. (2009). *Synlett* 13: 2109–2114.
45 Xi, J., Ma, M., and Hu, W. (2016). *Tetrahedron* 72: 579–583.
46 Ma, C., Xing, D., Zhai, C. et al. (2013). *Org. Lett.* 15: 6140–6143.

47 Ma, C., Chen, J., Xing, D. et al. (2017). *Chem. Commun.* 53: 2854–2857.

48 (a) Jing, C., Xing, D., Qian, Y. et al. (2013). *Angew. Chem. Int. Ed.* 52: 9289–9292. (b) Jing, C., Xing, D., and Hu, W. (2014). *Chem. Commun.* 50: 951–953.

49 Nicolle, S.M., Lewis, W., Hayes, C.J., and Moody, C.J. (2016). *Angew. Chem. Int. Ed.* 55: 3749–3753.

50 Devi, L., Shukla, R., and Rastogi, N. (2019). *Org. Biomol. Chem.* 17: 135–139.

51 Liu, S., Jiang, J., Chen, J. et al. (2017). *Chem. Sci.* 8: 4312–4317.

52 Zhou, C., Wang, J., Wei, J. et al. (2012). *Angew. Chem. Int. Ed.* 51: 11376–11380.

53 Ren, L., Lian, X., and Gong, L. (2013). *Chem. Eur. J.* 19: 3315–3318.

54 Guo, X., Huang, H., Yang, L., and Hu, W. (2007). *Org. Lett.* 9: 4721–4723.

55 (a) Yue, Y., Guo, X., Chen, Z. et al. (2008). *Tetrahedron Lett.* 49: 6862–6865. (b) Ji, J., Zhang, X., Zhu, Y. et al. (2011). *J. Org. Chem.* 76: 5821–5824.

56 Guo, X., Wang, J., Yang, L., and Hu, W. (2010). *Lett. Org. Chem.* 7: 106–109.

57 Han, X., Jiang, L., Tang, M., and Hu, W. (2011). *Org. Biomol. Chem.* 9: 3839–3843.

58 (a) Alcaide, B., Almendros, P., Aragoncillo, C. et al. (2009). *J. Org. Chem.* 74: 8421–8424. (b) Alcaide, B., Almendros, P., Aragoncillo, C. et al. (2012). *Eur. J. Org. Chem.*: 2359–2366.

59 Ji, J., Zhang, X., Jiang, L., and Hu, W. (2012). *Tetrahedron Lett.* 53: 182–185.

60 Lu, C., Liu, H., Chen, Z. et al. (2005). *Chem. Commun.* 41: 2624–2626.

61 (a) Zhang, X., Guo, X., Yang, L., and Hu, W. (2009). *Chin. Chem. Lett.* 20: 1299–1302. (b) Jing, C., Shi, T., Xing, D. et al. (2013). *Green Chem.* 15: 620–624.

62 Guo, X., Liu, W., and Hu, W. (2014). *Chem. Asian J.* 9: 117–120.

63 Nicolle, S.M., Lewis, W., Hayes, C.J., and Moody, C.J. (2015). *Angew. Chem. Int. Ed.* 54: 8485–8489.

64 Zhang, X., Huang, H., Guo, X. et al. (2008). *Angew. Chem. Int. Ed.* 47: 6647–6649.

65 Cai, J., Wang, X., Qian, Y. et al. (2019). *Org. Lett.* 21: 369–372.

66 Zhang, D., Kang, Z., Liu, J., and Hu, W. (2019). *iScience* 14: 292–300.

67 Li, M., Zan, L., Prajapati, D., and Hu, W. (2012). *Org. Biomol. Chem.* 10: 8808–8813.

68 (a) Jing, C., Xing, D., Wang, C., and Hu, W. (2015). *Tetrahedron* 71: 3597–3602. (b) Jia, S., Song, L., Lei, Y. et al. (2016). *Org. Biomol. Chem.* 14: 10157–10160.

69 Zhang, D. and Hu, W. (2018). *Adv. Synth. Catal.* 360: 2446–2452.

70 Jia, S., Lei, Y., Song, L. et al. (2017). *Adv. Synth. Catal.* 359: 58–63.

71 Medvedev, J.J., Galkina, O.S., Klinkova, A.A. et al. (2015). *Org. Biomol. Chem.* 13: 2640–2651.

72 Ma, X., Jiang, J., Lv, S. et al. (2014). *Angew. Chem. Int. Ed.* 53: 13136–13139.

73 Wei, Q., Li, Z., Yang, X. et al. (2019). *Mol. Diversity* 23: 997–1010.

74 Zhang, D., Zhou, J., Xia, F. et al. (2015). *Nat. Commun.* 6: 5801–5808.

75 Xu, X., Han, X., Yang, L., and Hu, W. (2009). *Chem. Eur. J.* 15: 12604–12607.

76 (a) Zhu, Y., Zhai, C., Yang, L., and Hu, W. (2010). *Chem. Commun.* 46: 2865–2867. (b) Han, X., Gan, M., Qiu, H. et al. (2011). *Synlett* 12: 1717–1722. (c) Liu, G., Yu, S., Hu, W., and Qiu, H. (2019). *Chem. Commun.* 55:

12675–12678. (d) Yu, S., Chen, J., Liu, G. et al. (2020). *Chem. Commun.* 56: 1649–1652.

77 Ma, C., Zhou, J., Zhang, Y. et al. (2018). *Chem. Asian J.* 13: 2549–2558.

78 (a) Liu, S., Yao, W., Liu, Y. et al. (2017). *Sci. Adv.* 3: e1602467. (b) Qiu, H., Liu, X., Xiao, G. et al. (2020). *Adv. Synth. Catal.* 362: 1961–1965.

79 Guan, X., Yang, L., and Hu, W. (2010). *Angew. Chem. Int. Ed.* 49: 2190–2192.

80 Li, M., Guo, X., Zheng, Q. et al. (2017). *J. Org. Chem.* 82: 5212–5221.

81 Alamsetti, S.K., Spanka, M., and Schneider, C. (2016). *Angew. Chem. Int. Ed.* 55: 2392–2396.

82 (a) Alavala, G.K.R., Sajjad, F., Shi, T. et al. (2018). *Chem. Commun.* 54: 12650–12653. (b) Reddy, A.G.K., Niharika, P., Zhou, S. et al. (2020). *Org. Lett.* 22: 2925–2930.

83 Chen, D., Zhang, C., Hu, Y. et al. (2015). *Org. Chem. Front.* 2: 956–960.

84 (a) Cao, Z., Zhao, Y., and Zhou, J. (2016). *Chem. Commun.* 52: 2537–2540. (b) Yang, Y., Ma, C., Thumar, N.J., and Hu, W. (2016). *J. Org. Chem.* 81: 8537–8543.

85 Li, M., Guo, X., Jin, W. et al. (2016). *Chem. Commun.* 52: 2736–2739.

86 (a) Zhang, D., Kang, Z., and Hu, W. (2017). *J. Org. Chem.* 82: 5952–5958. (b) Dong, K., Pei, C., Zeng, Q. et al. (2019). *Chem. Commun.* 55: 6393–6396.

87 Huang, H. and Hu, W. (2007). *Synlett* 8: 1314–1316.

88 (a) Xiao, G., Ma, C., Xing, D., and Hu, W. (2016). *Org. Lett.* 18: 6086–6089. (b) Xiao, G., Chen, T., Ma, C. et al. (2018). *Org. Lett.* 20: 4531–4535. (c) Xiao, G., Ma, C., Wu, X. et al. (2018). *J. Org. Chem.* 83: 4786–4791.

89 (a) Zhai, C., Xing, D., Qian, Y. et al. (2014). *Synlett* 25: 1745–1750. (b) Zhang, T., Hao, W., Wang, N. et al. (2016). *Org. Lett.* 18: 3078–3081. (c) Jia, S., Yang, X., Dong, G. et al. (2019). *Org. Lett.* 21: 4014–4018.

90 (a) Che, J., Reddy, A.G.K., Niu, L. et al. (2019). *Org. Lett.* 21: 4571–4574. (b) Lv, X., Kang, Z., Xing, D., and Hu, W. (2018). *Org. Lett.* 20: 4843–4847.

91 (a) Hari, D.P. and Waser, J.J. (2016). *Am. Chem. Soc.* 138: 2190–2193. (b) Hari, D.P. and Waser, J. (2017). *J. Am. Chem. Soc.* 139: 8420–8423.

92 Yuan, W., Eriksson, L., and Szabó, K.J. (2016). *Angew. Chem. Int. Ed.* 55: 8410–8415.

93 Yamamoto, K., Qureshi, Z., Tsoung, J. et al. (2016). *Org. Lett.* 18: 4954–4957.

94 Reddy, A.C.S., Choutipalli, V.S.K., Ghorai, J. et al. (2017). *ACS Catal.* 7: 6283–6288.

95 (a) Shi, T., Guo, X., Teng, S., and Hu, W. (2015). *Chem. Commun.* 51: 15204–15207. (b) Shi, T., Teng, S., Reddy, A.G.K. et al. (2019). *Org. Biomol. Chem.* 17: 8737–8744. (c) Lv, X., Yang, H., Shi, T. et al. (2019). *Adv. Synth. Catal.* 361: 1265–1270.

96 Bao, M., Qian, Y., Su, H. et al. (2018). *Org. Lett.* 20: 5332–5335.

97 Liu, K., Zhu, C., Min, J. et al. (2015). *Angew. Chem. Int. Ed.* 54: 12962–12967.

98 (a) Zhang, X., Ji, J., Zhu, Y. et al. (2012). *Org. Biomol. Chem.* 10: 2133–2138. (b) Li, M., Chu, R., Chen, J. et al. (2017). *Org. Lett.* 19: 1290–1293.

99 (a) Dawe, S.G., Kanchupalli, V., Lad, B.S. et al. (2014). *Org. Lett.* 16: 3700–3703. (b) Jiang, J., Ma, X., Ji, C. et al. (2014). *Chem. Eur. J.* 20: 1505–1509. (c) Qiu,

L., Guo, X., Ma, C. et al. (2014). *Chem. Commun.* 50: 2196–2198. (d) Kang, Z., Zhang, D., and Hu, W. (2017). *Org. Lett.* 19: 3783–3786.

100 (a) Jiang, L., Jin, W., and Hu, W. (2016). *ACS Catal.* 6: 6146–6150. (b) Tang, M., Xing, D., Huang, H., and Hu, W. (2015). *Chem. Commun.* 51: 10612–10615. (c) Shi, T., Teng, S., Wei, Y. et al. (2019). *Green Chem.* 21: 4936–4940. (d) Tang, M., Wu, Y., Liu, Y. et al. (2016). *Acta Chim. Sinica* 74: 54–60. (e) Arunprasath, D. and Sekar, G. (2019). *Org. Lett.* 21: 867–871.

101 (a) Ma, C., Xing, D., and Hu, W. (2016). *Org. Lett.* 18: 3134–3137. (b) Li, J., Ma, C., Xing, D., and Hu, W. (2019). *Org. Lett.* 21: 2101–2105. (c) Zheng, Y., Bian, R., Zhang, X. et al. (2016). *Eur. J. Org. Chem.* 2016: 3872–3877. (d) Dong, K., Yan, B., Chang, S. et al. (2016). *J. Org. Chem.* 81: 6887–6892. e) Zheng, Y., Dai, P., Gao, D. et al. (2020). *iScience* 23: 101168. (f) Jana, S., Yang, Z., Pei, C. et al. (2019). *Chem. Sci.* 10: 10129–10134. (g) Chen, J., Liu, S., Lv, X. et al. (2020). *J. Org. Chem.* 85: 13920–13928.

102 Qian, Y., Xu, X., Jiang, L. et al. (2010). *J. Org. Chem.* 75: 7483–7486.

103 Xing, D., Jing, C., Li, X. et al. (2013). *Org. Lett.* 15: 3578–3581.

104 Chen, D., Zhao, F., Hu, Y., and Gong, L. (2014). *Angew. Chem. Int. Ed.* 53: 10763–10767.

105 Lian, X., Meng, J., and Han, Z. (2016). *Org. Lett.* 18: 4270–4273.

106 Qiu, L., Su, M., Wen, Z. et al. (2019). *Eur. J. Org. Chem.* 2019: 2914–2918.

107 Lakshmi, N.V., Sivakumar, P.M., Muralidharan, D. et al. (2013). *RSC Adv.* 3: 496–507.

108 (a) Qiu, L., Gao, L., Tang, J. et al. (2014). *J. Org. Chem.* 79: 4142–4147. (b) Jing, C., Li, Z., Jia, K. et al. (2017). *Arch. Pharm. Chem. Life Sci.* 350: e1600173.

109 (a) Lei, R., Wu, Y., Dong, S. et al. (2017). *J. Org. Chem.* 82: 2862–2869. (b) Jia, K., Lv, X., Xing, D. et al. (2017). *RSC Adv.* 7: 23265–23271.

12

Transition-Metal-Catalyzed Cross-Coupling with Carbene Precursors

Kang Wang and Jianbo Wang

Peking University, College of Chemistry and Molecular Engineering, No. 292 Chengfu Rd., Haidian District, Beijing 100871, China

12.1 Introduction

Transition-metal-catalyzed cross-coupling reactions have been extensively explored as indispensable methodologies in modern organic synthesis, providing opportunities of convenient assembly of molecular complexity [1]. Although this field has witnessed significant progresses, it is still fascinating to explore novel cross-coupling partners. Recently, diazo compounds, *N*-tosylhydrazones as well as various non-diazo compounds have been employed as versatile coupling partners in transition-metal-catalyzed carbene coupling reactions [2–4]. The common feature of this type of cross-coupling reactions is the formation of metal carbene species and the carbene migratory insertion process.

Mechanistically, the transition-metal-catalyzed carbene cross-coupling reactions are initiated by classic cross-coupling steps, such as oxidative addition, transmetalation, C—H bond activation, or C—C bond activation. The resulted organometallic species **1** can react with carbene precursors to produce metal–carbene species **2**, which then undergoes rapid carbene migratory insertion to generate a new C—C bond. Subsequently, the resulted organometallic species **3** undergoes various transformations, including β–H elimination, protonation, reductive elimination, or other process to produce the corresponding coupling products (Scheme 12.1).

Due to the generality of carbene migratory insertion process, the catalytic carbene cross-coupling reactions have been established as versatile methodologies toward the construction of diverse chemical structures. This area is still evolving, in which various transformations based on new carbene precursors, different coupling partners, and novel cascade processes have been explored continuously. This chapter is aimed to describe the development of transition-metal-catalyzed cross-couplings with various carbene precursors involving carbene migratory insertion. The chapter is primarily organized based on the transition metal catalysts that are employed in the coupling reactions, mainly including palladium, copper, and rhodium. In addition to the catalytic carbene coupling reactions employing electrophilic or

Transition Metal-Catalyzed Carbene Transformations, First Edition.
Edited by Jianbo Wang, Chi-Ming Che, and Michael P. Doyle.

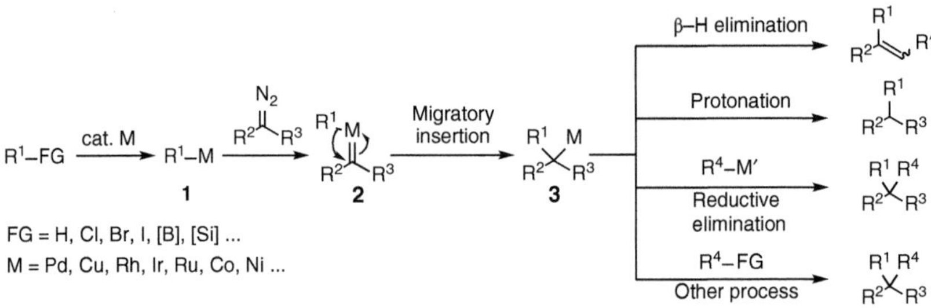

Scheme 12.1 General reaction pathways for carbene coupling reactions.

nucleophilic partners, the transition-metal-catalyzed C—H bond functionalization involving carbene migratory insertion is also discussed as an independent section (Section 12.5) in this chapter.

12.2 Palladium-Catalyzed Carbene Cross-Coupling Reactions

12.2.1 Diazo Compounds as Carbene Precursors

12.2.1.1 Reactions with Electrophiles

Diazo compounds are the most commonly used carbene precursors in transition-metal-catalyzed reactions. In carbene coupling reactions, they are considered as nucleophilic partners, thus reacting with various electrophilic partners to generate C—C bonds. The first Pd-catalyzed carbene cross-coupling reaction involving carbene migratory insertion was reported in 2001 by Van Vranken and coworkers, in which (trimethylsilyl)diazomethane ($TMSCHN_2$) was employed as the carbene precursor to react with benzyl halides to afford styrene derivatives [5].

Later, Yu et al. reported the Pd-catalyzed cross-couplings of benzyl bromides **4** with α-diazoesters **5**, producing various α,β-diarylacrylates **6** (Scheme 12.2) [6]. Mechanistically, the oxidative addition of benzyl bromides to Pd(0) species generates the benzyl-Pd(II) species **7**, which can decompose diazo compounds to

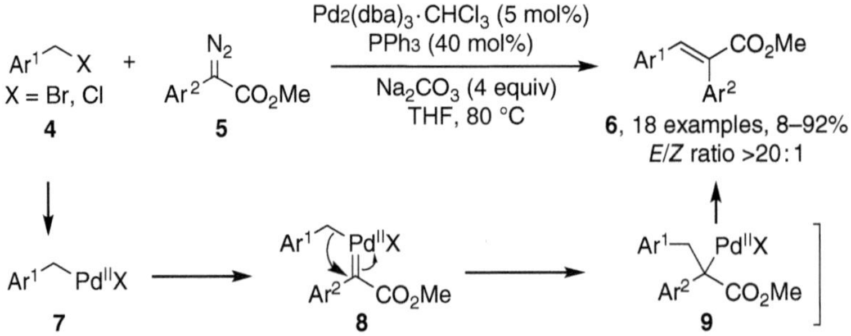

Scheme 12.2 Palladium-catalyzed coupling of aryldiazoesters with benzyl halides. Source: Modified from Yu et al. [6].

afford the Pd–carbene species **8**. The benzyl-Pd(II) species **9** resulted from carbene migratory insertion undergoes β–H elimination to afford the olefin products. Wang and coworkers developed the Pd(II)-catalyzed cross-coupling of trifluoromethyl- or phosphonate-substituted diazo compounds with benzyl halides, affording the trifluoromethyl- or phosphonate-substituted olefins in good yields [7, 8].

Allyl and aryl halides are also suitable electrophilic coupling partners. Wang and Chen developed the Pd-catalyzed cross-coupling of α-diazocarbonyl compounds with allyl halides **10**, affording polysubstituted 1,3-dienes **11** in good yields and *E*/*Z* selectivities (Scheme 12.3A) [9]. Additionally, the trifluoromethyl- or phosphonate-containing diazo compounds were also found as competent coupling partners [7, 8]. Later, they further employed aryl halides **12** as the electrophiles in Pd-catalyzed cross-coupling reaction, providing a series of α-aryl α,β-unsaturated carbonyl compounds **13** (Scheme 12.3B). The Ag_2CO_3 was required when α-diazoketones were employed as the carbene precursors [10].

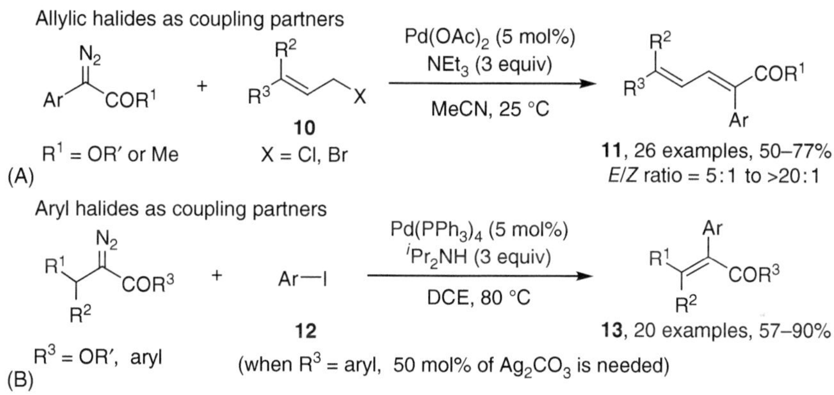

Scheme 12.3 Palladium-catalyzed couplings of diazo compounds with allyl halides and aryl iodides. Source: Modified from Chen and Wang [9].

12.2.1.2 Reactions with Nucleophiles

In 2008, Wang and coworkers reported the Pd-catalyzed oxidative cross-couplings of diazo compounds with organoboronic acids **14** (Scheme 12.4), which produced various α-aryl α,β-unsaturated carbonyl products **15** [11]. The reaction showed a wide scope of α-aryl and α-alkyl diazocarbonyl compounds, providing an alternative method for the preparation of electron-deficient olefins.

Mechanistically, this reaction is initiated by the formation of aryl-Pd(II) species **16** via transmetallation, which decomposes diazo compounds to afford Pd–carbene **17**. The subsequent carbene migratory insertion and β–H elimination then affords the alkene products and Pd(0) species. The Pd(0) species is oxidized by *p*-benzoquinone (BQ) to regenerate the Pd(II) catalyst. Subsequently, this strategy was further applied to the Pd-catalyzed oxidative cross-coupling reaction of vinyl boronic acid with diazo compounds. The reaction afforded a series of diene products bearing cyclic structures or lactone moieties [12].

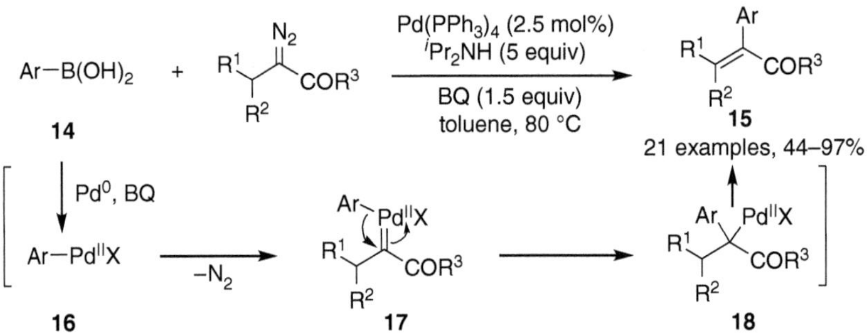

Scheme 12.4 Palladium-catalyzed oxidative coupling of diazo compounds and aryl boronic acids. Source: Modified from Peng et al. [11].

12.2.1.3 Palladium-Catalyzed Cascade Cross-Coupling Reactions

Prior to the formation of Pd–carbene species, the organopalladium species generated through oxidation addition may undergo various cascade reactions, including CO insertion, carbopalladation of unsaturated bonds, or other reactions to generate a new organopalladium species. Then, this organopalladium species decomposes diazo compound to furnish the carbene cross-coupling reaction. Alternatively, the organopalladium species generated from carbene migratory may also participate in other process before completing the catalytic cycle. Therefore, there are some possibilities to developed palladium-catalyzed cascade coupling reactions, in which carbene formation and migratory insertions are the key steps [13].

In 2010, Wang and coworkers reported a Pd-catalyzed four-component coupling reaction of aryl iodide, CO, α-diazoester, and triethylsilane **19**, which afforded a series of 1,3-dicarbonyl compounds **20** (Scheme 12.5) [14]. Mechanistically, the carbonylation affords a new acyl-Pd(II) species (**21**) to participate the catalytic cycle. The subsequent migratory insertion process of Pd–carbene **22** and the transmetalation with Et_3SiH produces a H-Pd(II) species **23**, which undergoes reductive elimination to afford the products.

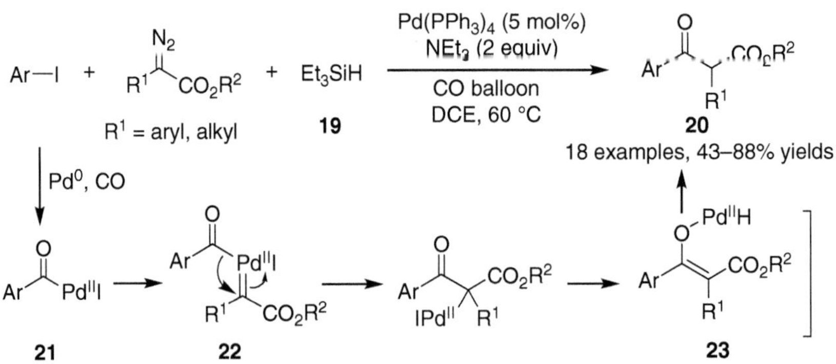

Scheme 12.5 Palladium-catalyzed four-component coupling of aryl iodide, CO, α-diazoester, and triethylsilane. Source: Modified from Zhang et al. [14].

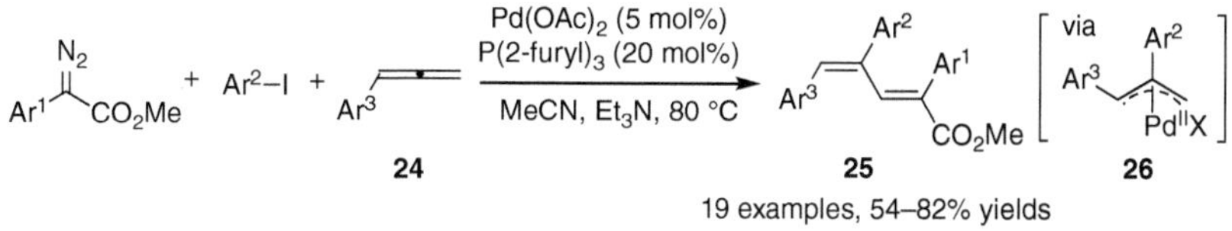

Scheme 12.6 Palladium-catalyzed cascade reaction of aryl iodide, allenes, and α-diazoester. Source: Modified from Xiao et al. [15].

The intermolecular carbopalladation of unsaturated bonds can also be integrated into the reaction. For example, Wang and coworkers reported a Pd-catalyzed cascade reaction of aryl iodide, allenes **24**, and diazo compounds, which can produce multi-substituted dienes **25** (Scheme 12.6) [15]. Mechanistically, the π-allylic palladium species **26** can be generated from the carbopalladation of allene **24** and aryliodide. This π-allylic palladium species further reacts with diazoacetate to afford 1,3-diene products **25** (Scheme 12.6) [15].

Intramolecular alkene carbopalladition can be employed to produce cyclic compounds. García-López and coworkers developed a Pd-catalyzed cascade coupling of α-diazoesters with bromo acylamides **27** [16]. This reaction proceeds through alkene carbopalladition and aryl C—H activation to afford a five-membered-palladacycle **29**, which can participate in the catalytic cycle involving carbene migratory insertion and reductive elimination to give the spiro-oxoindole products **28** (Scheme 12.7).

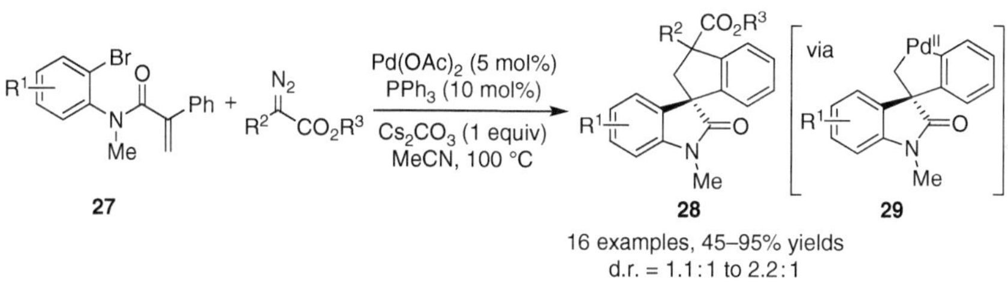

Scheme 12.7 Palladium-catalyzed cascade reaction involving intramolecular alkene carbopalladition. Source: Modified from Pérez-Gómez et al. [16].

The Pd-catalyzed C(sp^3)—H bond activation can also be merged into the Pd-catalyzed cross-coupling reaction. Martin and coworkers developed a Pd-catalyzed cascade cross-coupling of aryl bromides **30** with diazo compounds, affording polysubstituted indanes **31** as the products (Scheme 12.8) [17]. The proposed reaction mechanism involves oxidative addition step and intramolecular C(sp^3)—H bond activation to form the palladacycle **32**, which subsequently participates the catalytic cycle including migratory insertion to afford the final cyclization products **31**.

The organopalladium species resulted from carbene migratory insertion has rich downstream transformations. Especially, when the electrophilic partner or

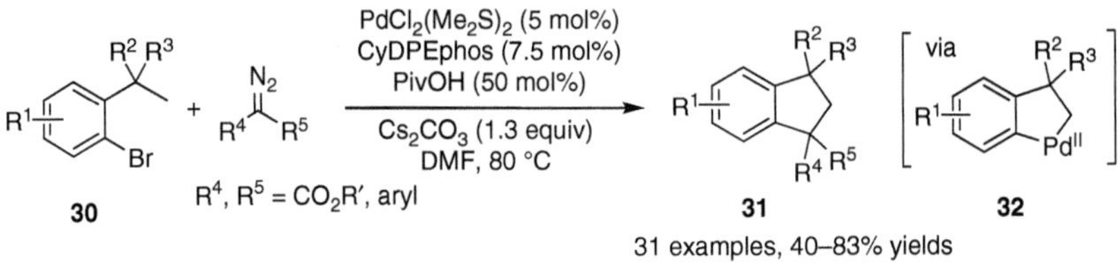

Scheme 12.8 Palladium-catalyzed $C(sp^3)$—H bond activation/carbene coupling reactions. Source: Modified from Gutiérrez-Bonet et al. [17].

carbene precursor contains an vinyl group, the resulted π-allyl-Pd(II) species can undergo allylic substitutions with internal or external nucleophiles. For example, Van Vranken and Devine developed the cross-coupling of diazo compounds, vinyl iodides **33** with external nitrogen- or carbon-based nucleophiles **34** or **36**, in which the regioselective allylic substitution afforded the allylic derivatives **35** or **37** (Scheme 12.9) [18, 19]. In addition, coupling partners bearing internal nucleophilic motifs can be employed to achieve the Pd-catalyzed cyclization to afford cyclic derivatives [4, 13].

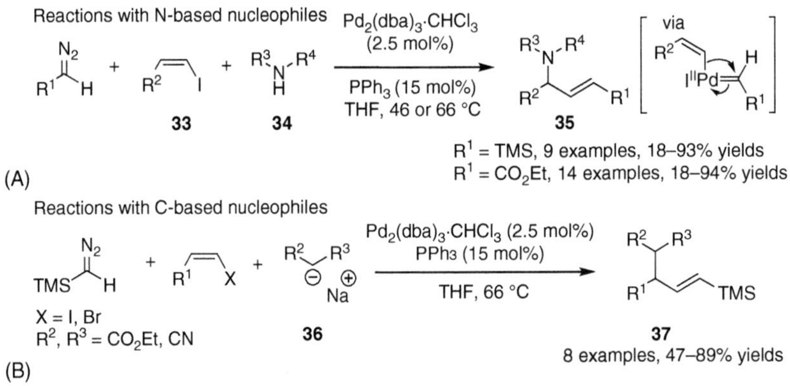

Scheme 12.9 Palladium-catalyzed carbene coupling cascade reactions employing external nucleophiles. Source: Based on Devine and Van Vranken [18, 19].

The alkyl-Pd(II) species generated via carbene migratory insertion may undergo intramolecular C—H activations to produce palladacycle species. In this context, Wang and coworkers developed the Pd-catalyzed coupling reaction of $TMSCHN_2$ with biaryl bromides or heteroaryl bromides **38** (Scheme 12.10A) [20]. In the proposed reaction mechanism, the benzyl-Pd(II) species generated through migratory insertion undergoes aromatic C—H activation to afford the palladacycle species **40**, which can undergo reductive elimination and desilylation to produce the fluorene derivatives **39**. Huang and coworkers developed the Pd-catalyzed intermolecular *O*-acylation of *ortho*-bromo benzaldehydes **41** with α-diazoesters (Scheme 12.10B). This reaction involved the activation of the aldehyde $C(sp^2)$—H bond [21]. After the carbene migratory insertion, the seven-membered palladacycle **43** generated via C—H activation undergoes C-O reductive elimination to afford the isocoumarin derivatives **42**.

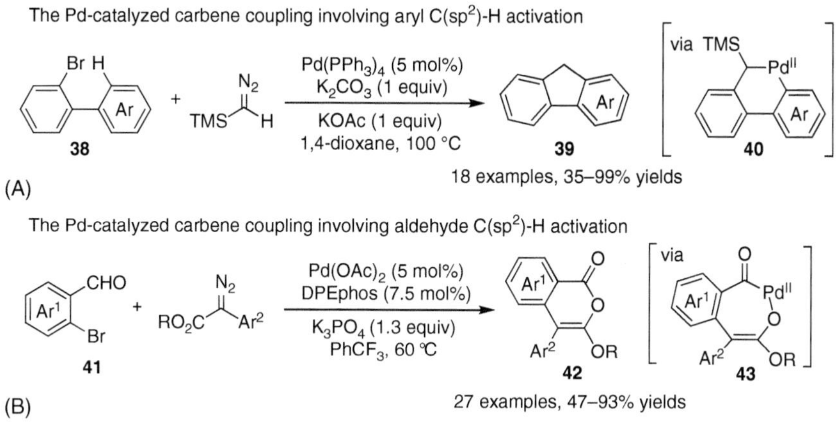

Scheme 12.10 Palladium-catalyzed carbene bridging C–H activation reactions. Source: (A) Modified from Xu et al. [21]. (B) Modified from Yu et al. [22].

12.2.2 *N*-Tosylhydrazones as Carbene Precursors

12.2.2.1 Reactions with Electrophiles

In general, diazo compounds bearing no electron-withdrawing substituents on the diazo carbon are unstable and thus difficult to handle. As an effective surrogate, *N*-tosylhydrazones, which can generate non-stabilized diazo compounds *in situ* via Bamford–Stevens reactions, have been employed as effective carbene sources in the carbene coupling reactions [4, 13]. In 2007, Barluenga and coworkers reported the seminal work by developing the Pd-catalyzed coupling reactions of *N*-tosylhydrazones **44** with aryl halides (Scheme 12.11) [22]. The reaction shows good substrate scope over a wide range of *N*-tosylhydrazones derived from ketones or aldehydes to afford polysubstituted alkenes **45**. They further disclosed that the polysubstituted olefins could also be obtained directly from ketones or aldehydes via the Pd-catalyzed one-pot coupling with aryl halides in the presence of tosylhydrazine ($TsNHNH_2$) [23].

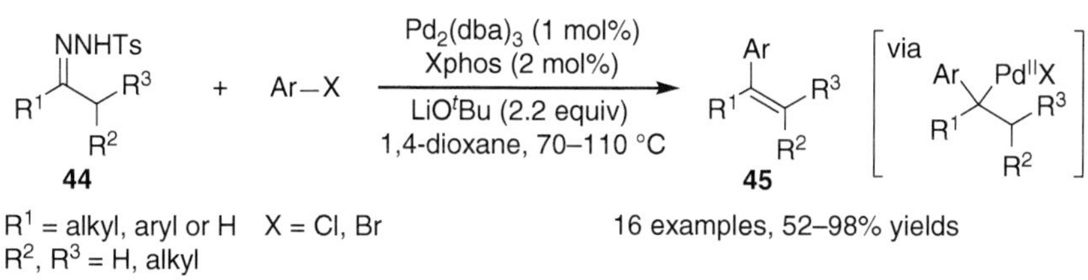

Scheme 12.11 Palladium-catalyzed carbene coupling reaction of *N*-tosylhydrazones with aryl halides. Source: Modified from Barluenga et al. [22].

Subsequently, the Pd-catalyzed cross-coupling of *N*-tosylhydrazones with aryl electrophiles have been extensively studied. For example, Wang and coworkers reported the Pd-catalyzed cross-coupling of cyclopropyl *N*-tosylhydrazones **46** with aryl halides (Scheme 12.12) [24]. A range of substituted *N*-tosylhydrazones were all

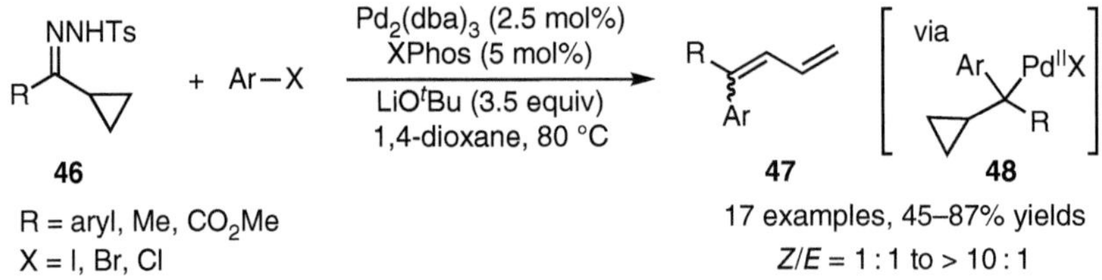

Scheme 12.12 Palladium-catalyzed carbene coupling reaction of cyclopropyl *N*-tosylhydrazones with aryl halides. Source: Modified from Zhou et al. [24].

tolerated to afford the 1,1-disubstituted 1,3-dienes **47**. In the reaction mechanism, the cyclopropylmethyl-Pd(II) species **48** generated via Pd–carbene migratory insertion undergoes β-carbon elimination to produce a homoallylic–Pd(II) species, which undergoes β–H elimination to afford the products. Later, they demonstrated that aryl fluorides were also suitable electrophiles via nitro group-promoted C—F bond oxidative addition [25].

In 2019, Wang and coworkers reported the synthesis of cross-conjugated polymers via Pd-catalyzed carbene coupling reactions (Scheme 12.13). Both AB type polymerization of Br-containing *N*-tosylhydrazones **49** and $A_2 + B_2$-type polymerization of *N*-tosylhydrazones **50** with aryl bromides **51** were conducted in the presence of $Pd_2(dba)_3$ and Xphos to afford poly(arylene-1,1-vinylidene)s (*iso*-PAVs) **52** [26]. These cross-conjugated polymers exhibited unique thermal stabilities, optical properties, and post-modifiabilities. In 2018, Chen and coworkers reported a $A_2 + B_2$ polymerization by employing heteroaryl bromides and *N*-tosylhydrazones as the monomers [27].

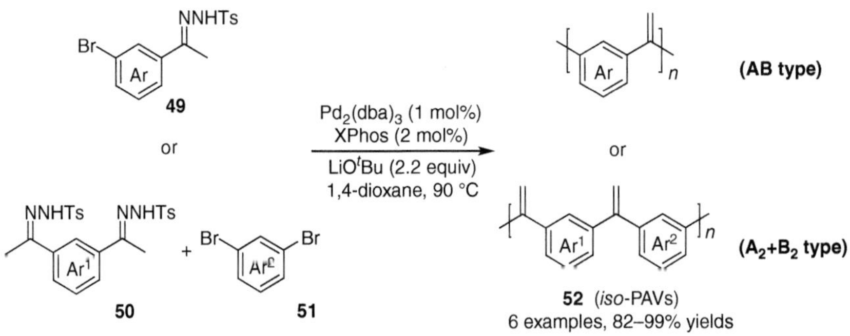

Scheme 12.13 Palladium-catalyzed synthesis of cross-conjugated polymers via carbene coupling reactions. Source: Modified from Zhou et al. [26]

When the β–H elimination is not feasible, the reductants can be introduced to achieve the reductive cross-coupling. In 2013, Wang and coworkers reported the Pd-catalyzed reductive coupling of *N*-tosylhydrazones with aryl halides in the presence of ammonium formate (HCO_2NH_4), producing a range of substituted triarylmethanes **53** (Scheme 12.14). Mechanistically, the alkyl-Pd(II) species generated from carbene migratory insertion reacts with the HCO_2NH_4 to form the H-Pd(II) species **54**, which can undergo reductive elimination to afford the products [28].

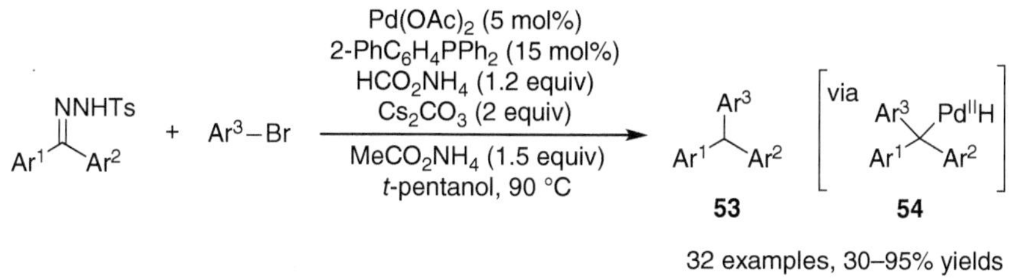

Scheme 12.14 Palladium-catalyzed reductive coupling of *N*-tosylhydrazones with aryl halides. Source: Modified from Xia et al. [28]

Activated alkyl electrophiles are also competent coupling partners. Wang and coworkers successively reported the Pd-catalyzed cross-coupling of *N*-tosylhydrazones with cyclopropyl [24], benzyl [29], or allyl [30] halides, providing efficient approaches toward substituted alkenes and dienes. *N*-tosylhydrazones bearing functional groups including CF_3- and $PO(OMe)_2$-groups can also be well tolerated [7, 8].

12.2.2.2 Reactions with Nucleophiles

The oxidative coupling of *N*-tosylhydrazones with various nucleophiles is also feasible. In this context, Wang and coworkers reported the Pd-catalyzed oxidative couplings of *N*-tosylhydrazones with arylboronic acids, in which the combination of catalytic amount of CuCl with O_2 was employed as oxidant [31]. They later reported the Pd-catalyzed oxidative coupling of terminal alkynes **55** with *N*-tosylhydrazones, leading to convenient synthesis of conjugated enynes **56** (Scheme 12.15). Terminal alkynes bearing aliphatic, aryl, heteroaryl, and silyl groups were all tolerated [32]. The mechanism involves the Pd–carbene migratory insertion to the alkynyl–Pd(II) bond and the subsequent β–H elimination of the resulted organopalladium species.

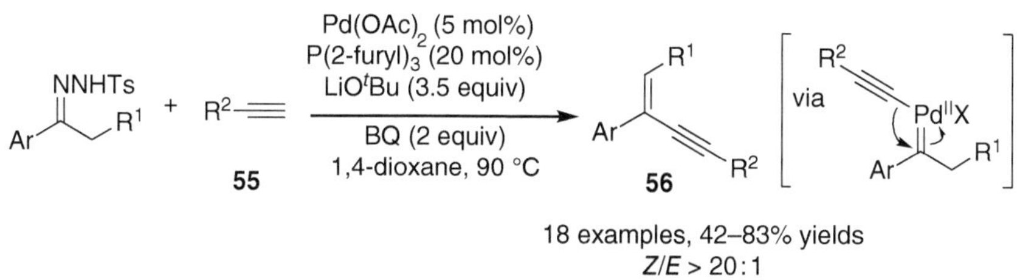

Scheme 12.15 Palladium-catalyzed oxidative coupling of *N*-tosylhydrazones with terminal alkynes. Source: Modified from Zhou et al. [32]

In 2015, Wang and coworkers reported the first Pd-catalyzed carbene Si—Si (or Sn—Sn) bond insertion reaction of disilane **57** or distannane **58** with *N*-tosylhydrazones (Scheme 12.16) [33]. This reaction is initiated by the oxidative addition of Si—Si (or Sn—Sn) bonds to Pd(0) species. The subsequent carbene migratory insertion of intermediate **61** and reductive elimination affords the geminal bis(silane) **59** or geminal bis(stannane) **60** products. They further developed the Pd-catalyzed Si—H bond carbene insertions of different silanes involving

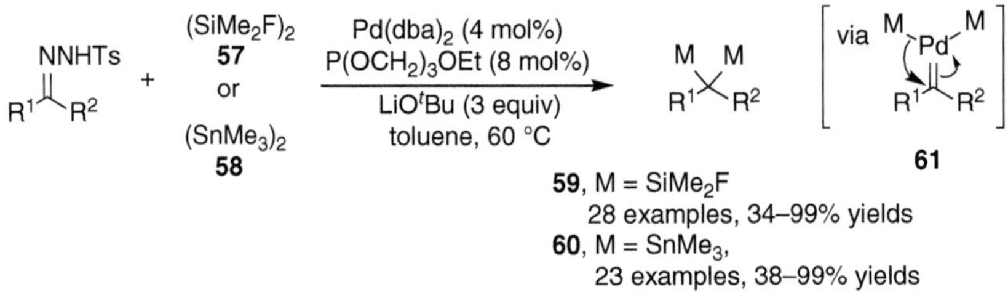

Scheme 12.16 Palladium-catalyzed carbene Si–Si and Sn–Sn bonds insertion reactions. Source: Modified from Liu et al. [33].

Pd–carbene migratory insertion, which is mechanistically different from the classic carbene Si—H bond insertions [34].

12.2.2.3 Palladium-Catalyzed Cascade Cross-Coupling Reactions

Different unsaturated bonds can be employed as the C2 building blocks in the Pd-catalyzed carbene cascade coupling reactions based on carbopalladation process. For example, Wang and coworkers introduced allenes into the carbopalladation/carbene coupling protocol, affording various multi-substituted dienes [15]. Later, they reported the three-component cross-coupling of aryl iodides, norbornene (NBE) **62**, and *N*-tosylhydrazones. The proposed reaction mechanism involves the intermolecular carbopalladation of the NBE with Ar–Pd(II) species, followed by Pd–carbene **64** migratory insertion to afford various carbocycles bearing exo-double bonds **63** (Scheme 12.17) [35].

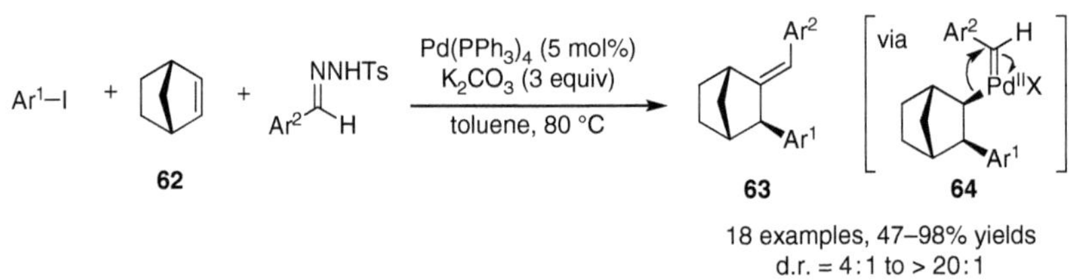

Scheme 12.17 Pd-catalyzed three-component coupling of aryl iodides, norbornene, and *N*-tosylhydrazones. Source: Modified from Hu et al. [35].

Cyclic products can be obtained through the cascade pathway involving intramolecular carbopalladation. Wang and coworkers reported the cascade reactions of alkyne-containing aryl iodides **65** with ketones or aldehydes in the presence of $TsNHNH_2$, which involves the carbopalladation of alkynes motif, Pd–carbene migratory insertion and the subsequent β-H elimination to afford substituted 3-vinylindoles and 3-vinylbenzofurans **66** (Scheme 12.18) [36].

The 1,5-Pd migration can also be employed to produce novel organopalladium species for carbene coupling reaction. Zhao and coworkers developed a Pd-catalyzed C(sp^3)—H bond olefination of silane-containing naphthyl bromides **68** with *N*-tosylhydrazones, which involves the aryl to alkyl 1,5-Pd migration (Scheme 12.19) [37]. The α-effect of the silicon atom was proposed to promote the

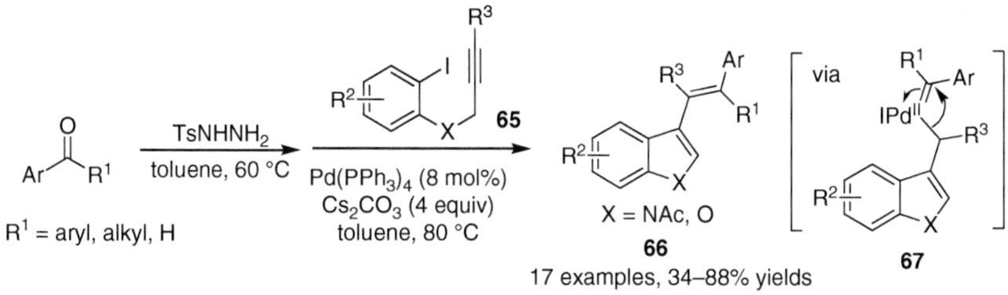

Scheme 12.18 Palladium-catalyzed cascade reactions of alkyne-containing aryl iodides with *N*-tosylhydrazones. Source: Modified from Liu et al. [36].

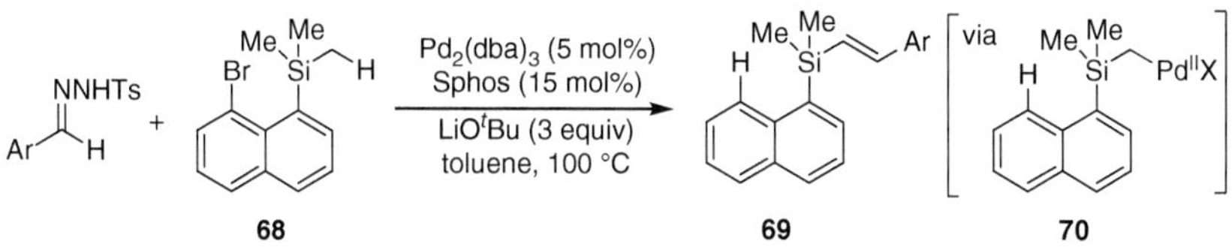

Scheme 12.19 Palladium-catalyzed $C(sp^3)$–H bond olefination of silane-containing naphthyl bromides with *N*-tosylhydrazones. Source: Modified from Han et al. [37].

novel 1,5-Pd/H shift, which can form the $C(sp^3)$–Pd species **70** to participate the following catalytic cycle.

Various cascade transformations after Pd–carbene migratory insertion can also be conducted. For example, the vinyl-group-containing electrophiles and *N*-tosylhydrazones can be employed to participate the Pd-catalyzed cyclization via allylic substitution [4, 13]. Wang and coworkers reported the Pd-catalyzed cross-coupling of salicylaldehyde derived *N*-tosylhydrazones **71** with β-bromostyrenes, which provided the 2-aryl substituted chromenes **72** (Scheme 12.20). Mechanistically, it has been proposed that the generated π-allyl-Pd(II) species **73** is trapped by the internal hydroxyl group to afford the oxacycle products [38].

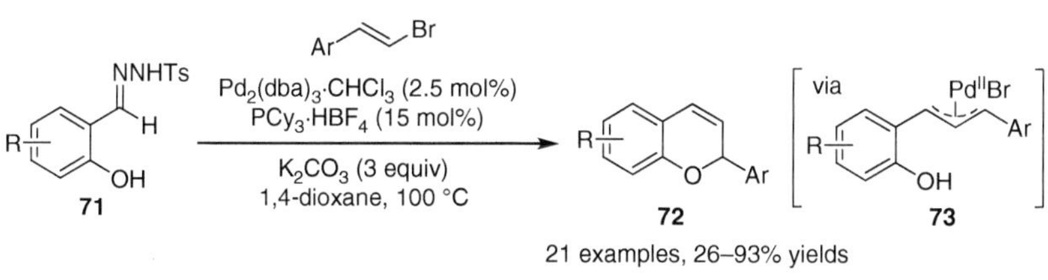

Scheme 12.20 Palladium-catalyzed cross-coupling of salicylaldehyde derived *N*-tosylhydrazones with β-bromostyrenes. Source: Modified from Xia et al. [38].

When the β–H elimination is not possible, the subsequent transmetalation and reductive elimination can lead to multicomponent products. Wang and coworkers employed terminal alkynes as the third partner in Pd-catalyzed cross-couplings of *N*-tosylhydrazones with aryl halides, which can afford the benzhydryl acetylene

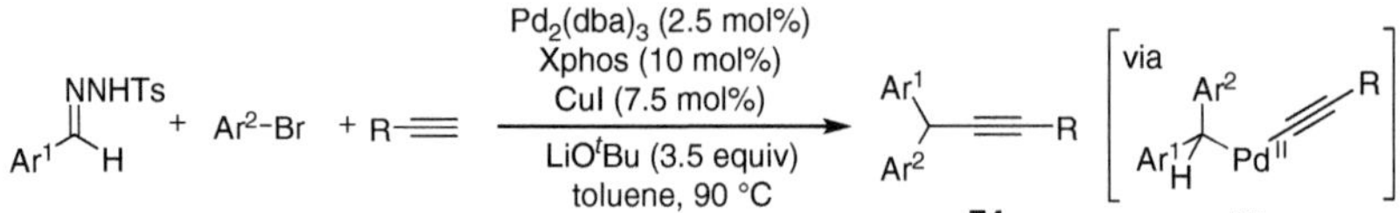

74 75

21 examples, 21–84% yields

Scheme 12.21 Palladium-catalyzed cross-couplings of *N*-tosylhydrazones with aryl halides and terminal alkynes. Source: Modified from Zhou et al. [39].

derivatives **74** (Scheme 12.21) [39]. The reaction proceeds with the following sequence. After the carbene migratory insertion, the alkynyl-Pd(II) species **75** is further generated through transmetallation, from which reductive elimination occurs to afford the final products **74**.

In 2020, Huang and coworkers reported the Pd-catalyzed synthesis of dibenzo-fused seven- or eight-membered lactones through the coupling of *N*-tosylhydrazones (**76**) with 2-bromobenzaldehydes **77** (Scheme 12.22) [40]. For the mechanism, the benzyl-Pd(II) species is proposed to be generated via carbene migratory insertion, giving a five-membered palladacycle bearing a H—Pd bond **79** via aldehyde C–H bond insertion of palladium. The hydride then transfers from the Pd center to the dibenzylic position to furnish the 1,4-Pd/H shift. The following ring closure by the tethered phenolic anion then affords the lactone products **78**.

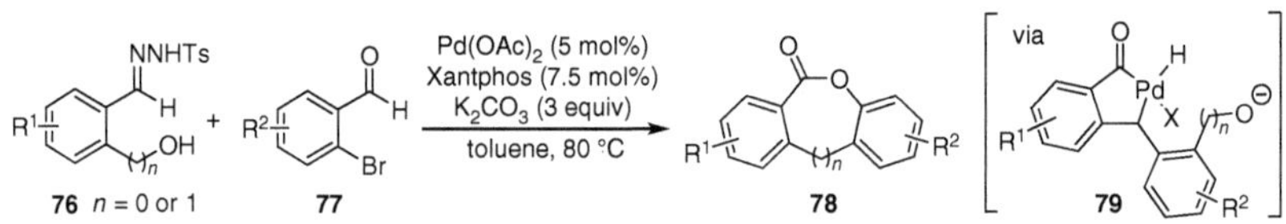

76 $n = 0$ or 1 77 78 79

82 examples, 31–92% yields

Scheme 12.22 Palladium-catalyzed cross-coupling of *N*-tosylhydrazones with 2-bromobenzaldehydes. Source: Modified from Yu et al. [40].

12.2.3 Non-Diazo Compounds as Carbene Precursors

It has been reported that metal carbene species generated via the 5-*exo*-dig cyclization of conjugated enynones can participate various classic carbene-based transformations. In 2013, Wang and coworkers reported the first Pd-catalyzed cross-coupling of conjugated enynones **80** with benzyl, aryl, and allylic halides, affording different furan-substituted olefins **81, 83,** and **84** (Scheme 12.23) [41]. Mechanistically, it has been proposed that the organopalladium(II) species is generated via oxidative addition coordinates to the enynones to trigger the 5-*exo*-dig cyclization. The resulted Pd–carbene **82** undergoes migratory insertion and β–H elimination to afford the products.

They further reported the Pd-catalyzed oxidative cross-coupling of enynones with aryl or vinyl boronic acids (**85, 86**), which afforded the furan-containing olefins **87** with good to excellent *E*-isomer selectivities (Scheme 12.24A) [42]. This strategy can be further extended to the oxidative coupling of enynones

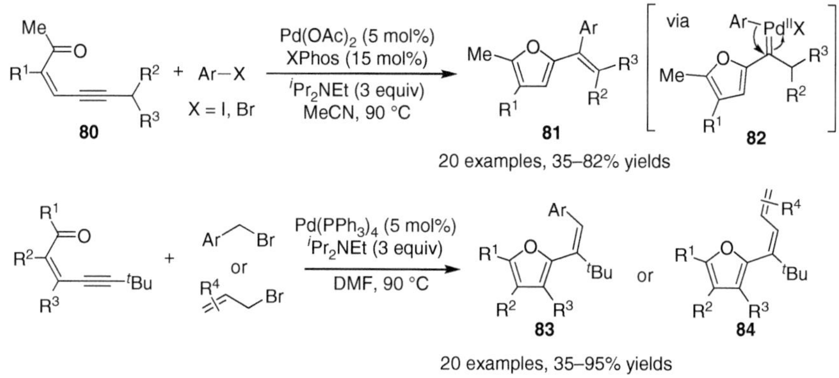

Scheme 12.23 Palladium-catalyzed coupling of conjugated enynones with organic halides. Source: Modified from Xia et al. [41].

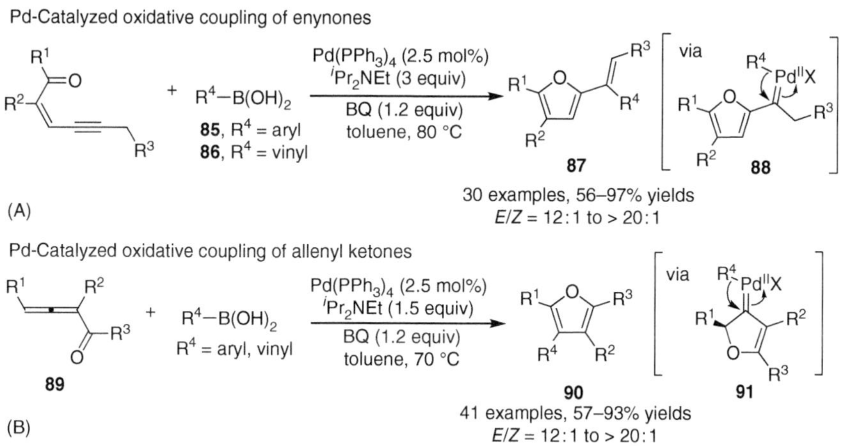

Scheme 12.24 Palladium-catalyzed oxidative coupling of enynones or allenyl ketones with organo boronic acids. Source: (A) Modified from Xia et al. [42]. (B) Modified from Xia et al. [45].

with terminal alkynes or diboron reagents, affording furan-substituted enynes or alkenylboronates [43, 44]. Additionally, allenyl ketones **89** are also suitable carbene precursors in Pd(II)-catalyzed oxidative cross-coupling with organoboronic acids (Scheme 12.24B), which can undergo the cycloisomerization to afford the Pd–carbene **91**. The subsequent migratory insertion/β-elimination then delivers the products **90** [45].

Alkynes can also be utilized as metal–carbene precursors with the assistance of an external nucleophile. In 2018, Wang and coworkers reported the Pd(II)-catalyzed oxygenative cross-coupling of ynamides (**92**) with benzyl bromides, in which the ynamides were employed as Pd–carbene sources in the assistance of 8-methylquinoline *N*-oxide (Scheme 12.25) [46]. A series of α,α-disubstituted amides **93** were obtained after the hydrogenation of the originally formed alkene

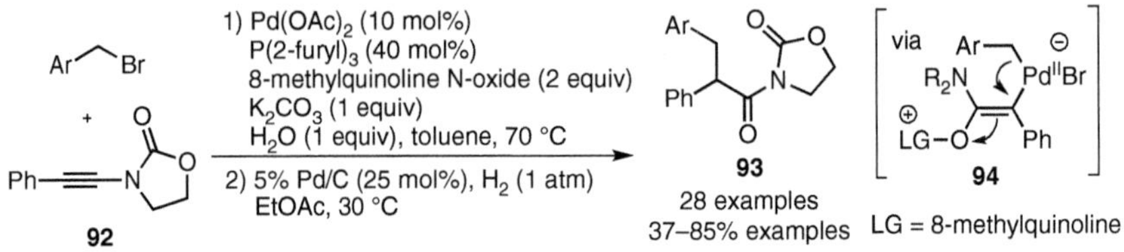

Scheme 12.25 Palladium-catalyzed oxygenative coupling of ynamides with benzyl bromides. Source: Modified from Gao et al. [46].

products. Mechnistically, the benzyl-Pd(II) species generated from oxidative addition activates the ynamide to promote the nucleophilic addition of quinoline *N*-oxides. The resulted vinyl Pd intermediate **94** undergoes γ-elimination to expel the quinoline part and generate the Pd–carbene, which then participates the following catalytic cycle.

The stable chromium(0) carbene complexes have been employed as versatile organometallic reagents, which can undergo carbene transfer with metal catalysts for further transformations. In 2017, Wang and coworkers reported the Pd-catalyzed [3+3] annulation of vinyl chromium(0) carbenes **95** with 2-iodophenols **96** to afford substituted flavonones **97**, which validated the possibility of combining carbene transfer with carbene migratory insertion (Scheme 12.26A). Different quinolones were afforded when 2-iodoanilines acted as coupling partners [47]. Mechanistically, the vinyl chromium(0) carbene undergoes carbene transfer with the aryl-Pd(II) species to afford the vinyl Pd–carbene **98**. The allylic-Pd(II) species generated via carbene migratory insertion is then trapped by the internal HO- or H_2N- groups to afford the products. Later, they reported the Pd-catalyzed reductive coupling of aryl chromium(0) complexes **99** with aryl iodides in the presence of hydrosilanes (Scheme 12.26B), affording a series of diarylmethyl ethers **100** [48]. Both mechanistic experiments and density functional theory (DFT) calculation support the

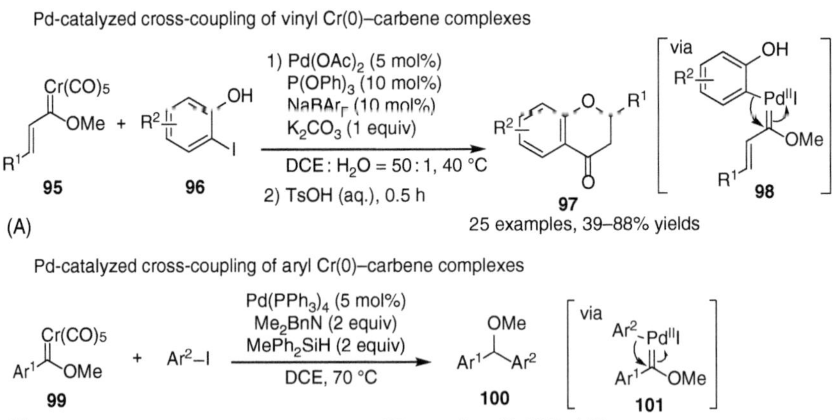

Scheme 12.26 Palladium-catalyzed coupling of chromium(0) carbenes with organic halides. Source: (A) Modified from Wang et al. [47]. (B) Modified from Wang et al. [48].

mechanism involving the formation of Pd–carbene **101** via carbene transfer and the subsequent migratory insertion process.

12.3 Copper-Catalyzed Carbene Cross-Coupling Reactions

12.3.1 Reactions with Terminal Alkynes

12.3.1.1 Multi-substituted Allenes as the Coupling Products

The organocopper species can also react with carbene precursors to afford Cu–carbene species. The new organocopper species resulted from Cu–carbene migratory insertion can undergo protonation, β–F elimination or react with electrophiles to afford different products.

In 2011, Wang and coworkers developed the Cu-catalyzed carbene coupling of *N*-tosylhydrazones with terminal alkynes (Scheme 12.27A), which provided the multi-substituted allenes (**103**) as the products in the presence of $Cu(MeCN)_4PF_6$ catalyst and bisoxazoline ligand (**102**) [49]. They further reported the synthesis of 1,3-disubstituted allenes (**104**) in a simplified CuI-catalyzed system (Scheme 12.27B). This Cu-catalyzed allene synthesis can also be conducted in a one-pot two-step protocol directly from aldehydes [50]. Mechanistically, these reactions involve the carbene migratory insertion to the alkynyl–Cu(I) bond, the resulted organocopper species (**105**) undergoes protonation at the alkynyl carbon to afford allene products.

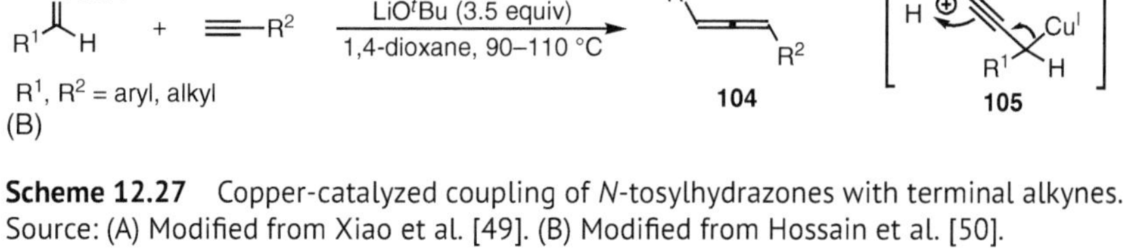

Scheme 12.27 Copper-catalyzed coupling of *N*-tosylhydrazones with terminal alkynes. Source: (A) Modified from Xiao et al. [49]. (B) Modified from Hossain et al. [50].

The Cu-catalyzed protocol of allene synthesis was further modified. The tri- or tetra-substituted allenes were obtained by employing allyl halides as a third component [51]. Additionally, various diazo compounds as well as conjugated enynones [52] were also proved as suitable carbene sources in Cu-catalyzed allene synthesis. On the other hand, the Cu-catalyzed tandem cyclizations through the *in situ* formation of allene intermediates have also been developed, affording a variety of carbo- and heterocyclic compounds [53].

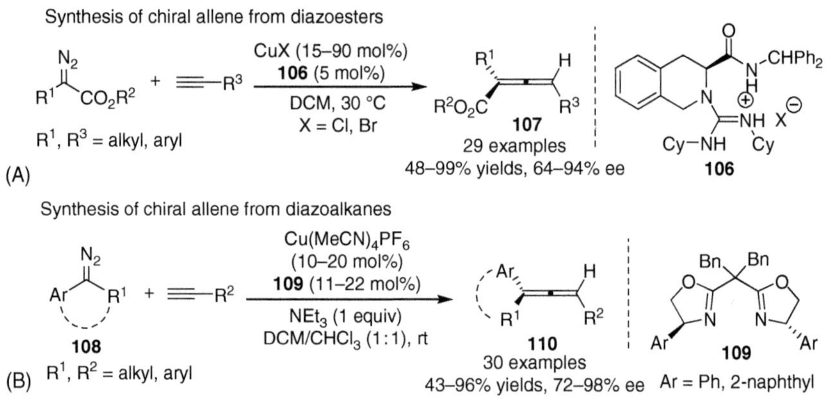

Scheme 12.28 Copper-catalyzed enantioselective synthesis of axially chiral allenes. Source: (A) Modified from Tang et al. [54]. (B) Modified from Chu et al. [55].

These coupling reactions can be employed to the enantioselective synthesis of axially chiral allenes, which has been a challenging topic in asymmetric catalysis. Feng and coworkers reported the Cu-catalyzed asymmetric synthesis of chiral 2,4-disubstituted allenoates (**107**) employing α-diazoesters in the presence of chiral guanidinium salts (**106**) (Scheme 12.28A) [54]. Wang and coworkers developed the Cu-catalyzed enantioselective synthesis of axially chiral trisubstituted allenes (**110**) by employing diazoalkanes (**108**) as the carbene precursors (Scheme 12.28B). A range of terminal alkynes and diazoalkanes were well tolerated in the presence of $Cu(MeCN)_4PF_6$ catalyst and the chiral bisoxazoline ligands (**109**) [55].

12.3.1.2 Internal Alkynes as the Coupling Products

The Cu-catalyzed carbene coupling with terminal alkynes can also afford internal alkynes. Wang and coworkers reported the Cu-catalyzed synthesis of internal alkynes (**112**) by employing *N*-tosylhydrazones and the trialkylsilyl substituted terminal alkynes (**111**) (Scheme 12.29). The trialkylsilyl group was proved to be crucial for the construction of the C(sp)—C(sp^3) bond. This reaction can also be conducted in the one-pot fashion directly from ketones or aldehydes [56]. Mechanistically, the propargyl–Cu(I) species (**113**) generated via carbene migratory insertion undergoes protonation at the carbenic carbon to afford the final products.

The Cu-catalyzed asymmetric cross-coupling of *N*-tosylhydrazones with trialkylsilyl ethynes were also developed by the same group, providing functionalized

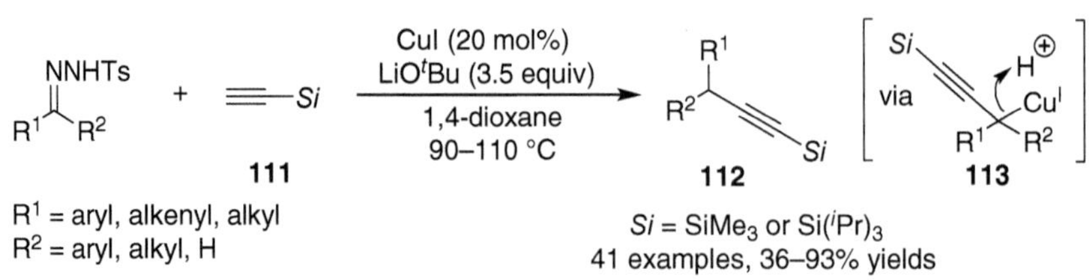

Scheme 12.29 Copper-catalyzed coupling of *N*-tosylhydrazones with trialkylsilyl substituted terminal alkynes. Source: Modified from Ye et al. [56]

alkynes in moderate to good enantioselectivities [57]. They further disclosed that the propargyl–Cu(I) species (**113**) can react with various electrophiles, including alkyl or benzyl halides, electron-deficient alkenes and aldehydes, to construct the all-carbon quaternary carbon centers or difluoro-conjugated enynes [58].

12.3.2 Reactions with Other Coupling Partners

Introduction of the fluoro-containing functional groups is of interest in organic chemistry. Hu and coworkers developed the Cu-catalyzed cross-coupling of diaryldiazomethanes with $TMSCF_3$ (**114**) (Scheme 12.30A), which involved the β-fluoride elimination of organocopper(I) species (**116**) to afford 1,1-difluoroalkenes (**115**) [59]. Goossen and coworkers reported the Cu-catalyzed trifluoromethylthiolation (or trifluoromethylselenation) of diazo compounds with Me_4NSCF_3 (**117**) (or Me_4NSeCF_3), which involved the protonation of Cu species (**119**) (Scheme 12.30B) [60].

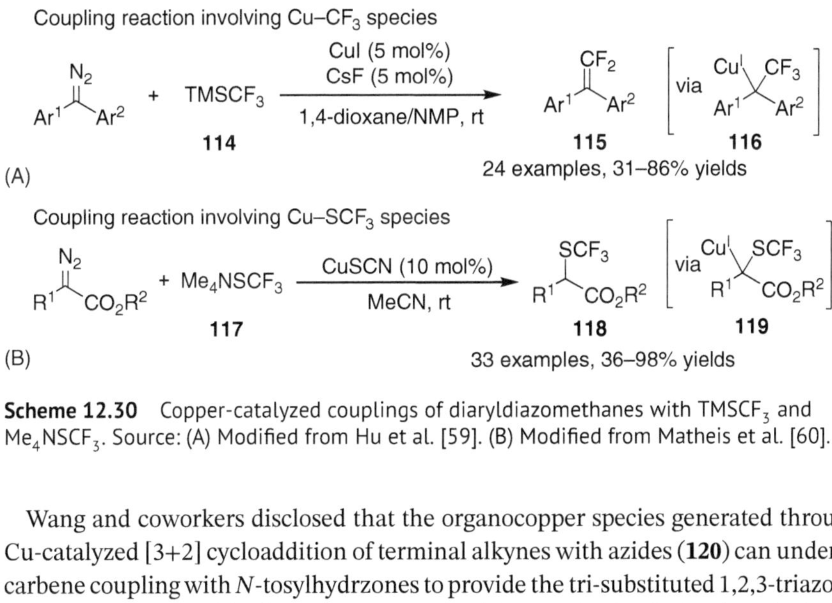

Scheme 12.30 Copper-catalyzed couplings of diaryldiazomethanes with $TMSCF_3$ and Me_4NSCF_3. Source: (A) Modified from Hu et al. [59]. (B) Modified from Matheis et al. [60].

Wang and coworkers disclosed that the organocopper species generated through Cu-catalyzed [3+2] cycloaddition of terminal alkynes with azides (**120**) can undergo carbene coupling with *N*-tosylhydrzones to provide the tri-substituted 1,2,3-triazoles (**121**) (Scheme 12.31) [61]. Later, they reported the synthesis of triazole-substituted difluoroalkenes by employing trifluoromethyl *N*-tosylhydrazones, which involved the β-fluoride elimination as the termination process [62].

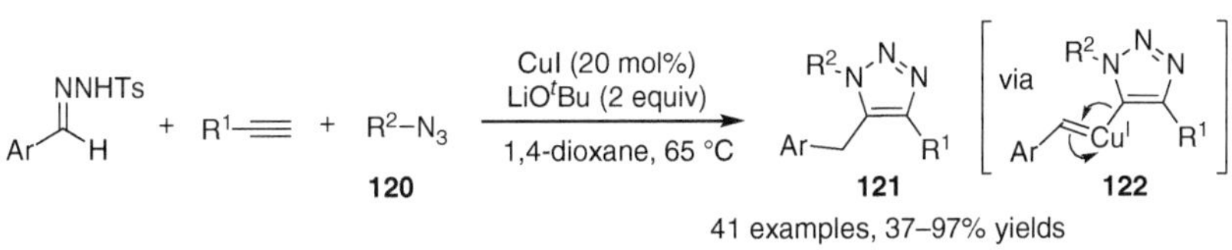

Scheme 12.31 Copper-catalyzed couplings of *N*-tosylhydrazones with alkynes and azides. Source: Modified from Zhang et al. [61].

12.4 Rhodium-Catalyzed Carbene Cross-Coupling Reactions

12.4.1 Generating Organorhodium Species Through Transmetalation

In 2011, Yu and coworkers reported the Rh-catalyzed carbene cross-coupling of α-diazoesters with arylboronates (**123**) and alkyl halides (**124**), providing the quaternary α,α-diaryl carboxylic esters as products (**125**) (Scheme 12.32) [63]. In this reaction, the aryl–Rh(I) intermediate (**126**) generated via transmetalation can afford Rh–carbene species (**127**) with diazo compounds. The oxa-π-allylrhodium complex (**128**) resulted from Rh–carbene migratory insertion undergoes the base-promoted alkylation with alkyl halides to afford the final products.

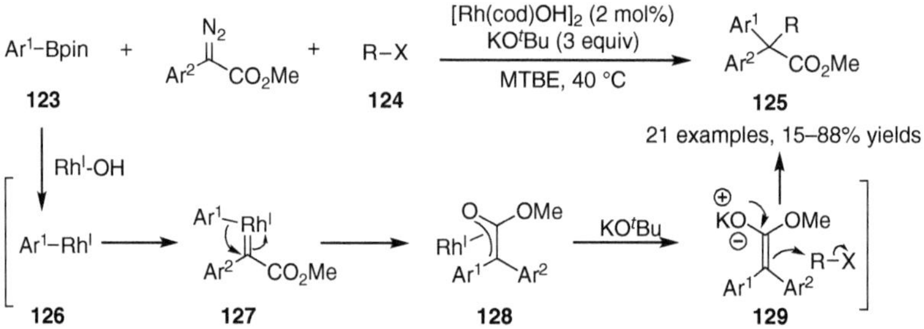

Scheme 12.32 Rhodium-catalyzed coupling of α-diazoesters with arylboronates and alkyl halides. Source: Modified from Tsoi et al. [63].

In 2016, Wang and coworkers reported the Rh(I)-catalyzed cross-coupling of trifluoromethyl *N*-tosylhydrazones with arylboronates (**130**), which involved the β–F elimination to afford 1,1-difluoro-2,2-diarylalkenes (**131**) (Scheme 12.33) [64]. Later, the Rh(I)-catalyzed cross-coupling of conjugated enynones with organic boronic acids was also developed, which can afford a series of furan-containing triarylmethanes [65].

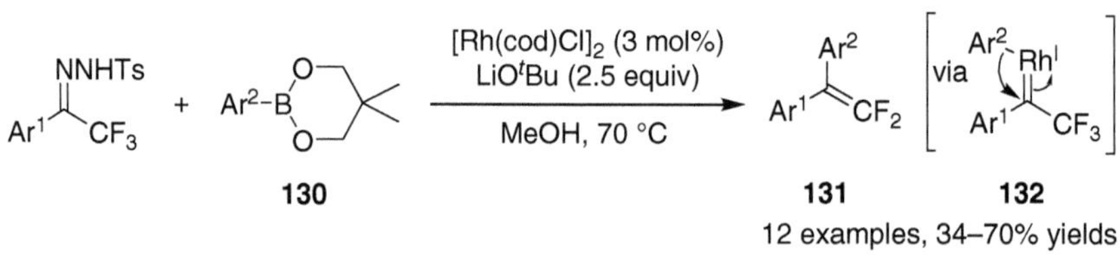

Scheme 12.33 Rhodium-catalyzed coupling of trifluoromethyl *N*-tosylhydrazones with arylboronates. Source: Modified from Zhang et al. [64].

In addition, other organometallic reagents have also been proved as competent substrates. Wang and coworkers reported the Rh-catalyzed cross-coupling of arylsiloxanes (**133**) with α-diazoesters (Scheme 12.34) [66]. The alkyl α-diazoesters which are problematic substrates in Rh-catalyzed coupling reactions with

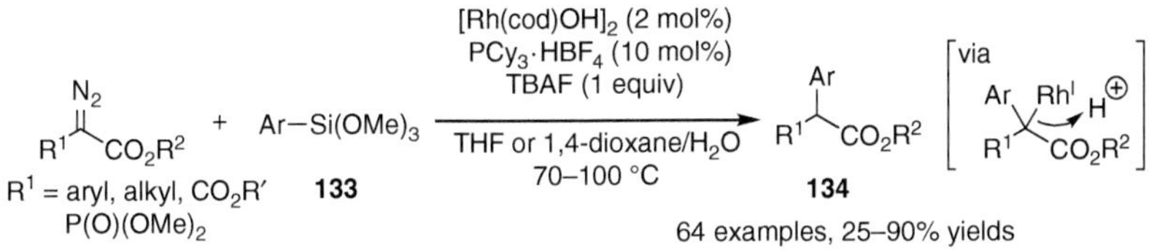

Scheme 12.34 Rhodium-catalyzed coupling of arylsiloxanes with α-diazoesters. Source: Modified from Xia et al. [66].

organoboron reagents can react with arylsiloxanes efficiently to afford the coupling products (**134**). Later, the organostannanyl reagents were also proved as suitable substrates [67].

12.4.2 Generating Organorhodium Species Through C—C Bond Cleavage

In 2014, Wang and coworkers reported a Rh(I)-catalyzed carbene insertion into the C—C bond of benzocyclobutenols (**135**) with α-diazoesters as carbene precursors, affording highly substituted indanol derivatives (**136**) as the coupling products (Scheme 12.35). Mechanistically, the aryl-Rh(I) species (**137**) generated via selective β-carbon elimination affords the Rh–carbene species (**138**) with diazo compounds, which is followed by migratory insertion and the intramolecular aldol reaction [68]. Meanwhile, Murakami and coworkers developed a similar Rh-catalyzed coupling reactions of *N*-tosylhydrazones involving the distal C(sp^3)—C(sp^3) bond cleavage of cyclobutanols [69].

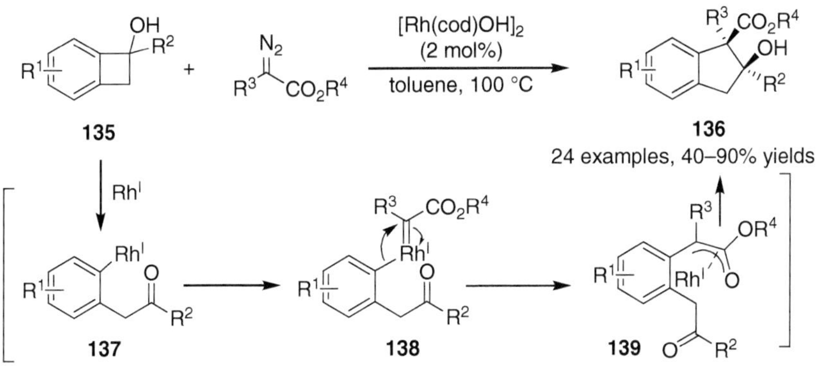

Scheme 12.35 Rhodium-catalyzed reaction of benzocyclobutenols with α-diazoesters. Source: Modified from Xia et al. [68].

The Rh(I)-catalyzed cleavage of C(sp)—C(sp^3) bonds of the *tert*-propargyl alcohols can also be incorporated into the carbene coupling reactions. In 2015, Wang and coworkers reported the Rh-catalyzed cross-coupling reactions of α-aryl α-diazoesters with *tert*-propargyl alcohols (**140**) and alkyl halides (Scheme 12.36) [70]. This reaction can construct the all-carbon quaternary center bearing four different carbon substituents (**141**). Mechanistically, the alkynyl-Rh(I) species

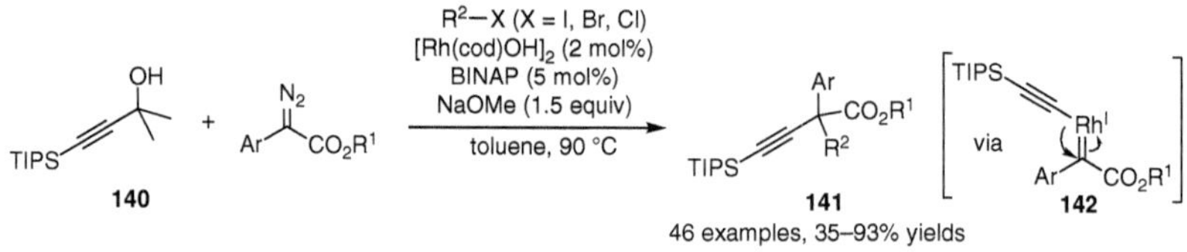

Scheme 12.36 Rhodium-catalyzed coupling of α-diazoesters with *tert*-propargyl alcohols and alkyl halides. Source: Modified from Xia et al. [70].

generated through β-alkynyl elimination affords the Rh–carbene species (**142**) with the diazo compound. The following Rh–carbene migratory insertion and the base-promoted alkylation then affords the final product.

Additionally, a Rh(I)-catalyzed C—C bond activation reaction of vinylcyclopropanes (**143**, VCPs) with α-diazoesters was developed by Wang and coworkers (Scheme 12.37A), which involved the cleavage of C2—C3 bond of the VCPs to afford 1,4-dienes (**144**) [71]. In the mechanism, the Rh–carbene insertion to the C2—C3 bond generates a new Rh–carbene species (**145**). The following carbene migratory insertion, β-H elimination, and reductive elimination then generate the products. They further reported that, by changing the catalyst to [Rh(cod)$_2$OTf], *E*-vinylcyclobutane (**146**) were formed with excellent chemoselectivities and in moderate to good yields (Scheme 12.37B), in which the C=C bond in VCPs was cleaved in the reaction mechanism involving intermediate (**147**) [72].

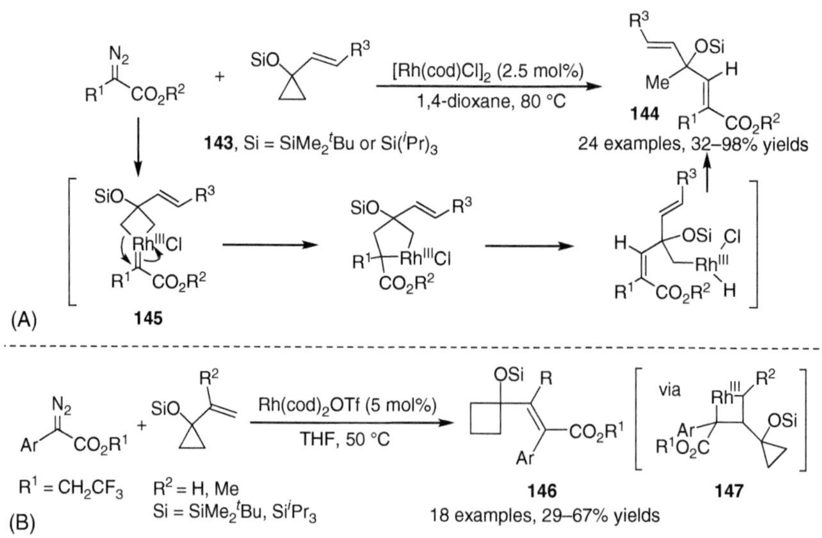

Scheme 12.37 Rhodium-catalyzed reactions of vinylcyclopropanes with α-diazoesters. Source: (A) Modified from Feng et al. [71]. (B) Modified from Feng et al. [72].

12.5 Transition-Metal-Catalyzed C–H Bond Functionalizations with Carbene Precursors

12.5.1 Non-Directing-Group-Assisted C–H Functionalizations

The Pd-catalyzed allyl C—H activation has become a powerful toolkit for the construction of allylic derivatives. In 2014, Gong and coworkers reported the Pd-catalyzed cross-coupling of diazo compounds with allyl hydrocarbons (**148**) via Pd-catalyzed allylic C—H bond activation (Scheme 12.38A). A series of aryl-, vinyl-, and alkyl-substituted allyl hydrocarbons were tolerated to provide the diene products (**149**) [73]. Later, Wang and coworkers employed conjugated enynones as the carbene precursors in Pd-catalyzed oxidative cross-coupling with allylarenes (**150**) via Pd-catalyzed allylic C—H activation, providing furan-substituted 1,3-dienes (**151**) in good yields and excellent *Z*-stereoselectivities (Scheme 12.38B) [74]. The catalytic amount of diphenylphosphate was necessary to promote the reaction.

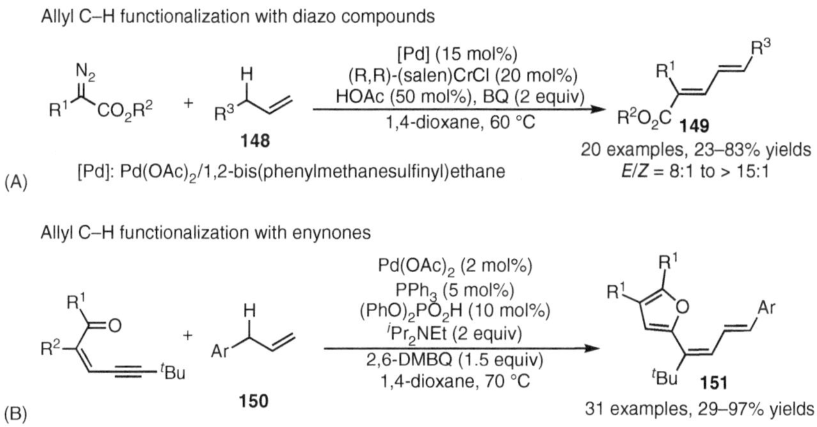

Scheme 12.38 Palladium-catalyzed allyl C–H activation involving carbene migratory insertion. Source: (A) Modified from Wang et al. [73]. (B) Modified from Ping et al. [74].

Arenes bearing acidic C(sp^2)—H bond are suitable coupling partners in Cu-catalyzed carbene cross-couplings. In 2011, Wang and coworkers developed a Cu-catalyzed C—H bond benzylation or allylation of 1,3-azoles (**152**) by employing *N*-tosylhydrazones as carbene precursors to provide C2-functionalized azoles (**153**) (Scheme 12.39) [75]. The sequence of Cu-catalyzed aryl C—H metallation, Cu-carbene migratory insertion, and protonation was proposed in the mechanism. In contrast to the classic carbene insertion into the electron-rich C—H bonds, this protocol provided the opportunities toward the functionalization of

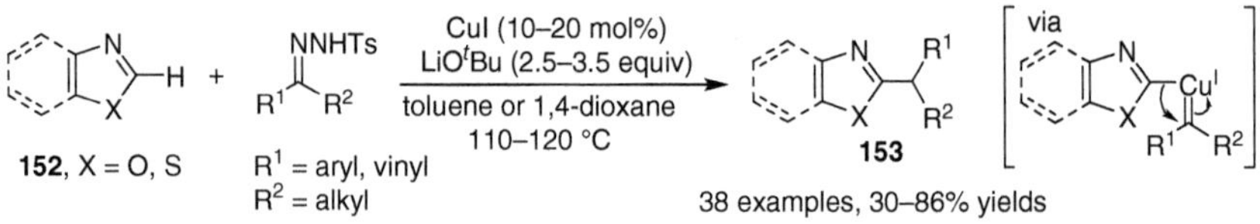

Scheme 12.39 Copper-catalyzed C–H bond functionalization of 1,3-azoles with *N*-tosylhydrazones. Source: Modified from Zhao et al. [75].

electron-deficient C—H bonds. Later, the bis(trimethylsilyl)diazomethanes was employed as carbene precursors. The resulted gem-disilylated oxazoles can be further converted to the methylation or olefination products [76]. Miura and coworkers reported the $NiBr_2$- or $CoBr_2$-catalyzed C2 alkylation of azoles with *N*-tosylhydrazones, providing a variety of azoles bearing unactivated secondary alkyl groups [77].

Recently, the Cu-catalyzed carbene coupling via C—H alkylation of other arenes and heteroarenes bearing acidic protons has been extensively studied [78]. For example, Wang and coworkers developed the C—H alkylation of polyfluoroarenes (**154**) with *N*-tosylhydrazones (Scheme 12.40A), in which 1,10-phenanthroline was employed as the effective ligand. A series of diaryldiazomethanes and α-diazoesters were also suitable substrates [79]. This reaction involves the Cu–carbene (**156**) migratory insertion and the subsequent protonation to afford the products (**155**).

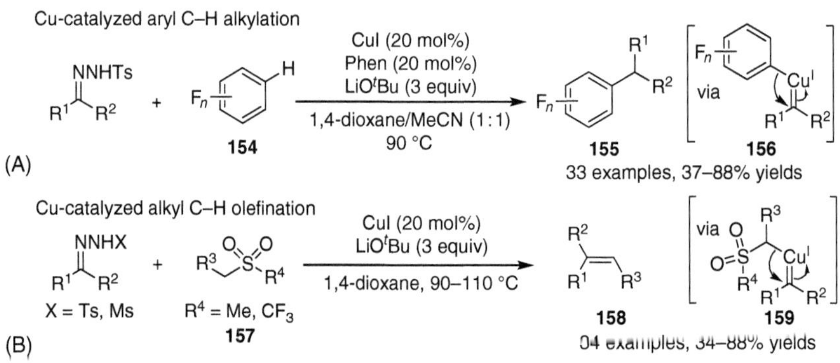

Scheme 12.40 Cu-catalyzed carbene coupling reactions via C–H functionalization. Source: (A) Modified from Xu et al. [79]. (B) Modified al. [80].

The acidic C(sp^3)—H bond can also be functionalized with the participation of *N*-tosylhydrazones. In 2016, Wang and coworkers reported the Cu-catalyzed olefination of *N*-sulfonylhydrazones with sulfones (**157**), affording a series of terminal olefins (**158**) (Scheme 12.40B) [80]. The trisubstituted olefins can be obtained by employing trifluoromethyl alkyl sulfones. Mechanistically, the alkyl-Cu(I) species generated through the base-promoted deprotonation/transmetalation of the sulfone (**157**) generates Cu–carbene (**159**) with *N*-sulfonylhydrazones. The following Cu–carbene migratory insertion and desulfination then affords the olefination products.

12.5.2 Directing-Group-Assisted C–H Bond Functionalizations

12.5.2.1 Generating Acyclic Products Through C–H Bond Activation

In 2012, Yu and coworkers reported the Rh(III)-catalyzed carbene cross-coupling of α-diazoesters with substituted arenes bearing directing-groups (**160**), which involves aryl C(sp^2)—H bond activation (Scheme 12.41) [81]. Various directing groups, including oximes, pyridines, carboxylic acids, and amines could promote the C—H alkylation. The mechanism involves the formation of aryl-Rh(III) species (**162**) via Rh-catalyzed electrophilic C—H metalation. After the Rh–carbene (**163**) generation, the following migratory insertion and protonation can produce the alkylation products (**161**).

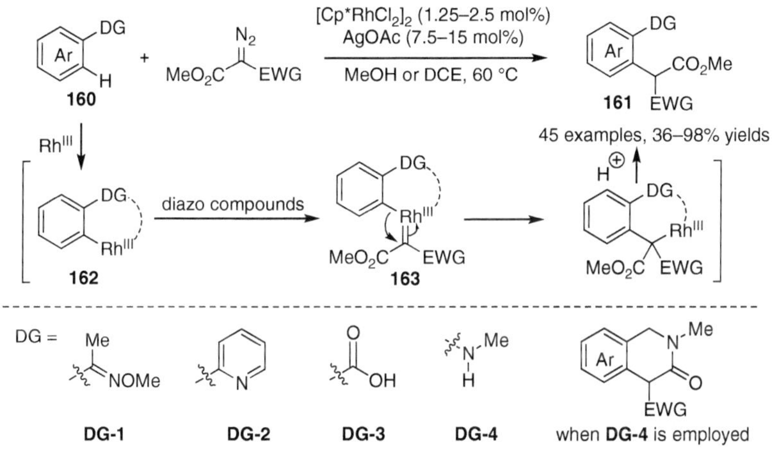

Scheme 12.41 Rhodium-catalyzed carbene coupling of α-diazoesters via C–H activation. Source: Modified from Chan et al. [81].

Recently, significant achievements have been made in the area of transition-metal-catalyzed carbene coupling via C—H bond activation [82]. A series of substrates bearing different directing groups have been reported to undergo C—H bond functionalization involving metal–carbene migratory insertion. Additionally, transition metals other than rhodium can also catalyze the carbene-based C—H functionalizations. For example, Wang and coworkers reported the Ir-catalyzed aryl C—H bond alkylation of **164** with α-diazomalonates (Scheme 12.42) [83]. Both mono- and bis-functionalized products (**165**) were obtained by adjusting the amount of diazo compounds. This reaction involved the sequence of Ir-catalyzed aryl C—H activation, Ir–carbene migratory insertion and the following protonation to afford the final products.

The alkyl–Rh species generated from the carbene migratory insertion can also undergo transformations other than protonation. For example, Wang and coworkers reported the Rh(III)-catalyzed *ortho*-alkenylation of *N*-phenoxyacetamides (**167**) with *N*-tosylhydrazones or diazo compounds, which provided *ortho*-alkenyl phenol products (**168**) (Scheme 12.43) [84]. Mechanistically, the rhodacyclic species (**169**)

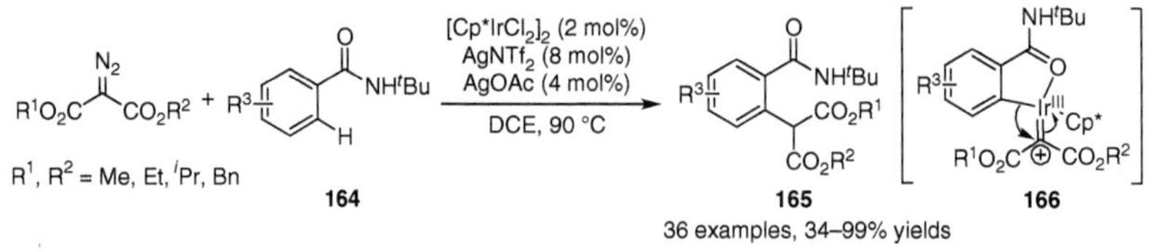

Scheme 12.42 Iridium-catalyzed aryl C—H bond alkylation of amides with α-diazoesters. Source: Modified from Xia et al. [83].

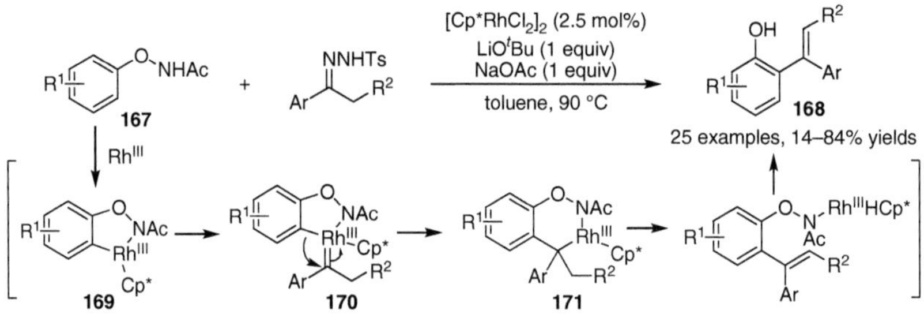

Scheme 12.43 Rhodium-catalyzed ortho-alkenylation of *N*-phenoxyacetamides via carbene migratory insertion. Source: Modified from Hu et al. [84].

generates the Rh–carbene (**170**) with carbene precursors, which can undergo migratory insertion and the following β–H elimination to afford the products. The *N*-oxyacetamide directing group can act as the internal oxidant to regenerate the Rh(III) catalytic species.

12.5.2.2 Generating Cyclic Products Through C—H Bond Activation

In the system of transition-metal-catalyzed carbene C—H activation, the organometallic species generated via carbene migratory insertion can also undergo annulation reaction through C–X (X = C, N, O…) reductive elimination, condensation, or addition process to provide a variety of cyclic compounds. In 2013, Rovis and coworkers reported the Rh(III)-catalyzed [4+1] annulation of *N*-pivaloyloxy benzamides (**172**) with diazo compounds, providing the isoindolone derivatives (**173**) bearing a quaternary carbon (Scheme 12.44) [85]. The mechanism involves the formation of Rh(III)–carbene species and the following carbene migratory insertion. The directing group can also act as the internal oxidant to regenerate the catalytic Rh(III) species after the C—N bond formation.

Various arenes and heteroarenes have been employed in the Rh-catalyzed annulation via C—H bond activation/C—N bond reductive elimination. For example, the Rh(III)-catalyzed asymmetric [4+1] annulation via C–N reductive elimination was later developed, in which the cyclopentadienyl ligand bearing chiral binaphthyl backbone was employed [86]. The conjugated enynones were also employed as carbene sources in this Rh-catalyzed [4+1] annulation to provide the furyl-substituted isoindolinones [87]. The annulation reaction via C—C or C—O bond formation was also feasible. For example, the Rh-catalyzed C—H activation of phenacyl ammonium salts (**174**) with α-diazomalonates can provide the α-indanones (**175**) via C—C

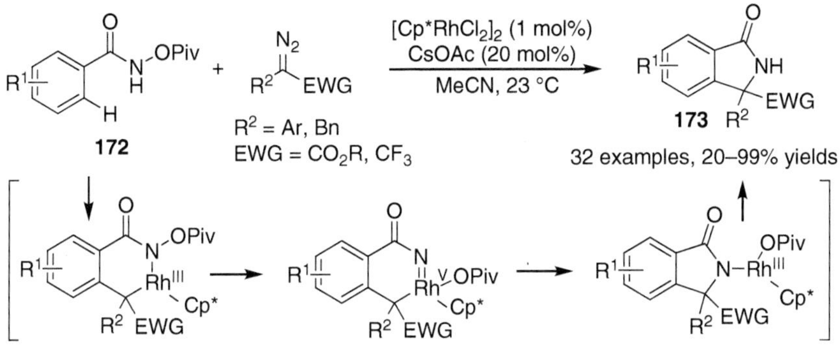

Scheme 12.44 Rhodium-catalyzed [4+1] annulation of *N*-pivaloyloxy benzamides with diazo compounds. Source: Modified from Hyster et al. [85].

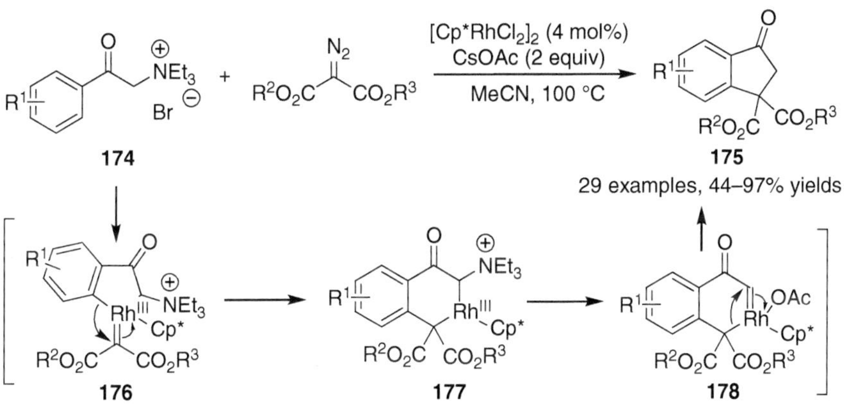

Scheme 12.45 Rh-catalyzed asymmetric [4+1] annulation via C–N reductive elimination. Source: Modified from Yu et al. [88].

bond formation (Scheme 12.45) [88]. After the Rh–carbene (**176**) formation and further migratory insertion, the rhodacycle (**177**) undergoes α-elimination of Et_3N to form another Rh–carbene (**178**). The second carbene migratory insertion and protonation can then afford the products.

Additionally, the condensation or addition reaction between the directing groups and the carbonyl moiety of the C—H activation products can lead to divergent cyclic products with the elimination of small molecules. Glorius and coworkers reported the Rh-catalyzed annulation of oximes (**179**) with diazo compounds, providing isoquinolines and pyridine *N*-oxides (**180**) (Scheme 12.46) [89]. This reaction was proposed to involve the condensation between the oxime and ketone groups in intermediate **181** or the 6π-electro-cyclization of the enol isomer **182**.

Wang and coworkers reported the Rh-catalyzed annulation of *N*-phenxoyacetamides (**183**) with cyclopropenes (**184**), affording 2*H*-chromene products (**185**) via C—O bond formation (Scheme 12.47A) [90]. Mechanistically, the cyclopropene is proposed as the vinyl carbene source to participate the annulation reaction. The rhodacyclic (**186**) generated through carbene migratory insertion and

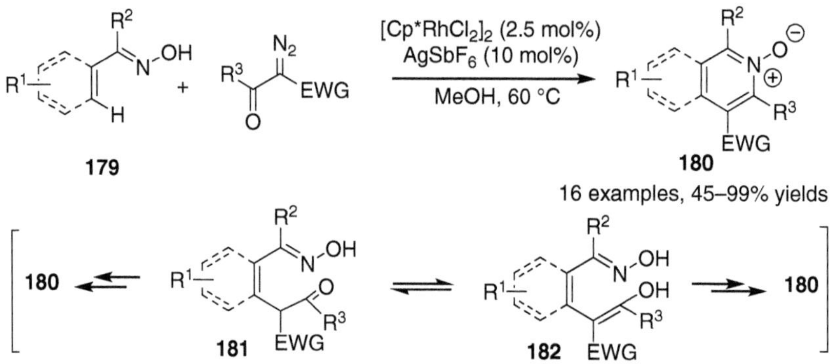

Scheme 12.46 Rh-catalyzed annulation of oximes with diazo compounds. Source: Modified from Shi et al. [89].

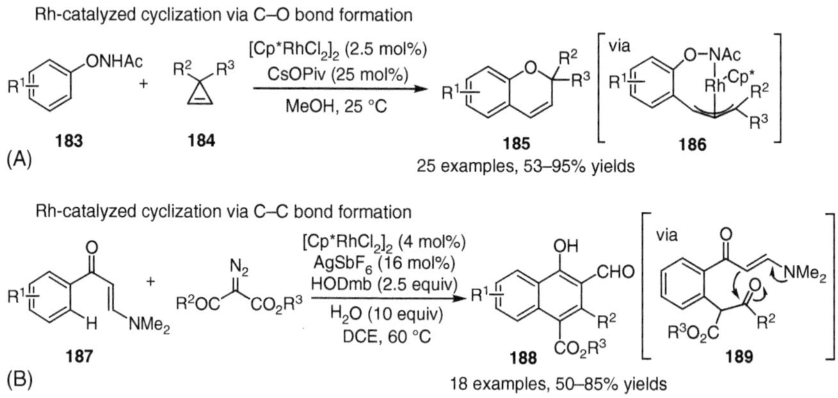

Scheme 12.47 Rhodium-catalyzed cyclization via C—O/C—C bond formation involving carbene migratory insertion. Source: (A) Modified from Zhang et al. [90]. (B) Modified from Zhou et al. [91].

the 1,3-Rh allylic migration then undergoes intramolecular substitution to form the C—O bond. Zhu and coworkers reported the Rh-catalyzed annulation of aryl enaminones (**187**) with α-diazo-β-ketoesters via C—C bond formation (Scheme 12.47B). Mechanistically, the C—H bond functionalization product (**189**) undergoes the acid-promoted condensation and hydrolysis to produce the naphthalene derivatives (**188**) [91].

12.6 Conclusion Remarks

As is demonstrated in this chapter, the transition-metal-catalyzed carbene cross-coupling reactions have evolved into a powerful methodology for the construction of diverse chemical structures. The typical features of the catalytic carbene cross-coupling reactions are the formation of metal carbene species and carbene migratory insertion process, which is followed by rich downstream transformations to afford various coupling products.

The catalytic carbene cross-coupling reactions have demonstrated the generality of carbene migratory insertion, in which a variety of carbene precursors can be tolerated to react with diverse electrophiles in a redox-neutral mode, or couple with different nucleophiles in the oxidative condition. Additionally, diverse cascade transformations can be merged into the carbene coupling reactions, bringing novel opportunities toward the feasible generation of molecular complexity.

Despite the significant achievements of the catalytic carbene cross-coupling reactions, several challenging issues still need to be addressed. For example, in contrast with the diverse transformations of diazo compounds or *N*-tosylhydrazones, some carbene precursors can only participate in limited types of coupling reactions. The discovery of new carbene sources and novel catalytic system is still desirable for exploring novel transformations. Additionally, the enantioselective carbene cross-coupling reactions, which are highly demanded in the synthesis of chiral molecules, have been less developed. Thus, further applications of the carbene cross-coupling chemistry in several areas, including material science, pharmaceutical synthesis, and polymer chemistry, can be anticipated in future.

Acknowledgment

This work is supported by grants from the Nature Science Foundation of China (No. 91956104) and Beijing National Laboratory of Molecular Sciences (BNLMS). We also acknowledge the funding support from the Health@InnoHK Program launched by Innovation and Technology Commission, The Government of Hong Kong Special Administrative Region of the People's Republic of China.

References

1 Negishi, E. -i. (2011). *Angew. Chem. Int. Ed.* 50: 6738.
2 Barluenga, J. and Valdés, C. (2011). *Angew. Chem. Int. Ed.* 50: 7486.
3 Xiao, Q., Zhang, Y., and Wang, J. (2013). *Acc. Chem. Res.* 46: 236.
4 Xia, Y., Di, Q., and Wang, J. (2017). *Chem. Rev.* 117: 13810.
5 Greenman, K.L., Carter, D.S., and Van Vranken, D.L. (2001). *Tetrahedron* 57: 5219.
6 Yu, W.-Y., Tsoi, Y.T., Zhou, Z., and Chan, A.S.C. (2009). *Org. Lett.* 11: 469.
7 Wang, X., Xu, Y., Deng, Y. et al. (2014). *Chem. Eur. J.* 20: 961.
8 Zhou, Y., Ye, F., Wang, X. et al. (2015). *J. Org. Chem.* 80: 6109.
9 Chen, S. and Wang, J. (2008). *Chem. Commun.* 2008: 4198.
10 Peng, C., Yan, G., Wang, Y. et al. (2010). *Synthesis* 2010: 4154.
11 Peng, C., Wang, Y., and Wang, J. (2008). *J. Am. Chem. Soc.* 130: 1566.
12 Xia, Y., Xia, Y., Liu, Z. et al. (2014). *J. Org. Chem.* 79: 7711.
13 Xia, Y., Zhang, Y., and Wang, J. (2013). *ACS Catal.* 3: 2586.
14 Zhang, Z., Liu, Y., Gong, M. et al. (2010). *Angew. Chem. Int. Ed.* 49: 1139.

15 Xiao, Q., Wang, B., Tian, L. et al. (2013). *Angew. Chem. Int. Ed.* 52: 9305.
16 Pérez-Gómez, M., Hernández-Ponte, S., Bautista, D., and García-López, J.A. (2017). *Chem. Commun.* 53: 2842.
17 Gutiérrez-Bonet, A., Juliá-Hernández, F., de Luis, B., and Martin, R. (2016). *J. Am. Chem. Soc.* 138: 6384.
18 Devine, S.K. and Van Vranken, D.L. (2007). *Org. Lett.* 9: 2047.
19 Devine, S.K. and Van Vranken, D.L. (2008). *Org. Lett.* 10: 1909.
20 Xu, S., Chen, R., Fu, Z. et al. (2017). *ACS Catal.* 7: 1993.
21 Yu, Y., Lu, Q., Chen, G. et al. (2018). *Angew. Chem. Int. Ed.* 57: 319.
22 Barluenga, J., Moriel, P., Valdés, C., and Aznar, F. (2007). *Angew. Chem. Int. Ed.* 46: 5587.
23 Barluenga, J., Tomás-Gamasa, M., Moriel, P. et al. (2008). *Chem. Eur. J.* 14: 4792.
24 Zhou, L., Ye, F., Zhang, Y., and Wang, J. (2012). *Org. Lett.* 14: 922.
25 Luo, H., Wu, G., Xu, S. et al. (2015). *Chem. Commun.* 51: 13321.
26 Zhou, Q., Gao, Y., Xiao, Y. et al. (2019). *Polym. Chem.* 10: 569.
27 Jiang, K., Zhang, L., Zhao, Y. et al. (2018). *Macromolecules* 51: 9662.
28 Xia, Y., Hu, F., Liu, Z. et al. (2013). *Org. Lett.* 15: 1784.
29 Xiao, Q., Ma, J., Yang, Y. et al. (2009). *Org. Lett.* 11: 4732.
30 Wang, K., Chen, S., Zhang, H. et al. (2016). *Org. Biomol. Chem.* 14: 3809.
31 Zhao, X., Jing, J., Lu, K. et al. (2010). *Chem. Commun.* 46: 1724.
32 Zhou, L., Ye, F., Ma, J. et al. (2011). *Angew. Chem. Int. Ed.* 50: 3510.
33 Liu, Z., Tan, H., Fu, T. et al. (2015). *J. Am. Chem. Soc.* 137: 12800.
34 Liu, Z., Huo, J., Fu, T. et al. (2018). *Chem. Commun.* 54: 11419.
35 Hu, F., Xia, Y., Liu, Z. et al. (2014). *Org. Biomol. Chem.* 12: 3590.
36 Liu, Z., Xia, Y., Zhou, S. et al. (2013). *Org. Lett.* 15: 5032.
37 Han, J.-L., Qin, Y., Ju, C.-W., and Zhao, D. (2020). *Angew. Chem. Int. Ed.* 59: 6555.
38 Xia, Y., Xia, Y., Zhang, Y., and Wang, J. (2014). *Org. Biomol. Chem.* 12: 9333.
39 Zhou, L., Ye, F., Zhang, Y., and Wang, J. (2010). *J. Am. Chem. Soc.* 132: 13590.
40 Yu, Y., Chakraborty, P., Song, J. et al. (2020). *Nat. Commun.* 11: 461.
41 Xia, Y., Qu, S., Xiao, Q. et al. (2013). *J. Am. Chem. Soc.* 135: 13502.
42 Xia, Y., Ge, R., Chen, L. et al. (2015). *J. Org. Chem.* 80: 7856.
43 Xia, Y., Liu, Z., Ge, R. et al. (2015). *Chem. Commun.* 51: 11233.
44 Ping, Y., Chang, T., Wang, K. et al. (2019). *Chem. Commun.* 55: 59.
45 Xia, Y., Xia, Y., Ge, R. et al. (2014). *Angew. Chem. Int. Ed.* 53: 3917.
46 Gao, Y., Wu, G., Zhou, Q., and Wang, J. (2018). *Angew. Chem. Int. Ed.* 57: 2716.
47 Wang, K., Ping, Y., Chang, T. et al. (2017). *Angew. Chem. Int. Ed.* 56: 13140.
48 Wang, K., Lu, Y., Hu, F. et al. (2018). *Organometallics* 37: 1.
49 Xiao, Q., Xia, Y., Li, H. et al. (2011). *Angew. Chem. Int. Ed.* 50: 1114.
50 Hossain, M.L., Ye, F., Zhang, Y., and Wang, J. (2013). *J. Org. Chem.* 78: 1236.
51 Ye, F., Hossain, M.L., Xu, Y. et al. (2013). *Chem. Asian J.* 8: 1404.
52 Hu, F., Xia, Y., Ma, C. et al. (2014). *Org. Lett.* 16: 4082.
53 Hossain, M.L. and Wang, J. (2018). *Chem. Rec.* 18: 1548.
54 Tang, Y., Chen, Q., Liu, X. et al. (2015). *Angew. Chem. Int. Ed.* 54: 9512.
55 Chu, W.-D., Zhang, L., Zhang, Z. et al. (2016). *J. Am. Chem. Soc.* 138: 14558.

56 Ye, F., Ma, X., Xiao, Q. et al. (2012). *J. Am. Chem. Soc.* 134: 5742.
57 Chu, W.-D., Guo, F., Yu, L. et al. (2018). *Chin. J. Chem.* 36: 217.
58 Wang, C., Ye, F., Wu, C. et al. (2015). *J. Org. Chem.* 80: 8748–8757.
59 Hu, M., He, Z., Gao, B. et al. (2013). *J. Am. Chem. Soc.* 135: 17302.
60 Matheis, C., Krause, T., Bragoni, V., and Goossen, L.J. (2016). *Chem. Eur. J.* 22: 12270.
61 Zhang, Z., Zhou, Q., Ye, F. et al. (2015). *Adv. Synth. Catal.* 357: 2277.
62 Zhang, Z., Zhou, Q., Yu, W. et al. (2017). *Chin. J. Chem.* 35: 387.
63 Tsoi, Y.-T., Zhou, Z., and Yu, W.-Y. (2011). *Org. Lett.* 13: 5370.
64 Zhang, Z., Yu, W., Zhou, Q. et al. (2016). *Chin. J. Chem.* 34: 473.
65 Xia, Y., Chen, L., Qu, P. et al. (2016). *J. Org. Chem.* 81: 10484.
66 Xia, Y., Liu, Z., Feng, S. et al. (2015). *Org. Lett.* 17: 956.
67 Liu, Z., Xia, Y., Feng, S. et al. (2015). *Aust. J. Chem.* 68: 1379.
68 Xia, Y., Liu, Z., Liu, Z. et al. (2014). *J. Am. Chem. Soc.* 136: 3013.
69 Yada, A., Fujita, S., and Murakami, M. (2014). *J. Am. Chem. Soc.* 136: 7217.
70 Xia, Y., Feng, S., Liu, Z. et al. (2015). *Angew. Chem. Int. Ed.* 54: 7891.
71 Feng, S., Mo, F., Xia, Y. et al. (2016). *Angew. Chem. Int. Ed.* 55: 15401.
72 Feng, S., Wang, K., Ping, Y., and Wang, J. (2020). *J. Am. Chem. Soc.* 142: 21032.
73 Wang, P.-S., Lin, H.-C., Zhou, X.-L., and Gong, L.-Z.,(2014). *Org. Lett.* 16: 3332.
74 Ping, Y., Zhang, S; Chang, T., and Wang, J., (2019). *J. Org. Chem.* 84: 8275.
75 Zhao, X., Wu, G., Zhang, Y., and Wang, J. (2011). *J. Am. Chem. Soc.* 133: 3296.
76 Wang, S., Xu, S., Yang, C. et al. (2019). *Org. Lett.* 21: 1809.
77 Yao, T., Hirano, K., Satoh, T., and Miura, M. (2012). *Angew. Chem. Int. Ed.* 51: 775.
78 Liu, Z. and Wang, J. (2013). *J. Org. Chem.* 78: 10024.
79 Xu, S., Wu, G., Ye, F. et al. (2015) *Angew. Chem. Int. Ed.* 54: 4669.
80 Xu, S., Gao, Y., Chen, R. et al. (2016) *Chem. Commun.* 52: 4478.
81 Chan, W.-W., Lo, S.-F., Zhou, Z., and Yu, W.-Y. (2012). *J. Am. Chem. Soc.* 134: 13565.
82 Hu, F., Xia, Y., Ma, C. et al. (2015). *Chem. Commun.* 51: 7986.
83 Xia, Y., Liu, Z., Feng, S. et al. (2015). *J. Org. Chem.* 80: 223.
84 Hu, F., Xia, Y., Ye, F. et al. (2014). *Angew. Chem. Int. Ed.* 53: 1364.
85 Hyster, T.K., Ruhl, K.E., and Rovis, T. (2013). *J. Am. Chem. Soc.* 135: 5364.
86 Ye, B. and Cramer, N. (2014). *Angew. Chem. Int. Ed.* 53: 7896.
87 Hong, S.Y., Jeong, J., and Chang, S. (2017). *Angew. Chem. Int. Ed.* 56: 2408.
88 Yu, S., Liu, S., Lan, Y. et al. (2015). *J. Am. Chem. Soc.* 137: 1623.
89 Shi, Z., Koester, D.C., Boultadakis-Arapinis, M., and Glorius, F. (2013). *J. Am. Chem. Soc.* 135: 12204.
90 Zhang, H., Wang, K., Wang, B. et al. (2014). *Angew. Chem. Int. Ed.* 53: 13234.
91 Zhou, S., Wang, J., Wang, L. et al. (2016). *Angew. Chem. Int. Ed.* 55: 9384.

Index

Transition Metal-Catalyzed Carbene Transformations, First Edition.
Edited by Jianbo Wang, Chi-Ming Che, and Michael P. Doyle.

g

h

i

q

r